# Advances in plant factories

## New technologies in indoor vertical farming

It is widely recognised that agriculture is a significant contributor to global warming and climate change. Agriculture needs to reduce its environmental impact and adapt to current climate change whilst still feeding a growing population, i.e. become more 'climate-smart'. Burleigh Dodds Science Publishing is playing its part in achieving this by bringing together key research on making the production of the world's most important crops and livestock products more sustainable.

Based on extensive research, our publications specifically target the challenge of climate-smart agriculture. In this way we are using 'smart publishing' to help achieve climate-smart agriculture.

Burleigh Dodds Science Publishing is an independent and innovative publisher delivering high quality customer-focused agricultural science content in both print and online formats for the academic and research communities. Our aim is to build a foundation of knowledge on which researchers can build to meet the challenge of climate-smart agriculture.

For more information about Burleigh Dodds Science Publishing simply call us on +44 (0) 1223 839365, email info@bdspublishing.com or alternatively please visit our website at www.bdspublishing.com.

**Related titles:**

Achieving sustainable cultivation of vegetables
Print (ISBN 978-1-78676-236-8); Online (ISBN 978-1-78676-238-2, 978-1-78676-239-9)

Achieving sustainable greenhouse cultivation
Print (ISBN 978-1-78676-280-1); Online (ISBN 978-1-78676-282-5, 978-1-78676-283-2)

Achieving sustainable urban agriculture
Print (ISBN 978-1-78676-316-7); Online (ISBN 978-1-78676-318-1, 978-1-78676-319-8)

Advances in horticultural soilless culture
Print (ISBN 978-1-78676-435-5); Online (ISBN 978-1-78676-437-9, 978-1-78676-438-6)

Advances in plant phenotyping for sustainable crop production
Print (ISBN 978-1-78676-856-8); Online (ISBN 978-1-78676-858-2, 978-1-78676-859-9)

Chapters are available individually from our online bookshop:
https://shop.bdspublishing.com

climate•SMART•publishing
IN AGRICULTURAL SCIENCE

BURLEIGH DODDS SERIES IN AGRICULTURAL SCIENCE
NUMBER 141

# Advances in plant factories

## New technologies in indoor vertical farming

Edited by Toyoki Kozai and Eri Hayashi

burleigh dodds
SCIENCE PUBLISHING

Published by Burleigh Dodds Science Publishing Limited
82 High Street, Sawston, Cambridge CB22 3HJ, UK
www.bdspublishing.com

Burleigh Dodds Science Publishing, 1518 Walnut Street, Suite 900, Philadelphia, PA 19102-3406, USA

First published 2023 by Burleigh Dodds Science Publishing Limited


Library of Congress Control Number: 2023940851

British Library Cataloguing in Publication Data
A catalogue record for this book is available from the British Library

ISBN 978-1-80146-316-4 (print)
ISBN 978-1-80146-318-8 (PDF)
ISBN 978-1-80146-317-1 (ePub)
ISSN 2059-6936 (print)
ISSN 2059-6944 (online)

DOI: 10.19103/AS.2023.0126

Typeset by Deanta Global Publishing Services, Dublin, Ireland

# Contents

## Part 3 Phenotyping



## Part 4 Spectral manipulations for controlling the growth and quality of leafy greens

# Series list

| Title | Series number |
|---|---|
| **Achieving sustainable cultivation of maize - Vol 1**<br>From improved varieties to local applications<br>Edited by: Dr Dave Watson, CGIAR Maize Research Program Manager, CIMMYT, Mexico | 001 |
| **Achieving sustainable cultivation of maize - Vol 2**<br>Cultivation techniques, pest and disease control<br>Edited by: Dr Dave Watson, CGIAR Maize Research Program Manager, CIMMYT, Mexico | 002 |
| **Achieving sustainable cultivation of rice - Vol 1**<br>Breeding for higher yield and quality<br>Edited by: Prof. Takuji Sasaki, Tokyo University of Agriculture, Japan | 003 |
| **Achieving sustainable cultivation of rice - Vol 2**<br>Cultivation, pest and disease management<br>Edited by: Prof. Takuji Sasaki, Tokyo University of Agriculture, Japan | 004 |
| **Achieving sustainable cultivation of wheat - Vol 1**<br>Breeding, quality traits, pests and diseases<br>Edited by: Prof. Peter Langridge, The University of Adelaide, Australia | 005 |
| **Achieving sustainable cultivation of wheat - Vol 2**<br>Cultivation techniques<br>Edited by: Prof. Peter Langridge, The University of Adelaide, Australia | 006 |
| **Achieving sustainable cultivation of tomatoes**<br>Edited by: Dr Autar Mattoo, USDA-ARS, USA and Prof. Avtar Handa, Purdue University, USA | 007 |
| **Achieving sustainable production of milk - Vol 1**<br>Milk composition, genetics and breeding<br>Edited by: Dr Nico van Belzen, International Dairy Federation (IDF), Belgium | 008 |
| **Achieving sustainable production of milk - Vol 2**<br>Safety, quality and sustainability<br>Edited by: Dr Nico van Belzen, International Dairy Federation (IDF), Belgium | 009 |
| **Achieving sustainable production of milk - Vol 3**<br>Dairy herd management and welfare<br>Edited by: Prof. John Webster, University of Bristol, UK | 010 |
| **Ensuring safety and quality in the production of beef - Vol 1**<br>Safety<br>Edited by: Prof. Gary Acuff, Texas A&M University, USA and Prof. James Dickson, Iowa State University, USA | 011 |
| **Ensuring safety and quality in the production of beef - Vol 2**<br>Quality<br>Edited by: Prof. Michael Dikeman, Kansas State University, USA | 012 |
| **Achieving sustainable production of poultry meat - Vol 1**<br>Safety, quality and sustainability<br>Edited by: Prof. Steven C. Ricke, University of Arkansas, USA | 013 |
| **Achieving sustainable production of poultry meat - Vol 2**<br>Breeding and nutrition<br>Edited by: Prof. Todd Applegate, University of Georgia, USA | 014 |

# Acknowledgements

The authors would like to thank deeply Francis Dodds and Amanda Renwick of Burleigh Dodds Science Publishing, Arun Rajakumar of Deanta and Hiroko Nakaoka of Japan Plant Factory Association for their excellent editorial assistance.

Cover image copyright:

- Cover image: Mugen Farm, created by Oishii, growing Omakase Berries in an indoor vertical farm. 2022. © Tsutomu Fujimoto.

# Part 1

## Introduction: backgrounds, concept and methodology of sustainable PFALs

# Chapter 1

# Characteristics, potential and challenges of plant factories with artificial lighting (PFALs): Introduction

*Toyoki Kozai and Eri Hayashi, Japan Plant Factory Association, Japan*



## 1 Introduction

This chapter provides a brief overview of opportunities and challenges for plant factories with artificial lighting (PFALs). Section 2 discusses:

1 global and local issues we are facing;
2 the contribution of PFALs;
3 reducing resource consumption and $CO_2$ emission by PFALs;
4 the aim of PFAL research and development; and
5 the airtight and thermally insulated scalable cultivation room with sensors.

Section 3 discusses some important challenges in PFAL research and development:

1 Initial investment cost and production capacity per floor area;
2 Virtual and dual PFALs;
3 Global and local features of PFALs;
4 Spatial distributions of environmental factors in a plant canopy;
5 Plant phenotyping; and
6 Expanding the target plant types for the PFALs.

For detailed information on each topic, including the references, refer to the corresponding chapters of this book.

http://dx.doi.org/10.19103/AS.2023.0126.01

# 2 Background and aim of plant factories with artificial lighting

## 2.1 Global and local issues currently being faced

We are facing:

1. an increasing urban population;
2. decreasing agricultural land and rural populations;
3. decreasing availability of river water and groundwater for irrigation;
4. degradation of agricultural soil (e.g. erosion, salt accumulation);
5. increasing extreme weather events causing droughts, floods and storm damage;
6. increasing demands for reductions in the consumption of fossil fuels, agrochemicals and chemical fertilizers;
7. an increasing need to reduce emissions of greenhouse gas (GHG) (e.g. $CO_2$, ethylene and nitrous oxide); and
8. occasionally unstable international trading conditions, disrupting stable supply of foods and resources for food production.

A diverse solution must be developed and applied to solve the aforementioned issues. PFALs are one such solution.

## 2.2 Contribution of plant factories with artificial lighting

An increasing number of people have recently become interested in PFALs or vertical farms, which are seen as contributing to solving the global and local issues mentioned earlier with respect to:

1. food safety and security;
2. restoring, protecting and enhancing the environment (and the ecosystem services it provides) through reductions in land area used for agriculture, use of water for irrigation and reduced use of fertilizers and pesticides;
3. increasing demands for higher quality, fresh, nutritious and pesticide-free foods with potential nutraceutical benefits;
4. reducing food mileage and food loss; and
5. increasing job opportunities in urban areas.

Compared to open-field production, PFALs can reduce:

1. water consumption per kg of produce by about 90%;
2. land area per ton of production by about 95%;
3. application of pesticides and herbicides by nearly 100%;

4 damage to produce due to adverse weather by 100%; and
5 reliance on agricultural machines and fuel for open-field cultivation by 100%.

## 2.3 Reducing resource consumption and $CO_2$ emission by plant factories with artificial lighting

Existing PFALs consume significant amounts of:

1 electricity for lighting, air-conditioning (cooling is always necessary during the photoperiod) and other electric equipment;
2 plastic products for cultivation beds and panels, trays, bags, boxes, films and strings;
3 chemical fertilizers for hydroponic nutrient solution; and
4 metals, cement and other materials for constructing the buildings and cultivation racks.

Additional $CO_2$ is also emitted to manufacture and transport these resources. Considerable reductions in consuming these resources and the use of alternative resources that emit much less $CO_2$ are therefore crucial to making the PFALs sustainable.

The most effective and immediate way to reduce $CO_2$ emissions per kg of produce is using electricity generated from natural energy sources such as solar energy, wind power and hydropower, instead of fossil fuels. The $CO_2$ emission factor of electricity generated by solar energy is around 50 $gCO_2$/kWh, which is about 1/15th of that of electricity generated by power plants using fossil fuels (Imamura et al., 2016; see Chapter 5 of this book). The cost of electricity generated by natural energy is currently comparable to or lower than that of fossil fuel. Electricity consumption per kilogram of produce for lighting and air-conditioning can be reduced considerably by improving lighting and air-conditioning systems (see Chapter 5 of this book), breeding cultivars suited to high $CO_2$ concentration and low photosynthetic photon flux density (PPFD) and uniform spatial environments in a plant community. Energy-autonomous PFALs powered only by natural energy are one of the essential goals of PFAL design and management.

The consumption and thus costs of chemical fertilizer, plastic products, metals and cement can also be reduced significantly by using plant-derived alternatives and improving the design and operation of PFALs. A hydroponic cultivation system with no drain of nutrient solution to the outside will reduce fertilizer consumption by around 20%, compared to the current hydroponic cultivation systems.

## 2.4 Aim of plant factories with artificial lighting research and development

PFAL research and development aims to establish basic concepts, methodologies and system modules for achieving ecologically, environmentally, economically and socially sustainable plant production systems. This goal can be realized by achieving the highest resource and monetary (or economic) productivities with the highest resource use efficiencies by realizing the highest annual yield with the highest quality, with minimum use of resources, and minimum emission of waste and GHGs, in the face of changing social, economic and technological conditions (Kozai, 2019).

PFALs can be seen to support the Sustainable Development Goals adopted by the United Nations (UN) in 2015, which the UN aims to achieve by 2030. Environment, Society and Corporate Governance (ESG) is another set of global standards used by socially conscious investors to analyze company performance on ethical and environmental grounds and to decide whether to invest. In addition to environmental benefits, PFALs can provide benefits to workers and other stakeholders. PFALs can offer full-time and part-time jobs opportunities for safe and less arduous work in relatively comfortable environments all year round to all adult citizens, including the disabled, elderly and young people. Routine jobs are conducted manually in small-scale PFALs, while these are mostly automated in large-scale PFALs. Mini-PFALs are used at schools, at home and in public spaces as an educational, self-learning or hobby tool to enjoy growing plants and learn the principle of agricultural ecosystems in a simplified form. Pest management in a large-scale PFAL for commercial production of leafy vegetables resembles that in the food factory rather than integrated pest management in the greenhouse. Pest management in a small-scale PFAL is not as strict as in food factories if produce is served after washing, boiling or heating.

## 2.5 Airtight, thermally insulated and scalable cultivation rooms with sensors

One of the core technologies of the PFAL derives from the use of an airtight and thermally insulated cultivation room with sensors for measuring all resource inputs, product outputs, environmental factors and plant traits or phenotype (e.g. plant canopy architecture and concentrations of chemical components; see Chapter 10 of this book). This means that plant environmental factors can be controlled at an optimum point with minimum resource inputs and waste outputs, regardless of local weather, soil and ecosystem conditions. Resource inputs, product outputs, environmental factors and plant phenotype are automatically recorded with high accuracy. In these cultivation room conditions, observability, controllability, traceability and reproducibility of most data are fairly guaranteed, and all data are digitally recorded for further utilization. This includes data are used in:

1 simulation using mechanistic, empirical and artificial intelligence (AI) models;
2 estimation of resource use efficiencies and resource productivities; and
3 detection of malfunctions of equipment and sensors.

Digital data can be shared worldwide via Social Network Services. In contrast, data sharing and analysis are less useful for growers cultivating crops in open fields and greenhouses under varying conditions (soil properties, weather conditions, etc.), which make comparison more challenging.

Most PFAL users want to start with a small-scale PFAL, such as a container-type PFAL or a PFAL with a floor area of 500 $m^2$ or less. Typical trouble when scaling up after success with a small PFAL is that the plants do not grow well in the scaled-up PFAL as they did in the small PFAL. This is attributable mainly to the changes in flows of air and nutrient solution within the plant community because these flows affect plant growth significantly even when variables such as air temperature, vapor pressure deficit (VPD), PPFD over the plant canopy, water temperatures, pH, dissolved oxygen and flow speed of nutrient solution at the entry point of cultivation beds remain the same (see Section 7.2 in Chapter 2 of this book). Another cause is changes in production process management. The design and management of scalable PFALs are essential to prevent these problems.

## 3 Challenges facing plant factories with artificial lighting

### *3.1 Initial investment cost and production capacity per floor area*

The initial investment costs per floor area of PFAL producing leafy vegetables are roughly tenfold more than a greenhouse equipped with environmental control units and roughly 100-fold more than open field production using standard agricultural machines and facilities. On the other hand, annual yield per land area of a PFAL with ten tiers is roughly 10-fold than that of a greenhouse and over 100-fold than that of an open field. The initial investment costs in terms of production capacity per floor or land area of a PFAL, greenhouse and open field are thus more or less the same.

The annual yield a PFAL can achieve in practice depends significantly on the skill of the PFAL manager and the hardware/software specifications of the PFAL. In contrast, the actual annual yield of an open field substantially depends on the weather, cultivar selection and the soil properties of the cultivation site. A greenhouse depends on both management skills and weather conditions.

Management skill in a PFAL can be acquired relatively quickly by analyzing time-series data on resource inputs, product outputs, management and environmental conditions. This is a big advantage of PFAL over greenhouse and open field cultivation where these measurements are harder to make and use in decision making, given more complex and variable conditions. Data analysis and use in management is well established in leafy vegetable production. Such analyses are necessary but more challenging for other products such as fruit vegetables and medicinal plants.

The aforementioned statements about the initial investments in the PFAL, greenhouse and open field are theoretical without experimental data. Experimental comparisons with risk analyses of plant growth and development in the aforementioned three plant production systems are essential to choosing the best production system under given conditions. Simulations using models are powerful tools to estimate costs for initial investment and environmental control. However, it is essential to compare simulated results of plant growth and development in the three plant production systems using the same simulation models. This is challenging since:

1 ecophysiological characteristics change in different ways as plants grow in the three different environments, even when using the transplants of the same cultivar with the same ecophysiological characteristics; and
2 optimal environments for the three production systems change as the plants grow, which in turn changes the ecophysiological characteristics of plants.

Cultivars suited to a PFAL differ from those suited to a greenhouse and an open field.

## *3.2 Virtual and dual plant factories with artificial lighting*

A virtual PFAL, or a PFAL simulator, can be a self-learning, training and educational tool. A virtual PFAL connected with a real PFAL would be a useful tool in improving skills in PFAL management. With this system, all input data to the virtual PFAL are transferred from the existing PFAL, and the outputs from the virtual and existing PFALs are compared to improve both PFALs.

Although PFALs with light-emitting diodes (LEDs) have great potential to achieve the goal of sustainable PFALs described earlier, current PFALs fall significantly short of the goal of sustainable PFALs mainly due to their recent development. A PFAL with red LEDs was first commercialized (not for research) in 1998 (Kozai et al., 2022), and a PFAL with white LEDs was first commercialized in around 2015. Many PFALs with white LEDs are currently introducing new advanced technologies such as AI, Internet of Things, robots, big data mining, networked smart micro-sensors, remote phenotyping (plant trait measurements), augmented reality, three-dimensional (3D) printers, genomic analysis and advanced breeding techniques (e.g. genome editing). However, only a few of them have been widely commercialized successfully as of 2022. At this early stage in PFAL development, sharing concepts and methodologies for achieving the goals described earlier is essential.

### 3.3 Global and local features common to smartphone and plant factories with artificial lighting technologies

The core technology of smartphone hardware and software is common globally, while their application software and peripheral devices are diverse, being adapted to users' age, needs, local culture and personal preferences. Application software developed by a third party can be installed on most smartphones under certain conditions, often free of charge. This feature, in principle, resembles PFAL technology. The core of PFAL technology is universal regardless of climate and local culture. At the same time, application software and hardware are diverse, being adapted to the user's needs, personality and locality. Application software and peripheral hardware developed by a third party can be integrated in a PFAL under certain conditions. PFAL technologies can be shared, improved and popularized rapidly, just like smartphones, although this is not yet the case in the existing PFALs as of 2022 (Lubna et al., 2022).

### 3.4 Spatial distribution of environmental factors in a plant community

In a densely populated plant community with a leaf area index (LAI; ratio of leaf area to cultivation bed area) of 1 or greater in a PFAL, PPFD and air current speed (or diffusion coefficients of $CO_2$ gas, water vapor and sensible heat energy) decreases sharply with depth from the canopy surface. The plant canopy (the aboveground portion of a plant community) restricts penetration of photosynthetic photons, diffusion of $CO_2$ gas and water vapor. PPFD and VPD thus decrease with depth in a plant community. $CO_2$ concentration during the photoperiod is lowest just below the canopy surface, because $CO_2$ is mostly absorbed there. Such spatially nonuniform environments reduce the plant community's net photosynthetic and transpiration rates with the increase in LAI. 3D uniformity of environmental factors [PPFD, VPD and air current speed (or diffusion coefficient of gases)] and $CO_2$ concentration at their optimal levels in the plant canopy will maximize net photosynthesis, growth of aerial parts of plants and 3D uniformity in the plant canopy.

Bottom-up or upward airflow in the plant canopy provides spatially uniform air current speeds, promoting net photosynthesis (refer to Chapter 8 in this book; Shibuya et al., 2006; Kitaya, 2016). Forced downward and sideward air flows would also promote the growth of the plant canopy to some extent (see previous references). Supplemental upward lighting with LEDs from the bottom of the plant canopy provides higher marketable lettuce leaf fresh weight by retarding the senescence of outer leaves (Zhang et al., 2015). Moreover, these uniform environments considerably simplify the relationships between the environment and plant growth/phenotype. This simplification is key to developing a practical growth model of the plant community as affected by the surrounding environment.

The transmittance of green leaves to ultraviolet, blue and red light is significantly lower than green and far-red light. The reflectance of green leaves to green and far-red light is higher than blue and red light (Saengtharatip et al., 2020; Amagai et al., 2022). Percentages of green and far-red light over light with a waveband of 300–800 nm increase with depth in a densely populated plant canopy. Light spectral distribution at the bottom of the canopy is thus significantly different from that at the canopy surface if no blue and red light is added in the middle and bottom of the canopy. Nonuniform vertical spectral distribution affects the phenotype of the plant canopy.

It is important to note that spatial distribution of phenotypes in a plant community may not be uniform even in uniform environments because of the complex interactions of matter and signals among plant organs. Environmental control based on nondestructive and continuous phenotyping will increase in importance in the future. The same applies to the rootzone environment. Environmental factors such as dissolved oxygen, nutrient solution flow speed, each ion concentration and microbes at the inlet and outlet of hydroponic cultivation beds often differ significantly from those in the rhizosphere (narrow region (2-3 mm) of the substrate), which are directly influenced by root secretions and the root microbiome. Optimal environmental parameters for commercial PFALs need to be determined not only by plant growth and development but also by production costs which are affected by prices of resources, market needs and prices, as well as allocation of available resources such as workers, cultivation area and time.

## *3.5 Plant phenotyping*

Having an almost airtight and well-insulated structure, PFALs can lead to accurate and straightforward data acquisition and environmental control, time series analysis and prediction regarding plant production, as discussed earlier. Variations in the plant phenotype of individual plants are mainly ascribed to three factors (Kozai et al., 2018; Hayashi et al., 2020; Hayashi et al., 2022):

1 Microenvironment (in the plant canopy) or environment (over the plant canopy or community);

2 Genotype (species and cultivar); and
3 Management (human and machine intervention).

Besides being impacted by genotype, a plant phenotype interacts with the microenvironment, wider environment and management. Using a phenotyping system, a tool for the noninvasive, nondestructive and continuous measurement of the development of plant phenotypes, time series interactions between phenotype, microenvironment, wider environment and management in PFALs can be revealed in real time during plant production. By considering plant phenotype, environmental control in PFALs can be improved together with improving plant productivity in PFALs. For details, refer to Chapter 10 of this book.

### 3.6 Expanding target plant types for commercial production

Currently, most commercial PFALs produce leafy vegetables, microgreens and leafy herbs. A limited number of PFALs produce transplants (e.g. seedlings, grafted transplants and rooted cuttings) (Park et al., 2022) for growing in the greenhouse or open fields. More PFALs are being developed for commercial production of:

1 vine crops such as tomatoes (Zelkind et al., 2022; Hisawa-Tanase et al., 2022) and cucumbers;
2 berries such as strawberries (see Chapter 18 of this book) and blueberries;
3 medicinal (see Chapter 15 of this book) and aromatic plants such as cannabis (Rodriguez-Morrison et al., 2021), ginseng, saffron (Dewir and Alsadon, 2022), Digitalis (*Digitalis purpurea*) (Verma et al., 2018) and vanilla;
4 head vegetables such as mini-cabbages and mini-Chinese cabbages (Kozai et al., 2022);
5 root vegetables such as carrots and turnips;
6 dwarf ornamentals (table or pot plants) such as orchids and roses and bulbous plants such as crocus; and
7 plant-derived pharmaceuticals for vaccines (Goto, 2020) and anti-cancer compounds.

These trends will become popular when new cultivars suited to PFALs are bred, and new cultivation systems suited to PFALs are developed. Some PFALs are now used for breeding and seed propagation of plants for cultivation in PFALs, greenhouses and open fields. PFALs for commercial production can be used in parallel for research and breeding because most data necessary for research, training, education and breeding are automatically collected. It is technically easy to grow staple crops such as wheat, rice and maize in a PFAL for research and breeding purposes. However, photosynthetic radiation energy and thus electric energy required per kilogram of produce is roughly tenfold, and economic value per kilogram of produce is roughly one-tenth compared with those of functional plants such as horticultural plants.

## 4 Scope of this book

This book reviews the wealth of research on optimizing PFALs as one potential solution to the challenge of genuinely sustainable agriculture. The book consists of seven parts, with a total of 21 chapters. Each part covers the following topics:

1. Part 1 covers the background, concept and methodology of sustainable PFALs.
2. Part 2 covers energy and other resource performance in relation to environmental assessment.
3. Part 3 covers phenotyping.
4. Part 4 covers spectral manipulations for controlling the growth and quality of leafy greens.
5. Part 5 covers spectral manipulations for ornamentals, fruit vegetables and herbs.
6. Part 6 covers PFAL case studies of businesses from North America, Europe and Asia, with different business models and technical features for commercial production of multiple crops.
7. Part 7 provides an overall conclusion to the book by highlighting the current challenges of sustainable PFALs in both food production and in well-being societies.

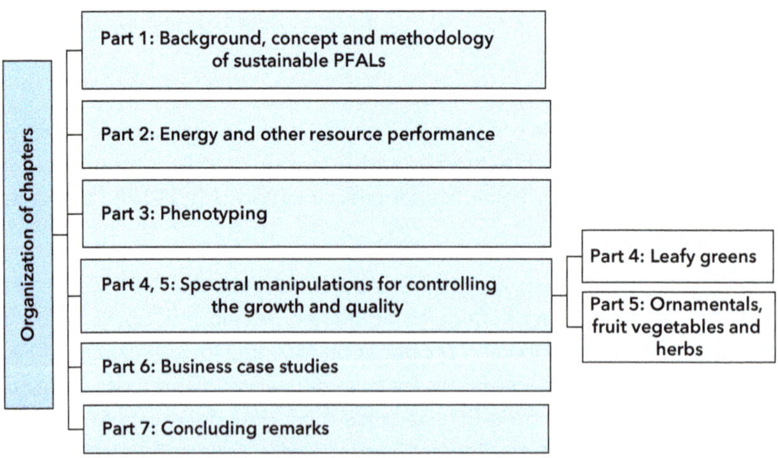


**Figure 1** Organization of this book.

This structure is shown in Fig. 1.

## 5 Conclusion

This chapter has provided an overview of the aims, characteristics and potential of PFALs. In achieving sustainable PFALs, challenges in PFAL research and development have been briefly introduced, including:

1 comparisons of initial investment per kilogram of produce among PFALs, greenhouses and open fields;
2 virtual and dual PFALs;
3 global and local features of PFALs;
4 spatial distributions of environmental factors in a plant canopy;
5 plant phenotyping; and
6 expanding the target plant types for PFALs.

This chapter has also outlined the scope of this book and it constituent parts.

## 6 Acknowledgement

The authors also would like to thank Dr. Hiroko Nakaoka for improving the quality of the manuscript.

## 7 References

Amagai, Y., Lu, N., Hayashi, E., Takagaki, M., Kikuchi, M., Ibaraki, Y. and Kozai, T. (2022). External green light as a new tool to change colors and nutritional components of inner leaves of head cabbages. *J. Food Meas. Charact.* 16: 269–280. https://doi.org/10.1007/s11694-021-01150-y.

Dewir, Y. H. and Alsadon, A. (2022). Effects of nutrient solution electrical conductivity on the leaf gas exchange, biochemical stress markers, growth, stigma yield, and daughter corm yield of saffron in a plant factory. Horticulturae 8(8). https://doi.org/10.3390/horticulturae8080673.

Goto, E. (2020). Chapter 18. Production of pharmaceuticals in a specially designed plant factory. In: *Plant Factory: an Indoor Vertical Farming System for Efficient Quality Food Production*, Kozai, T., Niu, G. and Takagaki, M. (Eds). Elsevier, 251–258.

Hayashi, E., Amagai, Y., Maruo, T. and Kozai, T. (2020). Phenotypic analysis of germination time of individual seeds affected by microenvironment and management factors for cohort research in plant factory. *Agronomy* 10(11): 1680. https://doi.org/10.3390/agronomy10111680.

Hayashi, E., Amagai, Y., Kozai, T., Maruo, T., Tsukagoshi, S., Nakano, A. and Johkan, M. (2022). Variations in the growth of cotyledons and initial true leaves as affected by photosynthetic photon flux density at individual seedlings and nutrients. *Agronomy* 12(1): 194. https://doi.org/10.3390/agronomy12010194.

Hisawa-Tanase, K., Kato, K. and Ezura, H. (2022). Chapter 20. Molecular breeding of miraculin-accumulating tomatoes with suitable traits for cultivation in plant factories with artificial lightings and the optimizations of cultivation methods. In: *Plant Factory: Basics, Applications and Advances*, Kozai, T., Niu, G. and Masabni, J. (Eds). Elsevier, 377-390.

Imamura, E., Iuchi, M. and Bando, S. (2016). Comprehensive Assessment of Life Cycle CO2 Emissions from Power Generation Technologies in Japan. *Socioeconomic Research Center Report (vol.Y06)*. 73pages. (in Japanese with English Abstract).

Kitaya, Y. (2016). Chapter 13. Air current around single leaves and plant canopies and its effect on transpiration, photosynthesis and plant organ temperatures. In: *LED Lighting for Urban Horticulture*, Kozai, T., Fujiwara, K. and Runkle, E. S. (Eds). Springer, 177-187.

Kozai, T. (2019). Towards sustainable plant factories with artificial lighting (PFALs) for achieving SDGs. *Int. J. Agric. Biol. Eng.* 12(5): 28-37.

Kozai, T., Lu, N., Hasegawa, R., Nunomura, O., Nozaki, T., Amagai, Y. and Hayashi, E. (2018). Chapter 26. Plant cohort research and its application. In: *Smart Plant Factory: the Next Generation Indoor Vertical Farms*, Kozai, T. (Ed.). Berlin, Germany: Springer, 413-431.

Kozai, T., Amagai, Y., Lu, N., Hayashi, E., Ibaraki, Y., Takagaki, M., Shinohara, Y. and Maruo, T. (2022). Chapter 23. Toward commercial production of head vegetables in plant factories with artificial lighting. In: *Plant Factory: Basics, Applications and Advances*, Kozai, T. Niu, G. and Masabni, J. (Eds). Elsevier, 417-434.

Lubna, F. A., Lewus, D. C., Shelford, T. J. and Both, A. (2022). What you may not realize about vertical farming. *Horticulturae* 8(4): 322. https://doi.org/10.3390/horticulturae8040322.

Park, Y., Gomez, C. and Runkle, S. E. (2022). Chapter 19. Indoor production of ornamental seedlings, vegetable transplants, and microgreens. In: *Plant Factory: Basics, Applications and Advances*, Kozai, T., Niu, G. and Masabni, J. (Eds). Elsevier, 351-375.

Rodriguez-Morrison, V., Llewellyn, D. and Zheng, Y. (2021). Cannabis yield, potency, and leaf photosynthesis respond differently to increasing light levels in an indoor environment. *Front. Plant Sci.* 12: 646020. https://doi.org/10.3389/fpls.2021.646020.

Saengtharatip, S., Goto, N., Kozai, T. and Yamori, W. (2020). Green light penetrates inside crisp head lettuce leading to chlorophyll and ascorbic acid content enhancement. *Acta Hortic.* (1273): 261-270. https://doi.org/10.17660/ActaHortic.2020.1273.35.

Shibuya, T., Tsuruyama, J., Kitaya, Y. and Kiyota, M. (2006). Enhancement of photosynthesis and growth of tomato seedlings by forced ventilation within the canopy. *Sci. Hortic.* 109(3): 218-222.

Verma, S. K., Gantait, S., Jeong, B. R. and Hwang, S. J. (2018). Enhanced growth and cardenolides production in *Digitalis purpurea* under the influence of different LED exposures in the plant factory. *Sci. Rep.* 8: 18009. https://doi.org/10.1038/s41598-018-36113-9.

Zelkind, M., Livingston, T. and Verlage, V. (2022). Chapter 16. Indoor production of tomatoes. In: *Plant Factory: Basics, Applications and Advances*, Kozai, T., Niu, G. and Masabni, J. (Eds). Elsevier, 295-305.

Zhang, G., Shen, S., Takagaki, M., Kozai, T. and Yamori, W. (2015). Supplemental upward lighting from underneath to obtain higher marketable, lettuce (*Lactuca sativa*) leaf fresh weight by retarding senescence of outer leaves. *Front. Plant Sci.* 6: 1110. https://doi.org/10.3389/fpls.2015.01110.

# Chapter 2

# Requirements and features of cultivation system modules in advanced plant factories with artificial lighting

*Toyoki Kozai and Eri Hayashi, Japan Plant Factory Association, Japan*



## 1 Introduction

This chapter discusses the environmental design and management principles of the cultivation room consisting of one or more airtight and thermally well-insulated cultivation system modules (CSMs) in advanced plant factories with artificial lighting (PFALs). The cultivation room aims to achieve the highest resource and monetary (or economic) productivities ($P_R$ and $P_M$) by realizing the highest annual yield at the highest quality or economic value, with the minimum use of resources and emission of waste, including greenhouse gases (GHG) during their construction and operation of PFALs. Emissions of GHG during the construction and operation of PFALs are analyzed and discussed based on life cycle assessment (LCA) by Molin and Martin (2018). The LCA in PFAL has been becoming increasingly important, as discussed by Martin and Orsini (2022) in Chapter 4 of this book.

http://dx.doi.org/10.19103/AS.2023.0126.03

The PFAL needs to be resource efficient, scalable, adaptable, and energy autonomous by using only electricity generated from natural energy to achieve the above goal. The use of electricity generated by natural energy in combination with improved resource use efficiency (RUE) and $P_R$ of the PFAL can achieve the energy autonomy of the PFAL. The energy autonomy will be enhanced by applying recent technologies and concepts, including plant phenotyping (trait measurements) and three-dimensional (3D) uniform distribution of environmental factors inside the plant community. This chapter provides a revised and updated methodology toward the above goal, based mainly on the energy and material balance equations of the CSM, initially developed by Kozai (2013) and Kozai et al. (2021).

# 2 Requirements of cultivation system modules

## *2.1 Basic requirements*

Each CSM is independent of (or has no interference with) other CSMs in energy and material (or substance) exchanges, while the CSM interacts with other CSMs via a local information network. Then, the cultivation room can be scaled up or down simply by changing the number of CSMs in the PFAL. Further, the CSMs can be standardizable, connectable, substitutable, compatible, open, low noise, and maintenance free. If the electricity generated and stored in batteries is insufficient to operate the whole PFAL, the PFAL with a reduced number of CSMs can be operated by the electricity generated by natural energy. The rest of the CSMs is operated by electricity from other energy sources. The basic requirements of CSMs for producing the highest yield and quality with minimum resource inputs include:

1. negligible influence of weather on environmental factors inside the CSM;
2. negligible loss of $CO_2$ and water vapor to the outside through air gaps;
3. negligible loss of water and nutrient components used for hydroponics and washing/or cleaning the floor and cultivation beds to the outside;
4. negligible condensation of water vapor on the inner surfaces of walls, floors, and ceilings;
5. relatively easy hygiene control of each CSM with no pesticide application; and
6. optimal environmental control for maximizing $P_R$ or $P_M$ with minimum electricity consumption.

The CSM consists of units of lighting, air conditioning, hydroponic cultivation, $CO_2$ enrichment, and measurement and control of environmental factors. In the case of multiple CSMs growing the same cultivar with the same crop calendar, a part of relatively expensive sensors in some CSMs, except the mother CSM, can be omitted to reduce the costs for sensors, assuming that all the plants are growing similarly in the other CSMs. The CSMs for a large-scale PFAL for commercial production require a mechanical ventilation unit with an automatic shutter for air exchange through fine-mesh filters, such as high-efficiency particulate air filters. This forced ventilation unit is only used in an emergency, such as fire and excessive accumulation of volatile organic gases, which may cause physiological disorders of plants and foul odors to humans or harm human health. Such gases are emitted by certain plants (e.g. ethylene, a plant hormone) and certain additives contained in plastic products, including paints on the surface of structural elements (e.g. trichloroethylene and methyl tert-butyl ether).

## 2.2 Measurements of resource inputs, product outputs, and environments

Figure 1 shows resource inputs ($R_I$), product outputs ($P_O$) from the CSM, and environmental control factors in the CSM (Kozai et al., 2021). In the CSM, $R_I$ and $P_O$ can be measured accurately, and all the environmental factors can be controlled at their set points. In other words, the CSM is characterized by high observability and controllability regarding $R_I$, $P_O$, and environments. In Fig. 1, the cameras for monitoring the plant phenotype (or traits) and machine operations are essential in a large-scale PFAL.

All the environmental sensors, flowmeters of $CO_2$, nutrient solution, wastewater, watt meters for the light source, air conditioners, and other electric equipment are relatively small, reliable, and inexpensive. The waste heat energy released from outer units (or condensers) of air conditioners for cooling can be estimated accurately as the sum of the electricity consumption of lighting systems, air conditioners, and other electric equipment such as water pumps and air circulation fans. The estimated waste heat energy under stable inside and outside air temperature conditions is accurate because of the highly airtight and thermally well-insulated CSMs. The airtightness of CSM can be expressed using the number of air exchanges per hour or hourly air infiltration rate divided by the room air volume of CSM, $N_h$ ($h^{-1}$). The thermal insulation characteristics of walls, floors, and ceilings are expressed using a heat transmission coefficient, K ($W/m^2{}^\circ C$).

The chemical composition and volume of a stock solution are recorded when prepared. Amounts of wastewater drained from hydroponic cultivation units and other drainages can be measured separately with filtered flowmeters. The number of seeds planted and locations of the seed trays are recorded with time. The area ratios of seeds, seedlings, and plants to the total cultivation area were calculated separately in PFAL. The fresh weight of plant residue, daily

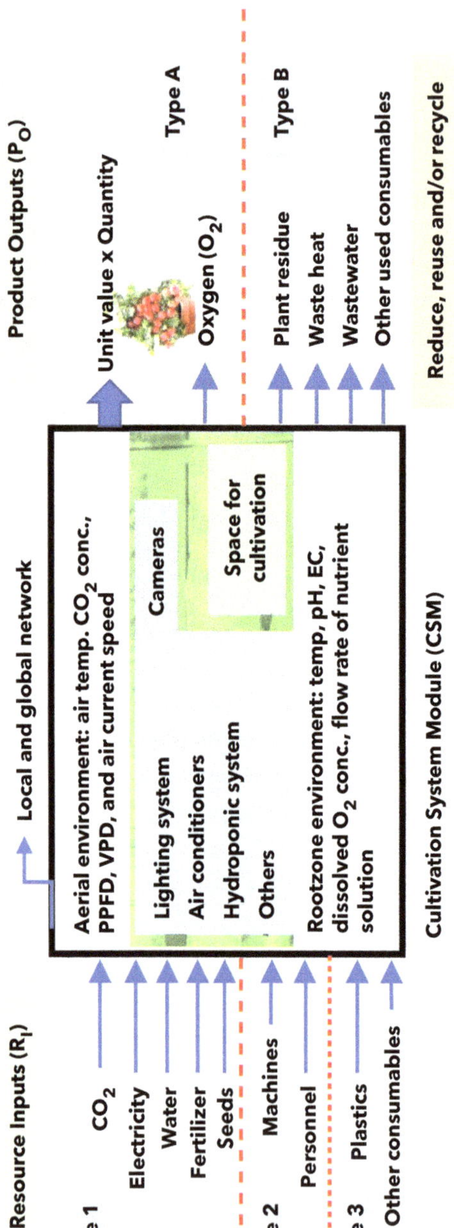


**Figure 1** Resource inputs ($R_I$) and product outputs ($P_O$) in the cultivation system module of plant factory with artificial lighting. Fertilizer includes nitrogen, potassium, and phosphate ions. Source: The revised version of Kozai et al. (2021).

mounts of the used substrate, and other consumables are easily measured or estimated using a weighing balance. The time of any person entering or getting out of the CSM from the air shower room (or entrance/exit of the cultivation room) can be automatically recorded. All the above variables are measured at time intervals of 0.1-24 h.

The data measured online shown in Fig. 1 are used for estimating hourly, daily, and weekly rates of net photosynthesis and transpiration, RUE, and $P_R$, as described in the next section. The PFAL gains high traceability, reproducibility, and predictability with time-series datasets. This feature can be a powerful tool to improve the productivity of each resource element of the CSM successively.

## 3 Net photosynthetic, transpiration and water uptake rates

### 3.1 Net photosynthetic rate of plant community

Under constant $CO_2$ concentration in the airtight CSM, the hourly net photosynthetic rate of the plant community in the CSM, $P_N$ ($kgCO_2/h$), is equal to the hourly $CO_2$ supply rate, $S_C$ (kg/h), namely $P_N = S_C$. This relationship is only valid if the hourly $CO_2$ exchange rate between the room air of the CSM and the following substances is negligible: (1) the air outside the CSM; (2) the substrate or medium for growing plants, and (3) the nutrient solution in the hydroponic cultivation system (saturated dissolved $CO_2$ concentration of nutrient solution decreases with increasing temperatures). In case any of the above exchange rates is not negligible, the $P_N$ is corrected by a corresponding correction factor.

In case $CO_2$ concentration at time t is X ppm higher than that at the time (t + 1 h), a correction factor of ($k_C \times V_R \times X/10^6$) needs to be added to the hourly net photosynthetic rate, as shown in Equation (1).

$$P_N = S_C + (k_C \times V_R \times X/10^6) \tag{1}$$

where $V_R$ is the room air volume ($m^3$) of the CSM, $k_C$ is a conversion factor from the volume to mass (1.96 $kg/m^3$ at 25°C), and $1/10^6$ is another conversion factor from ppm (parts per million) to kg/kg.

The respiration rate is the sum of photorespiration and dark respiration rates. The hourly dark respiration rate of the plant community (and microorganisms, if not negligible), $R_D$ ($kgCO_2/h$), can be estimated by Equation (2).

$$R_D = k_C \times V_R \times \delta CO_2/10^6 \tag{2}$$

where $k_C$ is a conversion factor from the volume to mass (=1.96 $kg/m^3$ at 25δ),$V_R$ is the air volume of the CSM ($m^3$), $\delta CO_2$ is an hourly increase in $CO_2$ concentration (ppm) under dark, and $1/10^6$ is a conversion factor from ppm to $m^3$.

In case any person works in the CSM, the amount of $CO_2$ released by them needs to be considered a $CO_2$ source for plants (around 40–45 g $CO_2$/h/person, which increases with total body weight and intensity of work) (Li et al., 2012; Kozai et al., 2020). $CO_2$ concentration in the CSM is often controlled at approximately 1000 ppm for two reasons. One reason is for human health regulation enacted by the government (its upper limit depends on the government and duration and type of work; however, 1000 ppm is the lowest upper limit in most countries). The other reason is that the net photosynthetic rate of plants generally increases with the increase of $CO_2$ concentration between 0 and 1500 ppm. Therefore, a $CO_2$ concentration higher than 1000 ppm may be better to increase the net photosynthetic rate of plants in case no person exists in the CSM. Besides, no person should enter the CSM for daily management, if possible, to maintain a hygienic environment and increase the floor area for the cultivation racks.

The effect of $CO_2$ concentration on the net photosynthetic rate depends on the cultivar, growth stage, leaf area index (LAI) of plants, and aerial and root-zone environmental factors other than $CO_2$ concentration. Therefore, the online accumulation of time-series data on the net photosynthetic rate, environmental factors, costs for resource supply, and others can be useful for successfully improving the resource and monetary productivities of PFALs (see Section 5 for more details).

## 3.2 Transpiration and water uptake rates of plant community

At a constant vapor pressure deficit (VPD), the hourly transpiration rate of the plant community in the airtight CSM is equal to the hourly water vapor condensation rate at the cooling panels of air conditioners for cooling because no water vapor is lost outside from the CSM and no condensation occurs on the inner surfaces of the thermally insulated CSM. If evaporation from the substrate is not negligible, the hourly water vapor condensation rate is considered the evapotranspiration rate. If the absolute humidity of the room air is Y kg/(kg dry air) higher at the time (t + 1) than at a time t, a correction factor of [Y × $V_R$× specific weight of dry air (kg/$m^3$ dry air)] is added to the hourly transpiration rate where $V_R$ is the volume of room air ($m^3$). The hourly rates of transpiration and evaporation are, respectively, proportional to the VPD regardless of air temperature. More strictly speaking, the transpiration rate is proportional to $VPD_L$, the difference between saturated vapor pressure at the leaf surface temperature and the vapor pressure of surrounding room air. The leaf surface temperature is often 1–2°C lower than the surrounding air temperature due to transpiration or 1–2°C higher due to radiative heating by the lamps.

Further, the hourly water uptake rate of the plant community in the CSM under constant environmental conditions can be estimated as the net supply rate of the nutrient solution to the cultivation beds. Suppose the mass of the nutrient solution in the hydroponic system is X kg greater at (t + 1 h) than at time t. In this case, the water uptake rate must be corrected by subtracting X kg from the above uptake rate, assuming that both the hourly transpiration and water uptake rates are measured accurately. Then, the hourly increase of the fresh weight of the plant community, $\Delta W_F$, is equal to the difference between hourly water uptake, $W_U$ and the sum of hourly transpiration rates, $W_T$, and the hourly dry weight increase rate, $\Delta W_D$, by hourly net photosynthesis (percent dry matter of increased fresh weight is roughly 5% or 0.05 $\Delta W_F$). Namely, $\Delta W_F = W_U - W_T + \Delta W_D$.

Water vapor condenses on the cooling panels of air conditioners when the panel temperature is lower than the dew-point temperature of room air and is converted into sensible heat energy and liquid water. The amount of condensed water is almost equal to that of the transpired water from plants, which is generally recycled as irrigation water after filtering and sterilization.

# 4 Online estimation of resource use efficiency and electric energy use efficacy

The term online refers to the estimation of relevant variables using data obtained through sensors connected to a computer. It does not necessarily mean 'real-time'. The term rate implies an amount of a resource such as water, $CO_2$, and energy per unit time interval (e.g. hour). The term efficiency is used when the denominator and nominator units of a ratio are the same (thus, the unit of efficiency is dimensionless). By contrast, the term efficacy is used when the units of denominator and nominator of a ratio are different.

## *4.1 Online estimation of resource use efficiency*

The RUE of type 1 in Fig. 1 is defined as the ratio of resource A ($CO_2$, electricity, water, fertilizer, or seeds) held or synthesized in plants to resource A supplied to the CSM. The theoretical maximum value of RUE for $CO_2$, water, fertilizer, and seeds is unity. Actual values of RUEs for $CO_2$, water, fertilizer, and seeds can be estimated online. Theoretical maximum values of RUEs for substrate and trays during one cultivation period are unity only if the trays containing plants together with the substrate are sold to the customer. The maximum theoretical value of RUE for electric (or electrical) energy use efficiency is much lower than 0.1, as is shown in Fig. 2. The online RUE estimation for each resource element is beneficial to improve RUEs successively using the time-series data on $R_I$, $P_O$, and environmental factors. It is also beneficial to monitor whether environmental control units are working well and the plants are growing normally.

Estimated RUEs for $CO_2$ and water are, for example, 0.87-0.89 and 0.95-0.98, respectively (Kozai et al., 2020). The value of RUE for seeds in most commercial

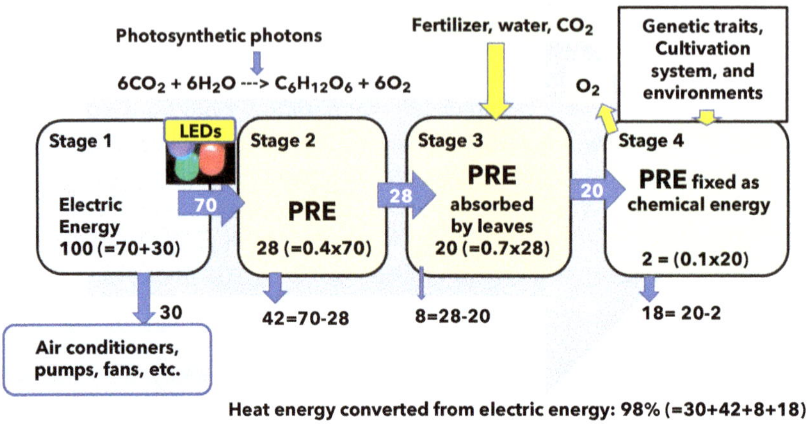


**Figure 2** Simplified conversion processes (stages 0–4) from electric to chemical and heat energy in the cultivation system module (CSM). PRE denotes photosynthetic radiation energy. The values shown are representative percentages of energy in total electric energy consumption in the CSM. Only 2% of the highest of the electric energy is fixed as chemical energy in plants; the rest of 98% (=30 + 42 + 8 + 18) is converted to heat energy or non-PRE.

PFALs in Japan is 0.95 or higher. The RUEs for $CO_2$ and irrigation water increase with the decrease in the number of air exchanges per hour, $N_h$, and the differences in $CO_2$ concentration and water vapor pressure between inside and outside the CSM. RUEs for irrigation water and chemical fertilizer increase with a reduction of the nutrient solution percentage drained to the outside.

The RUE for water used for washing/cleaning the floor, cultivation beds, and trays is zero if not recycled after filtering and sterilization. The RUE for irrigation water in greenhouses is 0.02–0.03 because all the transpired water vapor disappears into the atmosphere, while the RUE for PFALs is around 0.90 (Kozai, 2013). Therefore, the RUE for irrigation water is more than about 30 (0.90/0.03) times greater in PFALs than in greenhouses. Moreover, irrigation water is significantly greater in greenhouses than in PFAL because the greenhouse plants uptake the irrigated water partially, and the rest is evaporated from the soil or substrate surface or leached out from the root zone.

The RUE of each nutrient ion can be estimated online if the concentration of each ion in the nutrient solution is measured separately (Kozai, 2013). However, measuring the concentrations of a few specific ions (e.g. $PO^{4-}$) individually online is difficult. Also, the $NO_3^-$ ion in the nutrient solution, for example, is reduced to $NH_4^+$ ion when absorbed by the plant. Thus, it is a challenge to control each nutrient ion concentration independently and to keep the concentrations of harmful ions and organic acids at safe levels.

## 4.2 Online estimation of electric energy use efficacy

The hourly electric energy use efficacy of lighting for net photosynthesis (mol/kJ) is defined as the ratio of the hourly net photosynthesis rate of the plant community ($CO_2$ mol/h) to hourly electricity consumption for lighting (kJ/h), which can be estimated online. This efficacy is affected by the photosynthetic photon efficacy of light emitting diodes (LEDs (mol/kJ)), environmental factors, the ratio of photosynthetic photons emitted by LEDs to those received by leaves, physiological characteristics of the plant community, and genetic traits of plants. The 6 mol each of $CO_2$ and $H_2O$ in the presence of photosynthetic photons is converted in a leaf to 1 mol of carbohydrate $C_6H_{12}O_6$ and six moles of $O_2$. The net $CO_2$ absorption (mol) by leaves divided by photosynthetic photons (mol) absorbed by leaves is called quantum yield or photosynthetic photon yield. The carbohydrates are further converted to protein, lipid, and other secondary metabolites, such as pigments and vitamins (Fig. 2).

## 4.3 Offline estimation of electric energy use efficiency

Electric energy use efficiency of the CSM for dry mass increase is defined as the ratio of electric energy fixed as dry mass or chemical energy by plants to electric energy consumed in the CSM. It is known from the literature that chemical energy fixed in dry mass is approximately 20 MJ/kg and that the minimum amount of photosynthetic radiation energy (PRE) to fix 1 mol of $CO_2$ is 0.475 MJ/mol (Kozai, 2013). The electric energy use efficiency is approximately 0.02 in a well-designed CSM with suitable environmental conditions (Fig. 2). However, this efficiency is probably around 0.01 in most existing commercial PFALs. The remaining 98–99% of electric energy is converted into heat energy or non-PRE with a wavelength longer than 700 nm. Non-PRE is quickly converted into heat energy when absorbed by any object inside the CSM.

After all, 98–99% of electric energy is converted to heat energy in the CSM and is removed as sensible heat energy from the outer units of air conditioners to the outside. Therefore, it is not recommended to grow stable crops on a large scale by using the electricity generated from fossil fuel and nuclear power because staple crops require, in general, a higher photosynthetic photon flux density (PPFD), a longer cultivation period, and a larger cultivation area to grow per kilogram of produce compared to horticultural and herbal medicinal plants. The production of staple crops on a large scale can be justified only when the electricity is generated by natural energy and the heat energy removed from the PFAL is used to heat or dry an object in a closed space.

The monetary value per kilogram (dry mass) of horticultural crops (e.g. roughly US$6–8/kg fresh weight or US$120–160/kg dry mass for fresh tomatoes with 5% dry matter) is roughly 100-fold of staple crops [e.g. roughly US$1/kg for wheat flour with 85% dry matter or roughly US$1.2/kg dry mass (USDA Economic Research, 2022; Statista, 2022)]. The dry mass of plants, not the fresh mass (dry mass + water) of plants, increases with increasing photosynthetic photons received by the leaves, thus increasing the electricity consumption for lighting.

The electric energy use efficiency of the CSM for the increase in carbohydrates (or dry mass of primary metabolites), $E_{0-4}$, is calculated by multiplying four factors: $S_{0-1}$, $S_{1-2}$, $S_{2-3}$, and $S_{3-4}$ (Equation 3).

$$E_{0-4} = S_{0-1} \times S_{1-2} \times S_{2-3} \times S_{3-4} \tag{3}$$

where $S_{0-1}$ is the ratio of electric energy consumed by LEDs to the electric energy consumed by LEDs, air conditioners, pumps, and fans; $S_{1-2}$ is the ratio of PRE emitted by LEDs to electric energy consumed by LEDs; $S_{2-3}$ is the ratio of PRE absorbed by leaves to PRE emitted by LEDs; and $S_{3-4}$ is the ratio of PRE fixed as chemical energy in carbohydrates in leaves to the PRE received by leaves (called PRE yield). The PRE yield is affected by the spectral PRE distribution of the LED, surrounding environments, the ecophysiological status of plants, plant community architecture, genetic traits of plants, and optical characteristics of leaves. A certain portion of the carbohydrates in leaves is further translocated and/or converted to secondary metabolites in the plant at stage 5.

Currently, it is difficult to estimate the electric energy use efficiency for the dry mass increase online because the PRE use efficiency is too low to measure accurately in stage 4 in Fig. 2. CSM's highest electric energy use efficiency is a product of four expected highest factors (Fig. 2). They are approximately 0.90, 0.60, 0.80, and 0.12, respectively, thereby resulting in the highest electric energy use efficiency of 0.05 (=0.90 × 0.60 × 0.80 × 0.12). However, it is challenging to achieve this efficiency with the use of future technological innovations. It is also noted that only a portion of the plants is marketable or salable (e.g. about 85% for leafy vegetables and about 50% for fruiting vegetables).

## 5 Estimation of resource productivity and monetary productivity

### 5.1 Online estimation of resource productivity

$P_R$ is defined as the ratio of product amount, S, to the amount of resource supplied or applied, $R_{sp}$, as shown in Equation 4 [see Chapter 12 of Kozai et al. (2021) for details]. The $P_R$ is estimated in terms of electric energy (kg of produce per kWh), working hours (kg/h), cultivation bed area and time [kg/(m² h], PRE (kg/kWh), water, $CO_2$, fertilizer element, and other resource elements in a definite period (e.g. one lighting cycle, one day, and one cultivation period).

$$P_R = S/R_{sp} \tag{4}$$

where S is expressed as fresh or dry mass (or weight) or molar weight; $R_{sp}$ is expressed as the weight, amount of energy (kJ or kWh), mol, space ($m^2$ or $m^3$), or time (t or d). Then, the amount of resource supplied, $R_{sp}$, is expressed by Equation (5):

$$R_{sp} = R_{min}/RUE \tag{5}$$

where $R_{min}$ (= RUE × $R_{sp}$) is the minimum amount of resources required to produce S ($R_{sp} \geqq R_{min}$), the RUE is the use efficiency of the resource (= $R_{min}/R_{sp}$) in question with a maximum of 1.0, excluding the PRE use efficiency of approximately 0.1 maximum and the electric energy use efficiency of roughly 0.02 maximum (see Fig. 2; Kozai (2013); Kozai et al., 2013, 2018). Equations (4) and (5) may lead to Equation (6) for $P_R$:

$$P_R = RUE \times S/R_{min} \tag{6}$$

### 5.2 Online estimation of monetary productivity

$P_M$ is expressed by multiplying the selling unit price ($U_P$) in the numerator and the purchase unit price or cost, $U_C$, in the denominator of Equation 7. Therefore, $P_M$ can be estimated online only when the $U_P$ and $U_C$ are obtained online.

$$P_M = U_P \times S/U_C \times R_{SP} = P_R \times (U_P/U_C) = (U_P \times RUE \times S)/(U_C \times R_{min}) \tag{7}$$

On the other hand, not all of the product is sold in reality. Then, a more refined monetary productivity, $P_{MR}$, needs to be developed by considering the rate of the product sold, r. Furthermore, the $P_M$ for case r and RUE are less than 1.0, and $P_{MRP}$ needs to be developed by considering the unit cost of the product residue treatment, $U_R$, and the unit cost of applied resource residue treatment, $U_M$. Further, $P_M$ can be estimated by considering the unit cost of treating the unsold product (plant residue), the unit cost of the treatment of resource residue, and the cost of resources supplied (see Chapter 12 of Kozai et al., 2021 for more details). The $P_M$ of all resources applied equals the value of produce for sale divided by the cost of all the resources applied. The $P_M$ described above can be estimated online if the costs of each resource element and residue and sales prices of produces are obtained online.

## 6 Some practical features of cultivation system modules

### 6.1 Vapor pressure deficit as affected by lighting cycle

The VPD in a CSM with growing plants reaches nearly 0 kPa [or relative humidity (RH) 100%] half an hour after all the LEDs are turned off and during the dark period. This humid room air tends to cause the growth of fungi, stem (or internode) elongation of plants, and physiological disorders of plants. To avoid the low VPD during the dark period, LEDs are, for example, divided into three groups based on cultivation racks or cultivation tiers. Two groups selected from three groups of LEDs are always turned on. In case each photoperiod is 16 h per day, for example, the photoperiods of groups A, B, and C are 0–16 h, 8–24 h, and 16–8 h, respectively (the next day). Then, the CSM air is cooled all day and dehumidified at the same time by air conditioners to remove the heat energy generated by two groups of LEDs. With this lighting method, maximum electricity consumption for lighting and air conditioning can be reduced by one-third compared with three groups of LEDs turning on and off simultaneously. In this case, daily electricity consumption for lighting and air conditioning is almost the same. When plants are weakly sensitive to light and receive adverse pathomorphological effects from neighbor LEDs, the LED lamps must be covered with reflectors so that the light does not reach the plants growing in the neighbor cultivation racks and cultivation tiers. Moreover, many types of seedlings can grow normally under a 24-h photoperiod, although some plant species may suffer from physiological disorders such as necrosis, chlorosis, tipburn, and shortened stem.

Actual VPD and RH in the CSM are determined by the dynamic balance between the transpiration rate of plants and the condensation rate of water vapor at the cooling panel of air conditioners for cooling (Kozai et al., 2020). In the case of the cultivation beds with plants at an air temperature of approximately 20°C, the VPD (or RH) is generally about 0.5 kPa (or 80–85%). However, the VPD (or RH) is affected by the total leaf area of the plant community, stomatal conductance of leaves, air current speed over leaves, and the surface temperature of and airflow rate over the cooling panels. A PFAL user can ask a professional air conditioning engineer to reduce the surface temperature of cooling panels (i.e. dehumidification mode) to increase the condensation rate. The electricity consumptions of the CSM for lighting, air conditioning, and others (e.g. pumps and fans) are approximately 75, 20, and 5% in summer, respectively.

### 6.2 No heating is required throughout the year

If the lighting cycle described above is adopted (i.e. two-thirds of LEDs are turned on all day), heat energy generated from two-third of LEDs is more than sufficient to maintain the room air temperature in the CSM around 25°C even at the outside air temperature of around −30°C. On the other hand, heating may be required on cold days in mushroom and aquaponic cultivation systems because heat energy generated by lamps is much less than that in the CSM. The cooling load of the CSM is almost equal to the sum of sensible heat energy generated by LED lamps and other electric equipment, such as water pumps for nutrient solution delivery, air circulation fans, and electric transportation machine. Approximately 50% of sensible heat energy is converted to latent heat energy (or water vapor) in leaves with most stomata open during transpiration and on the wet floor, walls, substrate, and structural element surfaces during evaporation.

## 6.3 Recycling use of $CO_2$ in the cultivation system module

Under an 8-h shift of lighting cycle, the $CO_2$ concentration inside the CSM is kept at around 1000 ppm all day, approximately 600 ppm higher than outside (about 400 ppm). This high $CO_2$ concentration is one of several reasons why the CSM should not be ventilated even when the room air temperature is higher than its set point and air temperature outside. Under the lighting cycle mentioned above, two-third of plants are under the light and one-third of plants are under the dark. Then, plants under the light [PPFD higher than the light compensation point (approximately 5/μmol$^2$ s)] absorb $CO_2$ released by plants in the same CSM under dark or at PPFD lower than the light compensation point. Then, $CO_2$ is recycled in the CSM, and the daily $CO_2$ supply rate can be reduced by around 10% (the dark respiration rate at 10°C is around 10% of the gross photosynthesis rate).

## 6.4 $CO_2$ and water vapor losses due to air exchanges through air gaps

The cultivation room is assumed to consist of one CSM only to simplify the discussion below. Then, loss rates of $CO_2$ gas and water vapor (kg/h), i.e. $L_C$ and $L_W$, because of air exchanges (or air infiltration) from the CSM to the outside because of air gaps are expressed as follows, respectively:

$$L_C = k_C \times N_H \times V_R \times (C_{in} - C_{out})/10^6 \quad (8)$$

$$L_W = k_W \times V_R \times N_H \times (X_{Win} - X_{Wout}) \quad (9)$$

where $k_C$ is a conversion factor of $CO_2$ from volume to mass (1.96 kg/m$^3$ at 25°C); $k_W$ is a conversion factor of water vapor from volume to mass (0.80 kg/m$^3$ at 25°C); $N_H$ is the number of air exchanges per hour of CSM (1/h); $V_R$ is the air volume of CSM (m$^3$); $C_{in}$ and $C_{out}$ are $CO_2$ concentration (ppm) inside and outside the CSM, respectively; and $X_{Win}$ and $X_{Wout}$ are the absolute humidity of the air inside and outside the CSM [kg/kg (dry air)], respectively. It is desirable

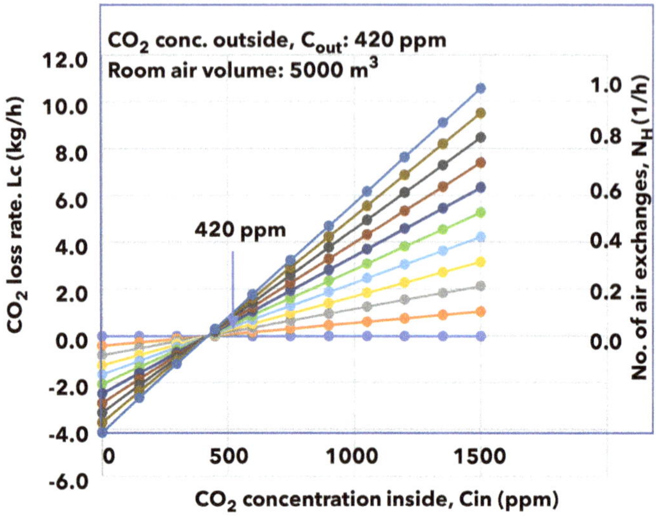


**Figure 3** $CO_2$ loss rate as affected by $CO_2$ concentration inside the cultivation room, $C_{in}$, and number of air exchanges per hour, $N_H$.

to estimate the $N_H$ immediately after the construction of the CSM without plants and maintain $L_C$ and $L_W$ below 1% of the maximum expected supply rates of $CO_2$ and water to the CSM, respectively. The $N_H$ is expected to be lower than 0.03–0.05/h.

The heat energy exchange rate between inside and outside the sidewalls and ceiling, $L_H$, is expressed by Equation (10):

$$L_H = K \times A_W \times (T_{in} - T_{out}) \tag{10}$$

where K is a heat transmission coefficient of sidewalls, ceiling, and floor, and the desirable K value is 0.1-0.2 W/m²/°C. Heat energy exchange through the floor must be considered, especially in cold regions. Wind speed and direction outside the building of the PFAL and other factors affect the K value.

Figure 3 shows the $CO_2$ loss rate, $L_C$, as a function of $CO_2$ concentration inside the CSM, $C_{in}$, and the number of air exchanges per hour, $N_H$. It is assumed that the $CO_2$ concentration outside the CSM, $C_{out}$, is 420 ppm, and the air volume of the CSM, $V_R$, is 5000 m³ ($L_C$ is proportional to $V_R$ and $N_H$). The average $C_{out}$ is 420 ppm as of 2021, although it varies in the range of 380–440 ppm depending on the season, time of day, weather, and region. Besides, the average $C_{out}$ has been increasing by more than 1 ppm each year lately. The recommended $N_H$ is 0.03–0.05/h or less. The $N_H$ is somewhat affected by wind speed and direction surrounding the CSM.

## 7 Challenges for realizing potential features of cultivation system modules

Challenges in implementing and upscaling vertical farming are well-reviewed by Van Delden et al. (2021). This section discusses a few selected issues related to the challenges.

### 7.1 Phenotype-based environmental control

The plant phenotype (or plant traits) refers to expressions of ecophysiological status, chemical composition, concentrations, the architecture of plant community affected by genetic traits of plants, management by humans, machines/cultivation systems, and environments (Hayashi et al., 2020; Hayashi et al., 2022). Figure 4 shows the plant phenotype-based environmental control of CSM (Kozai et al., 2021). With this system, set points of environmental factors are determined based on the multi-purpose function with target values of plant production provided by a PFAL manager concerning harvesting date, quantity and quality of produce, market information, and phenotype data. The market information includes demand, retail prices, and weather forecasts. The phenotype is primarily obtained automatically using various cameras and other remote sensors (see Fig. 1 for details of $R_I$ and $P_O$). It should be noted, in Fig. 4, that P, E, and M interact with each other.

The PFAL shown in Fig. 4 will be developed as the next-generation PFAL in the coming years (Kozai et al., 2020). The next-generation PFAL contains a PFAL simulator or a virtual PFAL containing a decision-making module with mechanical, multiple statistical, behavioral, and artificial intelligence (AI) models. Then, the simulator provides the PFAL manager with several production plans with the data on suggested environmental control set points and their expected costs to optimize the multi-objective function.

### 7.2 Uniform three-dimensional distributions of environmental factors within the plant community

Two hypotheses (A and B) described below are potential basic features of environmental control in the CSM to be introduced in the next-generation PFALs. These hypotheses are presented assuming that the spatial environmental factors inside the plant community, not those above or surrounding the plant community, need to be controlled.

#### 7.2.1 Hypothesis A

3D uniformity of environmental factors [PPFD, VPD, and air current speed (diffusion coefficient of gases)] and $CO_2$ concentration at optimal levels in the

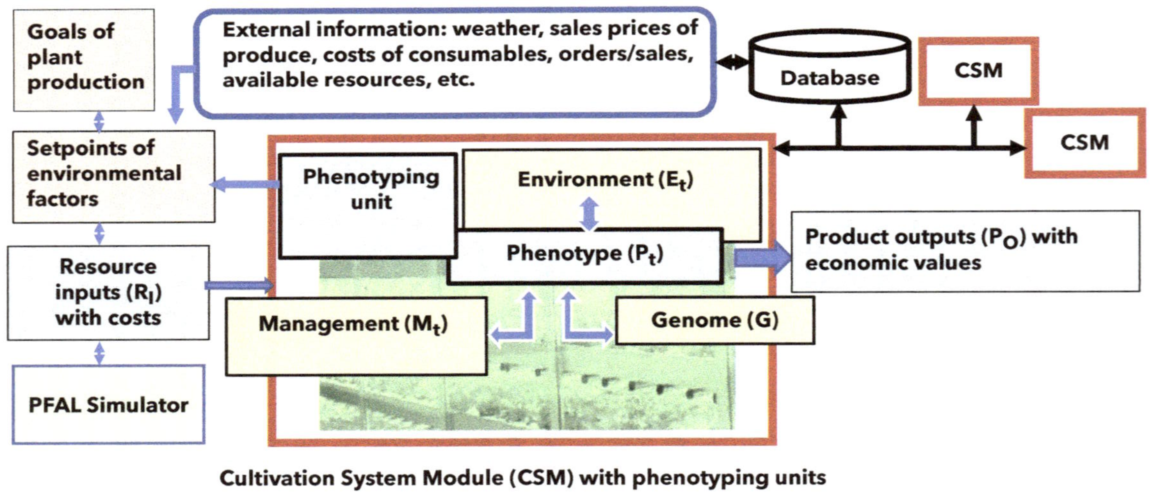


**Figure 4** Scheme showing phenotype-based environmental control of cultivation system module where set points of environmental factors are determined based on aims of plant production, market information, and the phenotype data, which are mostly obtained by use of various types of cameras and other types of sensors. The P at the time (t + 1) is expressed by Function($P_t$, $E_t$, $M_t$, G). See Fig. 1 for details of $R_I$ and $P_O$. Source: Substantially revised version of Kozai et al. (2021).

plant community maximizes the net photosynthesis, growth of aerial parts of plants, and 3D uniformity of phenotype in the plant community.

#### *7.2.1.1 Background*

Bottom-up or upward airflow in cultivation panels provide spatially uniform air current speeds, promoting the net photosynthesis of plant community (e.g. Shibuya et al., 2006; Kitaya, 2016). Supplemental upward lighting with LEDs from beneath at the bottom of the plant community provides higher marketable lettuce leaf fresh weight by retarding the senescence of outer leaves (Zhang et al., 2015). Moreover, these uniform environments considerably simplify the relationships between the environment and plant growth/phenotype. This simplification is key to developing a practical growth model of the plant community as affected by the surrounding environment.

In a densely populated plant community in the CSM, PPFD and air current speed (or diffusion coefficients of $CO_2$ gas, water vapor, and sensible heat energy) decrease sharply with depth. Then, the plant canopy restricts the penetration of photosynthetic photons, diffusions of $CO_2$ gas, and water vapor. Therefore, VPD decreases with depth in the plant community. Such environments reduce the net photosynthetic and transpiration rates by increasing plant community depth and growth. The term canopy means the aboveground portion of a plant community.

The transmittance of green leaves to blue and red lights is lower than green and far-red lights. The reflectance of green leaves to green and far-red lights is higher than blue and red lights (Saengtharatip et al., 2020; Amagai et al., 2021). Thus, percentages of green and far-red lights over visible light increase with depth in a densely populated plant canopy. Then, light spectral distribution at the bottom of the canopy is considerably different from that at the canopy surface if no blue and red lights are supplemented in the middle and bottom of the canopy. It is also noted that the spatial distributions of the phenotype in the plant community may not be uniform even in uniform environments because of the complex interactions of matter and signals between the plant and its organs.

Figure 5 represents the upward (or bottom-up), downward, and horizontal airflows in a closed sidewall cultivation tier (layer) in a CSM. The white-colored internal sidewall surfaces improve the uniformity and average PPFD at the canopy surface. Besides, closed sidewalls can avoid the effects of airflow just outside the layer on the airflow velocity in the growing area and the airflow direction.

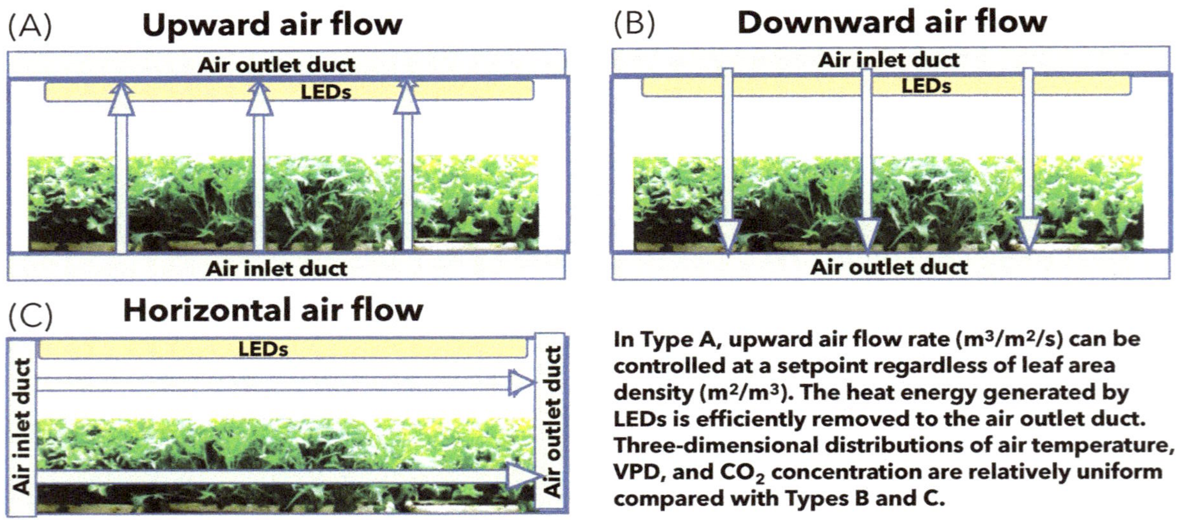


**Figure 5** Three patterns (A, B, and C) of airflow in a cultivation tier in the CSM. Internal surfaces of sidewalls and ceiling are white for high light reflection. Air temperature, VPD, and $CO_2$ concentration are controlled during air circulation. Arrows indicate the direction of airflow.

### 7.2.2 Hypothesis B

When the supply rate of water, nutrients, or dissolved oxygen is too low to maintain or grow the aerial parts, the root weight (or mass) tends to increase, or the aerial mass tends to decrease to balance the supply and consumption relationships. On the other hand, if roots (or rhizosphere) are provided with more than adequate water, nutrients, and dissolved oxygen by spatially uniform root-zone environments, the number and mass of secondary and hairy roots will not increase significantly (Kozai et al., 2021). Under such conditions, the root-zone mass is minimized without restricting the growth of aerial parts of plants, the harvest index (mass ratio of aerial parts to whole plants) of leafy vegetables is maximized, and the relationships between the root-zone environments and root phenotype are simplified. This simplification is key to developing a practical root-zone growth model of the plant community as affected by the surrounding root-zone environment.

#### 7.2.2.1 Background

Growing and actively metabolizing plants generally require high amounts of water, nutrients, and dissolved oxygen. In hydroponic cultivation beds with continuous and one-way nutrient solution supply, the nutrient solution flow around the crowded root mass of growing plants is restricted compared to the less crowded mass of young plants. In contrast, the crowded root mass needs greater amounts of water, nutrients, and dissolved oxygen for further growth. In this case, the supply of water, nutrients, and dissolved oxygen to the central part of the crowded root mass is restricted even when the total flow rate of nutrient solution in the cultivation bed is relatively high. Under such conditions, new roots emerge to get dissolved oxygen, while the old roots tend to be damaged or die. In other words, the roots tend to grow to uptake more water, nutrient elements, or dissolved oxygen only when they are insufficient. It should be noted that physiological status, the surface area of roots, the population of microbes/pathogens, bio-stimulants, and endophytes inside and outside the roots affect uptake rates of water, nutrient ions, and dissolved oxygen and that the roots can excrete chemical substances such as organic acids and produce phytohormones such as cytokinin.

### 7.3 Computational fluid dynamics of the air and nutrient solution flows inside the plant community

The computational fluid dynamics (CFD) software helps simulate airflow patterns in the CSM and nutrient solution flow patterns in the hydroponic cultivation beds (e.g. Zhang and Kacira, 2018). However, there is little research on the airflow patterns in the plant canopy as a function of plant architecture [leaf angle distribution, plant density ($m^2$ leaf area/$m^3$ cultivation space), leaf and stem elasticity causing complex 3D vibrations of leaves and stems, and gas diffusion coefficients] under given external air and energy flows around the plant canopy. The same applies to the nutrient solution flow patterns in the hydroponic cultivation beds. Therefore, CFD research is expected to simulate spatial airflow patterns in plant canopies and spatial nutrient solution flow patterns in the cultivation beds.

## 7.4 Automatic spacing

Prince and Bartok (1978) pointed out that automatic 2D and frequent or continuous spacing of seedlings/plants as they grow is critical for commercializing PFALs. However, this pressing issue remains unsolved technologically, which is a limiting factor in popularizing commercial PFALs.

In PFALs without automatic spacing, seedlings of leafy vegetables are transplanted manually or by robotic hands, mostly two times around 2 and 4 weeks after sowing. Then, the LAI is lower than 0.05 after the first transplanting and 0.10 after the second. On the other hand, at the third growth stage, the LAI increases quickly and often reaches two or greater several days before harvesting and three or greater at harvesting. An excess increase in the LAI causes mutual shading of leaves and, thus, an exponential decrease in PPFD with a depth of plant canopy. Also, the dense plant canopy restricts air movement in the lower part of the plant canopy. Thus, $CO_2$ (or photosynthesis) and water and nutrient uptake (or transpiration) can also be affected. The above restrictions often cause senescence, yellowing/etiolation, growth retardation, and death of lower leaves of the plant canopy, leading to low PPUE and relative growth rate of the plant community.

Automatic and frequent spacing is a key to improving the cultivation space (or area) use efficiency [CSUE; the ratio of the cultivation tray/bed space ($m^3$) or area ($m^2$) to the floor space/area of the cultivation room]. The CSUE is much greater in the PFAL with automatic spacing than the one without automatic spacing. Besides, working hour efficiency (the number of plants transplanted for spacing per unit time) is higher in the PFAL with automatic spacing.

Another advantage of automatic spacing is the high photosynthetic photon (or light energy) use efficiency (PPUE; the ratio of the net photosynthetic rate of the plant community, $P_N$, to the PPFD at the canopy surface of the plant community. The $P_N$ is equal to the $CO_2$ supply rate in case the plants are grown in an airtight growing space at a constant $CO_2$ concentration. The $P_N$ and, thus, PPUE are affected by the LAI (the area ratio of leaves to the cultivation bed/tray) and the growing-space environment for photosynthesis. The LAI can be controlled optimally by automatic spacing around unity at the first and second growth stages characterized by a relatively high planting density (or narrow cultivation area) and a low PPFD, and about 2 at the third (or last) growth stage (about 10 days before harvest in the case of leafy vegetables) characterized by a low planting density and a high PPFD. By doing so, the PPUE increases significantly.

In short, developing automatic 2D and frequent/continuous spacing units is a challenge for improving the CSUE, PPUE, working hour efficiency, and the relative growth rate in the next-generation PFALs and greenhouses and perhaps in the space farms on the Moon and Mars.

## 8 Conclusion

Illuminated by LEDs for the commercial production of leafy lettuce plants, PFAL has only 25 years of history since it was first built in 1998 (Kozai et al., 2021). Thus, its technology for commercialization is still immature in many aspects of productivity improvements, reduction in GHG or $CO_2$-equivalent emissions per kg of produce, and energy autonomy. For example, implementations of recent technologies to the PFALs have been limited concerning AI, the Internet of Things, image processing using small multifunctional cameras, and LED lighting systems. Moreover, substantial developments are necessary regarding the concept and methodology of the next-generation PFAL design and management. An ultimate goal of the next-generation PFAL is the energy-autonomous PFAL, which thinks on its own to optimize the multi-purpose under the given climatic, social, and economic conditions. This chapter is a small step toward the goal of energy-autonomous PFAL.

The airtight and thermally insulated CSM with the characteristics described in this chapter provides several advantages over conventional cultivation rooms. Using the CSMs, the estimation and analyses can be conducted online relatively accurately regarding environmental effects on net photosynthetic and transpiration rates of the plant community, RUE of each resource element, and $P_R$ in the CSM during actual commercial plant production processes. By adding a phenotyping unit and local/global network to the CSM, $P_M$ can be optimized online based on the automatic determination of environmental set points with optimal resource allocation. The above data can be analyzed online regarding interactions between the room environments and ecophysiological characteristics (LAI, plant height, leaf thickness, and pigment concentrations) of the plant community. The PFAL with the above capability will improve its sustainability significantly.

## 9 Acknowledgements

The authors would like to thank Dr. Francis Baumont de Oliveira for his valuable comments on improving the manuscript.

## 10 Where to look for further information

The following review provides an overview and features of advanced plant factories with artificial lighting.

It will provide a brief understanding of the requirements and features of the cultivation system modules in advanced plant factories with artificial lighting.

- Kozai, T. (2018) Benefits, problems and challenges of plant factories with artificial lighting (PFALs): a short review. Acta Horticulturae 1227: 25-30. DOI: 10.17660/ActaHortic.2018.1227.3. https://doi.org/10.17660/ActaHortic.2018.1227.3.

This article discusses the benefits and problems to be solved and the challenges for plant factories with artificial lighting (PFALs). It also includes the challenges for the next-generation smart PFALs, such as introducing advanced technologies and artificial intelligence using big data, genomics, and phenomics.

## 11 References

Amagai, Y., Lu, N., Hayashi, E., Takagaki, M., Kikuchi, M., Ibaraki, M. and Kozai, T. (2021). External green light as a new tool to change colors and nutritional components of inner leaves of head cabbages. *Journal of Food Measurement and Characterization* 16, 269–280 (2022). https://doi.org/10.1007/s11694-021-01150-y.

Hayashi, E., Amagai, Y., Maruo, T. and Kozai, T. (2020). Phenotypic analysis of germination time of Individual seeds affected by microenvironment and management factors for cohort research in plant factory. *Agronomy* 10(11), 1680. https://doi.org/10.3390/agronomy10111680.

Hayashi, E., Amagai, Y., Kozai, T., Maruo, T., Tsukagoshi, S., Nakano, A. and Johkan, M. (2022). Variations in the growth of cotyledons and initial true leaves as affected by photosynthetic photon flux density at individual seedlings and nutrients. *Agronomy* 12(1), 194. https://doi.org/10.3390/agronomy12010194.

Kitaya, Y. (2016). Air current around single leaves and plant canopies and its effect on transpiration, photosynthesis, and plant organ temperatures. Chapter 13 of *LED Lighting for Urban Agriculture*, Eds. Kozai, T., Fujiwara, K. and Runkle, E. S. 177–187. Springer Nature: Singapore

Kozai, T. (2013). Resource use efficiency of closed plant production system with artificial lighting: concept, estimation and application to plant factory. *Proceedings of the Japan Academy Series B* 89, 447–461.

Kozai, T., Niu, G. and Takagaki, M. (Eds.) (2020). *Plant Factory: an Indoor Vertical Farming System for Efficient Quality Food Production* (2nd edn.). Academic Press, 487 pages. Elsevier: UK.

Kozai, T., Niu, G. and Masabni, J. (Eds.) (2021). *Plant Factory: Basics, Applications and Advances*. Academic Press, 444 pages. Elsevier: UK

Li, M., Kozai, T., Ohyama, K., Shimamura, S., Gonda, K. and Sekiyama, T. (2012). $CO_2$ balance of a commercial closed system with artificial lighting for producing lettuce plants. *HortScience* 47(9), 1257–1260.

Martin, M. and Orsini, F. (2022). *Life cycle assessment of indoor vertical farms*. In: Kozai, T. and Hayashi, E. (eds.), *Advances in plant factories: New technologies in indoor vertical farming*, Burleigh Dodds Science Publishing, Cambridge, UK, 2023, (ISBN: 978 1 80146 316 4).

Molin, E. and Martin, M. (2018). Reviewing the energy and environmental performance of vertical farming systems in urban. IVL Swedish Environmental Research Institute, Report number C. 30 pages.

Prince, P. R. and Bartok Jr., J. W. (1978). Plant spacing for controlled environment plant growth. *Transactions of the ASAE. American Society of Agricultural Engineers* 21(2), 0332-0336. https://doi.org/10.13031/2013.35299.

Saengtharatip, S., Goto, N., Kozai, T. and Yamori, W. (2020). Green light penetrates inside crisphead lettuce leading to chlorophyll and ascorbic acid content enhancement. *Acta Horticulturae* 1273, 261–269.

Shibuya, T., Tsuruyama, J., Kitaya, Y. and Kiyota, M. (2006). Enhancement of photosynthesis and growth of tomato seedlings by forced ventilation within the canopy. *Scientia Horticulturae* 109(3), 218–222.

Statista (2022). *Retail Price of All Purpose White Flour in the United States from 1995 to 2021*. https://www.statista.com/statistics/236624/retail-price-of-white-flour-in-the-united-states/#:~:text=In%202021%2C%20the%20price%20for,to%20approximately%20 0.39%20U.S.%20dollar.

US Department of Agriculture. National retail report (2022). *Specialty Crops*, Page 11. https://www.ams.usda.gov/mnreports/fvwretail.pdf.

Van Delden, S. H., SharathKumar, M., Butturini, M., Graamans, L. J. A., Heuvelink, E., Kacira, M., Kaiser, E., Klamer, R. S., Klerkx, L., Kootstra, G., Loeber, A., Schouten, R. E., Stanghellini, C., van Ieperen, W., Verdonk, J. C., Vialet-Chabrand, S., Woltering, E. J., van de Zedde, R., Zhang, Y. and Marcelis, L. F. M. (2021). Current status and future challenges in implementing and upscaling vertical farming systems. *Nature Food* 2(12), 944–956. https://doi.org/10.1038/s43016-021-00402-w.

Zhang, Y. and Kacira, M. (2018). Analysis of environmental uniformity in a plant factory using computational fluid dynamics (CFD) analysis. *Acta Horticulturae* 1227, 607-614. https://doi.org/10.17660/ActaHortic. 2018.1227.77.

Zhang, G., Shen, S., Takagaki, M., Kozai, T. and Yamori, W. (2015). Supplemental lighting from underneath to obtain higher marketable lettuce (*Lactuca sativa*) fresh leaf weight by retarding senescence of outer leaves. *Frontiers in Plant Science* 6(1110), doi: https://doi.org/10.3389/fpls.2015.01110.

# Chapter 3

# Research and technology in plant factories with artificial lighting: past, present and future

*Ying Liu, Paul Kusuma and Leo F. M. Marcelis, Wageningen University & Research, The Netherlands*



## 1 Introduction

Technology in plant factories with artificial light (PFALs) is rapidly progressing. PFALs - also called vertical farms or indoor farms - are controlled environment systems where plants (and mushrooms) grow on a single layer, on stacked layers or on vertical walls in a building without the direct use of solar light for illumination. This chapter describes the past, present, and future of six PFAL technology categories (Fig. 1):

- Architecture and automation
- Lighting
- Climate control
- Sensors and algorithms
- Rootzone
- Genetics

http://dx.doi.org/10.19103/AS.2023.0126.04

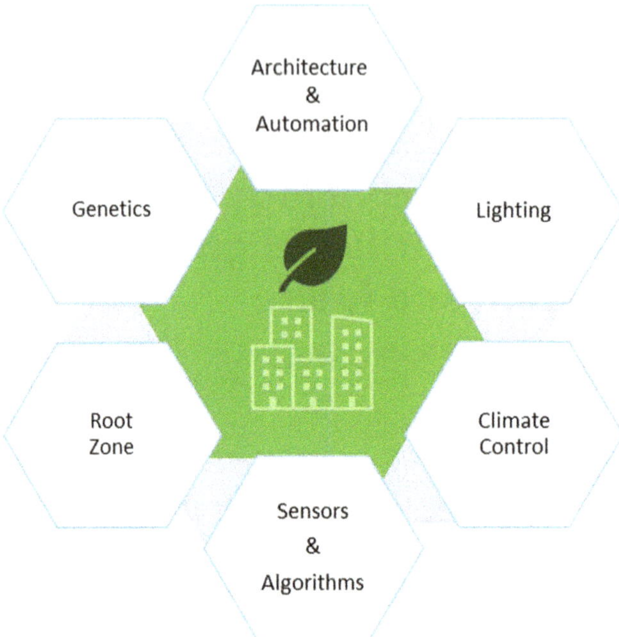


**Figure 1** Six broad technology categories in PFALs.

This chapter is partially based on a recent FAO (Food and Agriculture Organization of the United Nation) report on technologies in urban and peri-urban agriculture (FAO, 2023). Research in this field drives management innovations and technological developments (Van Delden et al., 2021), which further enhances yield, quality, and efficiency in these systems.

PFALs produce high yields year-round. The prevailing crops in these systems currently include leafy greens, herbs, and microgreens, although a diversity of crops is technically possible. Fruit crops like cherry tomatoes and berries are the targets for near-future development, but their economic suitability is still undetermined. Staple crops are not commercially grown in PFALs due to a lack of profitability at the current stage of technology (Pattison et al., 2018; Asseng et al., 2020). The potential yield for lettuce - and possibly other leafy greens - in PFALs was estimated to be 700 kg $m^{-2}$ $year^{-1}$ (Jin et al., 2022), but current yields are only a fraction of this. Lettuce and tomato yields in highly controlled PFAL environments are currently about double the yields of low- to mid-tech greenhouses and about 50 times the yields of open fields on a unit growing area basis (Fig. 2).

PFALs improve crop quality - including sensory, nutritional, and functional attributes - to a premium level. In addition to superior genetic material, this quality is the result of morphological and metabolic manipulation through

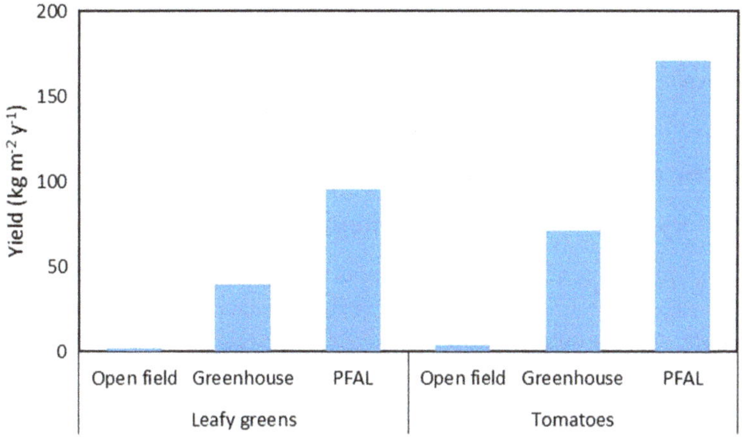


**Figure 2** Current crop yield (fresh weight per unit cultivation area per year) of leafy greens and tomato fruits in open field, greenhouse, and PFALs according to company reports and literature (Maureira et al., 2022; Heuvelink, 2018; FAO, 2023).

precise management and environmental control. For instance, high light intensity applied at the end of production has been shown to increase the nutritional quality of lettuce and basil (Min et al., 2021; Larsen et al., 2022). Beyond physical and chemical attributes of quality, these systems can also provide products that are pesticide and herbicide free. The clean environment avoids the need to wash produce to remove contaminants, which contributes to increased shelf life. Finally, although it is not directly related to the technology, the short supply chains of these systems can ensure fresher produce compared to field production. All these attributes contribute to the perceived increase in quality of PFAL products, putting them on a price point which is competitive with the premium paid for organic produce.

Finally, for equivalent yields, PFAL production uses fewer natural resources - including water and land area - compared to greenhouse and open field production, reducing environmental impact (Orsini et al., 2020). Water and nutrients can be recirculated in PFALs, which ensures their efficient use and prevents potential negative environmental effects. In open field production, water and nutrients are lost to the environment, leading to nutrient runoff and eutrophication (Romeo et al., 2018). Furthermore, high-quality water is becoming increasingly scarce, and agriculture is one of the primary consumers of water (Patle et al., 2020).

One concern about PFALs is that their high yield and high quality come at the cost of significant energy usage. With advances in technology, energy efficiency is increasing. This has largely been driven by the rapid increase in LED efficiency over the past decade. However, despite this improvement, energy use is still very high. With their high energy use, these systems may be the most economical or practical solution in extreme climates such as desert regions, arctic regions, or tropical regions, particularly compared to greenhouses, which are not well insulated.

## 2 Architecture and automation

### *2.1 History of controlled environments*

Documentation of controlled environments dates as far back as the first century CE, where *Cucumis* were grown for Emperor Tiberius in small structures, called *Specularia*, that had transparent stone roofs. In this system, *Cucumis* plants - likely *Cucumis melo* - were planted in containers on carts that could be rolled into the *Specularia* to be kept warm on cold days, allowing year-round growth (Janick and Paris, 2022). Over a millennium later, in the 1600s and 1700s, *Orangeries* with large southern facing windows were used to grow rare fruits and vegetables - primarily citrus - during winters in France and other European countries (Raviv et al., 2008; Nemali, 2022). Conservatories, with a higher fraction of their walls made of glass, were a further advance from *Orangeries*. An additional feature of conservatories was their sloped glass ceilings, which provided more light to the plants. *Orangeries* and conservatories are not currently suitable for commercial plant production, but they laid the groundwork for current plant factories by developing structures that began to control the local environment around the plants. Modern greenhouses, along with cheaper modern protected structures (e.g. shade houses, high tunnels, parral-style greenhouses, and solar greenhouses), are suitable for profitable crop production, but these systems lack the complete control of PFALs.

### *2.2 Current status of plant factories with artificial light*

Unlike other controlled environment structures with transparent walls and ceilings for the transmission of sunlight, PFALs are contained within a well-insulated building without any access to sunlight. All the photons for plant growth are provided by electric light sources, usually LEDs.

PFAL systems can range from mini-scale appliance farms to large-scale production systems (Fig. 3). Some PFALs focus primarily on large-scale production facilities in a single large room, while other PFALs take advantage of modularity. Modularity refers to smaller units - similar to container farms - that are repeated multiple times in the whole operation. The primary benefit of modularity is that it allows for functional zones. For example, crops with different temperature requirements, like lettuce and basil, can be grown in separate modules each at their optimum temperature. Another advantage of modularity is the reduced risk of pest and disease outbreaks.

**Figure 3** Four types of PFAL systems: crops are cultivated (a) in an in-store farm at retailers, (b) in a container, (c) in independent mega farms inside a building, and (d) directly in a building. Source: Photos are from Infarm (www.infarm.com), Urban Crop Solutions (www.urbancropsolutions.com), and Aerofarms (www.aerofarms.com).

**Figure 4** Two PFAL architectures: (a) stacked layers (Source: Aerofarms, www.aerofarms.com) and (b) vertical wall towers (Source: Plenty, www.plenty.ag).

Whether the system is modular or not, the internal architecture within the walls can be (Fig. 4):

- a single layer,
- stacked layers, and
- vertical walls.

The more traditional and more common stacked layer systems utilize shelving (Beacham et al., 2019), while less common vertical walls utilize growing towers with slots for plants. In the vertical wall system, lighting is also positioned in a vertical orientation, such that the whole system can be thought of as a stacked layer rotated 90°. Plants in vertical walls are not typically oriented at exactly 90° but angled slightly upward.

Plant mobility and variable spacing are possible in both architectures. Vertical towers can hang from a conveyor system, which allows them to easily move through the facility - from sowing to harvest. In stacked layers, lifts and mobile gutters can be used to move plants. Vertical walls have been suggested to have better environmental control because they allow for improved natural air movement and circulation (warm air rises). Stacked layers could theoretically be designed in such a way that also allows for this natural air movement, minimizing trapped air. Vertical wall systems allow fruits to hang in the air, similar to gutter systems in greenhouses, which is highly beneficial for crops like strawberry because it decreases disease pressure compared to fruits lying on a surface. Overall, vertical walls may have advantages over stacked layers, or vice versa, but these technologies are still young. It should be noted that if space is not a concern, growing on a single layer is also a viable common approach for high-output-controlled environment crop production.

## 2.3 Automation

The investment cost of automation is high, but this investment can significantly reduce labor costs, which constitute 25-45% of the operational cost (FAO, 2023). Furthermore, because cost of automation does not scale linearly with growing area, the relative investment into automation is lower for larger PFALs. Automated tasks include planting, transplanting, spacing, harvesting, and packaging. Automated machines require precision to perform these tasks without damaging plants, and this is an issue that requires continued development. These systems can include vision systems for higher precision, but this significantly increases costs.

Utilizing plant mobility has two benefits: variable spacing and decreased investment into other technologies. Variable spacing involves moving the plants farther apart as they grow, allowing for a high planting density at the beginning of growth, and a lower planting density by the end of growth. The result of variable spacing is a high light capture efficiency throughout the growth cycle without inducing plant competition. Plant mobility can decrease investment into other technologies by moving plants to measurement or treatment devices, rather than installing these devices above each plant. For example, ultraviolet (UV) light is sometimes applied to the plants at the end of the growth cycle to increase quality. Instead of installing expensive UV lamps above each plant, UV lamps can be installed in a specific treatment section where the plants are moved to at a specific time. Similarly, plants can be regularly moved to an imaging station for performance assessment.

Theoretically, the entire system can be automated, and many companies are moving in this direction. However, it is currently still common for workers to be involved in many processes. With minimal automation, a single worker is expected to maintain and monitor between 100 $m^2$ and 250 $m^2$ of growing area but, as automation increases, this capacity could increase to one worker per 1500–2000 $m^2$.

### 2.4 Future of architecture and automation

As previously mentioned, the superiority of either stacked layers or vertical walls over the other remains to be determined. Innovations are continuing in both systems and overall designs are continuously improving. It seems possible that specific technologies within each type of architecture could provide an advantage, but this remains to be seen. Perhaps an even better, more radical reimagining of PFAL architecture is still in the future.

Automation will continue to develop alongside system architecture. The towers in vertical wall systems are currently moved farther apart as the plants grow, but plant spacing on a single tower is fixed. Perhaps light use efficiency (LUE) in this type of system could be further improved by developing variable spacing in both directions, although it is always important to consider the trade-off between this kind of investment and payoff in economic terms. The precision of automation, especially regarding plant monitoring and harvesting, still requires further development, such as vision recognition, decision-making, and adapted robotics. It is important to restate that leafy greens are currently the primary focus of PFALs. This means that automated processes are currently being optimized for harvesting leaves and not fruits. As fruits become more economically viable in PFALs, systems for improving their automation will likely follow.

## 3 Lighting

### 3.1 History of artificial lighting for plant growth

The use of electric lamps to grow plants dates back to experiments in 1861 (Mangon, 1861; Wheeler, 2008), and references to their use commercially date as far back as 1893 when lettuce and cucumber were grown under arc lamps in a greenhouse to increase yields by 15% (Scientific American, 1893; Rawson, 1895; Spence, 1962). For much of the twentieth century, fluorescent lamps were paired with incandescent lamps to provide photons within the photosynthetically active radiation (PAR; 400–700 nm) and far-red (700-800 nm) regions (Downs, 1977). In the latter half of the twentieth century, gaseous discharge lamps, especially fluorescent and high-intensity discharge (HID)

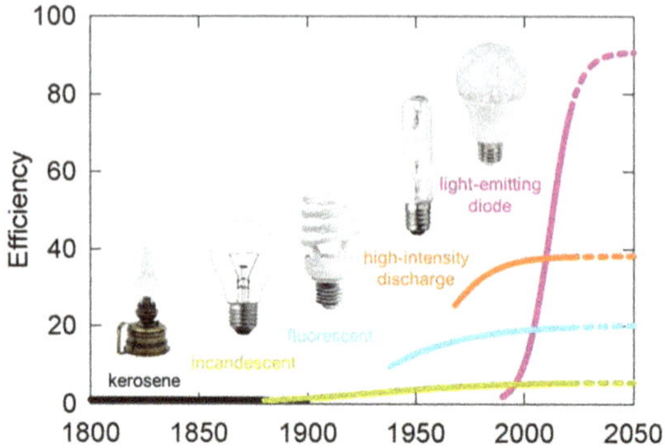


**Figure 5** Technology development of electrical lighting (Kusuma et al., 2021). Efficiency here refers to the energy in the emitted photons with wavelengths between 400 nm and 700 nm divided by the input electrical energy. Efficiency values do not include driver or ballast losses.

lamps, including metal halide and high-pressure sodium (HPS), became more common for both research and commercial use (Fig. 5).

Fluorescent lamps ionize mercury gas, and the emitted UV photons are converted into PAR photons using fluorescent materials called phosphors, which absorb the shorter wavelength photons and then reemit them at longer wavelengths (Fujiwara, 2020). Although they are inexpensive, have a long lifetime, and are relatively easy to obtain, fluorescent lamps are only about 20% efficient, with photosynthetic photon efficacies (PPEs) - calculated as µmol of photosynthetic photons produced per J of input energy required - of about 1 µmol $J^{-1}$ (Nelson and Bugbee, 2014; Shelford and Both, 2021). Despite their low efficiency by modern standards, fluorescent lamps used to be the preferred light for many indoor plant growth environments because of their low heat production, allowing them to be placed close to the plants without burning the leaves. Fluorescent lamps are still used in some Japanese plant factories to this day; however, their usage in new PFALs has mostly ended (Kozai, 2020).

HID lamps are similar to fluorescent lamps (a low-pressure discharge lamp), but they operate at much higher pressures, temperatures, and intensities. Although there are many types of HID lamps, the most common for urban agriculture are HPS. HPS are typically reported to have efficacies of about 1.7 µmol $J^{-1}$ (Nelson and Bugbee, 2014; Radetsky, 2018; Shelford and Both, 2021), but some reports suggest that they can achieve higher efficacies of 2.1 (Philips lighting, 2017) to 2.18 µmol $J^{-1}$ (Ushio, 2021) - although it should be noted that these latter two values do not include efficiency losses from the ballast. Significant further improvements are not expected to be feasible for HPS, which limits their future use in PFALs, but HPS lamps are still widely used in greenhouses.

The first electroluminescent materials were initially discovered about a century ago by Henry Round and Oleg Losev (Zheludev, 2007; Dupuis and Krames, 2008), but these devices were significantly different from modern LEDs (Dupuis and Krames, 2008; Tsao et al., 2015). The first LEDs only output infrared photons, followed decades later by LEDs that output red photons. Then, in the early 1990s, the Nobel Prize-winning work of Isamu Akasaki, Hiroshi Amano, and Shuji Nakamura led to the development of LEDs with substantial fluxes of blue photons (Tsao et al., 2015). LED efficiency has developed rapidly over the past two decades (Fig. 5) with an approximate increase in efficiency from around only 10% in the year 2000 to approximately 60–80% today. This rapid advancement in LED efficiency has been largely driven by the general illumination industry, which is a multibillion-dollar industry – in 2019, the global LED luminaire market revenue was about US$51 billion (Hansen et al., 2022). These advancements in white and blue LEDs from the lighting sector have made PFALs a feasible industry with 50–75% higher PPEs than HPS. The horticultural industry has driven the development of 660 nm red LEDs that have a further 50% higher efficacy than blue or white LEDs.

## *3.2 Current state of LED lighting technology*

LED luminaires can now achieve more than double the efficacy of other electric lamps. LEDs without down-conversion materials output a roughly Gaussian spectral photon distribution with a peak wavelength that can theoretically lie anywhere across the photobiologically active region of radiation (280–800 nm). The three most common single peak LEDs for horticulture are 450 nm blue, 660 nm red, and 730 nm far-red. Current PPEs of these LEDs under standard conditions[1] are 2.8 μmol $J^{-1}$ for 450 nm blue LEDs, 4.2 μmol $J^{-1}$ for 660 nm red LEDs, and 3.6 μmol $J^{-1}$ for far-red LEDs (Kusuma et al., 2020, 2022).

In addition to these single peak LEDs, phosphors are applied to 450 nm blue LEDs to create broad-spectrum white light; these white LEDs are often referred to as phosphor-converted white[2] (but here, phosphor-converted LEDs are referred to as white LEDs). Currently, two common phosphors are applied to the 450 nm blue LEDs to provide white light: a green-yellow phosphor with an emission peak at about 550 nm and a red phosphor with an emission peak at about 650 nm (Pattison et al., 2022). These phosphors can be combined in different ratios to provide white light with varying correlated color temperatures from about 2700–6500 K. The phosphors used in white LEDs ultimately results in a fundamental decrease in the energy efficiency of the LED, due to Stokes

[1] Standard conditions include a specified junction temperature and current density.
[2] Phosphor-converted white LEDs stand in contrast with color-mixed white LEDs, which create white light with LEDs that directly emit blue, green, and red photons without a phosphor. But, color-mixed white LEDs are not common in horticulture.

shift, but the efficiency on a photon emitted per photon absorbed basis is about 81-90% for the red phosphor and 88-98% for the green-yellow phosphor, depending on the junction temperature (Pattison et al., 2022).

Incorporating LEDs into a luminaire requires a consideration of current droop, thermal droop, driver efficiency, and optical losses. Droop refers to the decrease in efficacy as either the current (current droop) or the junction temperature (thermal droop) increases. Calculations to predict efficacy losses as LEDs are incorporated into luminaires are described in Kusuma et al. (2020, 2022) but, in general, 15-30% losses are expected. Currently, the best horticultural luminaires have efficacies between 3 μmol $J^{-1}$ and 4 μmol $J^{-1}$ and contain 60-95% red photons (Kusuma et al., 2020, 2022).

Many in the PFAL industry think that efficacy is overemphasized. Instead, reliability, longevity, and ease of installation ought to be considered as important characteristics of horticultural lighting. Many growers are looking for practical and technical improvements. However, global events, including the Covid-19 pandemic and the Russia-Ukrainian War have caused energy prices in Europe to rise by three- to five-fold from 2021 to 2022 (European Commission, 2022). The increased cost of energy will probably keep efficacy at the top of features required by the PFAL industry.

## 3.3 Quality, quantity and duration of light for plant growth

Effects of lighting on plant growth and development are summarized in numerous reviews, including Davis and Burns (2016), Demotes-Mainard et al. (2016), Lazzarin et al. (2021), and Zhen et al. (2022). Here, we briefly describe some considerations relating to photon quality, quantity, duration, and timing in PFALs.

Within the photobiologically active region of radiation (280-800 nm), UV-B can have beneficial effects on quality when applied in short durations, but it ought to be used with caution because it can cause physiological damage. Currently, UV-B LEDs have both low efficiency and low lifetimes. UV-A could play a role in plant coloration, but similar effects could be achieved with blue (450 nm) LEDs, which have higher efficiencies and are potentially more effective. Blue photons are applied with 450 nm blue LEDs or white LEDs. Many luminaires include some portion of blue (or white) LEDs because growing plants in the absence of blue photons appears to induce poor growth (Snowden et al., 2016). This requirement of blue photons has been apparent since the first studies using only red LEDs to grow plants at NASA (Bula et al., 1991; Yorio et al., 2001). Green photons are typically applied with white LEDs, with the potential role of alleviating plant compactness that can occur under only blue and red LEDs; however, many studies show a minimal effect of green photons on plant morphology (Zhen et al., 2022). Red photons from 660 nm red LEDs are both electrically and photosynthetically efficient (McCree, 1971). In general, the higher the fraction red in the spectrum, the higher the PPE. Far-red photons can increase leaf and stem elongation, potentially increasing growth rates (Jin et al., 2020), but some growers have stated that they prefer to avoid far-red because the longer and thinner leaves have been associated with decreased quality. Far-red has also been associated with increased biomass partitioning to fruits in some studies (Ji et al., 2020), but not in others (Chen et al., 2022).

Photosynthetic photon flux density (PPFD) requirements are species dependent. In general, the higher the PPFD, the faster the growth. However, higher PPFD comes at the cost of either increased investment cost or increased operational cost, or both. Some companies highlight the cost of lighting relative to all other costs and conclude that the increase in lighting costs at high PPFDs (above 500 μmol $m^{-2}$ $s^{-1}$) only marginally increases the total cost of the operation, while the yields are significantly increased. It is worth noting here that energy is often estimated to be about 20–40% of operational expenditure, and lighting is estimated to be about 40–80% of the energy cost (FAO, 2023). Using the mid-point of these two ranges, lighting is then estimated to constitute about 18% of the operational cost.

In some cases, high PPFD can be detrimental for plant growth. For lettuce, daily light integrals above 17 mol $m^{-2}$ per day (e.g. a PPFD of 295 μmol $m^{-2}$ $s^{-1}$ for 16 h per day) have been shown to induce tipburn in 'Ostinata' Bibb lettuce (Both et al., 1994); other lettuce cultivars could have different thresholds. Tipburn is a complex process, and it is not only caused by high PPFD. Frantz et al. (2004) successfully grew lettuce at a PPFD of 1000 μmol $m^{-2}$ $s^{-1}$ by aerating the meristem.

Photoperiod is the number of hours light is provided in a 24-h daily cycle. The ideal photoperiod would be continuous light for all 24 h because dark periods of plant growth are unproductive. However, many plant species, such as tomato, perform poorly under continuous light (Velez-Ramirez et al., 2014). Additionally, the induction of flowering in many plant species requires a dark period. Therefore, continuous lighting is only applied occasionally, but growers in PFALs will often apply the longest photoperiod they can.

Dynamic lighting refers to the modulation of spectrum, PPFD, and photoperiod throughout the day or throughout the whole life cycle. Some luminaires have dimming capabilities, which can facilitate dynamic lighting, but this feature increases the investment cost and can decrease luminaire efficacy. Automation and plant mobility may therefore be a less expensive way to accomplish dynamic lighting, although this likely depends on scale. In addition to dynamic lighting during growth, postharvest lighting treatments with high fractions of blue have been shown to maintain high anthocyanin levels in purple peppers (Liu et al., 2022).

## 3.4 Current and future developments in lighting technologies

At the LED package level, some continued minor advances are expected in 450 nm blue LEDs and 660 nm red LEDs. Over the past few years, the efficacy of the 660 nm red LEDs has improved by about 5–6% each year, driven either by an increase in the photon flux output or by a decrease in the forward voltage. As efficiency approaches 100%, continued gains are going to become increasingly difficult, and a 100% efficient device is not really possible. This is partially due to inherent internal resistances (e.g. wire bonds and pads) that impede the flow of current through the device.

The sub-efficiencies that determine the approximate 70% efficiency of 450 nm blue and 660 nm red LEDs – light extraction efficiency (photons that exit the LED per photon created), internal quantum efficiency (photon per electron), and electrical efficiency (energy in the photons divided by the voltage energy) – are each about 90% efficient (electrical efficiency is closer to 95%). Therefore, rather than a potential 30% improvement in efficiency, there is only about 10% more improvement possible in each of these sub-efficiencies. The current difficulty in increasing efficiency toward 100% is caused by impurities in the LED dye, but improvements in the manufacturing process can solve some of these issues. LED manufacturers are continuing to investigate new material systems that may be more efficient, and they are looking into cheaper materials that could achieve the same efficacies. This includes approaches like lamination of LEDs in an epoxy-based plastic package rather than a traditional ceramic package. With new materials and approaches, costs, reliability, and lifetime must be considered.

Although improvements in 450 nm blue and 660 nm red LEDs are expected to slow, improvements in other narrow peak LEDs will continue. Of particular interest are UV-B, green, and far-red LEDs. UV-B LEDs are currently less than 5% efficient, and they are comparatively expensive (Kusuma et al., 2022). Mercury vapor lamps are therefore currently widely used for UV-B (and UV-C) applications. However, LEDs are inherently more controllable, and therefore there is a demand for improvements in these LEDs. Green LEDs (peak at about 530 nm) could replace white LEDs in providing green photons because their theoretical maximum efficacy is higher, but currently their low efficacy of 1.3 $\mu mol\ J^{-1}$ limits their use (Kusuma et al., 2022). As green LEDs improve, they will be increasingly appropriate for many applications. However, because horticulture is so sensitive to efficacy as a measure of performance, it is unlikely that they will be used anytime soon. Indeed, even a substantial increase in green LED efficacy is not going to increase overall system efficiency by large margins. This is because white LEDs already have quite a high efficacy, and typical horticultural luminaires contain lower fractions of green in favor of red photons.

Far-red LEDs have seen significant increases in efficacy over the past few years, and many of the improvements that have been applied to 660 nm red LEDs could be applied to far-red LEDs. Over the next few years, far-red LEDs will likely reach efficiencies similar to blue and red LEDs, and due to the lower energy of far-red photons, this means the photon efficacies will be even higher than 660 nm red LEDs.

One of the continued areas of improvement for white LEDs focuses on the emission spectrum of the phosphors. White LEDs for general illumination provide more red photons than is necessary, considering the ability to directly emit red with 660 nm LEDs. LED package companies are working on developing phosphors that emit more green-yellow photons and fewer orange-red photons. In addition to developments in phosphor technology, advancements in other photon down-conversion devices, such as quantum dots, are also possible. However, the efficiency of quantum dots is currently lower than phosphors, and therefore they are not commonly used.

At the luminaire level, improvements can be made in current and thermal droops, optical efficiency, and driver efficiency. Optics are used to increase light uniformity across the growing surface, which increases the LUE of the entire canopy. Unfortunately, these optics (lenses and diffusers) usually decrease the total photon output, thus decreasing efficiency (Li et al., 2014; Zhen and Bugbee, 2019). Luminaire manufacturers can choose to minimize optics to maintain a lower cost and minimize optical losses, but this could decrease both the uniformity of output and the lifetime of the LED. Using a higher density of LEDs in a luminaire is another method of ensuring good uniformity of output. LED drivers play several important roles in LED luminaires: they convert AC to DC; they regulate voltage and current; and they incorporate dimming and control capabilities. Currently, the average driver for general lighting is about 88% efficient (Pattison et al., 2022). Higher-efficiency drivers are available, but they are more expensive. Despite this higher cost, horticultural luminaires are likely to include higher efficiency drivers. Driver efficiency is expected to increase slightly over the coming years, which will increase overall system efficiency. Finally, a significant component of luminaire PPE and initial cost is the specific choice of LEDs in the luminaire. Although many companies use white LEDs in their luminaires, some in the industry think that the use of white LEDs in PFALs may decrease in favor of luminaires with combinations of only blue and red LEDs.

To summarize, LED package and LED luminaire efficiency has increased rapidly over the past decade, and significant work has gone into designing the optics so that plants can use the photons most efficiently. However, because most of the sub-efficiencies are approaching 100%, orders of magnitude improvements in lighting technologies are no longer possible. There will be small, continued improvement in overall lighting system efficiency over the coming decade, but these improvements are expected to be more marginal. Investment cost into lighting is also generally expected to be marginally reduced in the coming years due to improvements in manufacturing. It is important to recognize that other components of lighting, such as durability, lifetime, and ease of installation, also play a significant role in the overall quality of these products.

# 4 Climate control

## *4.1 History of controlled environment agriculture*

Controlled environment agriculture (CEA) uses horticultural and engineering methods to provide growers with a high degree of control over the cultivation environment for increased crop yield, crop quality, and production efficiency (Albright, 1990). CEA dates back millennia to the first use of natural elements to protect crops (Mitchell, 2022), and the concept of greenhouse can be traced back to 30 CE (see Section 2.1; Janick and Paris, 2022). However, controlled, indoor environments, named phytotrons, first emerged in the mid-twentieth century for research purposes, with the goal of carefully examining how certain environmental factors - in isolation from other environmental factors - affected plant growth and development. The concept of phytotron[3] is very close to the modern commercial PFAL. PFALs have thus expanded from scientific research tools to economically viable commercial operations. This has been facilitated by technical advancements in equipment capacity and efficiency, a decline in energy requirements, and the advent of new technologies.

Climate control has been at the forefront of CEA since the days of growing *Cucumis* in *Specularia*, and the structures that followed - *Orangeries*, conservatories, and greenhouses - all trapped solar energy as heat during the colder seasons, which increased the temperature around the plants compared to the outdoor environment. However, the modern well-insulated PFAL has the opposite problem;[4] electric lighting and high transpiration rates in PFALs significantly heat these systems, and heating, ventilation, and air-conditioning (HVAC) is required for cooling and dehumidification.

Air-conditioning has developed through an improved understanding of thermodynamics, especially related to psychrometrics. Willis Carrier developed both the psychrometric chart and the first modern electrical air-conditioning unit in the first decade of the twentieth century (Simha, 2012). Carrier's original air-conditioning system used well water for cooling, but modern HVACs cool

[3] The *phytotron* design incorporated centralized environmental control facilities, including electricity, water, and HVAC (heating, ventilation, and air-conditioning) systems to serve individual plant growing compartments, enabling multiple growth rooms to function independently (Mitchell, 2022).

[4] Cooling is also performed in greenhouses with techniques including shading, natural ventilation, forced ventilation, and evaporative cooling (Nemali, 2022).

based on the principle of latent heat: refrigerants are actively compressed and passively expanded between liquid and gas phases in separate locations - respectively heating and cooling the surrounding air in these separate locations.

## 4.2 Key environmental factors

Temperature significantly affects the rate of plant growth and development. For example, the developmental rates of tomato plants and fruits linearly increase with temperature from 17°C to 27°C (De Koning, 1994). This is because biological reaction rates are generally slow at low temperatures and accelerate as temperatures rise within a reasonable range. Temperatures of 27-32°C have been found to hamper lycopene formation in tomato fruit (Gautier et al., 2008). At the other extreme, low night temperatures of 10°C caused sweet pepper fruit to be small and deformed (Tiwari et al., 2007). Decreasing temperature of the dark period can increase carbon use efficiency by a few percentage points, but the overall effect on yield has been shown to be insignificant (Frantz et al., 2004b). Similarly, fluctuations in temperature and PPFD have been shown to minimally affect total plant biomass accumulation - a response attributed to the storage of non-structural carbohydrates (Zepeda et al., 2022).

Lighting significantly heats PFALs, even when efficient LEDs are adopted. Heating systems are therefore unnecessary in thermally well-insulated PFALs, according to Kozai and Niu (2020). In fact, cooling is usually the only temperature control measure needed in PFALs. Plants lose heat energy through transpiration, convection, and long-wave radiation emission to the surroundings (Niu et al., 2020). Transpiration is a major cooling process for plants through latent heat transfer when liquid water converts to vapor. The transpiration rate is determined by the difference in vapor pressure between the plant leaf and the surrounding air, which is called the vapor pressure deficit ($VPD_{leaf\text{-}air}$). Plants also directly transfer heat to the air when the air temperature is lower than the plant surface, and this is accelerated with increased airflow. Heat convection can be free and forced, with free convection relying on the change in air density caused by plants warming the air and forced convection relying on air circulation driven by pumps, fans, or other external power sources. Forced convection is better for heat exchange compared to free convection due to increased air movement.

Air humidity is crucial for PFAL production because plant transpiration continuously adds water vapor to the air. However, humidity control is difficult, especially at the microclimate scale. The $VPD_{leaf\text{-}air}$ is the most appropriate metric for discussing effects of humidity on plant growth and performance, as it influences stomatal conductance, which further affects photosynthesis and transpiration. A low $VPD_{leaf\text{-}air}$ inhibits plant transpiration, which results in increased leaf temperature, and a higher likelihood of condensation on the leaves. A high $VPD_{leaf\text{-}air}$ stimulates stomatal closure to prevent excessive water loss, but as a result, photosynthesis is reduced. Both too high and too low $VPD_{leaf\text{-}air}$ may hamper water movement within plants and cause local $Ca^{2+}$ shortages, which leads to disorders like tipburn in young lettuce leaves and blossom-end-rot in tomato fruits. As reviewed by Rouphael et al. (2018), both increasing $VPD_{leaf\text{-}air}$ during the photoperiod or reducing $VPD_{leaf\text{-}air}$ during the night improved $Ca^{2+}$ distribution, minimizing tipburn.

Maintaining adequate air flow within PFALs ensures a uniform distribution of temperature, air humidity, and $CO_2$ concentration. As air speed increases across plants, the boundary layer is reduced, which increases plant photosynthesis, transpiration, and heat transfer (Kitaya et al., 2000). Sufficient airflow at the apical meristem of lettuce has been shown to decrease the incidence of tipburn by promoting transpiration of the apical meristem, and thus increases the $Ca^{2+}$ concentration at the tipburn prone areas (Goto and Takakura, 1992; Both et al., 1997; Frantz et al., 2004). Air speed should be properly controlled, as strong speeds can cause reduction in plant growth (Chehab et al., 2009).

$CO_2$ is required for photosynthesis. Its enrichment is a standard practice in PFALs because elevated $CO_2$ results in higher assimilation rates, faster growth rates, and higher yields. Among PFALs, $CO_2$ supply varies significantly, ranging from 500 ppm to 2000 ppm, depending on the crop and PPFD. In addition to the $CO_2$ concentration in the air, the diffusion of $CO_2$ from air to the internal leaf space also affects photosynthetic rates. This diffusion rate is influenced by the stomatal resistance and boundary layer resistance. Proper control of light, temperature, humidity, and airflow can enhance $CO_2$ diffusion by reducing stomatal resistance and the boundary layer thickness.

## 4.3 Climate control technologies and future directions

Climate management in PFALs consumes significant energy. The cost and efficiency of climate control in PFALs depends on the location, local climate, and the type of PFAL. Climate uniformity is challenging in PFALs because the distance between growing layers is small, and plants are grown at high densities. HVAC systems are currently the method used to dehumidify and cool indoor environments. HVAC systems cool the air below the dew point with a chilled coil. This causes condensation of water, which can be extracted and recirculated. After dehumidification, the cold air is reheated to prevent overcooling before entering the grow room. Both the insulation of the building and the local climate are important because they determine the size of the HVAC systems. Overall, both the CapEx and the OpEx of HVAC are high. PFALs historically used commercial HVAC equipment designed for office buildings. These commercial systems are suitable for heating, cooling, and dehumidification, but the main requirements of a PFAL HVAC are cooling and dehumidification

- with dehumidification playing the most significant role. By designing more appropriate systems, overall costs have been reduced. Besides HVAC, there are dehumidification systems that absorb moisture with desiccants, but these are uncommon. In addition, forced air circulation is critical in PFALs. For example, top-down ventilation systems have been embedded within the stacked layers of some PFALs to reduce tipburn. Parallel efforts on sensors and algorithms (Section 5) need to be emphasized to monitor and adjust climate conditions.

Compressed liquid $CO_2$ is the source for $CO_2$ supply in PFALs. $CO_2$ is collected from oil refineries or the incineration of residual wastes. Even though $CO_2$ extraction equipment is not installed in PFALs as part of the climate control, it is important to note that society is transitioning toward a climate-neutral, fossil fuel-free world. $CO_2$ in controlled environments will receive a lot of attention in the coming years, despite its current low price. The elimination of fossil fuels would mean the elimination of this cheap source of $CO_2$. Technologies that directly capture $CO_2$ from air are still under development, and while promising, they are costly.

## 5 Sensors and algorithms

### *5.1 History of sensors*

Scientific instruments have helped humanity explore, keep track of time, and understand the environment for thousands of years. Innovations such as magnetic compasses, clocks, hourglasses, scales, astrolabes, and more recently thermometers, barometers, and hydrometers, all facilitated human discovery, observations, record-keeping, and research (Gallicchio, 2017). Many of these early instruments were crude and contained errors based on an incomplete understanding of meteorological phenomena. For example, in the late sixteenth century, Gailileo Galilei built a temperature measurement device called a thermoscope, which could show variation in temperature based on the principle of expanding and contracting gases. However, these thermoscopes were sensitive to barometric pressure as well as temperature, and thus sealing the device allowed for improved measurements. However, even with a more precise device, these instruments preceded a standardization of temperature scales - such as Fahrenheit, Celsius, and Kelvin - which were not developed until the eighteenth and nineteenth centuries. During this time, most record-keeping had to be manual, but some automatic instruments were able to continuously record environmental parameters, such as barographs, which recorded atmospheric pressure.

Instrumentation took a giant leap forward during the first and second industrial revolutions when electronic versions of many of these instruments, like thermocouples, were invented. At this point, temperature recording was still performed on chart recorders, which are electromechanical devices that record the data - e.g. temperature and pressure - on either strip chats, circular charts, or roll charts. Chart recorders were used for environmental monitoring throughout most of the twentieth century. More recently, the ability to digitally store and process data has revolutionized data acquisition and processing for controlled environments. Electrical environmental sensors that change in voltage, resistance, capacitance, or amperage, with a change in some environmental parameter, can be easily paired with a datalogger, which collects, stores, and often processes the data.

Over the past two centuries, sensors related to environmental monitoring, such as temperature, humidity, and $CO_2$, have become relatively mature technologies. By contrast, sensors related to plant performance, and the algorithms used to interpret the data, are still being developed. These two technologies, sensors and algorithms, work in tandem to optimize the system, maximizing both yield and quality, while minimizing energy use. By incorporating feedback into the system (automated control), PFALs can be maximized for efficient, high-quality production.

## 5.2 Sensors

Sensors for environmental measurement are reasonably well developed. This includes sensors to measure light intensity, light spectrum, $CO_2$ concentrations, temperature, humidity, air pressure, wind speed, dissolved oxygen concentrations, pH, electrical conductivity, and volumetric water content. Additionally, some development has gone into sensors to monitor plant performance, including growth rates and water status. These include lysimeters or load cells, sap flow sensors, dendrometers, leaf thickness sensors, and plant height sensors.

Like other plant performance sensors, imaging sensors can be used to determine growth rates and water status, but they are also useful for measuring pigment concentrations (e.g. chlorophyll, anthocyanins, carotenoids), photosynthetic properties (e.g. photosynthetic efficiency), plant stress (biotic and abiotic), elemental concentrations (nitrogen and phosphorous), lignin, cellulose and hemicellulose concentrations, and protein concentrations (Mishra et al., 2017; Sarić et al., 2022). Imaging sensors include RBG cameras, chlorophyll fluorometers, thermal cameras, and multispectral sensors. Vision-based technologies are still a young technology for PFAL production, and continued research will reveal many promising opportunities and uses for these sensors.

Sensors are often limited to a single location, and therefore a high quantity of sensors is required for spatial measurements of both the environment and the plant. Typically, sensors for measuring the environment are much less expensive than sensors for measuring plant performance. Automation could provide an opportunity for reducing these costs by moving the plants to plant performance sensors, but the cost must be compared to the cost of automation itself. Non-destructive plant performance measurements allow for environmental adjustments between measurements. With this, the system can be optimized for efficient production, high yields, and maximum quality.

## 5.3 Algorithms

The environmental and plant performance data are processed through algorithms to create models that predict the optimal conditions for optimal growth and performance. These algorithms ideally integrate both existing knowledge about plant physiology (e.g. models of photosynthesis, transpiration, and compound synthesis) alongside machine learning to achieve this goal. At its core, machine learning is just advanced statistics. Machine learning allows systems to automatically learn from past data without specific programming. One of the goals of these algorithms is model-assisted decision-making, and eventually systems will move to full autonomous control (Van Delden et al., 2021). Ideally, these systems can be designed to react to unanticipated events.

In general, these data- and process-driven models for crop growth are in their infancy, but some recent work has shown promise in greenhouses. A good example of developing crop-based algorithms is the autonomous greenhouse challenge: a competition where teams remotely control a greenhouse climate and crop management with artificial intelligence (AI) techniques to maximize net profits. In the first autonomous greenhouse challenge, crop production was increased by 6% and net profit by 17% compared to reference growers, and in the second autonomous greenhouse challenge, two competitors achieved double the profit compared to the reference growers (Hemming et al., 2020). PFAL companies that are developing their own algorithms are performing similar tests.

## 5.4 Future prospects

Significant research is still required to improve sensors and algorithms to maximize profits and energy efficiency in PFALs. As previously mentioned, sensors for plant performance and the algorithms are still quite young technologies. These technologies will therefore likely see rapid advancements in the coming years. The algorithms ought to have a physiological basis, and therefore research from universities will likely have a large impact on these technologies. In the future, these algorithms may be able to determine production and efficiency optima for specific crops in PFALs, and because these environments are entirely controllable, once these optima are determined the investment into sensors may be reduced. In the nearer term, growers are not concerned with the nuances of the data collected from sensors, rather they want instructions for actions they should take. Currently, many of these technologies are being developed by experts working in PFALs, but as they become more ubiquitous, user interface will be important.

# 6 Rootzone

## 6.1 History of hydroponics

After domestication, irrigation was probably the first environmental modification in agriculture (Mitchell, 2022). Archeological evidence suggests that crop rootzones were modified with manure in the Neolithic era, over 5000 years ago (Vaiglova et al., 2014; Wang et al., 2018). This significant advancement in agriculture is also mentioned by Homer in the Odyssey, over 2500 years ago, where a pile of manure is mentioned to be ready to apply on the King's estate. Early examples of hydroculture include the Hanging Gardens of Babylon and the Chimanpas (floating gardens) of the ancient Aztecs.

Early documentation of liquid water culture comes from the experiments of the Naturalist John Woodward in 1699 – his experiments found that spearmint growing in water from the Thames River grew 50% more than plant growing in (distilled) rainwater (Woodward, 1699), a finding that contributed to the field of plant mineral nutrition. In 1933, Dennis Hoagland published a hydroponic nutrient solution recipe that is commonly referred to despite its many revisions (Hoagland and Arnon, 1950). Today, hydroponic systems include a wide array of technologies, including soilless media, liquid water culture with dissolved nutrients, and aeroponics.

## 6.2 Types of hydroponics

Hydroponic systems are reviewed in Van Delden et al. (2021), Mattson and Lieth (2019), and Van Os et al. (2019). Hydroponics can refer to water cultures, but the definition often includes soilless media (Figs. 6 and 7). Soilless media provide a structure for the roots. Ideally, they minimally bind water and nutrients so that these are available to the plant, while simultaneously maintaining ample oxygen. Soilless media include natural organic, natural inorganic, processed, and synthetic materials (Fig. 6).

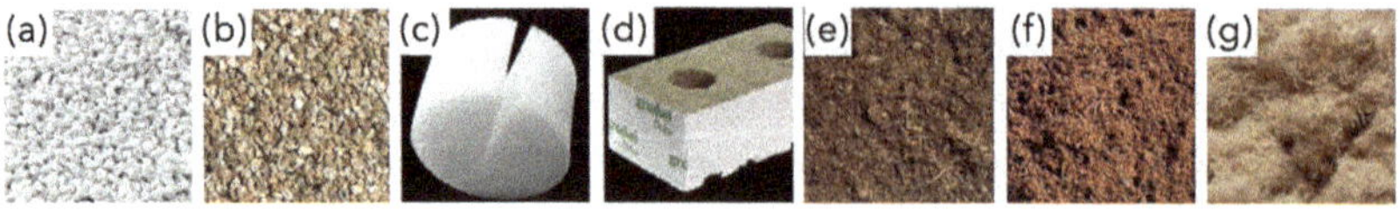


**Figure 6** Examples of some soilless media: (a) perlite, (b) vermiculite, (c) polyurethane, (d) stonewool, (e) peat, (f) coconut coir, and (g) hydrafiber (a processed wood-based media).

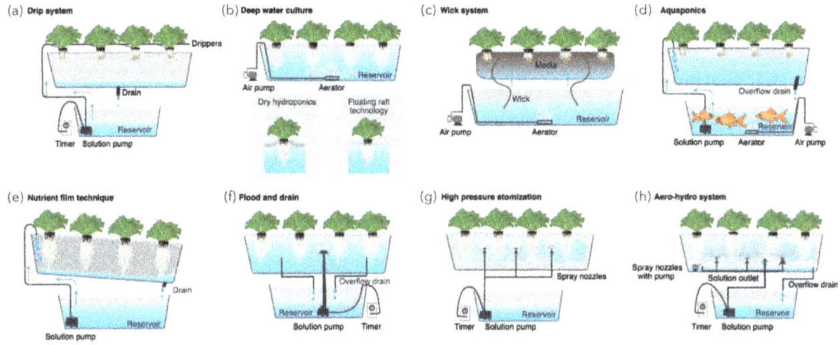


**Figure 7** Hydroponic culture systems (Van Delden et al., 2021).

Perlite and vermiculite are two common processed materials for soilless culture. They both come from natural materials that are heated to 1000°C until they expand. They are lightweight and highly porous, which allows them to absorb high amounts of water. Stonewool is also a processed material with natural origins: volcanic rock is heated to high temperatures (1600°C) to become molten, after which it is spun to form fibers. In addition to this raw fibrous material, binders and wetting agents are added to change the physical properties of the media. By changing the ratios of fibers, binders, and wetting agents, the stonewool media can vary in density and hydraulic properties. Stonewool is not a loose material, which can be beneficial compared to loose materials that can clog the irrigation systems.

Peat and coconut coir are among the most common organic soilless media. Schmilewski (2017) estimated that 75% of the total soilless media used in the European Union was peat. Other organic material constitutes 11% of the total; compost constitutes 8%; and mineral media (including stonewool) constitutes 6%. Peat is available, consistent, and generally has desirable physical and chemical characteristics - including high porosity, good aeration, and high water-holding capacity. The main drawback of peat is its unsustainability. Peat bogs are environmental carbon sinks, and thus draining the bogs to collect peat results in the release of this carbon sink. Many governments are taking steps to reduce peat use; for example, the United Kingdom is looking to end the use of peat in horticulture by 2028 (Department for Environment, Food and Rural Affairs, 2021). Coconut coir, which comes from coconut mesocarps, is another organic media. It is an abundant plant-derived organic waste material in many tropical countries. After an initial separation phase to remove fibers, the coconut pith is composted for several months. Similar to peat, coconut coir has high porosity, good aeration, and high water-holding capacity.

Soilless media can be used throughout the whole growing cycle or only in the propagation stage, after which the plug is transplanted into a system where the roots hang in water or in air. Drip irrigation is a common technique to apply water to soilless plant production. Here, nutrient solution is applied directly to the rootzone through tubes on top of the media. In ebb and flow (or ebb and flood) systems, tables are flooded so that the water wicks into either a pot with soilless media or a stonewool block. The table remains flooded for a short period and is then drained. The tables must be designed to adequately remove water during drainage to inhibit algae growth. Ebb and flood can also be used in a system without media.

The nutrient film technique (NFT) applies water to an inclined surface so it flows past the roots. The thin layer of water in this system ensures high levels of oxygenation in the rootzone. However, due to its low volume of water, rootzone temperature is more difficult to control. The deep flow technique or deep water culture refers to a system where the roots are constantly submerged in water. It is important to maintain ample oxygen in this system, at about 8 ppm (Langenfeld et al., 2022), which can be achieved with either injectors or pumps. Additionally, water circulation is required to avoid oxygen- and nutrient-deficient zones around the roots. However, in general, the large volume of water is well buffered for temperature, nutrients, and pH. The deep flow technique does not necessarily use more water than a system like NFT because water can be shielded from evaporation in both systems, and thus water loss primarily occurs through transpiration.

Aeroponic systems suspend roots in the air and are either continuously or intermittently sprayed with a nutrient filled mist. Other aeroponics systems use ultrasonic fogging systems with finer particles – this system is sometimes called fogponics. These systems allow for maximum oxygenation of the rootzone but, like NFT, it can suffer from a lack of a buffer for temperature. Aeroponics are sometimes considered the optimal rootzone system, but many PFAL companies consider aeroponics to be more trouble than they are worth – primarily due to the labor cost requirements to fix hardware failures. Furthermore, research has not consistently shown that aeroponics provide higher yields compared to other hydroponic systems (Li et al., 2018).

In general, water use efficiency in all hydroponic systems can be high, especially in controlled environments due to the recirculation of the drained water and because water that is lost as vapor through transpiration can be collected and recirculated into the nutrient solution. Recirculation of the nutrient solution ensures high nutrient use efficiency and minimizes (theoretically eliminates) nutrient runoff into the environment. Despite attempts to make PFALs clean and aseptic, many organisms can still enter these systems. Irrigation solution must be regularly purified and disinfected. According to Van Os (2019), heat treatment and UV radiation are the best options to disinfect recirculated solutions in nurseries larger than 2 ha cultivation area, and sand filtration is a good choice for nurseries smaller than 1 ha. Ozone and membrane filtration are also useful methods but demand high investment.

### 6.3 Future prospects

Nutrient solutions can be further optimized as many PFALs must dump their nutrient solutions when certain ions build up in the circulated solution, for example sodium and chlorine. This may be solved by choosing salts without these elements. There may also be a potential use for organic, rather than synthetic, sources of nutrients. Additionally, dynamic nutrient solution recipes that apply specific elemental ratios at specific growth stages could improve yields (Mattson and Lieth, 2019). Apart from nutrients, soilless media may see new developments with products that are more sustainable, while still providing adequate water, nutrients, and oxygen to the rootzone. Sensors that provide an indication of water content, use, and availability in the system can increase overall efficiency. Furthermore, the development of sensors that measure the concentrations of specific ions in the solution could improve nutrient use efficiency. Nutrients are a small portion of costs in PFALs and their optimization does not tend to be a primary focus, but nutrient and water availability can significantly affect the quality of the final product, so these aspects should not be overlooked.

## 7 Genetics and plant breeding

### 7.1 History of plant breeding

Domesticated crops appeared as early as 13000 years ago, with the evidence for domesticated wheat dating to 10500 years ago. Some of the evidence for this early domestication comes from a comparison of the spikelets of wild wheat to spikelets of charred wheat fragments collected at an archeological site. The spikelets of the domesticated wheat were difficult to remove, while the spikelets of the wild wheat were easily detached. The non-easily detached spikelet would have been evolutionarily unfavorable because it prevented natural seed dispersal (Balter, 2007). However, this trait benefited human collection of seeds (Schlegel, 2017).

As agriculture developed, farmers were the first plant breeders (Bradshaw, 2017). The pioneering work of Gregor Mendel, whose genetic inheritance studies, known as the laws of Mendelian inheritance, laid a foundation for scientific plant breeding. The application of Mendel's theory on heredity and knowledge about hybrids and mutations led to a significant advancement in crop breeding (Dudley, 1994; Murphy, 2007). After World War II, the Green Revolution greatly contributed to global food security by increasing yield, particularly by developing dwarf varieties (Evenson and Gollin, 2003). Plant breeding has now entered the stage of molecular breeding. Over the last 30 years, the ongoing development of plant biotechnology, molecular markers, and genomics have introduced powerful tools for creating, analyzing, and manipulating genetic variations in the development of new varieties. Developing a new variety by conventional plant breeding can take more than a decade but, through genetic engineering, this period is significantly reduced. However, genetically engineered crops are not allowed in all countries. Most cultivars are bred for open field conditions, but some are bred specifically for CEA, such as hydroponic lettuce. However, little development has been made to breed new cultivars specifically suited to PFALs.

## 7.2 The current status and focus of plant breeding in plant factories with artificial light

In contrast to the advancements in many of the technologies mentioned above, the development of the most important element of PFALs – the crop – is lagging behind. PFAL producers demand suitable genetics, but there are no cultivars on the market that were specifically bred for PFALs. This deficiency has been brought to the attention of the breeding industry, and a small proportion of breeding companies are beginning to fill this gap.

The development of suitable genetics for PFALs follows a biphasic approach. The first phase involves testing and trialing varieties under PFAL conditions to see how they perform. This initial screening will be done by seed companies with strong R&D. A selection of the highest-performing varieties will then be trialed by PFAL growers. Screening allows breeding companies to both identify eligible varieties and obtain an understanding of the genetics suitable for these systems. The second phase involves investigating the suitable genetics to breed new varieties. The general process of plant breeding is to cross selected parent plants with desirable traits and then to continue this process to generate new varieties with combined desired traits. A sound understanding of genetics at the first stage provides breeders with a blueprint for targeting specific traits and, ultimately, developing varieties that perform well.

Leafy greens are currently the breeding focus of PFALs but, with fruit crops receiving more attention, they will likely become a breeding target in the future. Desired traits in PFALs are different from desired traits in open field (Fig. 8). PFALs protect crops from biotic and abiotic stresses that are prevalent in outdoor environments, which means breeding for resistance is no longer a top priority. Instead, crops can be bred to maximize genetic plasticity and satisfy consumer expectations. Even though the ideal plant type for PFAL is still undetermined, new varieties will likely take advantage of the architecture and environment in PFALs, for example traits like dwarfism, rapid growth rates, low light tolerance, high plant density tolerance, and efficient nutrient uptake are desirable (Folta, 2019; Leong and Urano, 2018).

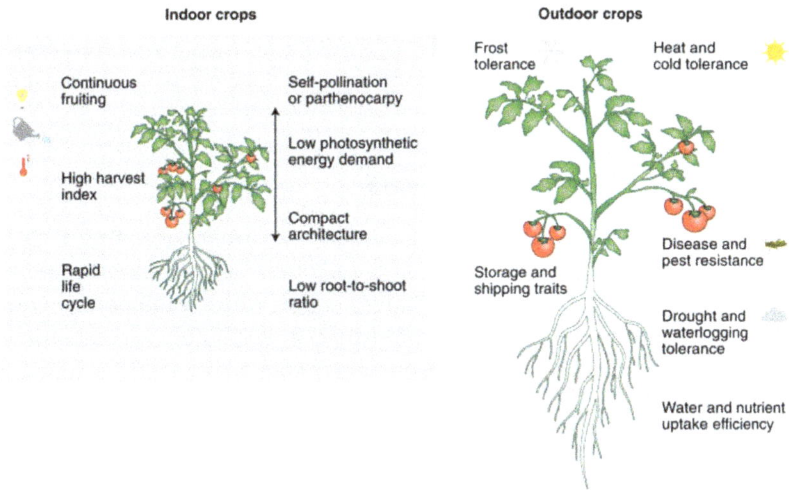


**Figure 8** Differences between indoor and outdoor farming (O'Sullivan et al., 2020).

## 7.3 Desired traits for plant factories with artificial light

The ideal crop architecture should be amenable to the production system, although it is currently difficult to conclude what the best crop traits are for PFALs. Leafy vegetables are relatively compact, with short growing cycles and high harvest index, which allow them to grow efficiently in PFALs. These characteristics make them more suitable to PFALs when compared to fruit crops.

### 7.3.1 Breeding for crop yield

Maximizing LUE is critical for increasing the yield per unit input. One potential way to increase LUE is by increasing the leaf area index, namely, aiming for faster leaf expansion. Overexpression of *LsaGRF5*, a growth-regulating transcription factor, in lettuce (*Lactuca sativa* L.) resulted in larger leaves (Zhang et al., 2021). Another method to increase LUE is to improve the light distribution within the canopy by finding a suitable plant architecture. Although dwarfism has been suggested as a desirable trait for PFALs (Folta, 2019), it is not necessarily ideal for light distribution. A spacious canopy structure with long internodes, elevated leaf angles, and long narrow leaves may be better for deeper light penetration (Sarlikioti et al., 2011). Even for tomato, there is a whole range of heights between a dwarf and a full vine. A balance must be reached between efficient space use, high yields, and high fruit quality. Finally, LUE may be improved by investigating the parameters of photosynthetic quantum yield (Simkin et al., 2019; Ding et al., 2016). A positive relationship between crop yield and photosynthetic efficiency has been demonstrated in several studies through genetic engineering (Simkin et al., 2019). For instance, photosynthetic carbon assimilation and yield in tomato were increased by overexpressing sedoheptulose-1,7-bisphosphatase (SBPase), an enzyme in the Calvin cycle (Ding et al., 2016).

### 7.3.2 Breeding for marketable yield

Marketable yield is the saleable produce with no browning or rotting, the right appearance (size, shape, and color), and a pleasant sensory experience. This can be maximized by increasing the harvest index, which is the partitioning of assimilates to the marketable organs. Vallarino et al. (2020) increased tomato fruit yield up to 23% by enhancing source strength with genetic approaches. Marketable yield can also be maximized by minimizing defects. Even though PFALs are pathogen- and insect-free environments, loss of produce can still occur, for example through the physiological disorder tipburn. Su et al. (2019) studied tipburn in Chinese cabbage by quantitative trait loci (QTL) mapping and identified a tipburn resistance gene, *BrCRT2$^R$*. By overexpressing *BrCRT2$^R$* in an Arabidopsis *crt2* mutant, $Ca^{2+}$ content increased and cell death decreased in the leaf tips under $Ca^{2+}$ deficient conditions.

### 7.3.3 Breeding for quality

Product quality includes attributes like flavor, aroma, texture, shape, color, antioxidant level, mineral content, nutritional value, medical value, and shelf life. PFALs should position themselves to focus on the high-end quality of the market. In fact, quality should be more important than yield because the high market value of premium products is currently the only way to cover the high cost in PFALs. For example, there is a trend to breed snacking type tomatoes, like cherry and cocktail, because of their good taste and easy consumption. It is also important for breeders to consider the genetic performance throughout the life of the crop - from seed to consumption. Therefore, postharvest qualities retained at the consumer level are also important traits (Hewett, 2019).

### 7.3.4 Breeding for automation

PFALs are becoming increasingly automated - from sowing to harvest. To facilitate this automation, the plant architecture should be uniform and accessible. For example, dwarf tomatoes would be easier to harvest if fruit trusses were clearly separated from leaves. Another significant factor is the uniformity of the fruits and the uniformity of the whole plant. For instance, fruit harvest would be facilitated by synchronous ripening, so that all the fruits can be harvested together. Crops, especially the marketable parts, should be resistant to injury, but this resilience should not come at the expense of quality characteristics. For instance, hard tomatoes may withstand robotic harvesting, but they deviate from consumer preferences. Finally, selecting self-pollinating and parthenocarpic varieties, and varieties with a flower type that is amenable to mechanical vibration, would make the next-generation pollination easier, especially in PFALs, where the unique light spectra cause issues in bumblebee pollination.

### 7.3.5 Breeding to reduce costs

Optimizing genetics to enable plants to grow in hotter and more humid conditions could reduce cost by requiring less resource allocation to environmental control. Given that PFALs vary significantly depending on their location, systems, access to energy, and energy costs, varieties that maintain yield and quality under a wide range of conditions may be desirable – especially in cases where energy prices spike. Varieties that mature faster could also decrease inputs while maintaining outputs, which potentially allows growers to produce more cycles per year and increase revenue.

### 7.4 Genome editing

Recently, the UK enacted new legislation that took effect in January 2022, which states that genome editing can be used to speed up the breeding of varieties that could otherwise be obtained more slowly by traditional breeding (Department for Environment, Food and Rural Affairs, 2022). Although this is still not allowed in the EU, genome editing is beginning to unlock opportunities for creating desirable varieties for PFAL. Genome editing can now involve CRISPR-Cas technology, which is simple, efficient, specific, and multiplexing (Ran et al., 2013). CRISPR-Cas9 technology has been applied in fruits and vegetables to enhance their resistance and quality (Brandt and Barrangou, 2019). As this tool is more widely permitted, breeding programs will advance more rapidly.

## 8 Conclusion

PFALs will continue to bring unprecedented crop yield and quality, but production is currently expensive. In Europe and North America, investment costs are estimated at €1500–€2500 per square meter of cultivation area, excluding the building – which, if included, could double this cost. These values have fallen from about €4000 in the past few years partially due to advances in research and technology. With the potential advantage of an efficient use of all resources, PFALs could benefit population, economy, and society.

## 9 Where to look for further information

The six technology categories outlined here are discussed for the broader urban agriculture landscape (including vertical farming) in an upcoming FAO report. This report also tends to focus more on associated costs. See Appolloni et al. (2022) for summary of most important findings.

For a short overview of trends in plant factories, see Ji et al. (2023). Good overviews of improvements in LED technology are provided by Pattison et al. (2018), Kusuma et al. (2020), and Kusuma et al. (2022).

Van Delden et al. (2021) provide comprehensive review on plant factories including a good overview of different aspects of rootzone technologies in plant factories, and it provides some examples of the use of sensors in these systems.

Finally, for more information breeding efforts in plant factories, see Folta (2019) or SharathKumar et al. (2020).

## 10 References


Albright, L. D. (1990). *Environment Control for Animals and Plants*. American Society of Agricultural Engineers.

Appolloni, E., Pennisi, G., Tonini, P., Marcelis, L. F. M., Kusuma, P., Liu, Y., Balseca, V., Jijakli, H., Orsini, F. and Monzini, J. (2022). How to make urban farming economically viable?. *Acta Horticulturae* (1356), 445–452.

Asseng, S., Guarin, J. R., Raman, M., Monje, O., Kiss, G., Despommier, D. D., Meggers, F. M. and Gauthier, P. P. G. (2020). Wheat yield potential in controlled-environment vertical farms. *Proceedings of the National Academy of Sciences of the United States of America* 117(32), 19131–19135.

Balter, M. (2007). Seeking agriculture's ancient roots. *Science* 316(5833), 1830–1835.

Beacham, A. M., Vickers, L. H. and Monaghan, J. M. (2019). Vertical farming: A summary of approaches to growing skywards. *The Journal of Horticultural Science and Biotechnology* 94(3), 277–283.

Both, A. J., Albright, L. D., Langhans, R. W., Reiser, R. A. and Vinzant, B. G. (1997). Hydroponic lettuce production influenced by integrated supplemental light levels in a controlled environment agriculture facility: experimental results. *Acta Horticulturae* (418), 45–52.


Both, A. J., Albright, L. D., Langhans, R. W., Reiser, R. A. and Vinzant, B. G. (1997). Hydroponic lettuce production influenced by integrated supplemental light levels in a controlled environment agriculture facility: experimental results. *Acta Horticulturae* (418), 45–52.


Bradshaw, J. E. (2017). Plant breeding: Past, present and future. *Euphytica* 213, 1–12.

Brandt, K. and Barrangou, R. (2019). Applications of CRISPR technologies across the food supply chain. *Annual Review of Food Science and Technology* 10, 133–150.

Bula, R. J., Morrow, R. C., Tibbitts, T. W., Barta, D. J., Ignatius, R. W. and Martin, T. S. (1991). Light-emitting diodes as a radiation source for plants. *Hortscience* 26(2), 203-205.

Chehab, E. W., Eich, E. and Braam, J. (2009). Thigmomorphogenesis: A complex plant response to mechano-stimulation. *Journal of Experimental Botany* 60(1), 43-56.

Chen, S., Marcelis, L. F. M. and Heuvelink, E. (2022). Far-red radiation increases flower and fruit abortion in sweet pepper (*Capsicum annuum* L.). *Scientia Horticulturae* 305, 111386.

Davis, P. A. and Burns, C. (2016). Photobiology in protected horticulture. *Food and Energy Security* 5(4), 223–238.

De Koning, A. N. (1994). *Development and Dry Matter Distribution in Glasshouse Tomato: A Quantitative Approach*. Wageningen University and Research.

Demotes-Mainard, S., Péron, T., Corot, A., Bertheloot, J., Le Gourrierec, J., Pelleschi-Travier, S., Crespel, L., Morel, P., Huché-Thélier, L., Boumaza, R., Vian, A., Guérin, V., Leduc, N. and Sakr, S. (2016). Plant responses to red and far-red lights, applications in horticulture. *Environmental and Experimental Botany* 121, 4–21.

Department for Environment, Food & Rural Affairs (2021). Plans to phase out the use of peat in the amateur horticulture sector. Available at: https://www.gov.uk/government/news/plans-to-phase-out-the-use-of-peat-in-the-amateur-horticulture-sector.

Department for Environment, Food & Rural Affairs (2022). New powers granted to research gene editing in plants. Available at: https://www.gov.uk/government/news/new-powers-granted-to-research-gene-editing-in-plants.

Ding, F., Wang, M., Zhang, S. and Ai, X. (2016). Changes in SBPase activity influence photosynthetic capacity, growth, and tolerance to chilling stress in transgenic tomato plants. *Scientific Reports* 6, 32741.

Downs, R. J. (1977). Incandescent lamp maintenance in plant growth chambers. *HortScience* 12(4), 330–332.

Dudley, J. W. (1994). Plant breeding–a vital part of improvement in crop yields, quality and production efficiency. In: Frey, K. J. (Ed.). *Historical Perspectives in Plant Sciences*. Iowa State University Press, Ames. pp. 163–177.

Dupuis, R. D. and Krames, M. R. (2008). History, development, and applications of high-brightness visible light-emitting diodes. *Journal of Lightwave Technology* 26(9), 1154–1171.

European Commission (2022). *Quarterly Report on European Electricity Markets*. Available at: https://ec.europa.eu/info/sites/default/files/energy_climate_change_environment/quarterly_report_on_european_electricity_markets_q1_2022.pdf (vol. 15).

Evenson, R. E. and Gollin, D. (2003). Assessing the impact of the Green Revolution, 1960 to 2000. *Science* 300(5620), 758–762.

FAO (2023). *Technologies in Urban Agriculture. Adapting Plant Cultivation Systems to the Urban Environment*.

Folta, K. M. (2019). Breeding new varieties for controlled environments. *Plant Biology* 21 (Suppl. 1), 6–12.

Frantz, J. M., Ritchie, G., Cometti, N. N., Robinson, J. and Bugbee, B. (2004a). Exploring the limits of crop productivity: Beyond the limits of tipburn in lettuce. *Journal of the American Society for Horticultural Science. American Society for Horticultural Science* 129(3), 331–338.

Frantz, J. M., Cometti, N. N. and Bugbee, B. (2004b). Night temperature has a minimal effect on respiration and growth in rapidly growing plants. *Annals of Botany* 94(1), 155–166.

Fujiwara, K. (2020). Light sources. In: *Plant Factory*. Academic Press, pp. 139–151.

Gallicchio, N. (2017). Inventions of weather instruments (1400–1800). *Evol. Meteorol. Look Past Present Future Weather Forecast*, 11–17.

Gautier, H., Diakou-Verdin, V., Bénard, C., Reich, M., Buret, M., Bourgaud, F., Poëssel, J. L., Caris-Veyrat, C. and Génard, M. (2008). How does tomato quality (sugar, acid, and nutritional quality) vary with ripening stage, temperature, and irradiance? *Journal of Agricultural and Food Chemistry* 56(4), 1241-1250.

Goto, E. and Takakura, T. (1992). Prevention of lettuce tipburn by supplying air to inner leaves. *Transactions of the American Society of Agricultural and Biological Engineers* 35, 641-645.

Hansen, M., Bardsley, N., Pattison, M. and Lee, K. (2022). 2022 DOE SSL manufacturing status & opportunities. Available at: https://www.energy.gov/sites/default/files/2022-02/2022-ssl-manufacturing-status-opportunities_0.pdf. United States Department of Energy.

Hemming, S., Zwart, F. D., Elings, A., Petropoulou, A. and Righini, I. (2020). Cherry tomato production in intelligent greenhouses–sensors and AI for control of climate, irrigation, crop yield, and quality. *Sensors* 20(22), 6430.

Heuvelink, E. (Ed.) (2018). *Tomatoes* (vol. 27). CABI Publishing.

Hewett, E. W. (2019). Preharvest factors affecting postharvest quality of horticultural products. *Acta Horticulturae* (1245), 11-20.

Hoagland, D. R. and Arnon, D. I. (1950). *The Water-Culture Method for Growing Plants without Soil*. Circular. Agricultural Experiment Station, CA. p. 347.

Janick, J. and Paris, H. (2022). History of controlled environment horticulture: ancient origins. *Hortscience* 57(2), 236-238.

Ji, Y., Kusuma, P. and Marcelis, L. F. M. (2023). Vertical farming. *Current Biology* 33(11), R471-R473.

Ji, Y., Nunez Ocana, D., Choe, D., Larsen, D. H., Marcelis, L. F. M. and Heuvelink, E. (2020). Far-red radiation stimulates dry mass partitioning to fruits by increasing fruit sink strength in tomato. *New Phytologist* 228(6), 1914-1925.

Jin, W., Urbina, J. L., Heuvelink, E. and Marcelis, L. F. M. (2020). Adding far-red to red-blue light-emitting diode light promotes yield of lettuce at different planting densities. *Frontiers in Plant Science* 11, 609977.

Jin, W., Formiga Lopez, D., Heuvelink, E. and Marcelis, L. F. M. (2022). Light use efficiency of lettuce cultivation in vertical farms compared with greenhouse and field. *Food and Energy Security* 12(1), e391.

Kitaya, Y., Tsuruyama, J., Kawai, M., Shibuya, T. and Kiyota, M. (2000). Effects of air current on transpiration and net photosynthetic rates of plants in a closed plant production system. In: *Transplant Production in the 21st Century*. Springer, Dordrecht, pp. 83-90.

Kozai, T. (2020). PFAL business and R&D in Asia and North America: status and perspectives. *Plant Factory: an Indoor Vertical Farming System for Efficient Quality Food Production* (2nd edn.). Academic Press, San Diego, CA, 39-51.

Kozai, T. and Niu, G. (2020). Role of the plant factory with artificial lighting (PFAL) in urban areas. *Plant Factory: an Indoor Vertical Farming System for Efficient Quality Food Production* (2nd edn.). Academic Press, San Diego, CA, 7-34.

Kusuma, P., Pattison, P. M. and Bugbee, B. (2020). From physics to fixtures to food: Current and potential LED efficacy. *Horticulture Research* 7, 56.

Kusuma, P., Fatzinger, B., Bugbee, B., Soer, W. and Wheeler, R. (2021). LEDs for extraterrestrial agriculture: tradeoffs between color perception and photon efficacy. Available at: https://ntrs.nasa.gov/citations/20210016720.

Kusuma, P., Pattison, P. M. and Bugbee, B. (2022). Photon efficacy in horticulture: Turning LED packages into LED luminaires. In: *Plant Factory Basics, Applications and Advances*. Academic Press, pp. 115–128.

Langenfeld, N. J., Pinto, D. F., Faust, J. E., Heins, R. and Bugbee, B. (2022). Principles of nutrient and water management for indoor agriculture. *Sustainability* 14(16), 10204.

Larsen, D. H., Li, H., van de Peppel, A. C., Nicole, C. C. S., Marcelis, L. F. M. and Woltering, E. J. (2022). High light intensity at End-Of-Production improves the nutritional value of basil but does not affect postharvest chilling tolerance. *Food Chemistry* 369, 130913.

Lazzarin, M., Meisenburg, M., Meijer, D., Van Ieperen, W., Marcelis, L. F. M., Kappers, I. F., Van der Krol, A. R., Van Loon, J. J. A. and Dicke, M. (2021). LEDs make it resilient: Effects on plant growth and defense. *Trends in Plant Science* 26(5), 496–508.

Leong, R. and Urano, D. (2018). Molecular breeding for plant factory: strategies and technology. In: *Smart Plant Factory*. Springer, Singapore, pp. 301–323.

Li, T., Heuvelink, E., Dueck, T. A., Janse, J., Gort, G. and Marcelis, L. F. M. (2014). Enhancement of crop photosynthesis by diffuse light: Quantifying the contributing factors. *Annals of Botany* 114(1), 145–156.

Li, Q., Li, X., Tang, B. and Gu, M. (2018). Growth responses and root characteristics of lettuce grown in aeroponics, hydroponics, and substrate culture. *Horticulturae* 4(4), 35.

Liu, Y., Schouten, R. E., Tikunov, Y., Liu, X., Visser, R. G. F., Tan, F., Bovy, A. and Marcelis, L. F. M. (2022). Blue light increases anthocyanin content and delays fruit ripening in purple pepper fruit. *Postharvest Biology and Technology* 192, 112024.

Mangon, M. H. (1861). Production of the green matter of leaves under the influence of the electric light. *The London, Edinburgh, and Dublin Philosophical Magazine and Journal of Science* 22(147), 327–328.

Mattson, N. and Lieth, J. H. (2019). Liquid culture hydroponic system operation. In: *Soilless Culture. Theory and Practice*. Elsevier, pp. 567–585.

Maureira, F., Rajagopalan, K. and Stöckle, C. O. (2022). Evaluating tomato production in open-field and high-tech greenhouse systems. *Journal of Cleaner Production* 337, 130459.

McCree, K. J. (1971). The action spectrum, absorptance and quantum yield of photosynthesis in crop plants. *Agricultural Meteorology* 9, 191–216.

Min, Q., Marcelis, L. F. M., Nicole, C. C. S. and Woltering, E. J. (2021). High light intensity applied shortly before harvest improves lettuce nutritional quality and extends the shelf life. *Frontiers in Plant Science* 12, 615355.

Mishra, P., Asaari, M. S. M., Herrero-Langreo, A., Lohumi, S., Diezma, B. and Scheunders, P. (2017). Close range hyperspectral imaging of plants: A review. *Biosystems Engineering* 164, 49–67.

Mitchell, C. A. (2022). History of controlled environment horticulture: indoor farming and its key technologies. *Hortscience* 57(2), 247–256.

Murphy, D. J. (2007). *People, Plants and Genes: The Story of Crops and Humanity*. University Press, Oxford.

Nelson, J. A. and Bugbee, B. (2014). Economic analysis of greenhouse lighting: Light emitting diodes vs. high intensity discharge fixtures. *PLoS ONE* 9(6), e99010.

Nemali, K. (2022). History of controlled environment horticulture: greenhouses. *Hortscience* 57(2), 239–246.

Niu, G., Kozai, T. and Sabeh, N. (2020). Physical environmental factors and their properties. In: *Plant Factory*. Academic Press, pp. 185–195.

O'Sullivan, C. A., McIntyre, C. L., Dry, I. B., Hani, S. M., Hochman, Z. and Bonnett, G. D. (2020). Vertical farms bear fruit. *Nature Biotechnology* 38(2), 160–162.
Orsini, F., Pennisi, G., Zulfiqar, F. and Gianquinto, G. (2020). Sustainable use of resources in plant factories with artificial lighting (PFALs). *European Journal of Horticultural Science* 85(5), 297–309.
Patle, G. T., Kumar, M. and Khanna, M. (2020). Climate-smart water technologies for sustainable agriculture: a review. *Journal of Water and Climate Change* 11(4), 1455–1466.
Pattison, M., Hansen, M., Bardsley, N., Thomson, G. D., Gordon, K., Wilkerson, A., Lee, K., Nubbe, V. and Donnelly, S. (2022). *Solid-State Lighting R&D Opportunities. 2022 Solid-State Lighting R&D Opportunities*. Available at: energy.gov (vol. 2022).
Pattison, P. M., Tsao, J. Y., Brainard, G. C. and Bugbee, B. (2018). LEDs for photons, physiology and food. *Nature* 563(7732), 493–500.
Philips lighting (2017). *Master Greenpower Plus 1000W EL/5X6CT*. Available at: https://www.lighting.philips.com/main/prof/conventional-lamps-and-tubes/high-intensity-discharge-lamps/hid-horticulture/horticulture/928196305116_EU/product.
Radetsky, L. C. (2018). *LED and HID Horticultural Luminaire Testing Report*. Lighting Research Center, Rensselaer Polytechnic Institute, Troy, NY.
Ran, F. A. F. A., Hsu, P. D., Wright, J., Agarwala, V., Scott, D. A. and Zhang, F. (2013). Genome engineering using the CRISPR-Cas9 system. *Nature Protocols* 8(11), 2281–2308.
Raviv, M., Lieth, J. H. and Bar-Tal, A. (2008). Significance of soilless culture in agriculture. In: *Soilless Culture. Theory and Practice*. Elsevier, pp. 1–11.
Rawson, W. W. (1895). Gardening under glass. *Twenty-Second Annual Report of the [New Jersey] State Board of Agriculture*, 1895, 161.
Romeo, D., Vea, E. B. and Thomsen, M. (2018). Environmental impacts of urban hydroponics in Europe: A case study in Lyon. *Procedia CIRP* 69, 540–545.
Rouphael, Y., Kyriacou, M. C., Petropoulos, S. A., De Pascale, S. and Colla, G. (2018). Improving vegetable quality in controlled environments. *Scientia Horticulturae* 234, 275–289.
Sarić, R., Nguyen, V. D., Burge, T., Berkowitz, O., Trtílek, M., Whelan, J., Lewsey, M. G. and Čustović, E. (2022). Applications of hyperspectral imaging in plant phenotyping. *Trends in Plant Science* 27(3), 301–315.
Sarlikioti, V., de Visser, P. H., Buck-Sorlin, G. H. and Marcelis, L. F. M. (2011). How plant architecture affects light absorption and photosynthesis in tomato: Towards an ideotype for plant architecture using a functional-structural plant model. *Annals of Botany* 108(6), 1065–1073.
Schlegel, R. H. (2017). *History of Plant Breeding*. CRC press.
Schmilewski, G. (2017). Growing media constituents used in the EU in 2013. *Acta Horticulturae* (1168), 85–92.
*Scientific American* (1893). The Photoret 69(26), 406. Available at: https://archive.org/details/scientific-american-1893-12-23/page/n5/mode/2up?view=theater.
SharathKumar, M., Heuvelink, E. and Marcelis, L. F. M. (2020). Vertical farming: moving from genetic to environmental modification. *Trends in plant science* 25 (8), 724-727.
Shelford, T. J. and Both, A. J. (2021). On the technical performance characteristics of horticultural lamps. *AgriEngineering* 3(4), 716-727.
Simha, R. V. (2012). Willis H carrier. *Resonance* 17(2), 117–138.

Simkin, A. J., López-Calcagno, P. E. and Raines, C. A. (2019). Feeding the world: Improving photosynthetic efficiency for sustainable crop production. *Journal of Experimental Botany* 70(4), 1119–1140.
Snowden, M. C., Cope, K. R. and Bugbee, B. (2016). Sensitivity of seven diverse species to blue and green light: Interactions with photon flux. *PLoS ONE* 11(10), e0163121.
Spence, C. C. (1962). Early uses of electricity in American agriculture. *Technology and Culture* 3(2), 142–160.
Su, T., Li, P., Wang, H., Wang, W., Zhao, X., Yu, Y., Zhang, D., Yu, S. and Zhang, F. (2019). Natural variation in a calreticulin gene causes reduced resistance to Ca2+ deficiency-induced tipburn in Chinese cabbage (*Brassica rapa* ssp. pekinensis). *Plant, Cell and Environment* 42(11), 3044–3060.
Tiwari, A., Dassen, H. and Heuvelink, E. (2007). Selection of sweet pepper (*Capsicum annuum* L.) genotypes for parthenocarpic fruit growth. *Acta Horticulturae* (761), 135–140.
Tsao, J. Y., Han, J., Haitz, R. H. and Pattison, P. M. (2015). The Blue LED Nobel Prize: historical context, current scientific understanding, human benefit. *Annalen der Physik (Leipzig)* 527, 1–9.
Ushio (2021). Ushio Germany presents most efficient HPS 1000 W lamp with 2180 μmol/s. Available at: https://www.blv-licht.com/ushio-germany-presents-another-product-innovation/.
Vaiglova, P., Bogaard, A., Collins, M., Cavanagh, W., Mee, C., Renard, J., Lamb, A., Gardeiseng, A. and Fraser, R. (2014). An integrated stable isotope study of plants and animals from Kouphovouno, southern Greece: A new look at Neolithic farming. *Journal of Archaeological Science* 42, 201–215.
Vallarino, J. G., Kubiszewski-Jakubiak, S., Ruf, S., Rößner, M., Timm, S., Bauwe, H., Carrari, F., Rentsch, D., Bock, R., Sweetlove, L. J. and Fernie, A. R. (2020). Multi-gene metabolic engineering of tomato plants results in increased fruit yield up to 23%. *Scientific Reports* 10(1), 17219.
Van Delden, S. H., SharathKumar, M., Butturini, M., Graamans, L. J. A., Heuvelink, E., Kacira, M., Kaiser, E., Klamer, R. S., Klerkx, L., Kootstra, G., Loeber, A., Schouten, R. E., Stanghellini, C., van Ieperen, W., Verdonk, J. C., Vialet-Chabrand, S., Woltering, E. J., van de Zedde, R., Zhang, Y. and Marcelis, L. F. M. (2021). Current status and future challenges in implementing and upscaling vertical farming systems. *Nature Food* 2(12), 944–956.
Van Os, E. A., Gieling, T. H. and Lieth, J. H. (2019). Technical equipment in soilless production systems. In: *Soilless Culture. Theory and Practice*. Elsevier. pp. 587–635.
Velez-Ramirez, A. I., Van Ieperen, W., Vreugdenhil, D., Van Poppel, P. M., Heuvelink, E. and Millenaar, F. F. (2014). A single locus confers tolerance to continuous light and allows substantial yield increase in tomato. *Nature Communications* 5, 4549.
Wang, X., Fuller, B. T., Zhang, P., Hu, S., Hu, Y. and Shang, X. (2018). Millet manuring as a driving force for the Late Neolithic agricultural expansion of north China. *Scientific Reports* 8(1), 5552.
Wheeler, R. M. (2008). A historical background of plant lighting: An introduction to the workshop. *Hortscience* 43(7), 1942–1943.
Woodward, J. (1699). II. Some thoughts and experiments concerning vegetation. *Philosophical Transactions of the Royal Society of London* 21(253), 193–227.

Yorio, N. C., Goins, G. D., Kagie, H. R., Wheeler, R. M. and Sager, J. C. (2001). Improving spinach, radish, and lettuce growth under red light-emitting diodes (LEDs) with blue light supplementation. *Hortscience* 36(2), 380–383.

Zepeda, A. C., Heuvelink, E. and Marcelis, L. F. M. (2022). Carbon storage in plants: A buffer for temporal light and temperature fluctuations. *Frontiers in Plant Science* 13, 968881.

Zhang, B., Tong, Y., Luo, K., Zhai, Z., Liu, X., Shi, Z., Zhang, D. and Li, D. (2021). Identification of GROWTH-REGULATING FACTOR transcription factors in lettuce (*Lactuca sativa*) genome and functional analysis of *LsaGRF5* in leaf size regulation. *BMC Plant Biology* 21(1), 485.

Zheludev, N. (2007). The life and times of the LED—A 100-year history. *Nature Photonics* 1(4), 189–192.

Zhen, S. and Bugbee, B. (2019). Diffuse light panels slightly increase canopy light use efficiency but decrease light intensity and canopy photosynthesis of lettuce. *Controlled Environments*. Available at: https://digitalcommons.usu.edu/cpl_env/12, Paper 12.

Zhen, S., Kusuma, P. and Bugbee, B. (2022). Toward an optimal spectrum for photosynthesis and plant morphology in LED-based crop cultivation. In: *Plant Factory Basics, Applications and Advances*. Academic Press, pp. 309–327.

# Part 2

## Energy and other resource performance

# Chapter 4

# Life cycle assessment of indoor vertical farms

*Michael Martin, IVL Swedish Environmental Research Institute and KTH Royal Institute of Technology, Sweden; and Francesco Orsini, University of Bologna, Italy*



## 1 Introduction

It is becoming increasingly important to find solutions for more resilient food production methods closer to urban environments with less vulnerability to supply-chain shocks (Benke and Tomkins, 2017; O'Sullivan et al., 2020; Pulighe and Lupia, 2020). Indoor vertical farming (IVF[1]) systems have emerged worldwide as a result of the need for more resilient food provisioning. IVF has been promoted for its potential to extend seasonal availability, produce more sustainable food, secure food supplies, and reduce pressure on agricultural land (Graamans et al., 2018; Martin and Molin, 2019; Thomaier et al., 2014; van Delden et al., 2021). Furthermore, IVF has seen a dramatic increase in recent years, attracting considerable interest and funding (Agritecture, 2022; Orsini et al., 2020; Weidner et al., 2019).

IVFs are relatively new in the context of food supply chains, and thus it is an expanding subject of inquiry. A large share of the scientific and grey literature promotes vertical farming as a sustainable solution for food provisioning

[1] The acronym IVF henceforth refers to the verb 'indoor vertical farming' to denote the practice and the noun 'indoor vertical farm' to denote the sites.

http://dx.doi.org/10.19103/AS.2023.0126.06

(Al-Chalabi, 2015; Benke and Tomkins, 2017; Despommier, 2011). However, assessments of the environmental implications of IVFs remain limited in scientific literature, with few cases applying systematic environmental assessments (Dorr et al., 2021; Martin and Molin, 2019; Martin et al., 2022; Romeo et al., 2018). A number of theoretical studies have assessed IVFs to compare their performance against competing systems such as open-field production and greenhouses, see e.g. Graamans et al. (2018) and Weidner et al. (2021, 2022). Nonetheless, empirical evidence from real case studies is lacking in the literature, which may be due in part to their novelty and evolving nature. As such, there are few studies that validate claims made by vertical farming of their resource efficiency and reduced environmental impacts, which are often focused primarily on the farm-level metrics.

Consumers, businesses, and decision-makers are becoming increasingly attentive to the use of feedback and information through credible systems to communicate and evaluate the environmental impacts of goods and services. In particular, the food sector has been increasingly employing life cycle assessment (LCA) for highlighting the 'footprint' of their products (Freidberg, 2014). Employing LCAs may be an important methodology for IVFs to meet the criticism of many of the claims in the industry and provide knowledge for working with sustainability more strategically, providing transparent and scientifically based metrics.

This chapter aims to provide insights on conducting an environmental sustainability assessment of an IVF employing LCA methodology and outline important considerations during the process. The chapter is designed to provide an overview of the method, and thereafter describes the different phases of conducting a LCA, providing guidance specifically for IVFs. It also outlined the limitations of employing LCA and provides knowledge on challenges, important aspects, and possibilities to improve the environmental performance of IVFs based on previous research. The methodology and insights are applicable to different forms of IVF. This includes assessments of, for example, the production of edible crops, mushrooms, and production of crops for other purposes (e.g. pharmaceutical applications).

## 2 Life cycle assessment and its application in indoor vertical farms

LCA is a broadly employed and accepted method in which the environmental impacts related to a product system or service are quantified and illustrated during its life cycle, i.e. from raw material extraction via production and use phases to waste management and transportation (Finnveden et al., 2009). The method has been used for several decades and has even been standardized by the International Standards Organization (ISO, 2006). It is used by organizations to identify so-called 'hot-spots' in their life cycle which can be used to improve processes and make strategic decisions on the best course of action to improve. LCA also provides information that can be used for labeling and decision-making.

According to ISO 14044 (2006), the process of conducting an LCA is based on four required steps (also referred to as phases), including (1) goal and scope definition, (2) inventory analysis, (3) environmental impact assessment, and (4) interpretation; see Fig. 1. These steps are specifically reviewed in subsequent sections with application to applying LCA of IVFs.

## 2.1 Goal and scope of the study

In the first step, the goal and scope of the LCA are defined to ensure that the outcome is consistent with the objectives, setting the context for the study. For this step, it is important that the purpose of the study is clearly defined in the documentation but also agreed upon between those conducting the LCA and the receiver (typically the client). For example, this can include the intended application of the study, such as communication, comparative assertions, product improvement, planning processes, strategic decisions, or policy-making. It is also important to highlight the conditions and assumptions for which the results of the assessment are valid, which are of utmost importance to the study.

In the scope description, the definition of the functional unit, i.e. the service delivered by the product system, is critical to agree upon. Specifically for IVFs this is typically associated with the production output, e.g. one kilogram of the edible portion (fresh weight) of a particular food available to consumers.

The definition of system boundaries should also be set. This includes what is to be included and what is to be omitted. Ideally, the system boundaries cover the full life cycle, with both upstream and downstream inputs and outputs, including all material, energy, and processes. However, often simplifications are done in order to match the scope of the study and reduce the complexity. In the

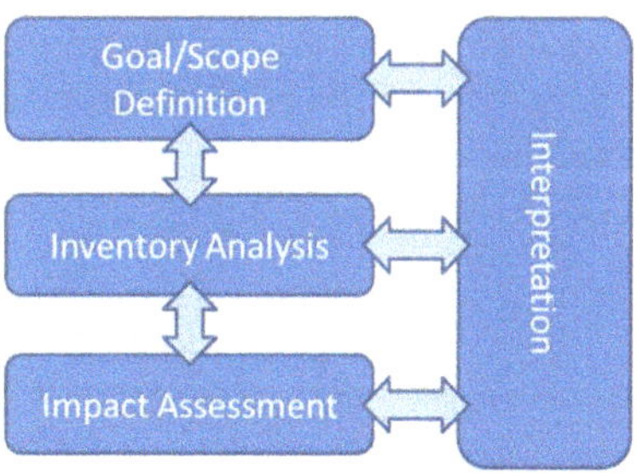


**Figure 1** Elements of the Life Cycle Assessment Method (based on ISO, 2006).

LCA field, the system boundaries are typically referred to as being e.g. a cradle-to-gate study versus a cradle-to-grave study.[2] Cradle-to-gate assessments include all inputs and processes up to the availability of a product at retail or availability at the consumer. Cradle-to-grave assessments include the same as the cradle-to-gate studies but also include the use and waste management of the product after consumption, e.g. the waste handling methods for the growing media and packaging. Figure 2 illustrates an example of the system boundaries for a cradle-to-gate and a cradle-to-grave perspective applicable for an IVF. Note that the waste handling from the farm is included in the cradle-to-gate boundaries and that retail is often not included. However, the 'gate' can also be extended to include the availability of the product at the retail location, but not the use and final disposal of the product. This should be clear in the system boundaries of the study. Typically, the retail phase for food products has only a minor contribution to the environmental performance compared to other processes along the life cycle and is often omitted given this motivation.

## 2.2 Life cycle inventory

The life cycle inventory (LCI) is used to collect data necessary for the quantification of the environmental impacts of the service or product. This is done by mapping the applicable material and energy flows and processes involved along the life cycle confined within the system boundaries set in the scope of the project.

For IVFs, data collection includes all material and energy inputs, all transportation requirements, processes, infrastructure, and processes required for cultivation and waste handling. Figure 2 outlines many of these for a typical IVF. This is often the most tedious part of conducting any LCA, and will require a number of iterations between the LCA practitioner and the IVF, i.e. if the LCA is performed by an external source.

In order to alleviate the data collection, especially for IVFs, often a timeframe is chosen to collect data, e.g. based on annual production[3] figures. For the LCI, data quality issues can also be noted. For example, data should be as accurate as possible. Nonetheless, assumptions may need to be made when the availability of accurate data is limited, which may require a sensitivity analysis later in the assessment to see the influence this has on the results. When assumptions are made by the LCA practitioner, these should be reviewed by the IVF representative involved with the LCA in order to assure the assumptions are realistic.

[2] In agricultural LCAs, cradle-to-gate can also be referred to as 'farm-to-gate' and the cradle-to-grave approach, referred to as a 'farm-to-fork' assessment.
[3] It is recommended to conduct the study using annual values if a screening of impacts from the system is to be done. If the production and energy demand are consistent, a monthly value may be used. If more details on how to improve seasonally are required, monthly data can be important.

Published by Burleigh Dodds Science Publishing Limited, 2023.

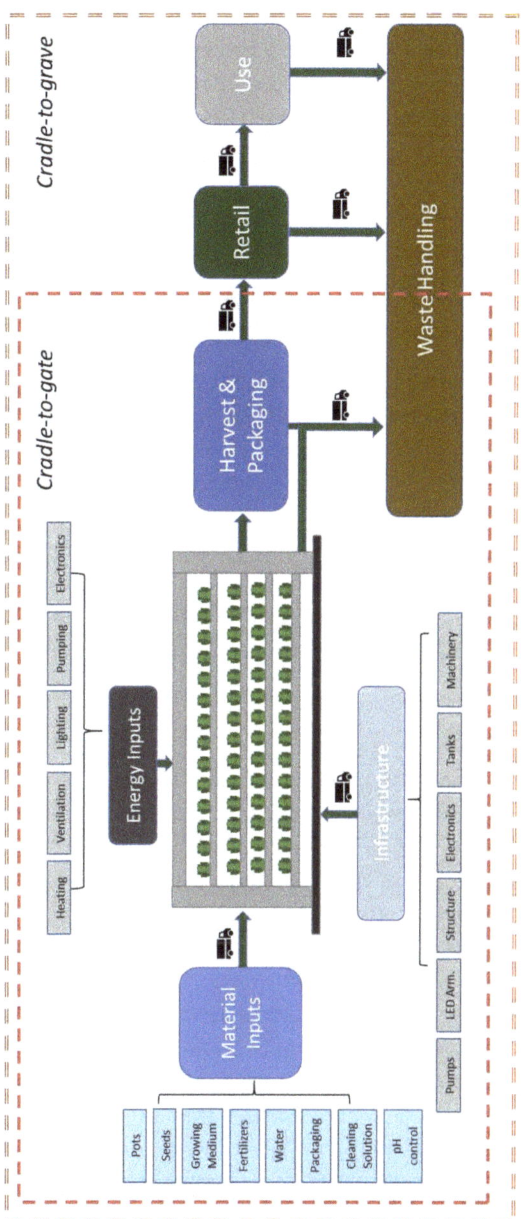


**Figure 2** System boundaries for the life cycle assessment of an IVF, including inputs, outputs, of the system. Based on Martin et al. (2022).

### 2.2.1 Material inputs

For IVFs, a number of material inputs, or consumables, are required. Depending upon the system, this can include all consumable inputs such as, packaging for inputs (e.g. cardboard, plastics, etc.), growing media, water, fertilizers, pH control, protective wear, seeds, $CO_2$ enrichment, cleaning supplies, other chemicals, and packaging for products (including pots, outer wrapping, cardboard, plastics). For some materials, there may be more thorough data available. Assumptions can be made and documented on the amounts employed based on the total outputs. Practically, it may also be possible to review expenses and make estimates on the amounts used annually.

### 2.2.2 Energy

For IVFs, the energy demand is of key importance to the environmental impacts. As such, it is crucial that data is collected on the amount of energy employed, in addition to its sourcing. For electricity, information on the source is important for the LCA and relevant information on certificates of origin for the electricity should be reviewed.

Energy consumption may be primarily related to electricity demand for LED lighting, pumps, HVAC[4] systems, monitoring equipment, and other processes. If external heating sources from other energy carriers than electricity are used, these should also be included in the study, e.g. natural gas or district heating.

### 2.2.3 Transportation

For the LCA, information on the transportation distance of all materials into the farm, and products and waste leaving the farm, are included in the assessment. Important information here includes the distances from the sourcing of the materials and supplies. While detailed information can be included, assumptions can be made on the distances. Furthermore, for transportation of the products to markets, a breakdown of different markets can be provided, or an average travel distance could be provided.

For LCAs, the transportation of materials and products is typically assessed as tonne-kilometers (tonne-km) as the LCI datasets for transportation logistics are provided in these units. As such, the distance is then multiplied by the mass (in tonnes) of the material or product. See, e.g. the row 'Transportation-Inputs' in Table 1, which sums up all the tonne-km for all material inputs; similarly this is also done for the outputs and infrastructure. If transportation of items to and from the farm is conducted in other forms, datasets, e.g. for driving a car or van, can also be included, and are typically related to the number of kilometers driven.

[4] Heating, ventilation, and air-conditioning (HVAC).

**Table 1** Simplified input-output table employed for the LCI in LCAs of IVFs. Values are arbitrarily included.

| Type | Category | Specific category | Inputs | Amount (annually) | Unit | Transport distance (km) |
|---|---|---|---|---|---|---|
| Inputs | ***Material inputs*** | **Growing medium** | Growing media | 10 | kg | 100 |
| | | **Fertilizers** | Fertilizer 1 | 1 | kg | 50 |
| | | | Fertilizer 2 | 2 | kg | 50 |
| | | **Seeds** | Seeds | 5 | kg | 100 |
| | | Water | Tap Water | 100 | $m^3$ | - |
| | | **Other** | $CO_2$ (enrichment) | 60 | kg | 45 |
| | | **Packaging** | Polystyrene (PS) | 3 | kg | 50 |
| | | | PET | 5 | kg | 50 |
| | | | Polypropylene | 5 | kg | 50 |
| | | | Cardboard | 20 | kg | 50 |
| | **Transportation-inputs** | | Material inputs (Truck) | 6 | tonne-km | - |
| | **Energy inputs** | | Electricity | 5000 | kWh | - |
| | | | Heat | 1000 | kWh | - |
| Outputs | **Production outputs** | | Plant (Type 1) | 400 | edible kg | 50 |
| | | | Plant (Type 2) | 500 | edible kg | 20 |
| | **Cultivation and packaging wastes** | | Packaging waste (Plastic) | 5 | kg | 50 |
| | | | Packaging waste (Cardboard) | 15 | kg | 50 |
| | | | Organic waste | 10 | kg | 50 |
| | **Transportation-outputs/ market** | | Market (Truck) | 30 | tonne-km | - |
| | | | Waste handling (Truck) | 1.5 | tonne-km | - |
| Infrastructure | **Infrastructure** | | Steel tray structures | 1500 | kg | 100 |
| | | | Plastic | 50 | kg | 100 |
| | | | Pipes (Polyethylene) | 15 | kg | 100 |
| | | | LED light fixtures | 50 | units @ 1.5 kg | 100 |
| | | | Tanks | 2 | units | 100 |
| | | | Seeding machine | 1000 | kg | 100 |
| | | | Other electronics | 25 | kg | 100 |
| | **Transportation-infrastructure** | | Truck | 359 | tonne-km | - |

### 2.2.4 Infrastructure

Infrastructure for IVF is an important input in the assessment of the system. For an IVF, this can include steel structures, tanks, tubing, pumps, electronics, machinery (seeding, packaging, etc.), HVAC equipment, LED armatures, etc. It is also important to take into account the lifetime of the infrastructure. As such, the entire infrastructure does not affect the environmental impacts of the vertical farm on one specific year but is distributed over the lifetime. An example is the assumed steel structures which have a long lifetime. As 1500 kg of steel is assumed to be included in the structure, with a lifetime of 25 years, the impact on the annual system would be related to employing 60 kg of steel. The infrastructure also may have varying durability, i.e. associated lifetimes. While steel structures may have an assumed long lifetime, LED lighting fixtures and other active equipment, such as pumps, tubing, and various plastics may have a much shorter lifetime due to their active use. Maintenance and added equipment and infrastructure can also be included if the information is available. The LCA practitioner and the IVF firm should make assumptions on the lifetime of all infrastructures based on evidence from both practical experience, information from the producer, scientific literature, and ensure to document this in the study.

### 2.2.5 Outputs

While some producers specialize in a single or a handful of outputs, others may have a large span of different products. These can be accounted for differently, depending on the type of output (e.g. salad, herbs, mushrooms, etc.). Furthermore, while some farms may sell harvested crops, others may sell as potted plants with packaging and growing media. As such it is important for an LCA to take into account the type of product sold. Typically this is done by accounting for the edible portion, e.g. in kilograms, of the product to retail. For potted plants, the total number annually can be used. Furthermore, waste from the farm should also be accounted for, see subsequent section.

### 2.2.6 Waste handling

Besides the edible portion of the products sold to market, there may be a number of waste and residual streams produced from the farm. This includes, but is not limited to, biowastes (e.g. growing media and plants), packaging wastes, wastewater to drain, equipment, and any other consumables (e.g. protective wear). For each of these flows, the amount and their method of waste handling should be accounted for. The waste handling methods for different waste streams may vary depending upon the location and IVF. This can include sending the waste for incineration, landfill, or to recycling for certain materials. If residual streams are included in the output, e.g. products or energy sold to other systems, these are important for the LCA. This can include residual heat which may be used in other applications, e.g. heating of the host building and other co-located spaces (Martin et al., 2022).

In studies with a cradle-to-grave perspective, an important aspect to take into account is the waste handling of the final product and packaging. As such, the materials and design of the packaging may be important at this stage. Once again, methods for the treatment of these waste flows should be accounted for. Practically, after the product leaves the farm this may not be clear, as information on consumer behavior, including how they employ the product, the share of waste, and what they do with the waste may not be available. As such, assumptions can be made based on local conditions and systems available in the region.

### 2.2.7 Connecting to LCI datasets

Once the inventory is complete, LCI data is gathered to match those flows and processes. Despite the comprehensive nature of LCI databases from sources such as Ecoinvent (Frischknecht et al., 2005) not all datasets are available to match with flows and processes outlined in the inventory. As such, representative datasets can be chosen, although assumptions should be documented. For example, there may not be datasets available for a specific component, e.g. a specific sensor. However, a dataset specifying an 'active electronic component' may be used as a proxy. At this stage, it is important to ensure that the LCI datasets and the inventory listing have similar units. As an example, in the case of the sensors, in the inventory, the number of sensors may be included. However, the LCI data may be provided for the mass of the electronic components, e.g. in kg. As such conversions will be required to allow for these to be used. When conducting an LCA, it is important to always list which LCI datasets we used, which are often listed as a table, in order to improve transparency.

## 2.3 Life cycle impact assessment

The next phase of conducting an LCA is the life cycle impact assessment (LCIA). In this phase, the environmental impacts associated with all inputs and processes are quantified and provided for different selected environmental impact categories. When conducting this phase, predefined LCIA methods included in LCA software are employed. These are used to aggregate the environmental impacts into different environmental impact categories.

LCIA methods are categorized as midpoint and endpoint approaches. Firstly, midpoint approaches provide quantitative modeling for equivalent emissions of substances. It stops at this point to reduce the uncertainties and does not include weighting. Examples of midpoint categories are global warming potential (kg $CO_2$-eq.) or acidification potential (e.g. measured in kg $SO_2$-eq.). Taking this a step further, endpoint approaches model damages caused by the release of different substances and emissions. An example of an endpoint impact includes the Disability-adjusted life years (DALY) impact category which takes into account years lost to premature death due to illness, disability, or early death. For IVFs, the midpoint methods are most appropriate. As an example, for the impact category, Global Warming Potential (also referred to as carbon footprint), an aggregation of all greenhouse gas emissions (GHGs; such as $CO_2$, $CH_4$, $N_20$, etc.) is conducted to assess their impacts on climate change. These are expressed as kg $CO_2$-equivalent units, denoted as kg $CO_2$-eq.

While it is often assumed that carbon footprints are LCAs, this is only one impact category in a much broader set of environmental impact categories. Various LCIA methods contain a large number of impact categories. Some of these included the ReCiPe method (Huijbregts et al., 2016) or the ILCD Environmental Footprint v. 3.0 method (ILCD, 2010) with a large range of environmental impact categories available. The ReCiPe method contains 18 environmental impact categories, including agricultural land occupation ($m^2a$), climate change (kg $CO_2$-eq.), fossil depletion (kg oil-eq.), freshwater ecotoxicity (kg 1,4-DCB-eq.), freshwater eutrophication (kg P-eq.), human toxicity (kg 1,4-DCB-eq.), ionizing radiation (kg U235-eq.), marine ecotoxicity (kg 1,4-DB-eq.), marine eutrophication (kg N-eq.), metal depletion (kg Fe-eq.), natural land transformation ($m^2$), ozone depletion (kg CFC-11-eq.), particulate matter formation (kg $PM_{10}$-eq.), photochemical oxidant formation (kg NMVOC-Eq), terrestrial acidification (kg $SO_2$-eq.), terrestrial ecotoxicity (kg 1,4-DCB-eq.), urban land occupation ($m^2a$), and water depletion ($m^3$ water-eq.).

LCAs of IVFs may choose to include all or a selected number of environmental impact categories. However, for IVFs, several authors have shown that important indicators from LCIA methods include those related to carbon footprint, resource consumption, water depletion, ecotoxicity, acidification, and eutrophication (Dorr et al., 2021; Martin et al., 2022). LCIA methods, however, are regionally specific. Many of the LCIA methods are based on European conditions. As such, regionally specific LCIA methods for different regions can also be used. For example, the ILCD or ReCiPe methods can be used for European conditions, while the TRACI LCIA method is more applicable to North American conditions.

## 2.4 Partitioning of environmental impacts in multi-functional processes

If an IVF produces several products, referred to as a multi-functional process, the allocation to the functional unit and other products from the system is necessary. The impacts are therefore partitioned to all products of the system, which is done by physical or economic allocation. Physical allocation is done by partitioning impacts to the different products by their physical properties of the overall output, e.g. by their mass or energy. For IVFs, the mass of the products is primarily used, with edible mass preferable. Economic allocation refers to partitioning impacts on the products based on their economic value (e.g. market price). The economic value inherent in the products and by-products may change, or be different depending on the location of an IVF, inhibiting comparisons (Cherubini, 2010; van der Voet et al., 2010).

In the ISO standards, it is recommended that allocation be avoided if possible (ISO, 2006). To do so, a method referred to as system expansion may be used to avoid allocation, by removing impacts from conventional products replaced from by-products of the system (Weidema, 2001). This is done by identifying the equivalent amount of conventional products or materials replaced by a by-product of the system and thereafter finding the environmental impacts of producing that product. In the calculations, the impacts from the avoided system are credited. The system expansion method is a form of consequential modeling, also referred to as partial-consequential modeling (Brander et al., 2009), for the avoidance of processes created in the system affecting markets outside the system (Zamagni et al., 2012).

An example of this can include a mushroom farm, which produces mushrooms as the main outputs, but the spent mushroom substrate is used for soil amendment. Assuming that replaces, e.g. peat, the equivalent amount of peat, either in kilograms of volume ($m^3$) can be avoided and credited to the mushroom farm. See examples of such calculations for IVFs in Fig. 3 and also in previous assessments of IVFs including by-products that provide credits to the system such as the use of residual heat from IVFs (Martin et al., 2019, 2022).

It should be noted that the allocation in multi-functional processes is a highly controversial topic in the LCA field. There is extensive literature devoted to this (see e.g. Ekvall and Finnveden, 2001; Wardenaar et al., 2012). Nonetheless, as Guinée et al. (2004) suggest, there is no 'correct' way to solve this problem in practice or theory. It should be apparent, that the chosen allocation method is consistent with the research questions addressed and the main methodological choices made. One way to avoid critique on the allocation issue is to conduct the study and illustrate the results by applying different approaches, i.e. including allocation (e.g. mass or economic) and system expansion, thus showing the influence this can have on the results.

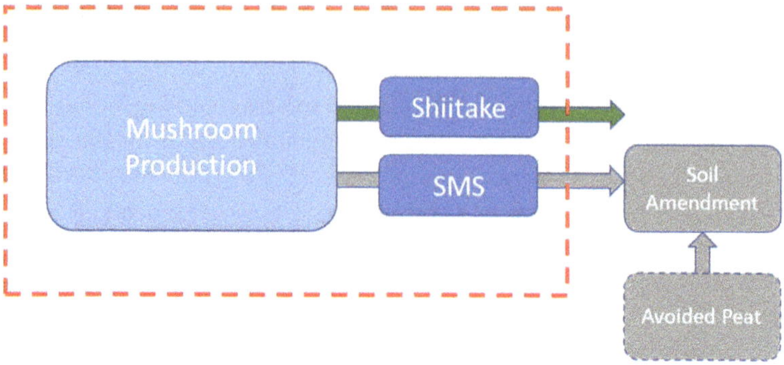


**Figure 3** Depiction of system-expansion method to show how the use of spent mushroom substrate (SMS) can lead to avoided peat use.

### *2.5 Interpretation*

The 'last' phase of an LCA is the interpretation phase. In this phase, the practitioner(s) interpret the inventory and impact results in order to assess the consistency, sensitivity, and significant issues to be used to formulate conclusions, recommendations, and limitations for the study. This is important for communication with external parties in order to highlight any limitations and uncertainties.

The process of interpretation is rather iterative in nature, requiring revisions to the model and data. A previous study by Lazarevic (2012, p. 3) describes the iterative nature of conducting the LCA as follows, 'the goal and scope of the study are defined, a life cycle model is developed, impact assessments produced, the goal and scope are then refined or revised if necessary, key data improved, impact assessment characterization factors are improved, results interpreted, reported and subjected to independent review if necessary.'

For LCA studies with IVFs, this may include reviewing the assumptions made on different processes, transportation distances, representative datasets employed, and making any necessary improvements to the model. Processes, inputs, and datasets which were found to have a large influence on the overall environmental impacts can also be analyzed further and sensitivity analyses can be conducted to show their influence. This is often done by changing the datasets, or showing how an increase or decrease in the specified amount can influence the results. As outlined as follows, for IVFs, one important aspect to conduct a sensitivity analysis is the electricity data employed.

## 3 Strengths and limits of life cycle assessments for indoor vertical farms

LCA's strength lies in its comprehensive approach to evaluate upstream and downstream flows of a product or service (Hermann et al., 2007; Finnveden et al.,

2009). This is important, as impacts from a product or service may have impacts unevenly distributed along the life cycle. For assessing the environmental implications of a product or service, LCA cannot completely be replaced, as other stools may not review a cradle-to-grave perspective (Finnveden, 2000).

However, LCA also has some limitations. The objectivity, methodological considerations, and completeness have been debated for decades (Arvidsson et al., 2018; Freidberg, 2014; Heiskanen, 1999). As previously outlined, allocation and other methodological choices have been subject to extensive scientific discussion in the field (Brandao et al., 2017; Ekvall and Finnveden, 2001; Plevin et al., 2014). Data availability can also be a limiting factor for conducting LCAs of IVFs, with limited data on specific materials and infrastructural components, e.g. substrates and components for the infrastructure.

LCA is also limited to environmental impacts. As such, it may be difficult to capture the benefits of IVFs for local food provisioning compared to conventional methods. Despite this, similar assessments can be made to assess the social and economic implications using other life cycle approaches, such as social LCA and life cycle costing. These can even be combined to provide a more holistic approach to assessing sustainability, using life cycle sustainability assessment (Sala et al., 2013; Zamagni et al., 2013). Furthermore, LCAs may not fully model rebound effects and future changes in technology, e.g. the claims by IVF firms that it 'frees' space for conventional agriculture.

## 4 Insights from life cycle assessments of indoor vertical farms

From previous studies applying LCA to IVFs, a number of processes and parameters are important, or sensitive, for the overall environmental performance. The following sections outline several of these important processes and parameters in order to provide information that can be used to improve the process and environmental performance of LCA practitioners and IVFs.

### *4.1 Electricity and climate control*

Similar to other forms of controlled environment agriculture, energy is of utmost importance for IVFs. From previous research, the largest share of environmental impacts from IVFs has been found to stem from energy demands for LED and HVAC systems; see e.g. findings in Graamans et al. (2018), Martin et al. (2022), and Weidner et al. (2022). Additionally, given the large share of emissions from energy sources, the results are highly sensitive to the source, and subsequent choice of LCI dataset, for electricity. It is advised that when conducting an LCA for IVFs, the regional energy mix, or mixes, should be used.

As an example, Martin et al. (2019, 2022) show the sensitivity of employing the Swedish electricity mix versus a Nordic electricity mix on the environmental performance of an IVF, confirming significant differences between the two options.

Furthermore, if an IVF purchases electricity with certificates of origin, e.g. from hydropower or wind, this should also be compared with the regional system. This is exemplified in a recent study for IVF in Sweden, where the choice of hydropower-based electricity led to lower environmental impacts compared to other studies from Sweden, see e.g. Milestad et al. (2020). As identified by Brander et al. (2018), it is not certain that the electricity is produced from their claimed origins or can lead to changes in the electricity mix of a given region or that the electricity used has the same profile. Additionally, different standards for conducting LCAs handle the use of energy sourcing differently, and it is advisable to show results based on different electricity sources and mixes in order to avoid criticism. In conclusion, it is advised when conducting an LCA to show the sensitivity of different electricity mixes on the results of the study.

## 4.2 Substrates

For IVFs, the choice of substrate can also have a significant influence on the environmental performance of an IVF. Previous research has shown that substrates such as perlite and peat may have large environmental impacts, while lower environmental impacts are found for by-products such as coir (Martin et al., 2019, 2022; Quantis, 2012; Toboso-Chavero et al., 2021; Vinci and Rapa, 2019). The use of peat also continues to be controversial (Chapman et al., 2003; Hojlund, 2008; Salomaa et al., 2018). However, further research should be conducted on the environmental performance of different substrates as the use of new materials evolves for applications in IVFs. These typically include blends of materials and there has been an increased influx of new substrate materials specifically designed for IVF applications.

Depending on the growing system adopted, the amount of substrate needed by the IVF may also significantly vary, or may even be absent (e.g. when aeroponic farms are using reusable sowing mats, obtained from recycled materials). Accordingly, strategies that account for substrate use reduction should also be envisaged, when comparative scenarios are elaborated.

It should also be noted that the inherent properties of different substrates may also affect their waste handling methods. For inorganic substrates, landfilling or incineration may be the only option, while for other bio-based materials, composting or recycling may be employed. This can also have an influence on waste management and should be considered in the design of the system, e.g. to promote more closed-loop or circular systems.

## 4.3 Infrastructure

The infrastructure has been shown to have a minor contribution to the overall environmental performance of IVFs. However, the infrastructure can contribute to as high as 10–15% of the overall GHG emissions, and equally contribute significantly to resource depletion, given the amount of metals and electronic components (Barge, 2020; Martin and Molin, 2019). This can be sensitive to the assumed lifetime of certain inputs and components. For example, as the industry is novel, there may not be a large base of experience to base assumptions on the lifetime of products. Creating new buildings for an IVF may have a large influence on the impact of the infrastructure, while employing residual or existing spaces may not require as many resources. However, the materials and processes needed to use existing spaces should also be taken into account. Again, for new structures or intermediate processing required to use residual and existing spaces, the assumed lifetime can have a large influence on the impact of the infrastructure. This suggests that, from a life cycle perspective, choices to improve the lifetime of the materials, have a benefit to the environmental performance of IVFs.

Furthermore, in previous assessments, the results were also found to be sensitive to datasets for the infrastructure, including, e.g. electronics, machinery, and metals (Barge, 2020). For infrastructure, it is crucial to carefully choose representative products in the LCI datasets. Additionally, for reviewers of the LCAs and for ensuring the scientific-based information provided by an LCA, transparently providing information on the assumed lifetimes in addition to the LCI datasets employed is of utmost importance for an LCA.

## 4.4 Packaging

In previous studies, packaging has been found to contribute only to a small share of the environmental impacts of IVFs. However, a large number of studies have also excluded packaging from their assessment (Graamans et al., 2018; Romeo et al., 2018; Weidner et al., 2022) motivating its minor share of the overall impacts. However, despite its relatively small share to the overall impacts, annually the total consumption of certain materials may contribute to a large environmental footprint (e.g. GHG emissions and resource consumption). Reducing the amount of material, e.g. plastics, and developing approaches to switch from conventional black-colored[5] plastics to other colors, or other bio-based materials, may allow for the possibility of recycling and composting. This is also regionally and context-dependent on the markets for IVF products. As such, the packaging should be carefully chosen as it can also influence its waste

[5] Black plastics are often difficult to recognize in sorting systems and often end up in incineration.

handling methods. Further studies are needed to understand this impact, especially as the IVF industry has been quick to address the use of plastics in their work with sustainability.

## 5 Life cycle management and increasing transparency

A large focus in the IVF industry has been geared to promote sustainability, although there is little documentation on how vertical farming companies work with sustainability beyond claims provided on the packaging and in media. For IVFs, it is important that claims made are substantiated. However, few transparent assessments and studies of their work are available; see e.g. (Agritecture, 2022). In previous research, it was found that IVFs consider environmental performance information important for benchmarking and communicating their environmental performance. However, few were willing to publicly provide the results of their assessments due to the fact that most systems are constantly improving and evolving. As such, the so-called 'snapshot' with an LCA may not be representative of a system in the near future. Claims and comparative assertions on the benefits of IVFs compared to conventional systems also contradict the original intent in the early development of LCA, making comparative assertions and suggesting one system is better than another. When drawing conclusions from LCAs, Finnveden (2000) suggested that conclusions cannot be made on which system is 'better,' though they can lead to decisions leading to a better course of action than would have been followed.

Using the information, however, can be essential to developing improvement options. However, developing prospective approaches can be a beneficial stance to proactively work with sustainability issues at an early stage of development to identify the best technologies for specific contexts, e.g. markets, geographic locations, and infrastructure available (Arvidsson et al., 2018; Martin et al., 2021).

## 6 Conclusion and future trends in research

The majority of current vertical farming systems employ linear approaches to their production. This entails that they employ imported (often virgin) materials from outside their immediate regions for all their resource and energy demands. In the future, employing recycled materials in addition to renewable and residual energy sources may reduce the environmental impacts of the consumables and other inputs. With cities becoming an important driving force for the circular economy and as a critical stakeholder for developing and improving food security, synergies between IVFs and their urban infrastructure are essential for understanding and planning for future urban food systems.

As such, sustainable solutions for more integrated food, water, energy, and transportation will become increasingly important (Specht et al., 2014; Martin et al., 2019, 2022; Rufi-Salis et al., 2020). There is currently a number of research projects exploring such developments ongoing worldwide, which have the potential to develop more circular approaches for the IVF industry and potentially improve the environmental performance of IVFs.

Furthermore, in the coming years, it is expected that IVF production will gain in terms of product diversification, moving from the current systems, which are mainly based on leafy vegetables, herbs, and microgreens, toward a wide range of agricultural goods. In the future, these are expected to include berries, edible flowers, potted seedlings, etc. For these products, comparative assessments versus conventional production systems (both in terms of cultivation and transportation/storage), may become more relevant to compare the implications of IVFs.

## 7 Where to look for further information

For further information about life cycle assessment and its application in general, there are a number of standards, guidelines, and helpful resources. A great introduction to LCA and its application can be found in the book by Tillman and Baumann, titled *The Hitch Hiker's Guide to LCA: An Orientation in Life Cycle Assessment Methodology and Application* (ISBN: 9789144023649). Further guidance on LCA methodology can be found from the European Commission's Joint Research Center in The International Reference Life Cycle Data System (ILCD) Handbook, document EUR 24378 EN-2010 (https://eplca.jrc.ec.europa.eu/ilcd.html).

Standards for LCA can also be reviewed, above all the International Standards Organisation (ISO) standards for LCA, through the ISO 14040 series. These provide relevant standards and guidelines on how to conduct LCA as per the industry standard and followed by much of the academic literature for applying LCA to products and services (https://www.iso.org/standard/38498.html).

A number of articles have also been published which assess the environmental performance of indoor vertical farming systems, several of which have been produced by the authors of this chapter, see e.g. Orsini et al. (2020), Martin et al. (2022) and Martin and Molin (2019).

There are also a number of current research projects related to indoor vertical farms and sustainability. The authors' own research projects provide a great reference for outputs and dissemination activities related to the sustainability of vertical and urban farms. These can be found through the Sustainable Urban Farming Lab (www.ivl.se/suf) and the FoodE project (www.foode.eu).

Finally, the Food and Agricultural Organization (FAO), the Association for Vertical Farming (AVF), and the International Society for Horticultural Sciences (ISHS) have developed reports, papers, and webinar series which address the sustainability of indoor vertical farming which are of interest.

## 8 Acknowledgements

The research leading to this publication was co-financed by the following funding sources: 1) Swedish Innovation Agency (Vinnova), within the Research Program, Innovations for a Sustainable Society, in the project 'Urban farming for resilient and sustainable food production in urban area,' grant code: 2019-03178, 2) the Swedish Research Council for Sustainable Development (FORMAS), within the Research Program, Increased mobility between academy and practice project 'Assessing and Improving the Sustainability of Urban Vertical Farming Systems,' grant code: 2019-02049, and 3) the Italian Ministry of Research and Education (MUR), within the Research Programmes of National Interest (PRIN) project 'Sustainable Vertical Farming (VFarm),' grant code: J33C20002350001.

## 9 References

Agritecture (2022). The 2021 global CEA consensus. Online, Available at: https://www.waybeyond.io/census, Accessed [2022 February 10].

Al-Chalabi, M. (2015). Vertical farming: skyscraper sustainability? *Sustain. Cities and Soc.* 18, 74–77.

Arvidsson, R., Tillman, A.-M., Sandén, B. A., Janssen, M., Nordelöf, A., Kushnir, D. and Molander, S. (2018). Environmental assessment of emerging technologies: recommendations for prospective LCA. *J. Ind. Ecol.* 22(6), 1286–1294.

Barge, U. (2020). *Analyzing the Environmental Sustainability of an Urban Vertical Hydroponic System*, UPTEC WUppsala University, Uppsala, p. 89.

Benke, K. and Tomkins, B. (2017). Future food-production systems: vertical farming and controlled-environment agriculture. *Sustain.: Sci., Pract. Policy* 13(1), 13–26.

Brandao, M., Martin, M., Cowie, A., Hamelin, L. and Zamagni, A. (2017). *Consequential Life Cycle Assessment: What, How, and Why?. Reference Module in Earth Systems and Environmental Sciences*. Elsevier, December 2017.

Brander, M., Tipper, R., Hutchison, C. and Davis, G. (2009). *Consequential and Attributional Approaches to LCA: A Guide to Policy Makers with Specific Reference to Greenhouse Gas LCA of Biofuels*. Technical Paper TP-090403-A.

Brander, M., Gillenwater, M. and Ascui, F. (2018). Creative accounting: A critical perspective on the market-based method for reporting purchased electricity (scope 2) emissions. *Energy Policy* 112, 29–33.

Chapman, S., Buttler, A., Francez, A.-J., Laggoun-Défarge, F., Vasander, H., Schloter, M., Combe, J., Grosvernier, P., Harms, H., Epron, D., Gilbert, D. and Mitchell, E. (2003). Exploitation of northern peatlands and biodiversity maintenance: a conflict between economy and ecology. *Front. Ecol. Environ.* 1(10), 525–532.

Cherubini, F. (2010). GHG balances of bioenergy systems - overview of key steps in the production chain and methodological concerns. *Renew. Energy* 35(7), 1565-1573.

Despommier, D. (2011). The vertical farm: controlled environment agriculture carried out in tall buildings would create greater food safety and security for large urban populations. *J.Verbraucherschutz Lebensmittelsicherheit* 6(2), 233-236.

Dorr, E., Koegler, M., Gabrielle, B. and Aubry, C. (2021). Life cycle assessment of a circular, urban mushroom farm. *J. Clean. Prod.* 288, 125668.

Ekvall, T. and Finnveden, G. (2001). Allocation in ISO 14041—a critical review. *J. Clean. Prod.* 9(3), 197-208.

Finnveden, G. (2000). On the limitations of life cycle assessment and environmental systems analysis tools in general. *Int. J. Life Cycle Assess.* 5(4), 229-238.

Finnveden, G., Hauschild, M. Z., Ekvall, T., Guinée, J., Heijungs, R., Hellweg, S., Koehler, A., Pennington, D. and Suh, S. (2009). Recent developments in life cycle assessment. *J. Environ. Manag.* 91, 1-21.

Freidberg, S. (2014). Footprint technopolitics. *Geoforum* 55, 178-189.

Frischknecht, R., Jungbluth, N., Althaus, H. J., Doka, G., Dones, R., Heck, T., Hellweg, S., Hischier, R., Nemecek, T., Rebitzer, G. and Spielmann, M. (2005). The ecoinvent database: overview and methodological framework (7 pp). *Int. J. Life Cycle Assess.* 10(1), 3-9.

Graamans, L., Baeza, E., van den Dobbelsteen, A., Tsafaras, I. and Stanghellini, C. (2018). Plant factories versus greenhouses: comparison of resource use efficiency. *Agric. Syst.* 160, 31-43.

Guinée, J. B., Heijungs, R. and Huppes, G. (2004). Economic allocation: examples and derived decision tree. *Int. J. Life Cycle Assess.* 9, 23-33.

Heiskanen, E. (1999). Every product casts a shadow: but can we see it, and can we act on it? *Environ. Sci. Policy* 2(1), 61-74.

Hermann, B. G., Kroeze, C. and Jawjit, W. (2007). Assessing environmental performance by combining life cycle assessment, multi-criteria analysis and environmental performance indicators. *J. Clean. Prod.* 15, 1787-1796.

Hojlund, B. A. (2008). Substitution of peat with garden waste compost in growth media preparation: a comparison from a LCA-modelling (EASEWASTE) perspective. Proceedings of the International Congress CODIS 2008. Solothurn, Switzerland.

Huijbregts, M. A. J., Steinmann, Z. J. N., Elshout, P. M. F., Stam, G., Verones, F., Vieira, M. D. M., Hollander, A., Zijp, M. and van Zelm, R. (2016). *ReCiPe 2016: A Harmonized Life Cycle Impact Assessment Method at Midpoint and Endpoint Level.* Report I: Characterization. RIMV Report 2016-0104.

ILCD (2010). *International Reference Life Cycle Data System (ILCD) Handbook—General Guide for Life Cycle Assessment—Detailed Guidance* (1st edn.).

ISO (2006). *ISO 14044:2006 Environmental Management -Life Cycle Assessment -Requirements and Guidelines*. International Standards Organisation (ISO).

Lazarevic, D. (2012). *Life Cycle Thinking and Waste Policy: Between Science and Society.* Royal Institute of Technology (KTH), Industrial Ecology.

Martin, M., Heiska, M. and Björklund, A. (2021). Environmental assessment of a product-service system for renting electric-powered tools. *J. Clean. Prod.* 281, 125245.

Martin, M. and Molin, E. (2019). Environmental assessment of an urban vertical hydroponic farming system in Sweden. *Sustainability* 11(15), 4124.

Martin, M., Poulikidou, S. and Molin, E. (2019). Exploring the environmental performance of urban symbiosis for vertical hydroponic farming. *Sustainability* 11(23), 6724.

Martin, M., Weidner, T. and Gullström, C. (2022). Estimating the potential of building integration and regional synergies to improve the environmental performance of urban vertical farming. *Front. Sustain. Food Syst.* 6, 849304.
Milestad, R., Carlsson-Kanyama, A. and Schaffer, C. (2020). The Högdalen urban farm: a real case assessment of sustainability attributes. *Food Secur.* 12(6), 1461-1475.
Orsini, F., Pennisi, G., Zulfiqar, F. and Gianquinto, G. (2020). Sustainable use of resources in plant factories with artificial lighting (PFALs). *European J. Hortic. Sci.* 85(5), 297-309.
O'Sullivan, C. A., McIntyre, C. L., Dry, I. B., Hani, S. M., Hochman, Z. and Bonnett, G. D. (2020). Vertical farms bear fruit. *Nat. Biotechnol.* 38, 160-162.
Plevin, R. J., Delucchi, M. A. and Creutzig, F. (2014). Using attributional life cycle assessment to estimate climate-change mitigation benefits misleads policy makers. *J. Ind. Ecol.* 18(1), 73-83.
Pulighe, G. and Lupia, F. (2020). Food first: COVID-19 outbreak and cities lockdown a booster for a wider vision on urban agriculture. *Sustainability* 12, 5012.
Quantis (2012). *Comparative Life Cycle Assessment of Horticultural Growing Media Based on Peat and Other Growing Media Constituents Final Report*. Quantis, Lausanne.
Romeo, D., Vea, E. B. and Thomsen, M. (2018). Environmental impacts of urban hydroponics in Europe: a case study in Lyon. *Procedia CIRP* 69, 540-545.
Rufí-Salís, M., Petit-Boix, A., Villalba, G., Gabarrell Durany, X. and Leipold, S. (2020). Combining LCA and circularity assessments in complex production systems: the case of urban agriculture. *Resour. Conserv. Recycl.* 166, 105359.
Sala, S., Farioli, F. and Zamagni, A. (2013). Progress in sustainability science: lessons learnt from current methodologies for sustainability assessment: Part 1. *Int. J. Life Cycle Assess.* 18(9), 1653-1672.
Salomaa, A., Paloniemi, R. and Ekroos, A. (2018). The case of conflicting Finnish peatland management - skewed representation of nature, participation and policy instruments. *J. Environ. Manage.* 223, 694-702.
Specht, K., Siebert, R., Opitz, I., Freisinger, U., Sawicka, M., Werner, A., Thomaier, S., Henckel, D., Walk, H. and Dierich, A. (2014). Urban agriculture of the future: An overview of sustainability aspects of food production in and on buildings. *Agric. Human Values* 31, 33-51.
Thomaier, S., Specht, K., Henckel, D., Dierich, A., Siebert, R., Freisinger, U. B. and Sawicka, M. (2014). Farming in and on urban buildings: present practice and specific novelties of zero-Acreage Farming (ZFarming). *Renew. Agric. Food Syst.* 30(1), 43-54.
Toboso-Chavero, S., Madrid-López, C., Villalba, G., Gabarrell Durany, X., Hückstädt, A. B., Finkbeiner, M. and Lehmann, A. (2021). Environmental and social life cycle assessment of growing media for urban rooftop farming. *Int. J. Life Cycle Assess.* 26(10), 2085-2102.
van Delden, S. H., SharathKumar, M., Butturini, M., Graamans, L. J. A., Heuvelink, E., Kacira, M., Kaiser, E., Klamer, R. S., Klerkx, L., Kootstra, G., Loeber, A., Schouten, R. E., Stanghellini, C., van Ieperen, W., Verdonk, J. C., Vialet-Chabrand, S., Woltering, E. J., van de Zedde, R., Zhang, Y. and Marcelis, L. F. M. (2021). Current status and future challenges in implementing and upscaling vertical farming systems. *Nat. Food* 2(12), 944-956.
van der Voet, E., Lifset, R. J. and Luo, L. (2010). Life-cycle assessment of biofuels, convergence and divergence. *Biofuels* 1(3), 435-449.
Vinci, G. and Rapa, M. (2019). Hydroponic cultivation: life cycle assessment of substrate choice. *Br. Food J.* 121(8), 1801-1812.

Published by Burleigh Dodds Science Publishing Limited, 2023.

Wardenaar, T., Van Ruijven, T., Beltran, A. M., Vad, K., Guinée, J. and Heijungs, R. (2012). Differences between LCA for analysis and LCA for policy: A case study on the consequences of allocation choices in bio-energy policies. *Int. J. Life Cycle Asses* 17, 1059-1067.

Weidema, B. (2001). Avoiding co-product allocation in life-cycle assessment. *J. Ind. Ecol.* 4(3), 11-33.

Weidner, T., Yang, A., Forster, F. and Hamm, M. W. (2022). Regional conditions shape the food-energy-land nexus of low-carbon indoor farming. *Nat. Food* 3(3), 206-216.

Weidner, T., Yang, A. and Hamm, M. W. (2019). Consolidating the current knowledge on urban agriculture in productive urban food systems: learnings, gaps and outlook. *J. Clean. Prod.* 209, 1637-1655.

Weidner, T., Yang, A. and Hamm, M. W. (2021). Energy optimisation of plant factories and greenhouses for different climatic conditions. *Energy Convers. Manage.* 243, 114336.

Zamagni, A., Guinée, J., Heijungs, R., Masoni, P. and Raggi, A. (2012). Lights and shadows in consequential LCA. *Int. J. Life Cycle Assess.* 17(7), 904-918.

Zamagni, A., Pesonen, H.-L. and Swarr, T. (2013). From LCA to life cycle sustainability assessment: concept, practice and future directions. *Int. J. Life Cycle Assess.* 18(9), 1637-1641.

# Chapter 5

# Reducing carbon emissions from plant factories with artificial lighting

*Toyoki Kozai, Japan Plant Factory Association, Japan*



## 1 Introduction

Industries worldwide are aiming to reduce greenhouse gas (GHG) emissions to mitigate climate change and its impacts (Gates, 2021). Many national governments have made the decision to halve GHG emissions by 2030 and eliminate them by 2050. Another global concern is to secure safe, healthy, fresh, locally produced plant-based foods. Plant factories with artificial lighting (PFALs) are expected to contribute to achieving this goal because PFALs have many advantages over greenhouses and open fields, as described in Chapter 1. However, PFALs consume significant amounts of electric energy per kilogram of produce ('product' is also used hereinafter) for lighting and air conditioning (Kozai et al., 2021). Furthermore, PFALs emit considerable quantities of $CO_2$-eq ($CO_2$-equivalent: $CO_2$, methane ($CH_4$); nitrous oxide ($N_2O$); and fluorinated gases, such as hydrofluorocarbons (HFCs) and perfluorocarbons (PFCs) used as refrigerants in air conditioners) during construction and operation (Martin and Orsini, Chapter 4 in this book). This dilemma must be overcome by developing

http://dx.doi.org/10.19103/AS.2023.0126.02

resource-efficient PFALs with their own sources of locally generated electricity using natural energy sources such as solar energy, wind power, hydropower, and biomass energy. This chapter discusses methods of reducing the emissions of $CO_2$-eq, representing the primary GHGs emitted in the construction and operation of PFALs, and achieving the most efficient use of resources per kilogram of produce.

Methods of reducing $CO_2$-eq emissions from resources used for the construction and operation of PFALs include the use of:

1) electricity generated using natural energy sources,
2) efficient controllers of electricity use and environmental control,
3) hydroponic cultivation systems without draining of nutrient solution containing fertilizer from the system (generating waste and a pollution risk),
4) organic fertilizer derived from plant residue,
5) paper-made bags, boxes, and trays as simpler, more sustainable alternatives to plastic packaging and other products,
6) minimum volume of substrate (culture medium) made from plant-derived materials or no use of the substrate,
7) reinforced wooden components as simpler, sustainable alternatives to the use of metals,
8) porous concrete and wooden floors,
9) reuse of water and wastewater by recycling after proper filtering and sterilization.

Methods for efficient use of electricity include:

1) an airtight and thermally insulated cultivation room for minimum loss of water vapor and $CO_2$, minimum fluctuation of room environment due to outside weather, and minimum invasion of pest insects, pathogens, and dust (Refer to Chapter 2 of this book for the definitions of airtightness (or the rate of air exchange) and thermal insulation),
2) photosynthetic photon (PP)-efficient lighting systems by using LEDs with optimal spectral distribution and high PP efficacy (μmol/J),
3) energy-efficient air conditioners with a high coefficient of performance (COP),
4) environment controls based on achieving multiple objectives of a high growth rate, high economic value, minimum resource inputs, and waste outputs,
5) breeding of PP-efficient cultivars with a high growth rate at high $CO_2$ concentrations and a low photosynthetic photon flux density (PPFD), and high content of intended chemical components (e.g. those linked to optimum sensory and nutritional quality).

Some of these methods to reduce $CO_2$-eq emissions are discussed in more detail in the following section.

## 2 Environmental impacts on global sustainability

The following environmental impacts decrease global sustainability (Ritchie et al., 2020):

1) GHG emissions, measured in the $CO_2$-eq emission rates,
2) Acidification (decrease in the pH of oceans, soil, rainwater, etc.) measured in sulfur dioxide ($SO_2$)-eq emission rates,
3) Eutrophication (nutrient-induced increase in phytoplankton productivity), measured in phosphate ion ($PO_4$)-eq emission rates,
4) Human toxicity, measured in the 1,4-dichlorobenzene (DCB)-eq emission rates of DCB, sulfur oxides ($SO_x$), nitrogen oxides ($NO_x$), carbon monoxide (CO), and particulate matter smaller than 10 μm in diameter (PM10),
5) Abiotic resource depletion (inorganic natural resources such as iron ore and crude oil), measured in megajoule equivalent.

Methane ($CH_4$) is a major GHG. $CH_4$ emissions from crop and livestock production and forestry are a key challenge to be solved in agriculture (Ritchie et al., 2020). The normal digestive process of domestic livestock, such as cattle, produces $CH_4$ through their belches and flatulence. Animal manure also emits $CH_4$ during storage. This means, e.g. that cattle are the largest $CH_4$ emission source in the United States which has a particularly large livestock sector (US EPA, 2022). $CH_4$ is also produced by land use changes and land management activities (e.g. forest and grassland fires used to control vegetation), decomposition of organic matter in coastal wetlands, and other events.

Carbon emissions from livestock production can be reduced by techniques such as hydroponically produced fodder with improved feed quality and water use efficiency (Newell et al., 2021, https://graze-it.com/). A feed supplement based on a variety of red seaweed is claimed to reduce $CH_4$ emissions from cows by nearly 90% by blocking one of the enzymes needed by $CH_4$-producing bacteria in the rumen (https://www.voltagreentech.com/solution/). The impact of these technologies could be substantial but they are currently under-researched at public institutions.

$CH_4$ and $N_2O$ emissions from PFAL construction and operation are considerably lower than $CO_2$ emissions. This is partly because national environmental regulations strictly control the collection and recycling of fluorinated gases used as refrigerants in air conditioners (a major source of $CH_4$). This chapter therefore discusses methods of reducing $CO_2$-eq emissions (or the carbon footprint) of PFALs only. Analyses of the effects of resource consumption on acidification, eutrophication, human toxicity, and abiotic resource depletion based on life cycle assessment (LCA) methods are also not covered (LCA of PFALs is discussed in Chapter 4 in this book by Martin et al.).

# 3 Resource inputs and product/waste outputs of plant factories with artificial lighting systems

## *3.1 $CO_2$ emission factor*

The $CO_2$ emission factor of electricity generation is expressed in grams of $CO_2$ per kilowatt-hour (or megajoules). Those of chemical fertilizer, plastic products, metals, and cement are expressed in grams of $CO_2$ per kilogram of product. These resources are the leading causes of $CO_2$ emissions in PFAL construction and operation (Molin and Martin, 2018; Martin and Molin, 2019; Kikuchi and Kanematsu, 2020), although the $CO_2$ emission factor of a product varies with its raw materials, scale of and manufacturing method/system (IEA, 2021; US EPA, 2022). Life cycle inventory databases, such as the ecoinvent database (2021), provide datasets of various types of sustainability assessments (https://ecoinvent.org/the-ecoinvent-database/; including a greenhouse dataset) but do not contain PFAL datasets.

## *3.2 Resource inputs*

To evaluate $CO_2$-eq emissions from PFALs, inputs from electrical energy, materials, infrastructure, and outputs such as waste are measured at the system boundaries in the LCA (Molin and Martin, 2018; Martin and Molin, 2019). Figure 1 shows $CO_2$ emissions at the construction, operation, and postharvest stages in response to resource inputs and product/waste outputs at the system boundaries of a PFAL.

The system boundaries of the PFAL for LCA are shown by the boxes of resource inputs and the cultivation room. Types of input are:

1) Electrical energy inputs: these include lighting, air conditioning (primarily for cooling with dehumidification), precooling of produce, sterilization, air circulation (fans), pumping for water and nutrient solution delivery, and vehicles with electric batteries for transport. Fossil fuel is not used in PFALs, although it is often used for vehicle transportation after shipping.
2) Material inputs: these include seeds, fertilizer, substrate (growing medium), and $CO_2$ for promoting plant photosynthesis; water for irrigation and washing/cleaning; cultivation and transportation trays;

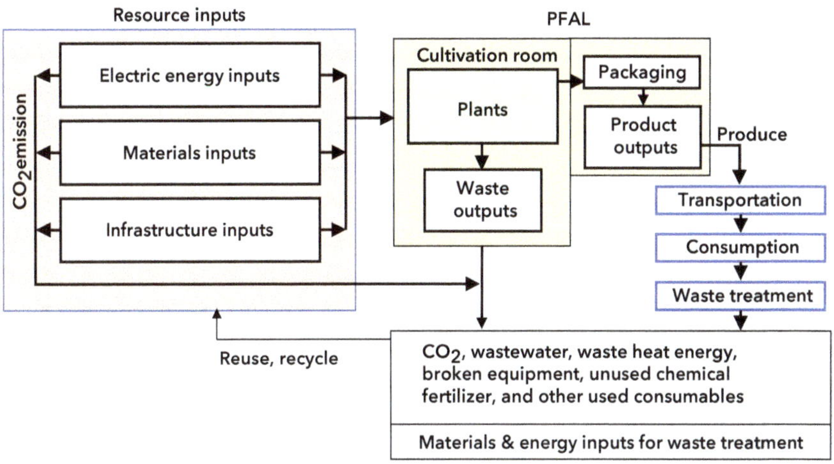


**Figure 1** Scheme showing $CO_2$-eq emissions at the construction, operation, and postharvest stages in response to resource inputs and product/waste outputs.

boxes and bags for packaging and shipping; boxes for waste collection; disinfected gloves, head caps, masks, boots, and clothes for workers; and films and papers for wrapping, wiping, cleaning, disinfecting, and other purposes.

3) Infrastructure inputs include the following: steel and other metals used as structural elements for buildings; cultivation racks for multilevel cultivation; hydroponic cultivation units with pumps, tubes/pipes, and nutrient solution tanks; LED lighting fixtures with wires and switches; air conditioners (or heat pumps) with air circulation fans for controlling temperature, vapor pressure deficit (VPD) or relative humidity, $CO_2$ concentration, and air current speed; root zone or nutrient solution environmental control units; air shower room units; spaces for trimming, grading, weighing, packing, packaging, and shipping; spaces for worker meetings and relaxation; and dressing rooms, concrete floors, weighing balances, vacuum cleaners, washing machines, dehydrators, and others.
4) Waste outputs include: plant residue; wastewater; used plastic films, bags, substrates, and boxes; waste heat energy released from air conditioner outer units (condensers) to the outside; and waste fertilizer drained from hydroponic cultivation units.

## 3.3 $CO_2$ absorption by photosynthesis and $CO_2$ fixation by afforestation

To keep the PFAL sustainable, the amount of $CO_2$ fixed by photosynthesis in the product during its life span should be greater than that emitted by the PFAL during its life span. $CO_2$ fixed by plants is neglected in LCA because almost all the carbon fixed in plants is rereleased as $CO_2$ to the atmosphere within a few days or weeks after harvesting (due to their decomposition by microorganisms with the consumption of $O_2$), except woody transplants (e.g. trees). Since one mole or 44 (=12+2×16) g of $CO_2$ is converted by photosynthesis to 30 (=12+2×1+16) g of $(CH_2O)n$ (carbohydrates) and fixed in a plant for a certain period of time, efficient use of carbohydrates and their derivatives as foods, wooden products, and trees are crucial to minimize or stop the increase in $CO_2$ concentration in the atmosphere.

Woody transplants for afforestation and reforestation continue to fix $CO_2$ for around 100 years (carbon offset) during their growth. Generally, woody transplants for forestation reach their maximum annual net $CO_2$ fixation rates 40-50 years after transplanting. Afterward, the annual net $CO_2$ absorption rate decreases until it reaches zero around 100 years later. The amount of $CO_2$ absorbed by trees aged 30-100 years used as wooden products for housing and furniture is much greater than that emitted to produce transplants and grow them into harvestable trees.

A closed woody transplant production system with artificial lighting was developed for commercialization in 2000 (Kozai and Chun, 2002). With this system, disease-, insect-, pesticide-free elite woody transplants can be produced efficiently with no harm from outside weather, soil, and pests. These transplants can be preacclimatized to harsh climates in the closed system before being taken outside, and grow rapidly after being transplanted. Production of high-quality woody transplants using this type of PFAL at sites of forestation and reforestation is expected to be applied on a large scale.

## 4 Actions required to reduce $CO_2$-eq emissions

Production costs for electricity, personnel (labor), depreciation (initial investment), and others account, respectively, for about 20%, 30%, 30%, and 20% of total production costs for PFALs in Japan as of 2021 (Kozai et al., 2021). Such percentages vary among countries or localities. However, electricity consumption for lighting per kilogram of a particular product, which accounts for 75-80% of the total electricity consumption in a cultivation room, is similar across countries (electricity consumption for air conditioning varies with the outside air temperature, etc.). The electricity consumption per kg of produce as of 2022 for lighting can be reduced by nearly half in the coming decades globally using the methods described in this chapter and other chapters in this book. The main reason for this is that current lighting technology in PFALs is still in an early stage and immature. On the other hand, PFAL technology is moving toward unified software and hardware systems with increasing diversity, complexity, and interdisciplinarity. Most of the methods for reducing electricity consumption described below can be applied in any locality because most of the software and hardware used in PFALs are common worldwide. Electrically self-sufficient/autonomous PFALs with minimum $CO_2$ emissions would be available in many regions by 2030–2035.

The technology for electrically autonomous PFALs is divided into four groups:

1) efficient use of electric energy (production per electricity input) for lighting and air conditioning under given conditions for high PFAL productivity;
2) electricity generation using local renewable sources such as solar energy, wind power, and biomass energy (although electricity generated by non-natural sources, such as fossil fuel, can be used as an emergency backup);
3) a PFAL operated by direct current (DC) to minimize the loss of electric energy by conversion from DC to AC (alternative current), the complexity and thus cost of electric circuits of LEDs. This is because electricity generated by solar panels/film is DC, and that LEDs are operated by DC;
4) a PFAL with a battery backup unit; and
5) breeding of light energy-efficient cultivars suited for PFALs.

It is noted that optimal PPFD, photon spectral distribution, and lighting cycle are affected by the genetic traits of target plants, the purpose of production, production costs, sale prices, and other factors.

Since electric power generation by solar energy and wind power is changeable over time, a PFAL with batteries can be connected to a local electric grid/supply and demand (consumption) network. Surplus electricity generated by the PFAL can be supplied to the network for consumption by local households, matching supply with demand. Since demand is also changeable with time, a system for adjusting demand and supply in the network is required (Kirkerud et al., 2021).

Actions required to reduce $CO_2$ emissions per kilogram of produce fall into the following groups:

1) Electric energy inputs: Minimize the electric energy consumption per kilogram of produce, maximize the use of electric energy generated by natural energy, and cease or minimize electricity generated by fossil fuel and nuclear power.
2) Material inputs: Minimize the amounts of material inputs per kilogram of produce and maximize the use of materials manufactured with minimum

$CO_2$-eq emissions (e.g. plant-made, reused, and recycled materials). Use genetically and physiologically high-quality seeds.

3) Infrastructure inputs: Minimize infrastructure inputs by adopting high-quality design, layout, and management. Use reinforced, lightweight, noninflammable timber materials with long life spans. Use recycled materials manufactured with minimum $CO_2$-eq emissions.
4) Waste outputs: Minimize the waste outputs per kilogram of produce with the efficient use of electrical energy and materials.
5) Product outputs: Maximize or optimize the yield and quality of products with minimum resource inputs using high-quality environmental control, cultivars, cultivation systems, and management.

The above actions need to be adjusted to meet technological advances and social requirements. A procedure for estimating the $CO_2$-eq emission rate per kilogram of produce needs to be developed to evaluate the contribution of each of the abovementioned actions to $CO_2$-eq emissions. A similar procedure has been developed to calculate the resource productivity of each input, measured in kilograms of produce per kilowatt-hour (electricity), per working hour, and per cultivation area and hour (Kozai et al., 2021). A combination of these procedures ($CO_2$-eq emissions and resource productivity) can be a valuable tool for estimating the effects of each action in reducing $CO_2$-eq emissions and increasing resource productivity.

## 5 Minimizing electricity consumption

In almost airtight and thermally insulated cultivation rooms, electricity is consumed for lighting ($E_L$), air conditioning ($E_A$), and operation of other electric equipment ($E_M$) such as water pumps, circulation fans, motors, and battery-powered vehicles. The percentages of electricity consumption for $E_L$, $E_A$, and $E_M$ of the total consumption ($E_L + E_A + E_M$) in temperate regions are roughly 75%, 20%, and 5% in summer, and 80%, 15%, and 5% in winter, respectively (Kozai, 2020). Efficient electricity use for lighting is thus the most critical requirement for reducing total electricity consumption and $CO_2$-eq emissions per kilogram of produce or sale price. The $E_A$ is lower in winter than in summer because the COP of air conditioners is higher in winter than in summer (see Section 6.2 in this chapter). In contrast, $E_L$ and $E_M$ remain almost the same throughout the year assuming monthly plant production remains the same.

### 5.1 Energy use ratio of the lighting system

Table 1 shows energy use ratios in Steps 2-6 of the energy conversion process in electric lighting (Kozai, 2011, 2013). In Table 1, electricity consumption for

**Table 1** Examples of energy use ratios in Steps 2-6 of the energy conversion process in electric lighting. The ratio given in each step is a representative one that varies with several factors (see the text). The $E_S$ of 1.00 in Step 1 is a reference value. The electric energy consumption is in kilowatt-hour or megajoule ((1 kWh = 1 kJ/s × 3600 s/h) = 3600 kJ = 3.6 MJ) for a time period of 1 h, day, week, cultivation period, and so on. Kozai (2011, 2013)

| Step | Energy use ratio in each step | Definition and representative value | Ratio to $E_S$ |
|---|---|---|---|
| 1 | The sum of electricity consumption for lamps ($E_L$), air conditioners ($E_A$), and other electric equipment ($E_M$), $E_S = E_L + E_A + E_M$ | $E_S$ 1.00 | 1.00 |
| 2 | The ratio of $E_L$ to $E_S$ | $E_L/E_S$ 0.70 | 0.70 ($E_L/E_S$) (= 0.70/1.00) |
| 3 | The ratio of photosynthetic radiation energy (PRE) emitted by lamps ($E_P$) to $E_L$ | $E_P/E_L$ 0.4 | 0.28 ($E_P/E_S$) (=0.4 × 0.7) |
| 4 | The ratio of $E_P$ received by plant canopy ($E_R$) to $E_P$ | $E_R/E_P$ 0.7 | 0.20 ($E_R/E_S$) (= 0.7 × 0.28) |
| 5 | The ratio of preabsorbed by leaves of plant canopy ($E_A$) to $E_R$ | $E_A/E_R$ 0.8 | 0.16 ($E_A/E_S$) (= 0.8 × 0.20) |
| 6 | The ratio of chemical energy fixed in the leaves ($E_C$) to $E_A$ | $E_C/E_A$ 0.1 | 0.016 ($E_C/E_S$) (=0.1 × 0.16) |

lighting ($E_L$) is assumed to account for 70% of the total electricity consumption ($E_S$) in Step 2; the remaining 30% is consumed by air conditioners ($E_A$) and other electric equipment ($E_M$). In Step 3, 40% of $E_L$, $E_P$, is converted into photosynthetic radiation energy (PRE). In Step 4, 70% of PRE is received by leaves. In Step 5, only 16% of $E_S$ is absorbed by leaves as PRE; the remaining 84% is converted into heat energy. In Step 6, only 10% of the 16% of $E_S$, or 1.6% of $E_S$, is fixed as chemical energy and incorporated into carbohydrates for growth (Eight moles of PPs are required, at least, to capture one mole of carbon dioxide and significant amounts of $CO_2$ is released to the room air by dark and photo respiration); the remaining 14% (= (1.0 − 0.1) × 16) of $E_S$ is converted into heat energy. The sum of the percentages of heat energy (98% = 84% + 14%) needs to be transferred outside by air conditioners to keep the room air temperature constant. In short, electrical energy fixed as chemical energy in plants via photosynthesis is generally as low as 1.6% of the total electricity consumption in the above example, which needs to be improved by increasing the energy use ratios in Steps 2-6. Around 50% of sensible heat energy and PRE transferred to the transpiring leaves with the most stomata open is converted into latent heat energy (contained in water vapor).

All the energy use ratios in Steps 2-6 are variable. In Step 2, $E_L/E_S$ can be increased by using air conditioners with high COPs to reduce $E_A$. In Step 3, $E_P/E_L$ or PRE efficiency can be improved by technical advances in LED fixtures.

$E_R/E_P$ in Step 4 is affected by the plant community architecture (shape, area, angle, and the number of leaves), the light reflectance and geometry of the inside surfaces of the cultivation space, angular or directional PRE distribution of lamps, lamp layout inside the cultivation space, and light reflectance of inside surfaces (including the plant canopy surface) of the cultivation space. The ratio $E_A/E_R$ in Step 5 is determined primarily by the spectral absorptance of leaves at the plant canopy surface. $E_C/E_R$ in Step 6 is determined by the surrounding aerial and root zone environments and the physiological status of plants. With a 10% improvement in the energy use ratios in Steps 2-4 and 6, for example, $E_C/E_S$ in Step 6 increases from 0.016 (=0.70 × 0.40 × 0.7 × 0.8 × 0.1) to 0.023 (=0.77 × 0.44 × 0.77 × 0.80 × 0.11) or increases by 44% (100 × 0.023/0.016).

## 5.2 Electric energy use efficacy of marketable produce

Table 2 shows formulas estimating the electric energy use efficacy of fresh mass ($W_{FM}$) and dry mass ($F_{DM}$) production of marketable produce, measures of resource/monetary productivity, and cost-benefit ratios. For Formulas 1 and 2, $E_L$ can be measured with a wattmeter or calculated as a product of the rated power consumption of lamps (1 W = 1 J/s) and the lighting period (s). The values in Formulas 3-7 are influenced by the genetic traits of cultivars, aerial and root zone environments, and the cultivation system adopted.

**Table 2** Formulas for estimating the efficacies and productivities of electricity consumption for lighting ($E_L$) and cost-benefit ratio for the fresh mass ($W_{FM}$) and dry mass ($W_{DM}$) production of marketable produce.

| | Coefficient/parameter | Symbol | Unit |
|---|---|---|---|
| | The unit price of fresh mass produce | $C_F$ | $/kg |
| | The unit price of dry mass produce | $C_D$ | $/kg |
| | The unit cost of electricity | $C_E$ | $/kWh |
| **No.** | **Variable** | **Formula (Definition)** | **Unit** |
| 1 | Electric energy use efficacy of $E_L$ (lighting) for $W_{FM}$ production | $W_{FM}/E_L$ | kg/kWh |
| 2 | Electric energy use efficacy of $E_L$ (lighting) for $W_{DM}$ production | $W_{DM}/E_L$ | kg/kWh |
| 3 | Harvest Index (HI) | $W_{FM}/W_{FW}$ | dimensionless |
| 4 | Monetary $W_{FM}$ productivity of lighting | $P_F = C_F \times (W_{FM}/E_L)$ | $/kWh |
| 5 | Monetary $W_{DM}$ productivity of lighting | $P_D = C_D \times (W_{DM}/E_L)$ | $/kWh |
| 6 | The cost-benefit ratio of $W_{FM}$ | $P_F/C_E$ | Dimensionless |
| 7 | The cost-benefit ratio of $W_{DM}$ | $P_D/C_E$ | Dimensionless |

The estimated efficacies, HI, and measures of productivity can be correlated with the environmental and design factors of the lighting system to reduce $CO_2$-eq emissions per kilogram of produce. A similar estimation can be conducted for the efficacies of PPs using the values of the PP efficacy of lamps (e.g. 2.5–3.5 μmol/J) instead of PRE.

## 5.3 Leveling electricity consumption over time

In many existing PFALs, about two-thirds of lamps are turned on throughout the day. This lighting cycle is achieved by dividing the lamps into three groups with a 16-h daily photoperiod. For example, groups A, B, and C are 12:00 am to 4:00 pm, 8:00 am–12:00 pm, and 4:00 pm to 8:00 am (next day). The lamps' electricity consumption rate (kilowatts) is thus constant all day. Furthermore, the daily maximum electricity consumption rate for lighting and the cooling load of air conditioners is reduced by one-third compared with those in a scenario where all the lamps are turned on and off simultaneously. The initial costs of the power supply unit and the air conditioners are reduced due to the reduction in daily maximum electricity consumption. If the basic monthly charge increases with the maximum electricity consumption rate, the basic monthly charge is also reduced.

Furthermore, since the air temperature is lower at night than in the daytime, the electricity consumption rate of air conditioners is lower at night than during the daytime due to a higher COP. One issue in keeping two-thirds of lamps always turned on is that the air temperature setpoints during the photoperiod and dark periods need to be the same in the same cultivation space. Fortunately, the air temperature is 1–2°C higher under light than in the dark at the same air temperature setpoint. Thus, the leaf temperature tends to be higher during the photoperiod than during the dark period. However, its degree of temperature rise during the photoperiod decreases with the transpiration rate.

## 5.4 $CO_2$ emission factors for different electricity generation systems

Table 3 shows averaged direct, indirect, and total (direct + indirect) $CO_2$-eq emission factors ($gCO_2$-eq/kWh) for different electricity generation systems in Japan (Imamura et al., 2016). Direct $CO_2$-eq emission refers to emissions associated with the combustion of fossil fuels in power plants. Indirect $CO_2$-eq emissions are associated with construction, operation, maintenance, and decommissioning. The total $CO_2$-eq emission factor is 7–85 times greater in coal-, oil-, and LNG-fired power plants than in nuclear and natural energy power plants except for offshore floating-type wind power plants. Thus, it is highly desirable not to use the electricity generated by fossil-fuel-fired power plants to

**Table 3** Averaged direct, indirect, and total $CO_2$-eq emission factors ($gCO_2$-eq/kWh) for different electricity generation systems in Japan (Imamura et al., 2016). LNG and MW denote liquid natural gas and megawatts, respectively.

| | | $CO_2$-eq emission factor | | |
|---|---|---|---|---|
| No. | Type of system | Direct | Indirect | Total |
| 1 | Coal-fired | 864 | 79 | 943 |
| 2 | Oil-fired | 695 | 43 | 738 |
| 3 | LNG-fired | 341 | 89 | 430 |
| 5 | Nuclear (Pul-thermal) | - | 19 | 19 |
| 6 | Hydro (waterway) | - | 11 | 11 |
| 7 | Geothermal | - | 13 | 13 |
| 8 | Wind (1-2 MW) (land) | - | 28 | 28 |
| 9 | Wind (40 MW) (offshore, floating/ implantation) | - | 99<br>26 | 99<br>26 |
| 10 | Solar-photovoltaic (2-10 MW) | - | 59 | 59 |

minimize $CO_2$-eq emissions per kWh. It is also noted that nuclear power plants might cause destructive long-term radioactive contamination when the plants are heavily damaged by unpredictable events such as severe earthquakes, volcanic eruptions, tsunamis, military or terrorist attacks, and other natural and artificial disasters.

The $CO_2$ emission factors shown in Table 3 are estimated based on several assumptions regarding lifetimes, electricity generation capacities, and technological levels in Japan as of 2009. In Table 3, lifetimes were assumed to be 40 years for all plants, and thermal efficiency (percentage of electric energy generated on heat energy contained in the fuel) of an oil-fired plant was assumed to be 38–39%. In reality, most existing PFALs purchase their electricity from a commercial power supplier procuring electricity from different types of power plants via a power grid network.

## 5.5 Electricity generation using natural energy

Technological advances and governmental policies have recently decreased the cost of generating electricity using natural energy which will continue to fall in the future. In 2020 levelized costs of electricity (LCOE) for solar energy and wind power were comparable to those of nuclear power, natural gas, and coal (Fig. 2). Solar energy's world electricity generation capacity has increased and will continue to rise in the coming decades (Fig. 3). However, the LCOE of solar energy varies by country (Fig. 4). Solar panel and wind power technologies have recently been advancing (Dambhare et al., 2021). The perovskite solar

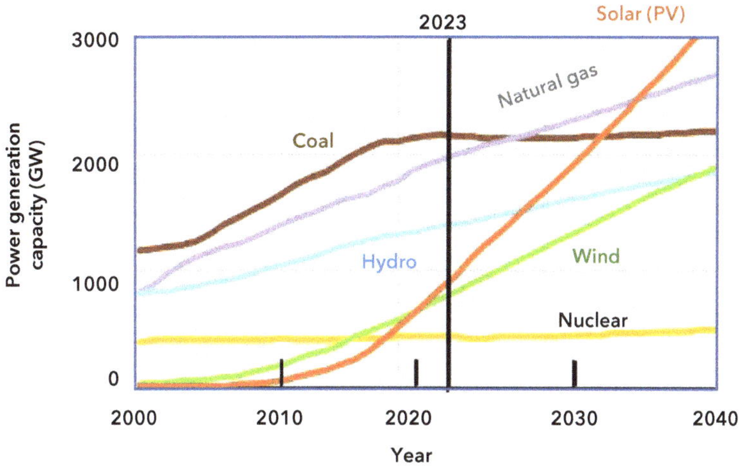


**Figure 2** Trends of electricity generation capacity by energy source (https://www.iea.org/reports/world-energy-outlook-2019).

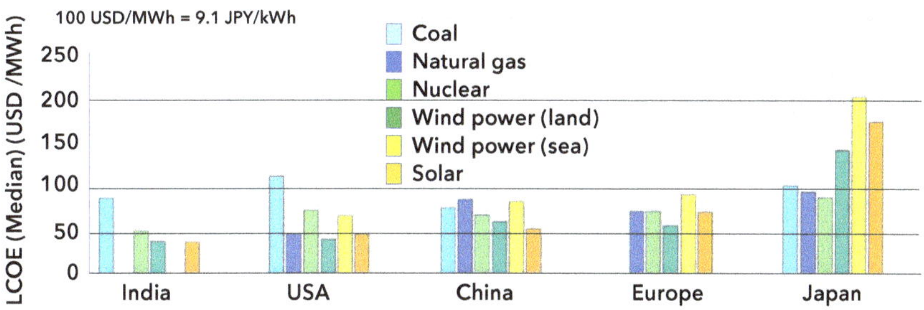


**Figure 3** Medians of LCOE by country and energy source. The LCOE is the sum of the average capital costs, decommissioning, fuel, fixed and variable operations, maintenance, financing, and an assumed utilization rate divided by a discounted sum of the actual energy amounts delivered. https://www.iea.org/reports/projected-costs-of-generating-electricity-2020.

cells have driven interest in the development of commercial thin and flexible solar film for power generation (Kim et al., 2020). The power of electricity generated by solar and wind power energy varies with the time of day and weather, so batteries are required for smoothing the available electricity. Because of recent technological reductions in batteries, PFALs powered solely by natural energy are now commercially available. PFALs able to generate their own electrical power autonomously and produce on a commercial scale have become a reality. Electricity generation by use of geothermal and ocean/tidal energy is also promising in those cases where PFALs are situated near these power sources (Baumont de Oliveira et al., 2023).

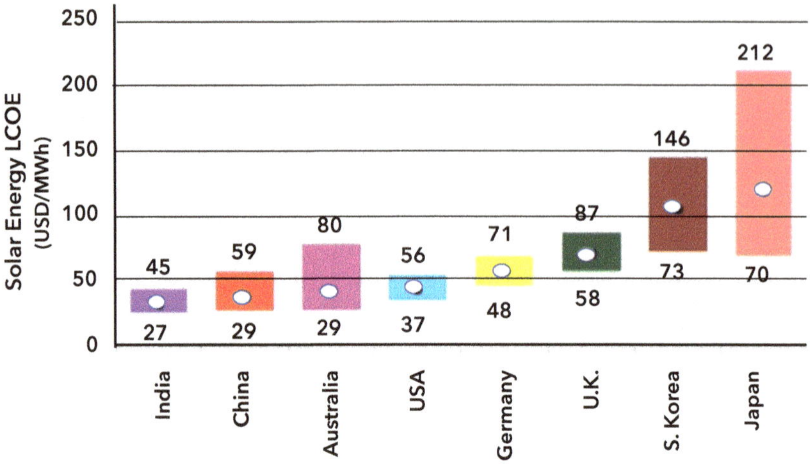


**Figure 4** LCOE for solar panels by country (https://www.econ.kyoto-u.ac.jp/renewable_energy/stage2/contents/column0210.html).

# 6 Air conditioners

## 6.1 Heat pumps

A heat pump transports heat energy for a short distance (around 0.5–15 m in most cases). Heat pumps are used in refrigerators, freezers, ice makers, water heaters, water coolers (chillers), dehumidifiers, driers, and air conditioners. Figure 5 shows a heat pump consisting of a compressor for contracting refrigerant gas, a condenser for heat energy release, an evaporator for heat energy collection, an expansion valve for controlling the amount of refrigerant released into the evaporator, and a three-way valve (not shown in the figure) for changing the flow direction of the refrigerant to switch the functions of the condenser and the evaporator. In this chapter, the medium in the heat exchangers of the evaporator and the condenser is assumed to be air (i.e. air-to-air heat pump), although water-to-air heat pumps are also widely used. A typical refrigerant is R32 emitting much less GHG than R22 which was widely used in the past.

In most cases, a heat pump transports heat energy against the temperature gradient (from low to high temperatures). However, the air conditioners installed in cultivation rooms remove heat energy, even when the inside air temperature is higher than that outside. The reason is that the cultivation room cannot be ventilated (in order to prevent the entry of small insects and dust, the loss of $CO_2$ and water vapor from the cultivation room to the outside, and to avoid the outside weather affecting the room environment). Nonetheless, a forced ventilation unit must be installed for use in emergencies, such as fire, extreme

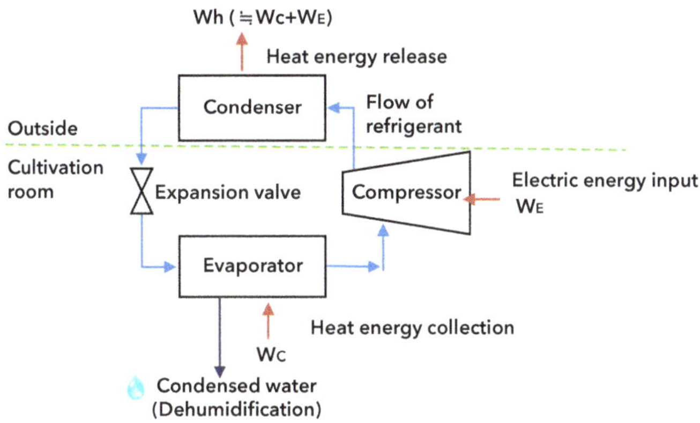


**Figure 5** Main components and flows of heat energy and refrigerant of the electric heat pump in cooling mode. The ratio ($Wc/W_{E)}$ expresses the COP of cooling ($COP_C$), and the ratio $W_h/W_E$ ($=W_C + W_E = COP_C + 1$) expresses the COP of heating ($COP_H$) (see also Figure 6).

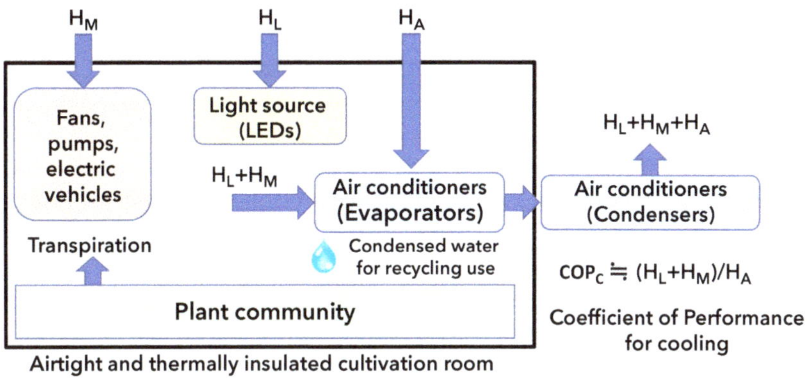


**Figure 6** Simplified energy balance of the airtight, thermally insulated cultivation room with a light source, air conditioners, and other electric equipment (e.g., air circulation fans and pumps). $H_L$, $H_A$, and $H_M$ denote the electric energy consumption of light sources, air conditioners for cooling and dehumidification, and other electric equipment, respectively. $H_L$ and $H_M$ are converted into sensible or latent heat energy in the cultivation room and removed with $H_A$ as sensible heat to the outside. $COP_C$ is expressed as $(H_L + H_M)/H_A$.

rise in $CO_2$ concentration (e.g. 5000 ppm or higher), and volatile compound gases harmful to plants and humans.

When cooling, the water vapor in the moist room air is condensed into liquid water on the cooling panels of the evaporator and generates sensible heat of condensation (2.27 kJ/kg). The room air is then dehumidified (Fig. 6). The resultant VPD of the room air is determined by the dynamic balance mainly between the transpiration rate of plants and the dehumidification rate of heat pumps. Where the cultivation beds are mostly covered with plants and about two-thirds of LEDs are turned on, the VPD of the room air naturally reaches around 1.0 kPa (or relative humidity of 80%) at 20°C. Where the piping from the cooling panels to the nutrient solution tank is clean, the water condensed at the cooling panels, originating from the transpired water vapor, will be clean enough for use as irrigation water. In reality, water is filtered and often sterilized before use for irrigation.

## 6.2 COP of air conditioners

In airtight, thermally insulated cultivation rooms, the $COP_C$ of air conditioners can be estimated as the ratio of the heat energy removed from the cultivation room ($H_L + H_M$) to the electric energy consumed by the air conditioners ($H_A$) or $(H_L+H_M)/H_A$. The $COP_H$ is expressed by $(H_L + H_M + H_A)/H_A = COP_C + 1$ (Fig. 6). The $COP_C$ and the $COP_H$ can thus measure electrical energy use efficiencies in air cooling and heating, respectively.

The COP generally increases with the difference in air temperature between the inside and outside of the cultivation room, which roughly ranges between 3°C and 12°C in temperate regions (Kozai et al., 2020, 2021). For example, if the percentages of $E_L$, $E_A$, and $E_M$ are 80%, 15%, and 5%, respectively, the $COP_C$ will be 5.7 (= (80 + 5)/15). Air conditioners generally provide the maximum $COP_C$ or $COP_H$ when the percentage of actual cooling load is 70–80% of their cooling capacity (Kozai et al., 2020). The $COP_C$ is affected by the two abovementioned factors and by the cooling load ratio of latent heat to sensible heat, the degree of maintenance to ensure optimum performance, the refrigerant gas pressure in the pipe, and specifications of air conditioners, among others. A worker in the cultivation room generates around 0.1 kW (0.1 kJ/s) of heat energy, whereas the heat energy generated by plant respiration is negligible. Continuous monitoring of $COP_C$ using the equation in Fig. 6 should then help identify any change in the above factors.

## 6.3 Reusing heat energy generated in the cultivation room

As described in Section 5.1, over 98% of the electric energy consumed by the lighting system ($E_L$), air conditioners ($E_A$), and other electric equipment ($E_E$) is converted into heat or radiation energy. When leaves or wet surfaces absorb heat and radiation energy, around half of the energy is converted to latent energy (water vapor). The sensible and latent heat energy in the room air is then absorbed in the cooling panels of air conditioners and released as sensible heat energy to the outside at the condenser. In cases where air-to-air air conditioners are used for cooling, the air temperature at the air outlet of the condenser is often 40–45°C. This warm air can be used to dry wet materials (e.g. plant residue), raise the water temperature for use in hydroponics, or for heating office buildings. Reusing the heat energy generated in the cultivation room is as important as reducing $CO_2$ emissions from electricity generation. When electricity is generated by solar energy, there is no net increase in heat energy because solar energy is converted somewhere on the earth into heat energy. By contrast, when electricity is generated by fossil fuel, heat energy is generated as net increases both at the power plant and the PFALs, in addition to $CO_2$ emissions at the power plant.

### 6.4 Heating, ventilation and air conditioning

The term HVAC is an abbreviation for heating, ventilation, and air conditioning. HVAC is frequently used in the field of environmental control of PFALs. This term has been used for many years in the design and environmental control of residential structures, industrial and office buildings, and vehicles such as cars and trains. In these fields, heating, and ventilation are generally as important as air conditioning. In using this term in the case of PFALs, it is important to note that the airtight and thermally insulated cultivation room in a PFAL is neither ventilated nor heated except in emergencies such as fire and rapid accumulation of gases harmful to humans and plants (The needs for airtight conditions and thermal insulation are described in detail in Chapter 2). This is why the term air conditioning is used instead of HVAC in this chapter and Chapter 2. However, control of light and root zone environments is as important as air conditioning.

## 7 Photosynthetic photon efficacy and electrical energy use efficiency

### 7.1 Photosynthetic photon efficacy

Pattison et al. (2018) proposed the following equation to evaluate the PP efficacy of dry mass (DW) production ($E_{dry}$) in the unit of grams of dry mass produced per mole of PPs:

$$E_{dry} = \text{Fa} \times \text{QY} \times \text{CUE} \times \text{HI} \times k \quad (1)$$

where Fa is the fraction of PPs absorbed by leaves; QY is the quantum yield (or PP yield), an essential physiological parameter of photosynthesis (moles of carbon fixed per mole of PPs absorbed); CUE is the carbon use efficiency (moles of carbon incorporated into plant biomass per mole of carbon fixed; a constant of 0.65 is assumed below, as given in the original paper); HI is the harvest index (moles of carbon in the edible or usable part of the product per mole of carbon in plant biomass); $k$ (30 = 12 + 2 + 16) is a constant that represents the mass of $CmH_2nOn$ (carbohydrates) as grams per mole of carbon in an edible or usable product.

The QY of a single leaf can be measured manually using a small assimilation chamber relatively quickly. However, it is not easy to measure the time course of the average QY of a plant canopy as it grows. Pattison et al. (2018) showed some numerical examples of $E_{dry}$, namely, 1.33 gDW/mol for leafy microgreens, assuming Fa = 0.95, QY = 0.08, and HI = 0.9; 0.80 gDW/mol for lettuce, assuming Fa = 0.65, QY = 0.07, and HI = 0.9; and 0.35 gDW/mol for tomatoes, assuming Fa = 0.60, QY = 0.05, and HI = 0.6. A QY of 0.08 means that 12.5 (=1.0/0.08) mol of PPs are required to fix 1 mol of carbon.

### 7.2 Electrical energy use efficacy

Using the data given by Pattison et al. (2018), the electric energy use efficacy of dry matter production can be calculated, for example, as 3.99 g/MJ (= 3 mol/MJ × 1.33 gDW/mol) or 1.11 gDW/kWh (1 kWh = 3.6 MJ) for leafy microgreens using the value of $E_{dry}$, assuming a PP efficacy of 3 μmol/J (or 3 mol/MJ) for LEDs. The electric energy use efficacies of lettuce and tomato are 0.67 and 0.29 gDW/kWh, respectively. In the case of a dry matter ratio ($W_{DM}/D_{FM}$) of 0.05, the electric energy use efficacy of fresh microgreens ($E_{fresh}$) is 22.2 (=1.11/0.05) gDW/kWh, and that of fresh lettuce is 13.4 (=0.67/0.05) gDW/kWh.

However, the electric energy use efficacy of fresh leafy lettuce in commercial PFALs in Japan is reported to be 110–140 gFW/kWh (Kozai et al., 2021). This efficacy is 8–10 times greater than that given by Pattison, et al. (2018). Possible reasons explaining this difference in PFALs are: high $CO_2$ concentration (1000 ppm), a photoperiod of 16 h $d^{-1}$, and low PPFD (150-200 μmol $m^{-2}$ $s^{-1}$). Further research is needed to elucidate the reason for the big difference between the two. Jin et al. (2022) compared the light use efficiency and related parameter values of lettuce cultivation in PFAL, greenhouse, and field, using Pattison's equation (Eq. 1).

## 8 Technical terms related to physiological and physical characteristics

In the design and control of lighting systems for PFALs, technical terms related to the physiological and physical characteristics of plants, LEDs, and light environments are frequently used. The technical terms shown below need to be used with caution so they are not misunderstood by people with different academic backgrounds (e.g. light as an illumination source for workers and visitors is out of the scope of this book). The following notes explain key terms and the relevant background for understanding how they are used in this chapter.

1) A plant receives light as photons at various photoreceptors during photosynthesis and photomorphogenesis. Photosynthesis is an energy

conversion process using photons with a waveband of 400–700 nm (blue, green, and red), whereas photomorphogenesis is a signal conversion process using photons with a waveband of 300–800 nm (ultraviolet, blue, green, red, and far-red). Photosynthesis and photomorphogenesis often occur concurrently in a plant. The required quantity of photons to grow plants is roughly 10–100 times greater in photosynthesis than in photomorphogenesis. Generally, photosynthesis increases the quantity (i.e. dry weight) of produce, whereas photomorphogenesis affects the quality (e.g. shape, color, taste, texture, nutrition, etc.) of produce positively or negatively. The photomorphogenetic response depends considerably on the wavelength or spectral distribution, while the photosynthetic response is hardly affected by the spectral distribution in the range of 400–700 nm (this is why the term PPFD [400–700 nm] is widely used).

2) The energy per photon (J μmol$^{-1}$), $e_\lambda$, is inversely proportional to the wavelength. For example, the $e_\lambda$ for a wavelength of 450 nm (blue: 400–500 nm) is 1.44 (=650/450) times greater than that at 650 nm (red: 600–700 nm). In other words, the red photon is 1.44 times more energy-efficient than the blue photon concerning photosynthesis, although plants growing under only red photons often show morphological disorders. Photosynthetic photons per energy (micromoles per joule), $M_\lambda$ (reciprocal of $e_\lambda$), are expressed as a function of wavelength $\lambda$ as shown in Equation 2 (Fujiwara, 2021). The $M_\lambda$ for $\lambda$ = 400, 500 nm, 600 nm, and 700 nm are 3.344 μmol J$^{-1}$, 4.180 μmol J$^{-1}$, 5.017 μmol J$^{-1}$, and 5.852 μmol J$^{-1}$, respectively.

$$M_\lambda = \lambda/119.6 \quad (2)$$

3) The photosynthetic photon efficacy of an LED fixture (μmol J$^{-1}$) means the conversion factor of the LED from the supplied electric energy to the emitted photosynthetic photons. The $M_\lambda$ is given in a product catalog for each product name provided by the manufacturer. The efficacy of LEDs currently commercialized for PFALs ranges mostly between 2.0 μmol J$^{-1}$ and 3.7 μmol J$^{-1}$ (Kusuma et al., 2021).
4) When discussing the effects of the blue/red (B/R) ratio on plant physiology and growth, it must be stated whether the ratio is photon ratio or energy ratio because of the reason mentioned above. The same applies to the ratio of other wavebands.
5) Photon flux (μmol s$^{-1}$) and radiant flux (W or J s$^{-1}$) are used to express, respectively, the photon and radiant emitting power of a light source. Photon flux density (μmol m$^{-2}$ s$^{-1}$) and radiant flux density (W m$^{-2}$) are

used to express the photon flux and radiant flux received by or crossing a surface.

6) The term light intensity is often used to mean PPFD (μmol $m^{-2}$ $s^{-1}$) or PAR flux density W $m^{-2}$). The light intensity is used also to mean illuminance (lux) or solar radiation flux density (W $m^{-2}$; 250–2500 nm). Daily light integral is used to mean daily PPFD integral, daily integral of PAR flux density, or daily solar radiation integral. To minimize the misunderstanding and confusion by people with different academic backgrounds, it is recommended not to use the term light intensity to mean the PPFD and PAR flux density. (At least, its unit needs to be added when the light intensity is used.) On the other hand, the word light can be used as a common noun without a unit, such as light environment, light source, and lighting system. It is also noted that the term intensity, used as radiant, luminous, and photon intensities, is defined differently from the above by the International Electrotechnical Commission (Fujiwara, 2021).
7) The B/R photon flux ratio is a property of a light source, while the B/R photon flux density ratio at a plant canopy surface is determined not only by the B/R photon flux ratio of the light source but also by the spectral reflectance of inner surfaces of cultivation space (ceiling, sidewalls, an upper part of the plant canopy, and substrate) due to the multiple reflections between the inner surfaces (Akiyama and Kozai, 2016). A green leaf reflects around 20% of green photons and around 40% of far-red photons and absorbs around 90% of blue and red photons. Thus, for example, the blue/green photon flux ratio of the light source differs from the blue/green photon flux density ratio at the canopy surface. Due to the multiple reflections between the inner surfaces mentioned above, the PPFD at the plant canopy surface decreases or increases by around 10% with increasing the leaf area index, even under a fixed photosynthetic photon flux of LEDs, when the reflectance of leaves is considerably different from that of the substrate. The PPFD increases with time in the case of the black substrate and decreases in the case of the white substrate (Hayashi et al., 2022a).
8) A white LED emits blue, green, and red photons. Some white LED products emit a small percentage (around 1%) of ultraviolet and/or far-red photons, which may affect the photomorphogenesis of plants. The percentage of green photons to the total photosynthetic photons largely varies with the product, ranging between 5% and 45%. The role of green light in plant physiology and production in PFALs with white LEDs is thus attracting attention (Folta and Maruhnich, 2007).

9) The spectral distribution of photons in the densely populated plant canopy is not spatially uniform because the transmittance and reflectance of leaves are wavelength dependent. Green and far-red photons penetrate deeper into the plant canopy because their transmittance and reflectance are significantly higher than blue and red photons. As a result, green photons are beneficial for increasing the photosynthesis of lower leaves in the densely populated plant canopy (Kozai et al., 2020). This spatially nonuniform spectral distribution inside the plant canopies in PFALs causes various physiological phenomena in plants, many of which remain unknown.

## 9 Reducing $CO_2$ emissions of resources other than electricity

In PFAL construction and operation, resources other than electricity are also consumed and emit significant $CO_2$ during manufacturing. These resources include growing medium (or substrate); fertilizer (e.g. nitrogen [N], phosphate [P], and potassium [K]); plastic products (e.g. pots, trays, and packaging materials); and construction elements (e.g. iron, steel, aluminum, and cement). Consumption of these resources needs to be minimized while optimizing productivity.

### *9.1 Water for irrigation and washing in hydroponic cultivation*

In the cultivation room, 90–95% of the water absorbed by the plant roots transpires from leaves; the remaining 5-10% is kept in the plants to increase their fresh weight (Kozai et al., 2020). A PFAL, consisting of an airtight, thermally insulated cultivation room, consumes only 5–10% of irrigation water per kilogram of produce compared with a greenhouse because water vapor transpired from plants in a PFAL is condensed and collected at the cooling panels of air conditioners and recycled, whereas the transpired water vapor from plants in a greenhouse is not recycled for irrigation (Kozai et al., 2020).

The recycling use of transpired water in a PFAL is technologically simple. In reality, however, transpired water is recycled in only 10% of PFALs in Japan, where the cost of high-quality water accounts for only about 2% of production costs (JGHA, 2022). On the other hand, recycling irrigation water is crucial in arid and very cold regions where the quality or availability of water is often low. In addition to irrigation water, existing PFALs consume significant amounts of water (probably comparable to net irrigation water consumption) for cleaning floors, cultivation panels, and beds often with pressurized or steaming water. This water can be reused after proper filtering and sterilization, although this is not done in practice in most PFALs in Japan.

Although there are no reliable statistics, estimates show that 10–20% of the nutrient solution in a hydroponic cultivation system is drained to the outside weekly in most PFALs to avoid ion imbalance and excess accumulation of unfavorable ions (e.g. chloride [Cl] and $PO_4$), as well as avoid dead and living algae, dead fine roots, unfavorable organic acids exuded from roots, pathogens, and extreme pH values. A hydroponic cultivation system that does not drain nutrient solution needs to be developed because drained nutrient solution contaminates the surrounding environment as well as losing nutrients that could potentially be reused. A hydroponic cultivation system with one-way nutrient solution delivery (no circulation of nutrient solution) is the next development target. This would eliminate the need for circulating pipes, sterilization and filtering units, and large storage capacity (which increases the risk of pathogens developing and spreading in the system). Such a system would need technologies such as ion sensors, IoT (Internet of Things), deep learning methods using AI, phenotyping using various types of small cameras, as well as simulation models for predicting nutrient/water uptake and plant growth.

If no water is lost from the cultivation room to the outside, the required amount of irrigation water equals the amount of water kept in plants up to harvest. Tap water or reverse osmosis membrane water can then be used as cost-effective irrigation water in PFALs. With this system, controlling nutrient solution composition in hydroponics is easier than the use of groundwater containing ions of natural minerals. When a reverse osmosis membrane is used for groundwater, $CO_2$ is emitted to generate the electricity used for desalination. In arid regions where water is scarce, groundwater is often pumped up from 1000 m or lower depths. In this case, the $CO_2$ emission factor is high, so the recycling use of transpired water in PFALs is advantageous over greenhouses. Besides, a significant amount of water is consumed for evaporative greenhouse cooling in arid regions. On the other hand, significant amounts of heat energy are required to melt ice for irrigation in cold regions.

## 9.2 Chemical fertilizer

N and P fertilizers become environmental pollutants when drained from PFALs to the outside, and they may cause the eutrophication of lakes and rivers. Moreover, the raw materials for P and K fertilizers (rock phosphate/guano and potash ore) and fossil fuels are depleting natural resources. $CO_2$ is also emitted to manufacture chemical fertilizers from raw materials. Drainage of nutrient solutions containing N, P, and K to the outside thus needs to be minimized. In a hydroponic cultivation system, nutrient supply is often stopped a few days before harvest. Ion concentrations in nutrient solution are drastically reduced before the drainage because most of the nutrient ions in the solution are absorbed by plant roots within a day or two after their supply is stopped.

In a hydroponic cultivation system, most nutrient ions absorbed by plant roots remain in either the produce or the plant residue. When plant residue is decomposed into organic fertilizer in the presence of microorganisms, the ion composition of the inorganic fertilizer should be more or less the same as that of plant residue, although it varies depending on several environmental, phytological, and microbiological factors. For example, the ion composition of produce (e.g. leaves with stems) might be different from that of plant residue (e.g. roots). Using the organic fertilizer originating from the plant residue will significantly help reduce the consumption of inorganic fertilizer, thus decreasing $CO_2$ emissions during manufacturing. Further research on this topic is required.

## 9.3 Plastic products, artificial media and their alternatives

Plastic products often used in PFALs include trays, pots, bags, boxes, containers, tanks, films, sheets, tapes, string, and pipes. Most of these products need to be replaced with alternatives made of reinforced paper or plant-based biodegradable plastics. Boxes, bags, and films made of semi-transparent paper are available for packaging and wrapping. Artificial media (or substrate) include chemically inert supports (e.g. formed and foamed urethane cubes) and chemically active supports (e.g. Rockwool cubes and mixtures of peat moss, vermiculite, and perlite). $CO_2$ emissions from pots with substrate can be significantly reduced by changing from gardening (natural and organic) soil to coir (fiber from the outer husk of coconut) and, to a lesser extent, from plastic to paper pots (Martin and Molin, 2019). A paper tray with cells containing no media (each cell itself acts as a support) would be ideal for next-generation PFALs.

## 9.4 Metals and cement and their alternatives

Metals (e.g. steel, aluminum, and alloy) and cement used for concrete floors and pathways are materials that produce significant amounts of $CO_2$ during the manufacturing, transportation, construction, and decommissioning of a PFAL (Kikuchi and Kanematsu, 2020). Significant amounts of metals are used as structural elements in conventional PFAL buildings as well as multilevel cultivation racks and machines/facilities inside buildings. Metals used in PFALs need to be reduced considerably by better design and the use of alternatives. Recycled metals and reinforced wooden structural elements can help mitigate $CO_2$ emissions during the PFAL life cycle.

Machines are made mostly of metals and plastics and consume electricity. In PFALs with a robotic machine for seeding and transplanting at a fixed position, trays with seeds, transplants, and plants ready for harvest are transported by an automatic or manual carrier to a designated position. Where trays do not move, robotic machines travel to trays for transplanting, spacing, or harvesting. The development of automatic plant spacing and allocation to maximize the resource and monetary productivity is a challenge for the next-generation PFAL. Other machines used in PFALs include air conditioners, heat pumps for heating or cooling of nutrient solutions, water pumps, air circulation fans, and sterilization units for nutrient solutions and room air. $CO_2$ emissions per kilogram of produce are thus affected by the amounts and types of metals used in machines, layouts of cultivation racks, machines and other facilities, and their operations. Hence, careful design (e.g. weight reduction, miniaturization, and multi-functionalization) and intelligent operation management are required to minimize $CO_2$ emissions per kilogram of produce.

Alternatives to metals include cross-laminated timber (CLT) and other reinforced wooden sheets/boards (Brandner et al., 2016; Molin and Martin, 2018; Pierobon et al., 2019). Porous concrete reduces cement consumption by about 25%, and CLT reduces the consumption of metals by nearly 100%. Fabricated, modularized, and scalable cultivation system modules would also reduce $CO_2$ emissions from the manufacture, transportation, construction, renovation, scaling, and demolition of PFALs (Zeidler and Shubelt, 2017).

## 9.5 Land area

The land area required to build a PFAL with ten layers (tiers) is roughly 1/100th of that required by open-field cultivation of vegetables to achieve the same yield (amount of produce per land area) as the PFAL. This 100-fold yield is due to the increase in yield (e.g., 10-fold by ten tiers, 2-fold by year-round production, 2-fold by a higher growth rate under optimal and stable environments for photosynthesis, 2-fold by no damage by pest insects, harsh weather, etc., and 1.5-fold by less plant residue due to a higher harvest index, less damage at harvesting, uniform plant growth, etc.) (i.e. $120 = 10 \times 2 \times 2 \times 2 \times 1.5$). High yields are a major advantage of building PFALs in a city with local production for local consumption to secure safe, healthy, plant-based food. A PFAL can be built on any small land area, such as shaded, nonfertile land, wasteland, and even vacant rooms and buildings. It can contribute to decreasing $CO_2$ emissions in terms of food mileage (losses of produce during transportation, fuel for transportation, and freshness of produce). It will also create job opportunities in cities.

## 9.6 $CO_2$ emissions per kilogram of produce for PFAL- and greenhouse-grown crops

Orsini et al. (2020) reviewed the use efficiencies of PFALS in water, land surface, and energy use with different types of lighting (lamp types, spectrum, and PPFD), cultivation/growing systems (substrate, watering, aeroponics, and spacing), and types of environmental control, and compared their performances with those of greenhouse and open-field cultivation. This type of research has been increasingly important in developing a sustainable plant production system in or near a city.

In the Netherlands, 0.92 kg of $CO_2$-eq is emitted to produce 1 kg of fresh tomatoes in a Dutch-style greenhouse heated with natural gas; the consumption of natural gas for heating and other purposes accounts for 72% (0.66 kg) of $CO_2$-eq (Ritchie and Roser, 2020; Hayashi, 2022b). In a PFAL, 0.27-0.74 kg $CO_2$-eq is emitted per edible plant material to produce leafy greens and herbs (e.g. basil) (Martin and Molin, 2019). LCA on different energy sources for electricity generation (Baumont de Oliveira et al., 2023) and plant production systems under various conditions is essential for developing an optimal strategy for reducing the $CO_2$ emissions per kilogram of produce.

## 10 Conclusion

The number of PFALs producing horticultural plants and herbs has been steadily increasing in Europe, North America, Asia, and some other regions since 2016 because of the advantages of PFALs over greenhouses and open fields. However, there are concerns about the ecological, environmental, and economic sustainability of PFALs because existing PFALs consume significant amounts of electricity, chemical fertilization, plastic products and cultivation mediums (substrate) for operation, and metals and cement for construction.

This chapter discusses reducing the consumption of the above resources and thus $CO_2$ emissions by improving resource use efficiencies, using alternatives, and reusing/recycling resources. This reduction is often accompanied by a reduction in production costs per kg of produce without reducing but even increasing the productivity of PFALs. Expected reductions in resource consumption are summarized below. These need to be experimentally and theoretically validated carefully because they are based on limited experiments and theoretical assumptions. On the other hand, rapid recent technological advances and potential future innovations in PFALs are remarkable and difficult to predict, so reductions in resource consumption and $CO_2$ emissions might exceed the expectations that follow:

1) Forty percent reduction in electricity consumption for lighting per kg of produce by intelligent design and management of lighting and cultivation systems with energy-efficient cultivars (Sections 5.1-5.4). By doing so, the electricity consumption of air conditioners per kg of produce is also reduced by 40% (Section 6).

2) Eighty percent reduction of $CO_2$ emission factor per kWh for electricity generation by use of natural energy only, compared to electricity generation systems only using fossil fuels (Section 5.4).
3) By combining (1) with (2), $CO_2$ emissions per kg of produce with electricity consumption is reduced by 88% (=100 - (100 - 40) × (1.0 - 0.8) = (100 - (60 × 0.2)) = 88).
4) Twenty percent reduction in chemical fertilizer consumption per kg of produce by not draining nutrient solution with the use of a one-way delivery method (Section 9.2), and with the use of organic fertilizer decomposed by microorganisms from plant residue.
5) Thirty percent reduction in consumption of plastic products by replacing them with plant-based products and simplifying production and marketing/sales processes (no experimental evidence. Just an expectation) (Section 9.3).
6) Thirty percent reduction in weight of metals and cement for constructing a PFAL per kg of produce (Section 9.4).
7) Twenty percent reduction in transport and delivery costs through simplified packaging and less loss of produce (Martin and Molin, 2019).
8) Fifty percent increase in worker productivity by improving flows of materials and workers and introducing AI and robot-assisted management (no experimental evidence. Just an expectation).

## 11 Acknowledgements

The author would like to thank Dr. Michael Martin, Dr. Francis Baumont de Oliveira, Dr. Eri Hayashi, and Dr. Hiroko Nakaoka for their valuable comments on improving the manuscript. The authors also would like to thank Dr. Hiroko Nakaoka for improving the quality of the manuscript.

## 12 References

Akiyama, T. and Kozai, T. (2016). Light environment in the cultivation space of plant factory with LEDs. Chapter 7. In: *LED Lighting for Urban Agriculture*, Kozai, T., Fujiwara, K. and Runkle, E. S. (Eds). Springer. 91–109.

Baumont de Oliveira, F., Bannon, S., Evans, L., Anderson, L., Myers, P. and Thomas, J. M. H. (2023). Pathways to net-zero farming: a carbon footprint comparison of vertical versus traditional agriculture. *Acta Horticulturae* 1369. 125–132. doi: 10.17660/Acta Hortic.2023.1369.15.

Brandner, R., Flatscher, G., Ringhofer, A., Schickhofer, G. and Thiel, A. (2016). Cross laminated timber (CLT): overview and development. *European Journal of Wood and Wood Products* 74.331-351.

Dambhare, M. V., Butey, B. and Moharil, S. V. (2021). Solar photovoltaic technology: a review of different types of solar cells and its future trends. *Journal of Physics: Conference Series* 1913(1), 16 pages. doi: 10.1088/1742-6596/1913/1/012053.

Folta, K. M. and Maruhnich, S. A. (2007). Green light: a signal to slow down or stop. *Journal of Experimental Botany* 58(12). 3099–3111. doi: 10.1093/jxb/erm130.

Fujiwara, K. (2021). Photonmetric quantities and their application. Chapter 5. In: *Plant Factory: Basics, Applications and Advances*, Kozai, T., Niu, G. and Masabni, J. G. (Eds). Elsevier. 83–99.

Gates, B. (2021). *How to Avoid a Climate Disaster. The Solutions We Have and the Breakthroughs We Need.* Random House Large Print, pp.384. ISBN: 978-0-593-21577-7.

Hayashi, E., Amagai, Y., Kozai, T., Maruo, T., Tsukagoshi, S., Nakano, A. and Johkan, M. (2022a). Variations in the growth of cotyledons and initial true leaves as affected by photosynthetic photon flux density at individual seedling and nutrients, *Agronomy*. 12(1). doi: 10.3390/agronomy12010194.

Hayashi, K. (2022b). *Life Cycle Assessment (LCA) in Greenhouse Horticulture. The 43rd (Webinar) Seminar on Greenhouse Horticulture Organized by the Japan Greenhouse Horticulture Association. The 16th Slide of Oral Presentation Material with 36 Slides* (in Japanese).

IEA (2021). *Emission Factors 2021: Annual GHG Emission Factors for World Countries from Electricity and Heat Generation*. Available at: https://www.iea.org/data-and-statistics/data-product/emissions-factors-2021.

Imamura, E., Iuchi, M. and Bando, S. (2016). *Comprehensive Assessment of Life Cycle $CO_2$ Emissions from Power Generation Technologies in Japan*. Socioeconomic Research Center Report (vol. Y06). 73 pages. (in Japanese with English Abstract).

JGHA (Japan Greenhouse Horticulture Association) (2022). *Annual Business Report Separate Volume No.1: Survey and Case Study on Greenhouses and Plant Factories.* 70 pages. (in Japanese). Available at: https://jgha.com/wp-content/uploads/2022/04/TM06-03-bessatsu1.pdf.

Jin, W., Lopez, D. F., Heuvelink, E. and Marcelis, L. F. M. (2022). Light use efficiency of lettuce cultivation in vertical farms compared with greenhouse and field. *Food and Energy Security*. e391. 10 pages. doi: 10.1002/fes3.391.

Kikuchi, Y. and Kanematsu, Y. (2020). Chapter 27. In: Plant factory: *An Indoor Vertical Farming System for Efficiant Quality Food Production.* Kozai, T. et al. (Eds). Elsevier. 383–395.

Kim, J. Y., Lee, J. W., Jung, H. S., Shin, H. and Park, N. G. (2020). High-Efficiency Perovskite Solar Cells. *Chemical Reviews* 120(15), 7867–7918. doi: 10.1021/acs.chemrev.0c00107.

Kirkerud, J. G., Nagel, N. O. and Bolkesjø, T. F. (2021). The role of demand response in the future renewable northern European energy system. *Energy* 235(15), 121336.

Kozai, T. (2011). Improving light energy utilization efficiency for a sustainable plant factory with artificial light. *Proceedings of the Green Lighting Shanghai Forum* 2011, 375–383.

Kozai, T. (2013). Resource use efficiency of closed plant production system with artificial light: Concept, estimation, and application to plant factory. *Proceedings of the Japan Academy Series B* 89, 447–461.

Kozai, T. and Chun, C. (2002). Closed systems with artificial lighting for the production of high-quality transplants using minimum resource and environmental pollution. *Acta Horticulturae* 578. 27–33. doi:10.17660/actahortic.2002.578.2.

Kozai, T., Niu, G. and Masabni, J. (Eds.) (2021). *Plant Factory: Basics, Applications and Advances*. Elsevier, 444 pages.

Kudo, R. and Yamamoto, K. (2016). Induction of plant disease resistance and other physiological responses by green light illumination LED lighting for urban horticulture. Chapter 19. In: *LED Lighting for Urban Agriculture*. Kozai, T., Fujiwara, K. and Runkle, E. S. (eds ), Springer, 261-274.

Kusuma, P., Pattison, P. M. and Bugbee, B. (2021). Photon efficacy in horticulture: turning LED packages into LED luminaires. Chapter 7. In: *Plant Factory: Basics, Applications and Advances*. Kozai, T., Niu, G. and Masabni, J. G. (Eds). Elsevier. 115–128.

Molin, E. and Martin, M. (2018). Reviewing the energy and environmental performance of vertical farming system in urban. IVL Swedish Environmental Research Institute, Report number C 298. ISBN 978-91-88787-34-7. 31 pages.

Martin, M. and Molin, M. (2019). Environmental assessment of an urban vertical hydroponic farming system in Sweden. *Sustainability* 11(15). 4124. 14 pages. doi: 10.3390/su11154124.

Martin, M. and Orsini, F. (2023). Life cycle assessment of indoor vertical farms. Chapter 4. In: *Advances in plant factories: New technologies in indoor vertical farming*. Kozai, T. and Hayashi, E. (eds). *Burleigh Dodds Science Publishing*, doi: 10.19103/AS.2023.0126.06.

Newell, R., Newman, L., Dickson, M., Vanderkooi, B., Fernback, T. and White, C. (2021). Hydroponic fodder and greenhouse gas emissions: a potential avenue for climate mitigation strategy and policy development. *FACETS* 6. 334–357 | doi: 10.1139/facets-2020-0066.

Orsini, F., Pennisi, G., Zulficar, F. and Glanquinto, G. (2020). Sustainable use of resources in plant factories with artificial lighting (PFALs). *European Journal of Horticultural Science* 85(5). 297–309 | ISSN 1611-4426. doi: 10.17660/eJHS.2020/85.5.1.

Pattison, P. M., Tsao, J. Y., Brainard, G. C. and Bugbee, B. (2018). LEDs for photons, physiology and food. *Nature* 563(7732). 493–500. doi:10.1038/s41586-018-0706-x.

Pierobon, F., Huang, M., Simonen, K. and Ganguly, I. (2019). Environmental benefits of using hybrid CLT structure in midrise non-residential construction: an LCA based comparative study in the U. S. Pacific Northwest. *Journal of Building Engineering* 26. 100862 doi.10.1016/J.jobe.2019.100862. 14 pages.

Ritchie, H., Rosado, P. and Roser, M. (2020). Environmental impacts of food production. *Published OurWorldInData.org* Available at: https://ourworldindata.org/environmental-impacts-of-food [Online Resource].

United States Environmental Protection Agency (2022). Greenhouse gas emissions Available at: https://www.epa.gov/ghgemissions/overview-greenhouse-gases.

Zeidler, C., Shubelt, D. and Vrakking, V. (2017). Vertical Farm 2.0: Designing an Economically Feasible Vertical Farm = A Combined European Endeavor for Sustainable Urban Agriculture Final Report. Institute of Space Systems. 81 pages. Available at: https://www.researchgate.net/publication/321427717.

# Chapter 6

## Optimizing energy and other resource use in vertical farms

*Francesco Orsini, Laura Carotti, Mohammad Kazem Souri, Giuseppina Pennisi and Giorgio Gianquinto, University of Bologna, Italy*



### 1 Introduction

Food security faces several challenges, including population growth, urbanization, and climate change and its consequences, such as extreme drought, agricultural soil degradation, and salinization (Sanyé-Mengual et al., 2019; Orsini et al., 2020). While data points to increasingly reduced resource availability, it is estimated that 10–20% more water and other resources will be needed, including cultivation land, fertilizers, and pesticides to maintain current levels of food production through conventional farming (De Pascale et al., 2016; Michelon et al., 2020). In the context of sustainable development, defined as 'development that meets the need of the present without compromising the ability of future generations to meet their own needs' (WCED, 1987), crop production in plant factories with artificial light (PFALs) has the potential to improve resilience to climate change and the sustainable use of resources (Orsini et al., 2020).

Resource use efficiency (RUE) is defined as the ratio between the amount of resource fixed or utilized by the plants and the amount of resource supplied to the system (Kozai et al., 2015). In order to quantify RUE, recent technological advances enable continuous and accurate measurement of resource inputs and outputs at a relatively low cost (Kozai et al., 2014).

http://dx.doi.org/10.19103/AS.2023.0126.07

The advanced and differentiated technologies in PFALS (e.g. hydroponic and aeroponic systems, LED (light-emitting diode) lighting, $CO_2$ enrichment, and closed fertigation units) allow for efficient use of resources (Orsini et al., 2020). However, there is still room for improvement, and substantial gains in RUE are expected from the greater knowledge of plant physiology, sensing, artificial intelligence, and climate management (Van Delden et al., 2021; de Carbonnel et al., 2022). In addition, the design of the system (e.g. the building characteristics and its engineering and materials) and crop selection will also play important roles. A lack of detailed engineering analysis of the system design can lead to inefficient resource use and consequent side effects.

Energy use is a major research topic, associated with high costs, significant carbon emissions, and environmental impacts (Avgoustaki and Xydis, 2020). Recent research estimates that the electricity used for LED lighting and heating, ventilation, and air-conditioning (HVAC) systems contributes to more than 50% of greenhouse gas emission impacts of PFALs (Martin et al., 2022). These results emphasize the urgent need to optimize the use of resources across different integrated technologies. Moreover, the integration of PFALs with city infrastructure, creating a mutually beneficial system and strengthening resource flow within the urban district (Gentry, 2019; Martin et al., 2022), is an additional perspective.

Accordingly, the objective of this chapter is to present the various strategies for maximizing resource (mainly energy, land, and water) use efficiency within PFALs, highlighting the latest and most innovative technologies.

## 2 Energy use efficiency

The various components of the PFAL system, such as artificial lighting, climate management, irrigation, and fertigation, require electrical energy. It has been estimated that energy use efficiency (EUE) in a PFAL ranges from 1 g fresh weight (FW) $kWh^{-1}$ to 140 g FW $kWh^{-1}$, with the degree of variation depending mainly on the technology used and the lighting features (Orsini et al., 2020). Kozai and Niu (2020) estimated the percentage of electricity consumption for lighting, air conditioning, and other electric equipment (fertigation pumps, air circulation fans, sterilization units, floor and cultivation panel cleaning units, and miscellaneous) to be split at 75–80%, 15–20%, and 5%, respectively. Nonetheless, when inefficient lamps are adopted, energy dissipation in the form of heat can also result in increased energy use by the climate control unit (Orsini et al., 2020).

Substantial energy reduction and optimization can come from lighting and climate management, use of alternative energy sources (such as renewable energies), and integration into the urban fabric in order to improve energy flows. All these aspects will be discussed and addressed in the following sections.

### 2.1 Light use efficiency

The efficient use of light as the most energy-consuming component in a PFAL is of great importance as any improvement could significantly influence the efficient and economic performance of the system. Light use efficiency (LUE) can be defined in several ways, according to different perspectives. From a plant physiology point of view, LUE is defined as the ratio between plant dry weight per incident light ($LUE_{inc}$), alternatively considering as incident light the photosynthetic photon flux density (PPFD, 400–700 nm) or the photon flux density (400–800 nm) (Jin et al., 2020). According to a recent study, the average $LUE_{inc}$ for lettuce cultivation in vertical farms is 0.55 g $mol^{-1}$, with the maximum value reaching 1.63 g $mol^{-1}$ (Jin et al., 2023). The same study also identified the maximum theoretical value, ranging from 1.26 g $mol^{-1}$ to 1.81 g $mol^{-1}$. The gap between the value obtained and the maximum theoretical value can be filled by applying strategies for improved light distribution within the crop, to increase the fraction of light intercepted by the crops, optimize leaf photosynthesis, and increase the biomass allocated to the leaves (Van Delden et al., 2021).

From an environmental and economic point of view, LUE [defined as the yield per unit of electricity consumed (L-EUE, g FW $kWh^{-1}$)] is a more interesting parameter to assess the overall economic performance of the system (Orsini et al., 2020). In this sense, the conversion factor of electrical energy into photons that can be used for plants' photosynthesis primarily affects L-EUE. Significant improvement can come from optimizing the spectral effects on plant morphology (which is often species-specific) and physiology (Kusuma et al., 2021), as well as from the optimization of light intensity (Orsini et al., 2020; Kong et al., 2019). In terms of L-EUE performance, especially when compared to other systems such as a technological greenhouse with artificial lighting, the geographical dimension of the area should also be taken into account to determine feasibility (Graamans et al., 2018). According to Pennisi et al. (2020b), $LUE_{inc}$ is directly connected to the L-EUE, and accordingly strategies leading to an increase in the L-EUE can improve $LUE_{inc}$ and vice versa. In the next section, an overview of these strategies is presented.

#### 2.1.1 LED lighting

The starting point for the optimization of LUE is the adoption of LED lighting. The fundamental role that this technology has played in the development of vertical farming (VF) has been highlighted several times (Paucek et al., 2020). According to an extensive study that analyzed the current global LED industry, based on firm-declared data, the greatest part of LED lighting systems has a photosynthetic photon efficacy (the lamp's efficacy of electricity-to-light conversion) ranging from 2 μmol $J^{-1}$ to 3 μmol $J^{-1}$ (Paucek et al., 2020). However,

LED lamp technology, and therefore their efficacy, is evolving fast. Different color LEDs have different photon efficacy, with the best performance identified by red diodes (Park and Runkle, 2018; Blanken et al., 2013). However, as reported by Pennisi et al. (2019b), in response to increased junction temperature, blue diodes may increase their performance in terms of photon efficacy, while red diodes decrease it. According to a recent calculation, potential performances for white + red fixtures can reach 3.4 μmol $J^{-1}$ and increase up to 4.1 μmol $J^{-1}$ for red + blue LEDs (Kusuma et al., 2020). Thus, the upper limit of fixture efficacy is determined by the choice of LEDs and operating conditions, but several factors can also contribute, such as drive current, junction temperature, driver efficiency, and optical losses in the fixture (Kusuma et al., 2020). Furthermore, as described in Section 2.2, the efficacy of LED lighting systems has an impact on climate management, since the energy not converted into light enters the system as heat and must be removed by the HVAC system.

### *2.1.2 Light management*

The use of highly technological and efficient lighting is essential to reach high L-EUE and LUE, to increase the sustainability of the system. Concurrently, application of advanced cultivation techniques is essential in order to fulfill the potential of LED technology. The ability to focus primarily on managing environmental parameters rather than relying on genetics will greatly enhance crop production in VF (SharathKumar et al., 2020). In that sense, light management plays a major role through knowledge and management of photobiology. Use of LED lighting can enable manipulation of the light spectrum, light intensity, and photoperiod, as well as light pulsation and direction. It is important to consider all these parameters holistically, as individually they have specific effects that vary when the other parameters change. This section aims to give an overview of the main light management strategies for maximizing the LUE.

#### *2.1.2.1 Management of the light spectrum*

In the light spectrum, the application of red (600-700 nm) and blue (400-500 nm) radiation (the most used wavelengths in indoor agriculture) in a ratio of red to blue between 2 and 4, (considered as a photon-based ratio), allows the optimization of L-EUE and $LUE_{inc}$ in different crops (Pennisi et al., 2019a,b; Yan et al., 2020).

Plants can, however, respond to a wider range of wavelengths, which can be fine-tuned in order to reach the highest LUE. According to Smith et al. (2017), green radiation (400–500 nm) added to a dichromatic red and blue spectrum can play a strategic role in photosynthesis at the leaf and canopy level, thereby significantly contributing to carbon assimilation. This appears to be a critical strategy, especially in a dense canopy environment, in which the addition of green radiation could efficiently increase L-EUE and $LUE_{inc}$, compared to an red and blue spectrum. The benefits of including green radiation in a customized spectrum for plant production would need to be evaluated with regard to the energy required to generate green wavelengths. For this reason, to be efficient in terms of energy use, the fraction of green radiation present in the spectrum must be tightly regulated. Indeed, fresh weight (FW) of lettuce increased by 48–57% with a proportion of green radiation in a white full spectrum, compared to the FW of lettuce grown with a green radiation (58–63%), indicating that an excessive amount of green radiation could reduce growth and be less energy-efficient (Nozue et al., 2018). Shade-avoidance syndrome can also be exploited to increase the fraction of light intercepted by the crop canopy. Shade-avoidance syndrome consists of a specific set of responses in plant development, triggered by the presence of far-red radiation, especially with a high far-red fraction (far-red red + far-red; considered as a photon-based ratio) (Kusuma and Bugbee, 2021). For instance, the addition of far-red radiation to a spectrum composed of red and blue radiation has been reported to increase leaf and dry weight in several species (Tan et al., 2021) while also increasing the plant leaf area, likely influencing light interception. This is critical as, according to Van Delden et al. (2021), in a leaf canopy, the maximum light interception is reached when the leaf area index (LAI, $m^2$ of leaf area per $m^2$ of floor) reaches values between 3 and 4. The addition of far-red radiation has the potential to increase light interception of the canopy, even when greater values of LAI are reached (Van Delden et al., 2021). In addition to the effect on plant morphology, far-red radiation has been reported to play an active role in photosynthesis, when added to a spectrum with shorter wavelengths (Zhen and Bugbee, 2020; Zhen and van Iersel, 2017). Accordingly, far-red radiation has been proposed to be included in the definition of photosynthetically active radiation. The integration of far-red radiation is a critical strategy to increase LUE, even if these effects need to be balanced with crop quality preservation, which may be negatively affected by the presence of far-red in the spectrum. Besides an increase in plant biomass, what is also interesting in a PFAL is a major allocation of the biomass into the leaves, or the edible portion of the plant in general. In this regard, dynamic management of light parameters offers great potential for VF. Indeed, by understanding the circadian cycle effect and its requirements, it is possible to regulate photomorphogenesis, secondary metabolites, and photosynthesis tailored to the needs of the different plants during the growth cycle (SharathKumar et al., 2020). The dynamic conditions of these parameters can take place on different days of the growth cycle or even throughout one day, involving 24-hour management, creating a great opportunity for the efficient use of system resources. This is in fact what happens in a natural environment, in which plants are subjected to numerous variations, even within the same day. There are two typologies of plant response under a fluctuating light environment – transient, which are reversed after light conditions change, and persistent, which must be fine-tuned, especially in the early stages of growth (Meng and Runkle, 2020). One strategy is to apply a spectrum in the early stages of growth that could allow greater leaf area development to increase light interception, and likely the biomass directed to the leaves, and then apply a spectrum to trigger specific responses in the secondary metabolites profile. In the case of lettuce, for instance, Meng and Runkle (2020) suggest the use of monochromatic blue radiation, applied at the end of the treatment, to enhance anthocyanin accumulation.

#### 2.1.2.2 Management of other light parameters

In addition to optimal spectrum management to achieve high RUE, the photoperiod (the duration of light per day) has a central role in affecting various plant mechanisms such as flowering and plant metabolism, together with the development of morphological characteristics and biomass accumulation (Pennisi et al., 2020a). Light duration can also affect the energy demands of the system. For leafy greens, such as lettuce and chicory, the adoption of a photoperiod up to 16 h day$^{-1}$ (DLI of 14.4 mol m$^{-2}$ day$^{-1}$) leads to an increase in biomass, but also greater use of resources, increasing energy, light, and water use efficiency (WUE), compared to a longer photoperiod (Pennisi et al., 2020a). The daily light integral (DLI) is the product of the PPFD and photoperiod (equal to the total sum of radiation in a 24-h period). It appears that at the same (and high) DLI, a low PPFD and a long photoperiod lead to an increase in fresh and dry weight in two different lettuce cultivars. Indeed, once the saturation point is reached, the increased photoperiod allows an increase in the net daily photosynthesis, using the light provided in a more efficient way (Kelly et al., 2020). As pointed out by Yan et al. (2020), the effects of light quality on lettuce growth and morphology have been shown only at high DLI, indicating that the effect of light quantity is stronger than that of light quality. Light intensity is estimated to have a significant effect on crop growth. Considering average values, it is estimated that a 1% increase in light could increase the yield by 0.8–1% in different crops (Marcelis et al., 2006). On the other hand, Yan et al. (2020) highlight a linear decrease in the $LUE_{inc}$ and L-EUE with increased light intensity, making the management (and increase) of the photoperiod a better strategy to yield improvement in PFALs, without negatively affecting the economic and environmental sustainability of the system. In this context, precise and efficient photoperiod management is an excellent strategy. It was also found that plants grown with a shorter photoperiod and intermittent light had a higher fresh biomass with reduced energy input. Another important strategy to increase $LUE_{inc}$ and L-EUE in VFs is the use of pulsed light, which is based on specific characteristics of LED lamps (Kanechi et al., 2016). Indeed, LED lamps can be switched on and off very rapidly, at extremely short intervals (hundreds of times per second). By exploiting the specific rates of reactions occurring in the photosystem, it is possible to maximize photosynthetic performance, reducing energy requirements (Davis and Burns, 2016). By adjusting the frequency and the duty ratio, different studies have shown a reduction in energy consumption for the same amount of biomass obtained, and even increased biomass (Kanechi et al., 2016; Carotti et al., 2021b; Song et al., 2019). According to Avgoustaki and Xydis (2020), it is possible to reduce the cost of indoor lighting of a VF via the adoption of a fluctuating lighting scheme based on the price of the electricity market on an hourly basis, without compromising the quantity and the quality of the products. To apply this strategy, it is necessary for the data concerning hourly arrangements to be fully accessible, as in the above-mentioned case carried out in Denmark. The approach used in the work of Li et al. (2014) for the reduction of energy associated with the lighting system was based on a strategy to reduce unnecessary illumination between adjacent crops, in this case lettuces. In the study, lettuces grown under conventional LED were compared to those under LED with zoom lenses, with the goal to cut off irradiation in the growing space not occupied by the plants. This approach reduced electricity use by 52%, increasing the L-EUE by 56%. However, the reduction in energy led to a decrease in temperature, which showed negative effects for leaf fresh and dry weight. As further reported subsequently, a reduction in the energy wasted between adjacent plants can be reached using dynamic plant densities. Finally, considering the intricate molecular and physiological mechanism by which plants respond to various light characteristics, the application of artificial intelligence with machine learning, deep learning, and the Internet of Things (IoT) has the potential to predict plant growth and design an optimal growing environment (Siregar et al., 2022).

### 2.1.3 Light use efficiency besides light

In addition to light, the precise management of other environmental parameters can also influence the LUE. PFALs are normally enriched in $CO_2$ content, which is an important measure for increasing indoor productivity. In a VF, $CO_2$ enrichment is generally accompanied by a low ventilation rate, as $CO_2$ can be mainly utilized by plants, without being released into the environment. A high $CO_2$ concentration, up to the saturation point, has the potential to increase photosynthetic performances and plant biomass, but also to increase LUE. As reported by Huber et al. (2021) for four different cultivars of tomato seedlings, enriching $CO_2$ from 1000 μmol mol$^{-1}$ up to 1600 μmol mol$^{-1}$ increases the $LUE_{inc}$ by 38–44%. Moreover, the same study highlighted how the cost of increasing $CO_2$ is minimal compared to other techniques that growers could apply in a PFAL. Therefore, growing tomato seedlings with a high DLI (13 mol day$^{-1}$) and a substantial increase in $CO_2$ can increase both the environmental and economic sustainability of production. Managing plant density is an important cultivation technique aimed at increasing the fraction of light intercepted, in order to avoid too much light ending up on the floor instead of being used for biomass production. As previously anticipated, maximum values of light interception are reached with an LAI of 3–4 (Van Delden et al., 2021). Beyond this level, the absorption curve flattens out and there are no longer major gains in LUE. According to the growth model proposed by Ioslovich and Gutman (2000), when the environment is constant and stable, in a VF the optimal spacing between growing plants is such that the crop density is constant over time, meaning that as the crop grows, the space must increase. These results increase the validity of agronomic techniques within VFs that enable dynamic density management. Management of plant density, together with harvest time, is the main technique that can be controlled in an open field in order to reach high LUE. In a VF, these parameters interact to a large extent with other controllable environmental factors. Jin et al. (2020) studied the interaction between far-red radiation and plant density, observing a significant increase in LUE by the addition of far-red radiation, which was greater when the planting density was lower. Increased PPFD, if supported by an adequate concentration of $CO_2$, higher than the environment (450 ppm), can lead to an increase in biomass, as demonstrated in lettuce with intensities of 200, 400, and 750 μmol $m^{-2}$ $s^{-1}$ (Carotti et al., 2021a). Indeed, by increasing PPFD, light is used less efficiently, hence the need to always frame the value of cultivation to calibrate both economic and environmental costs. As reported by Carotti et al. (2021a), root zone temperature management had little influence on dry matter accumulation but affected LUE (if considered as grams of FW per mole of light). Furthermore, besides the technological and management advances resulting from a deeper understanding of crop physiology, there is a need to delineate optimal conditions for efficient use of energy, even in crops other than leafy greens (Orsini et al., 2020). Jayalath and van Iersel (2021) compared the growth of lettuce and mizuna in a vertical farm grown with six different PPFDs, ranging from 50 μmol $m^{-2}$ $s^{-1}$ to 425 μmol $m^{-2}$ $s^{-1}$, with $CO_2$ enrichment, targeting the identification of the underlying physiological components that could explain differences in crop growth. They identified that mizuna can convert incident light into biomass more efficiently than lettuce (considering the amount of light reaching the canopy of the crop instead of the total amount of light provided), especially when plants are grown at low PPFDs. The fastest growth is the result of a higher chlorophyll content index and projected canopy size, which allows mizuna to capture more light and consequently to increase the $LUE_{inc}$ when compared to lettuce. The study highlighted that projected canopy size and $LUE_{inc}$ are good parameters for an adequate selection of crops with fast growth, and therefore suitable for VF.

There is great potential for non-destructive methods such as canopy imaging to screen crops adapted to VF.

## 2.2 Climate management

Due to the airtight conditions of the growing environment, the high planting density, and the use of artificial lighting as the sole source of light, continuous dehumidification and cooling of the environment is essential in a PFAL. In this sense, outdoor ventilation is not a common strategy as it could reduce $CO_2$ use efficiency and allow the entry of pathogens and insects (Kozai, 2013). Climate and microclimate management is therefore carried out by the installation of an HVAC system, generally consisting of a heat pump used for heating or cooling the environment when needed. An HVAC system is also necessary to maintain the appropriate levels of humidity and to allow an adequate evapotranspiration rate (Engler and Krarti, 2021). It has been estimated that the amount of energy used by the HVAC system can account for around 30% of the total energy used in a PFAL (Orsini et al., 2020). Distributed energy systems (e.g. combined heat and power, or combined cooling, heating, and power) can increase the energy use efficiency of a PFAL, particularly when integrated within the residential and industrial sector, thus allowing excess heat to be shared (Kikuchi et al., 2016). According to Yokoyama (2019), substituting the heat pump with co-generation equipment in a PFAL can lead to a 30% reduction in energy use for climate management.

Regarding temperature, Zhang and Kacira (2020) showed that, for lettuce cultivation, increasing the photoperiod temperature (in this case to 25°C) and reducing the dark period temperature (16°C) can help save energy for the HVAC system in four different climatic zones. On the other hand, from the point of view of EUE, the optimal day/night temperature combination had a 4-5°C difference, namely, 24°C/19°C (Zhang and Kacira, 2020). Considering the high amount of energy needed for cooling, exploring climate conditions atypical of traditional growing environments appears to be a reasonable strategy (Carotti et al., 2021a). Breeding for higher plant tolerance to air humidity and temperature could reduce energy consumption. Facility design and the building envelope should also be considered in order to make the use of energy as efficient as possible (Weidner et al., 2021; Zhang and Kacira, 2020). According to Graamans et al. (2020), the specific facade design of a PFAL can significantly contribute to a reduction in energy demand, with significant effects in different climate zones. Specifically, they found that an opaque facade with high *U*-values and optimized albedo could reduce energy demand by 6.1, 12.5, and 9.5% for PFALs located in the United Arab Emirates, the Netherlands, and Sweden, respectively. When electricity use is considered, the most efficient facade is transparent, which led to a decrease in electricity use ranging from 7.4% to 9.4%, according to the region analyzed.

### 2.3 Renewable energies

In addition to the optimal management of environmental parameters, a strategy to improve the sustainability of the PFAL system is the use of energy produced by renewable sources. According to a study carried out by Hyun et al. (2014), the coefficient of performance of a heat pump used in a PFAL was highest using power plant waste heat as compared to air source, sea water, river, or geothermal heat. The extent of electricity generation from renewable sources and its stability depends on geographical location, climate, and weather conditions. It is expected that the share of renewable energies in electricity generation will increase. However, as reported by Van Delden et al. (2021), the use of solar panels to cover a PFAL's energy needs requires a high amount of land area. It has been estimated that 1 $m^2$ of cultivated area of a vertical farm requires 28.2 $m^2$ of photovoltaic cell arrays. The calculation considers only the lighting energy requirements, since the need for climate management could change according to the specific location and the level of insulation of the chamber. It appears clear, therefore, that considering the contribution of the HVAC system, the estimated surface area for photovoltaic cell arrays could increase considerably. In another study (Kobayashi et al., 2022), the land area requirement for solar energy for the production of lettuce, tomato, potato, and wheat in Spain and Sweden in PFALs was assessed. Specifically, it was investigated whether the soil saved through multi-level vertical production was sufficient to produce energy using photovoltaic cells to cover the needs of the PFAL. The results highlighted the role of technology in determining the feasibility of the process. Indeed, with current technology, the land saved to produce energy from photovoltaic cells was only sufficient for lettuce production. However, with improved technology, in terms of both PFALs and photovoltaic energy production, the saved land could be sufficient to produce all the crops, with the exception of tomato in Sweden. Further analysis was also conducted into wind energy. In this case, even assuming the two different levels of technology, the only potential production that would not require additional land use was that of lettuce (Kobayashi et al., 2022).

## 3 Water use efficiency

The adoption of modern and high-tech growing strategies has contributed to improving WUE (g FW $L^{-1}$ $H_2O$) in PFALs. As reported by Orsini et al. (2020), the WUE for lettuce production in a PFAL could reach 80 g FW $L^{-1}$ H20, a greater value than those obtained in open field (3–20 g FW $L^{-1}$ $H_2O$) and greenhouse (5–60 g FW $L^{-1}$ $H_2O$) cultivation. These impressive values are not limited to lettuce (which is generally regarded as a model plant for controlled environment agriculture) but were also confirmed for basil and rocket. The relevance of these figures is further validated, taking into account that the agriculture sector uses almost 70% of the world's freshwater. The introduction of a precise environmental control system through the use of sensors to monitor crop water status, the use of advanced cultivation techniques, and the optimization of the growing system could lead to a further increase in WUE (Tavan et al., 2021; Orsini et al., 2020). An additional advantage of soilless cultivation, which goes beyond the efficient use of water, is the potential to produce a cleaner product without soil residues or pathogens (Van Delden et al., 2021). The following sections focus on the influence of specific growing systems, and the management of climate and other environmental parameters to increase WUE.

### 3.1 Growing systems and nutrient use efficiency

Generally, soilless systems are used in PFALs, allowing a high level of automatization and optimization of both irrigation and fertilization. There are different typologies of soilless system, divided into substrate culture and hydroponic culture. To simplify, while in hydroponic systems the plant roots grow directly in contact with the nutrient solution (as in the case of deep-water culture and aeroponic cultivation), in the substrate culture the plants are hosted in pots and supported by an inert substrate and the nutrient solution is continuously or periodically pump through the system. An example of substrate culture is the ebb-and-flow cultivation. The common feature of these types of cultivation is the fact that water supplies nutrients to the root zone (Orsini et al., 2020). A closed-loop circulation system is generally adopted, in which water not absorbed by plants is automatically checked for pH and electrical conductivity (EC), which is restored to set point values and then returned to the plants. This system increases water and nutrient efficiency and avoids water enriched in nitrates and phosphates from entering the environment (Jung et al., 2019). As pointed out by Orsini et al. (2020), a major limitation of this system stems from the fact that EC gives accurate information on the total nutrient content, but not on the concentration of individual ions. This problem can be overcome through the application of technologies allowing the assessment of individual nutrient ions in a nutrient solution - colorimetric, spectrophotometric, ion-selective electrodes, and emission spectroscopy (Langenfeld et al., 2022). In the case of ion-selective electrodes, the application of this technology is hindered by the limited stability and life of the sensors in the hydroponic environment (Gieling et al., 2005). Recently, Jung et al. (2019) showed the validity of this technology in conjunction with an ion-based nutrient dosing algorithm in an ebb and flow hydroponic system for lettuce cultivation. WUE (defined in that case as the ratio of dry mass to water transpired) can also be a fundamental parameter to model the concentration of each nutrient based on the mass balance, minimizing the need to monitor nutrient solutions at ion level (Langenfeld et al., 2022). According to this model, the nutrient concentration in the irrigation solution is the product of the optimal nutrient plant tissue concentration (easier to assess and characterize) and WUE, which is largely affected by the environmental conditions of a PFAL (Langenfeld et al., 2022).

When compared to the widely used commercial soilless systems, such as the ebb-and-flow one, an aeroponic system is generally associated with higher WUE (Patle et al., 2020), with the potential to better manage the plant's physiological constraints (due mainly to low oxygen availability and high root zone temperatures) normally associated with hydroponic cultivation (Eldridge et al., 2020). A comparative assessment of yield and WUE of lettuce (*Lactuca sativa* cv. Canasta) grown in ebb-and-flow and aeroponic systems under the same environmental conditions was carried out. Results showed no significant differences in yield between the two systems, but the WUE of aeroponics were 2.1 times higher as compared to ebb-and-flow (Carotti et al., 2023). Considering only the water used for irrigation as a reference, the hydroponic system reached a WUE of 32.5 g FW $L^{-1}$, while the aeroponic system reached 69.7 g FW $L^{-1}$ (Carotti et al., 2023). Furthermore, with regard to aeroponics, which is a relatively newer soilless technology, optimization of the system can come from improving the pressure and the frequency used to deliver the nutrient solution to the plants, together with droplet size (Li et al., 2018). However, as highlighted by Eldridge et al. (2020), aeroponics require greater control technology and more expensive farm infrastructure, which could make this system less compatible with certain business models and crops.

### 3.2 Management of climate and environmental parameters

As explained in the previous section, continuous dehumidification is necessary within a PFAL to allow proper plant evapotranspiration. A critical enhancement can come from dehumidifying the internal air to recover water transpired by plants and used for climate control (Orsini et al., 2020). Pacak et al. (2020) analyzed the potential of water recovery from exhaust air in a vertical farm, highlighting the effectiveness of the system, meaning that air is a stable source of water recovery. In the aforementioned study it has been reported that around 67% of the water used as an input could be recovered by adopting this strategy, increasing the WUE of ebb-and-flow and aeroponics to 86.0 g FW $L^{-1}$ and 161.9 g FW $L^{-1}$, respectively. Applying this strategy within a PFAL, the only water lost is that used for cleaning and managing operations (Carotti et al., 2023). According to Van Delden et al. (2021), by recycling water transpired by plants, water used would be reduced by 90% compared to greenhouses and

99% compared to open fields. The management of environmental parameters affects not only LUE but also WUE. WUE is a function of transpiration rate and stomatal conductance. Vapor pressure deficit (VPD) is one of the main factors driving transpiration. As reported by Langenfeld et al. (2022), decreasing the VPD decreases the transpiration rate by 50%. On the other hand, several aspects influence stomatal conductance, including $CO_2$ concentration and light environment. For instance, light spectrum management may modify stomatal conductance, morphology, and density (Orsini et al., 2020). To this end, various studies have been carried out on lettuce, basil, and cucumbers (Hogewoning et al., 2010; Pennisi et al., 2019a,b). Considering lettuce cultivation, the use of a photon ratio between red and blue equal to 3 led to an increase in WUE of 63% compared to a spectrum richer in blue radiation (red and blue ratio equal to 0.5). As reported by Smith et al. (2017), the enrichment of the red and blue spectrum with green radiation can lead to optimizing the whole-canopy WUE, due to the stimulation of stomata closure, which occurs in the presence of changes in the blue and green ratio, without compromising biomass accumulation. Furthermore, a study into lettuce showed that the use of a white spectrum enriched in red and blue radiation increased WUE as compared to a spectrum of red and blue only, in green lettuce but not in a red cultivar, suggesting, again, species-specific and even cultivar-specific behavior (Pennisi et al., unpublished data). Stomatal conductance is also influenced by the $CO_2$ level. In the presence of high $CO_2$ levels, stomatal conductance is reduced, resulting in decreased transpiration and a consequent increase in WUE and growth (Langenfeld et al., 2022).

Root zone management is also important as it can primarily influence the dissolution of nutrients in the solution provided by plants, thereby influencing plant growth and the contents of metabolites, contributing to the enhancement of crop quality (Avgoustaki and Xydis, 2020). As explained by Eldridge et al. (2020), high temperatures at root levels are not recommended because they may inhibit growth. The use of aeroponics, however, allows water supply temperature to be adjusted, and also provides additional $CO_2$, which, as shown in various studies, can lead to a different mineral uptake and higher nutrient content (He et al., 2010; He et al., 2013). The ability of aeroponics systems to specifically control the time and frequency of nutrient solution application allows the dosage to align very specifically with the light spectrum, which, as briefly described, can greatly influence water and nutrient use (Eldridge et al., 2020). Furthermore, the application of precise sensor technology can increase WUE, thereby minimizing the impact on the environment, as well as on yield, of excessive and poorly managed irrigation. Tavan et al. (2021) demonstrated that moisture sensors can be successfully used for automatic irrigation of microgreens, without compromising yield and quality.

### 3.3 Substrate

The choice of the substrate to be used for substrate culture is fundamental to the success of the system, since it influences crop performance and the irrigation and fertigation regime to be applied, contributing therefore to the WUE of the system (Khandaker and Kotzen, 2018). Furthermore, the substrate can also influence the quality of the products, as shown by the higher nitrate level in red and green basil microgreens grown with vermiculite or jute fiber, compared to coconut fiber (Bulgari et al., 2021). According to Du et al. (2022), the quality and suitability of growing media can be evaluated according to (1) dry bulk density, (2) total pore space, (3) particle density, (4) air capacity, and (5) water-holding capacity. The use of the most suitable substrate according to cultivation needs can lead to an increase in the overall sustainability of the system, as shown by Martin and Molin (2019) in the case of substituting gardening soil with coir, but also in the WUE affecting the amount of water retained by the media that is therefore no longer available for reuse within the system. However, there are additives that can be added to the substrates (e.g. biochar) that, by improving water retention capacity and moisture content, could positively affect crop production, together with nutrient and WUE (Wang et al., 2019; Fischer et al., 2019).

## 4 Land surface use efficiency

Land surface use efficiency (SUE, g FW $m^{-2}$ $day^{-1}$) is one of the most interesting and best performing aspects of cultivation in PFALs, mainly due to the ability to exploit the vertical dimension for plant cultivation (Al-Kodmany, 2018). As reported by Orsini et al. (2020), SUE values are generally between 10 and 300 times higher when compared to open field and greenhouse values for the same crops. Examples include lettuce cultivation, for which SUE values for field and greenhouse are between 10 g FW $m^{-2}$ $day^{-1}$ and 300 g FW $m^{-2}$ $day^{-1}$ and that for PFALs between 1300 g FW $m^{-2}$ $day^{-1}$ and 3300 g FW $m^{-2}$ $day^{-1}$. A similar trend can be observed for other leafy vegetables (e.g. basil, rocket, and chicory). The application of specific and advanced cultivation strategies, such as specific planting density, transplanting strategy, and the level of automation, can have a significant influence on SUE values in a PFAL (Orsini et al., 2020). In addition, the use of aeroponic cultivation systems, which lead to a smaller volume of circulating water, has the potential to further improve the use of the vertical dimension. According to a model for lettuce growth developed by Ioslovich and Gutman (2000) to optimize grower profit, the optimal spacing strategy in an environment in which the climate variables can be considered constant is to also maintain constant the planting density, meaning that the spacing between plants should increase as crops grow. In this sense, dynamic plant spacing in a PFAL (for instance, dedicating each cultivation layer to a specific crop growth stage) may also provide advantages in terms of air recirculation at canopy level (thereby decreasing the need for ventilation and de-humidification), but the feasibility of this technique depends on the level of automation of the system (Orsini et al., 2020; Van Delden et al., 2021). According to Van Delden et al. (2021), an increase in the automation level within the PFAL could also bring other advantages, such as the potential to use higher $CO_2$ concentration, higher humidity, higher temperature ranges, and UV radiation. In addition, substituting at least part of the human work by robot can lead to an even greater SUE by better exploiting spaces (both in the horizontal and vertical dimensions) that would otherwise be unfeasible or unsafe for human workers. Moreover, in the case of SUE, proper light management can lead to increased values. As demonstrated by Pennisi et al. (2019a) for basil cultivation, optimizing the photon ratio of red to blue radiation in the spectrum (in this specific case identified by a ratio between 2 and 4) can lead to a significant increase in the SUE, compared, for example, to a ratio of 0.5-1. From the point of view of land use, PFALs can also be implemented within existing or vacant urban spaces, so as to avoid construction of new buildings that may cause substantial environmental impacts (Orsini et al., 2020; Van Delden et al., 2021), although constraints may emerge when anti-seismic regulations need to be met, or rational space management of cultivation chambers is hindered by the presence of internal columns.

## 5 Integration of vertical farms within district heating

A strategy to improve the efficiency of PFALs can be based on integrating the growing system within public and private infrastructures, in order to establish a closed-loop system of energy, $CO_2$, and waste (Pimentel and Friedler, 2022). From the point of view of energy efficiency, the association between PFAL and district heating (DH) is interesting. DH is a centralized production of heat and hot water that can be used in the civil sector – essentially, a network of pipes used to distribute excess fuel or heat resources to the local market that would otherwise be wasted (Werner, 2017). This technology has several advantages with regard to practical, environmental, and safety aspects for the production of thermal energy. The DH network reduces pollutants and thermal emissions in the urban environment, removes the combustion system at the final user stage, removes fuel transportation in the urban area, increases the safety of the process, and lowers heating costs when fuel prices are particularly high (Ancona et al., 2014; Werner, 2017).

Integrating PFALs and DH creates a unique opportunity to improve the circularity of the system, increasing the efficiency of both systems (Gentry, 2019). As explained by Gentry (2019) for a case study in Sweden, the DH network could supply PFALs with their heat requirements, creating in this way a new customer for the DH that could benefit from (1) the supply of low-temperature buildings in the DH network, (2) lowering the DH temperature in the network, and (3) providing an outlet for waste products.

Accordingly, PFALs could potentially be integrated into every DH network. Moreover, this association could also establish DH in areas where it would otherwise not be possible due to low-heat demand. This symbiosis needs to be explored and modeled in more detail for various climatic regions while considering the increasing interest from the scientific community, industry, and policy organizations worldwide. The European Union, for example, has emphasized the importance of reducing the high-energy demands of the residential building sector, issuing specific directives that should be followed by each Member State (Kampelis et al., 2017). It should be noted, however, that often a major energy component in VF climate control is associated with cooling (as a consequence of heat emission by lamps); therefore, further technologies (such as heat exchangers) may be required.

## 6 Conclusion and future trends

This chapter has identified the major sustainability advantages offered by PFALs to the agricultural sector at large. Although the feasibility of crop production may still be hindered by the large energy demands of the technology, the potential for groundbreaking performances in terms of water, nutrient, and land use efficiencies is evident. In the coming years, technology innovation and the definition of crop-specific management protocols will likely pave the way for substantial improvements in energy use efficiency. This will necessarily build on enhanced lighting fixtures and management strategies, dynamic plant spacing, and optimized climate control technologies. Thanks to the potential of the technology to adapt to dense urban districts, advantages may also come from successful integrations within city infrastructure, merging resource flows and enabling increased circularity in sustainable food production.

## 7 Acknowledgements

The research leading to this publication was financed by the Italian Ministry of Research and Education (MUR), within the Research Programme of National Interest (PRIN) project 'Sustainable Vertical Farming (VFarm)', grant code: J33C20002350001.

## 8 References

Al-Kodmany, K. (2018). The vertical farm: a review of developments and implications for the vertical city. *Buildings* 8(2), 24. doi:10.3390/buildings8020024.
Ancona, M. A., Bianchi, M., Branchini, L. and Melino, F. (2014). District heating network design and analysis. *Energy Procedia* 45, 1225–1234. doi:10.1016/j.egypro.2014.01.128.

Avgoustaki, D. D. and Xydis, G. (2020). Plant factories in the water-food-energy Nexus era: A systematic bibliographical review. *Food Security* 12(2), 253–268. doi:10.1007/s12571-019-01003-z.

Blanken, W., Cuaresma, M., Wijffels, R. H. and Janssen, M. (2013). Cultivation of microalgae on artificial light comes at a cost. *Algal Research* 2(4), 333–340. doi:10.1016/j.algal.2013.09.004.

Bulgari, R., Negri, M., Santoro, P. and Ferrante, A. (2021). Quality evaluation of indoor-grown microgreens cultivated on three different substrates. *Horticulturae* 7(5), 96. doi:10.3390/horticulturae7050096.

Carotti, L., Graamans, L., Puksic, F., Butturini, M., Meinen, E., Heuvelink, E. and Stanghellini, C. (2021a). Plant factories are heating up: hunting for the best combination of light intensity, air temperature and root-zone temperature in lettuce production. *Frontiers in Plant Science* 11, 592171. doi:10.3389/fpls.2020.592171.

Carotti, L., Potente, G., Pennisi, G., Ruiz, K. B., Biondi, S., Crepaldi, A., Orsini, F., Gianquinto, G. and Antognoni, F. (2021b). Pulsed LED light: exploring the balance between energy use and nutraceutical properties in indoor-grown lettuce. *Agronomy* 11(6), 1106. doi:10.3390/agronomy11061106.

Carotti, L., Pistillo, A., Zauli, I., Meneghello, D., Martin, M., Pennisi, G., Gianquinto, G. and Orsini, F. (2023). Improving water use efficiency in vertical farming: Effects of growing systems, far-red radiation and planting density on lettuce cultivation. *Agricultural Water Management* 285, 108365. https://doi.org/10.1016/j.agwat.2023.108365.

Davis, P. A. and Burns, C. (2016). Photobiology in protected horticulture. *Food and Energy Security* 5(4), 223–238. doi:10.1002/fes3.97.

de Carbonnel, M., Stormonth-Darling, J. M., Liu, W., Kuziak, D. and Jones, M. A. (2022). Realising the environmental potential of vertical farming systems through advances in plant photobiology. *Biology* 11(6), 922. doi:10.3390/biology11060922.

De Pascale, S., Maggio, A., Orsini, F. and Barbieri, G. (2016). Cultivar, soil type, nitrogen source and irrigation regime as quality determinants of organically grown tomatoes. *Scientia Horticulturae* 199, 88–94. doi:10.1016/j.scienta.2015.12.037.

Du, M., Xiao, Z. and Luo, Y. (2022). Advances and emerging trends in cultivation substrates for growing sprouts and microgreens toward safe and sustainable agriculture. *Current Opinion in Food Science* 46, 100863. doi:10.1016/j.cofs.2022.100863.

Eldridge, B. M., Manzoni, L. R., Graham, C. A., Rodgers, B., Farmer, J. R. and Dodd, A. N. (2020). Getting to the roots of aeroponic indoor farming. *New Phytologist* 228(4), 1183–1192. doi:10.1111/nph.16780.

Fischer, B. M. C., Manzoni, S., Morillas, L., Garcia, M., Johnson, M. S. and Lyon, S. W. (2019). Improving agricultural water use efficiency with biochar – A synthesis of biochar effects on water storage and fluxes across scales. *Science of the Total Environment* 657, 853–862. doi:10.1016/j.scitotenv.2018.11.312.

Gentry, M. (2019). Local heat, local food: integrating vertical hydroponic farming with district heating in Sweden. *Energy* 174, 191–197. doi:10.1016/j.energy.2019.02.119.

Gieling, T., Vanstraten, G., Janssen, H. and Wouters, H. (2005). ISE and Chemfet sensors in greenhouse cultivation. *Sensors and Actuators. Part B: Chemical* 105(1), 74–80. doi:10.1016/S0925-4005(04)00113-3.

Graamans, L., Baeza, E., van den Dobbelsteen, A., Tsafaras, I. and Stanghellini, C. (2018). Plant factories versus greenhouses: comparison of resource use efficiency. *Agricultural Systems* 160, 31–43. doi:10.1016/j.agsy.2017.11.003.

Graamans, L., Tenpierik, M., van den Dobbelsteen, A. and Stanghellini, C. (2020). Plant factories: reducing energy demand at high internal heat loads through façade design. *Applied Energy* 262, 114544. doi:10.1016/j.apenergy.2020.114544.

He, J., Austin, P. T. and Lee, S. K. (2010). Effects of elevated root zone CO2 and air temperature on photosynthetic gas exchange, nitrate uptake, and total reduced nitrogen content in aeroponically grown lettuce plants. *Journal of Experimental Botany* 61(14), 3959–3969. doi:10.1093/jxb/erq207.

He, J., Qin, L. and Lee, S. K. (2013). Root-zone $CO_2$ and root-zone temperature effects on photosynthesis and nitrogen metabolism of aeroponically grown lettuce (*Lactuca sativa* L.) in the tropics. *Photosynt.* 51(3), 330–340. doi:10.1007/s11099-013-0030-5.

Hogewoning, S. W., Trouwborst, G., Maljaars, H., Poorter, H., van Ieperen, W. and Harbinson, J. (2010). Blue light dose-responses of leaf photosynthesis, morphology, and chemical composition of Cucumis sativus grown under different combinations of red and blue light. *Journal of Experimental Botany* 61(11), 3107–3117. doi:10.1093/jxb/erq132.

Huber, B. M., Louws, F. J. and Hernández, R. (2021). Impact of different daily light integrals and carbon dioxide concentrations on the growth, morphology, and production efficiency of tomato seedlings. *Frontiers in Plant Science* 12, 615853. doi:10.3389/fpls.2021.615853.

Hyun, I., Lee, J., Yoon, Y., Lee, K. and Nam, Y. (2014). The potential and utilization of unused energy sources for large-scale horticulture facility applications under Korean climatic conditions. *Energies* 7(8), 4781–4801. doi:10.3390/en7084781.

Ioslovich,I. and Gutman, P. O. (2000). Optimal control of crop spacing in a plant factory. *Automatica* 36(11), 1665–1668.

Jayalath, T. C. and van Iersel, M. W. (2021). Canopy size and light use efficiency explain growth differences between lettuce and mizuna in vertical farms. *Plants* 10(4), 704. doi:10.3390/plants10040704.

Jean, P. and Ferenc, F. (2022). Synthesis of integrated vertical farming systems with multiperiodic resource availability. *Chemical Engineering Transactions* 94, 1039-1044. doi:10.3303/CET2294173.

Jin, W., Formiga Lopez, D., Heuvelink, E. and Marcelis, L. F. M. (2023). Light use efficiency of lettuce cultivation in vertical farms compared with greenhouse and field. *Food and Energy Security* 12(1). doi:10.1002/fes3.391.

Jin, W., Urbina, J. L., Heuvelink, E. and Marcelis, L. F. M. (2020). Adding Far-red to red-blue light-emitting diode light promotes yield of lettuce at different planting densities. *Frontiers in Plant Science* 11, 609977. doi:10.3389/fpls.2020.609977.

Jung, D.-H., Kim, H.-J., Cho, W.-J., Park, S. H. and Yang, S.-H. (2019). Validation testing of an ion-specific sensing and control system for precision hydroponic macronutrient management. *Computers and Electronics in Agriculture* 156, 660–668. doi:10.1016/j.compag.2018.12.025.

Kampelis, N., Gobakis, K., Vagias, V., Kolokotsa, D., Standardi, L., Isidori, D., Cristalli, C., Montagnino, F. M., Paredes, F., Muratore, P., Venezia, L., Dracou, M. K., Montenon, A., Pyrgou, A., Karlessi, T. and Santamouris, M. (2017). Evaluation of the performance gap in industrial, residential & tertiary near-Zero energy buildings. *Energy and Buildings* 148, 58–73. doi:10.1016/j.enbuild.2017.03.057.

Kanechi, M., Maekawa, A., Nishida, Y. and Miyashita, E. (2016). Effects of pulsed lighting based light-emitting diodes on the growth and photosynthesis of lettuce leaves. *Acta Horticulturae* (1134), 207–214. doi:10.17660/ActaHortic.2016.1134.28.

Kelly, N., Choe, D., Meng, Q. and Runkle, E. S. (2020). Promotion of lettuce growth under an increasing daily light integral depends on the combination of the photosynthetic photon flux density and photoperiod. *Scientia Horticulturae* 272, 109565. doi:10.1016/j.scienta.2020.109565.

Khandaker, M. and Kotzen, B. (2018). The potential for combining living wall and vertical farming systems with aquaponics with special emphasis on substrates. *Aquaculture Research* 49(4), 1454–1468. doi:10.1111/are.13601.

Kikuchi, Y., Kanematsu, Y., Sato, R. and Nakagaki, T. (2016). Distributed cogeneration of power and heat within an energy management strategy for mitigating fossil fuel consumption: cogeneration for mitigating fossil fuel consumption. *Journal of Industrial Ecology* 20(2), 289–303. doi:10.1111/jiec.12374.

Kobayashi, Y., Kotilainen, T., Carmona-García, G., Leip, A. and Tuomisto, H. L. (2022). Vertical farming: A trade-off between land area need for crops and for renewable energy production. *Journal of Cleaner Production* 379, 134507. doi:10.1016/j.jclepro.2022.134507.

Kobayashi, Y., Kotilainen, T., Carmona-García, G., Leip, A. and Tuomisto, H. L. (2022). Vertical farming: A trade-off between land area need for crops and for renewable energy production. *Journal of Cleaner Production* 379, 134507. doi:10.1016/j.jclepro.2022.134507.

Kong, Y., Nemali, A., Mitchell, C. and Nemali, K. (2019). Spectral quality of light can affect energy consumption and energy-use efficiency of electrical lighting in indoor lettuce farming. *HortScience* 54(5), 865–872. doi:10.21273/HORTSCI13834-18.

Kozai, T. (2013). Plant factory in Japan-Current situation and perspectives. *Chronica Horticulture* 53, 8.e11.

Kozai, T. and Niu, G. (2020). Role of the plant factory with artificial lighting (PFAL) in urban areas. In: *Plant Factory*, pp. 7–34. Academic Press.

Kozai, T., Kubota, C., Takagaki, M. and Maruo, T. (2015). Greenhouse environment control technologies for improving the sustainability of food production. *Acta Horticulturae* (1107), (1–14).

Kozai, T., Niu, G. and Takagaki, M. (2015). *Plant Factory: an Indoor Vertical Farming System for Efficient Quality Food Production*. London: Academic Press, 423 pp.

Kusuma, P. and Bugbee, B. (2021). Far-red fraction: an improved metric for characterizing phytochrome effects on morphology. *Journal of the American Society for Horticultural Science* 146(1), 3–13. doi:10.21273/JASHS05002-20.

Kusuma, P., Pattison, P. M. and Bugbee, B. (2020). From physics to fixtures to food: current and potential LED efficacy. *Horticulture Research* 7, 56. https://doi.org/10.1038/s41438-020-0283-7.

Kusuma, P., Swan, B. and Bugbee, B. (2021). Does green really mean go? Increasing the fraction of green photons promotes growth of tomato but not lettuce or cucumber. *Plants* 10(4), 637. doi:10.3390/plants10040637.

Langenfeld, N. J., Pinto, D. F., Faust, J. E., Heins, R. and Bugbee, B. (2022). Principles of nutrient and water management for indoor agriculture. *Sustainability* 14(16), 10204. doi:10.3390/su141610204.

Li, K., Yang, Q.-C., Tong, Y.-X. and Cheng, R. (2014). Using movable light-emitting diodes for electricity savings in a plant factory growing lettuce. *HortTechnology* 24(5), 546–553. doi:10.21273/HORTTECH.24.5.546.

Li, Q., Li, X., Tang, B. and Gu, M. (2018). Growth responses and root characteristics of lettuce grown in aeroponics, hydroponics, and substrate culture. *Horticulturae* 4(4), 35. doi:10.3390/horticulturae4040035.

Marcelis, L. F. M., Broekhuijsen, A. G. M., Meinen, E., Nijs, E. M. F. M. and Raaphorst, M. G. M. (2006). Quantification of the growth response to light quantify of greenhouse crop production. *Acta Horticulturae* (711), 97–104. doi:10.17660/ActaHortic.2006.711.9.

Martin, M. and Molin, E. (2019). Environmental assessment of an urban vertical hydroponic farming system in Sweden. *Sustainability* 11(15), 4124. doi:10.3390/su11154124.

Martin, M., Weidner, T. and Gullström, C. (2022). Estimating the potential of building integration and regional synergies to improve the environmental performance of urban vertical farming. *Frontiers in Sustainable Food Systems* 6, 849304. doi:10.3389/fsufs.2022.849304.

Meng, Q. and Runkle, E. S. (2020). Growth responses of red-leaf lettuce to temporal spectral changes. *Frontiers in Plant Science* 11, 571788. doi:10.3389/fpls.2020.571788.

Michelon, N., Pennisi, G., Myint, N. O., Dall'Olio, G., Batista, L. P., Salviano, A. A. C., Gruda, N. S., Orsini, F. and Gianquinto, G. (2020). Strategies for improved yield and water use efficiency of lettuce (*Lactuca sativa* L.) through simplified soilless cultivation under semi-arid climate. *Agronomy* 10(9), 1379. doi:10.3390/agronomy10091379.

Nozue, H., Shirai, K., Kajikawa, K., Gomi, M., Nozue, M. and White, L. E. D. (2018). White LED Light with wide wavelength spectrum promotes high-yielding and energy-saving indoor vegetable production. *Acta Horticulturae* 1227(1227), 585–592. doi:10.17660/ActaHortic.2018.1227.74.

Orsini, F., Pennisi, G., Zulfiqar, F. and Gianquinto, G. (2020). Sustainable use of resources in plant factories with artificial lighting (PFALs). *European Journal of Horticultural Science* 85(5), 297–309. doi:10.17660/eJHS.2020/85.5.1.

Pacak, A., Jurga, A., Drąg, P., Pandelidis, D. and Kaźmierczak, B. (2020). A long-term analysis of the possibility of water recovery for hydroponic lettuce irrigation in indoor vertical farm. Part 1: Water recovery from exhaust air. *Applied Sciences* 10(24), 8907. doi:10.3390/app10248907.

Park, Y. and Runkle, E. S. (2018). Spectral effects of light-emitting diodes on plant growth, visual color quality, and photosynthetic photon efficacy: white versus blue plus red radiation. *PLoS ONE* 13(8), e0202386. doi:10.1371/journal.pone.0202386.

Patle, G. T., Kumar, M. and Khanna, M. (2020). Climate-smart water technologies for sustainable agriculture: A review. *Journal of Water and Climate Change* 11(4), 1455–1466. doi:10.2166/wcc.2019.257.

Paucek, I., Appolloni, E., Pennisi, G., Quaini, S., Gianquinto, G. and Orsini, F. (2020). LED lighting systems for horticulture: business growth and global distribution. *Sustainability* 12(18), 7516. doi:10.3390/su12187516.

Pennisi, G., Blasioli, S., Cellini, A., Maia, L., Crepaldi, A., Braschi, I., Spinelli, F., Nicola, S., Fernandez, J. A., Stanghellini, C., Marcelis, L. F. M., Orsini, F. and Gianquinto, G. (2019a). Unraveling the role of red:blue LED lights on resource use efficiency and nutritional properties of indoor grown sweet basil. *Frontiers in Plant Science* 10, 305. doi:10.3389/fpls.2019.00305.

Pennisi, G., Orsini, F., Blasioli, S., Cellini, A., Crepaldi, A., Braschi, I., Spinelli, F., Nicola, S., Fernandez, J. A., Stanghellini, C., Gianquinto, G. and Marcelis, L. F. M. (2019b). Resource use efficiency of indoor lettuce (Lactuca sativa L.) cultivation as affected by red:blue ratio provided by LED lighting. *Scientific Reports* 9(1), 14127. doi:10.1038/s41598-019-50783-z.

Pennisi, G., Orsini, F., Landolfo, M., Pistillo, A., Crepaldi, A., Nicola, S., Fernández, J. A., Marcelis, L. F. M. and Gianquinto, G. (2020a). Optimal photoperiod for indoor cultivation of leafy vegetables and herbs. *European Journal of Horticultural Science* 85(5), 329–338. doi:10.17660/eJHS.2020/85.5.4.

Pennisi, G., Pistillo, A., Orsini, F., Cellini, A., Spinelli, F., Nicola, S., Fernandez, J. A., Crepaldi, A., Gianquinto, G. and Marcelis, L. F. M. (2020b). Optimal light intensity for sustainable water and energy use in indoor cultivation of lettuce and basil under red and blue LEDs. *Scientia Horticulturae* 272, 109508. doi:10.1016/j.scienta.2020.109508.

Sanyé-Mengual, E., Specht, K., Grapsa, E., Orsini, F. and Gianquinto, G. (2019). How can innovation in urban agriculture contribute to sustainability? A characterization and evaluation study from five Western European cities. *Sustainability* 11(15), 4221. doi:10.3390/su11154221.

SharathKumar, M., Heuvelink, E. and Marcelis, L. F. M. (2020). Vertical farming: moving from genetic to environmental modification. *Trends in Plant Science* 25(8), 724–727. doi:10.1016/j.tplants.2020.05.012.

Siregar, R. R. A., Seminar, K. B., Wahjuni, S. and Santosa, E., Seminar (2022). Vertical farming perspectives in support of precision agriculture using artificial intelligence: A review. *Computers* 11(9), 135. doi:10.3390/computers11090135.

Smith, H. L., McAusland, L. and Murchie, E. H. (2017). Don't ignore the green light: Exploring diverse roles in plant processes. *Journal of Experimental Botany* 68(9), 2099–2110. doi:10.1093/jxb/erx098.

Song, S., Kusuma, P., Carvalho, S. D., Li, Y. and Folta, K. M. (2019). Manipulation of seedling traits with pulsed light in closed controlled environments. *Environmental and Experimental Botany* 166, 103803. doi:10.1016/j.envexpbot.2019.103803.

Tavan, M., Wee, B., Brodie, G., Fuentes, S., Pang, A. and Gupta, D. (2021). Optimizing sensor-based irrigation management in a soilless vertical farm for growing microgreens. *Frontiers in Sustainable Food Systems* 4, 622720. doi:10.3389/fsufs.2020.622720.

van Delden, S. H., SharathKumar, M., Butturini, M., Graamans, L. J. A., Heuvelink, E., Kacira, M., Kaiser, E., Klamer, R. S., Klerkx, L., Kootstra, G., Loeber, A., Schouten, R. E., Stanghellini, C., van Ieperen, W., Verdonk, J. C., Vialet-Chabrand, S., Woltering, E. J., van de Zedde, R., Zhang, Y. and Marcelis, L. F. M. (2021). Current status and future challenges in implementing and upscaling vertical farming systems. *Nature Food* 2(12), 944–956. doi:10.1038/s43016-021-00402-w.

Wang, Y., Zhong, B., Shafi, M., Ma, J., Guo, J., Wu, J., Ye, Z., Liu, D. and Jin, H. (2019). Effects of biochar on growth, and heavy metals accumulation of moso bamboo (Phyllostachy pubescens), soil physical properties, and heavy metals solubility in soil. *Chemosphere* 219, 510–516. doi:10.1016/j.chemosphere.2018.11.159.

Weidner, T., Yang, A. and Hamm, M. W. (2021). Energy optimisation of plant factories and greenhouses for different climatic conditions. *Energy Conversion and Management* 243, 114336. doi:10.1016/j.enconman.2021.114336.

Werner, S. (2017). International review of district heating and cooling. *Energy* 137, 617–631. doi:10.1016/j.energy.2017.04.045.

World Commission on Environment and Development (WCED) (Ed.). (1987). *Our Common Future*. Oxford, New York: Oxford University Press.

Yan, Z., He, D., Niu, G., Zhou, Q. and Qu, Y. (2020). Growth, nutritional quality, and energy use efficiency in two lettuce cultivars as influenced by white plus red versus red plus blue LEDs. *International Journal of Agricultural and Biological Engineering* 13(2), 33–40. doi:10.25165/j.ijabe.20201302.5135.

Yokoyama, R. (2019). Energy consumption and heat sources. In *Environmental Disruption and Clues to Agricultural Innovation*, Anpo, M., Fukuda, H. and Wada, T. (Eds). Amsterdam, The Netherlands: Elsevier, p. 177–184. doi:10.1016/B978-0-12-813973- 8.00016-6.

Zhang, Y. and Kacira, M. (2020). Enhancing resource use efficiency in plant factory. *Acta Horticulturae* (1271), 307–314. doi:10.17660/ActaHortic.2020.1271.42.

Zhen, S. and Bugbee, B. (2020). Far-red photons have equivalent efficiency to traditional photosynthetic photons: implications for redefining photosynthetically active radiation. *Plant, Cell and Environment* 43(5), 1259–1272. doi:10.1111/pce.13730.

Zhen, S. and van Iersel, M. W. (2017). Far-red light is needed for efficient photochemistry and photosynthesis. *Journal of Plant Physiology* 209, 115–122. doi:10.1016/j.jplph.2016.12.004.

# Chapter 7

## Energy consumption in plant factories with artificial lighting: concepts and pathways toward a sustainable future

*Michael Eaton and Neil Mattson, Cornell University, USA*



## 1 Introduction

Energy intensity of crop production in plant factories with artificial lighting (PFALs) is regarded as a key obstacle for the indoor agriculture industry to overcome (Avgoustaki and Xydis, 2020; Eaton et al., 2023; Kozai, 2018; Lubna et al., 2022). However, production facilities can minimize their total energy consumption, energy costs, and carbon impact by implementing energy-conservation measures (ECMs) (Eaton et al., 2023; Engler and Krarti, 2021; Graamans et al., 2020; Wang and Iddio, 2022) such as installing high-efficacy lighting fixtures, supplementing carbon dioxide to the production space to aid in photosynthesis, and utilizing outdoor air for cooling. Building energy models can inform designers, investors, and managers about the expected economic performance and environmental impacts of a proposed facility or design. This chapter reviews the published literature related to PFAL energy consumption, provides details on the concepts around energy-consuming systems, and gives example case studies with scenario analysis to illustrate the impacts of making energy-efficient decisions in PFALs.

http://dx.doi.org/10.19103/AS.2023.0126.32

## 2 The state of knowledge on plant factories with artificial lighting energy consumption

PFAL owners do not normally publish data regarding the energy intensity of production (EIP), though this would significantly assist in benchmarking industry energy demands, estimating the value of proposed efficiency measures (Eaton et al., 2023), and creating better models of economic performance and environmental sustainability (Nicholson et al., 2020). Due to a scarcity of industrial energy consumption data, most of the datapoints in this section come from small-scale experiments using growth chambers or shipping container farms and modeling studies. Growth chamber experimental data are quite valuable for the purpose of characterizing crop productivity under a variety of climatic conditions and providing information on equipment energy consumption. However, there is still a question of how well this information can scale to reflect commercial-scale plant production. Models can help bridge this information gap and provide insight into the biggest opportunities for improving efficiency and their economic value.

### *2.1 Scope*

While the term *energy modeling* can describe a variety of modeling scopes, this review focuses on PFAL models which use building energy modeling (BEM) techniques, a sub-domain of building performance simulation (Hensen and Lamberts, 2019). BEM techniques typically consider dynamic building models that simulate heat and mass flow throughout an indoor environment as well as the behavior and performance of energy-consuming equipment such as lighting fixtures, air-conditioning equipment, fans, and pumps. Though PFALs are crop production facilities, most equipment employed to control the environmental parameters in PFALs are common to other types of commercial facilities like warehouses and office buildings. This makes the scope of BEM convenient for modeling the PFAL environment because popular BEM software like EnergyPlus™ ('EnergyPlus Version 9.5.', 2021), IES ('IES Whole Building Simulation Virtual Environment', n.d.), and DOE-2 can manage and simulate complex building systems. Moreover, there are vast quantities of software documentation and large libraries of thermodynamic properties of construction materials, heating, ventilation, and air-conditioning (HVAC) configurations, and energy-efficiency measures in complementary software like OpenStudio (Guglielmetti et al., 2011) and DesignBuilder ('DesignBuilder', n.d.). Limiting our review to this subset of modeling techniques reduces the variety of assumptions to compare, aligns with our proposed focus of examining annual energy consumption and productivity data, and increases the availability of model component documentation.

Before summarizing the scientific literature, we will give a brief outline of the BEM process.

### 2.1.1 Building energy modeling process

Typical BEM input procedure involves defining the building geometry and makeup, providing local weather data, specifying HVAC system equipment and performance characteristics, defining control strategies, defining schedules of room equipment and occupants, and defining central energy delivery to the building (i.e. natural gas boilers, photovoltaic panels, and district energy systems). After the building model is completely specified, simulation parameters such as numerical technique, time-step size, and simulation period are selected ('EnergyPlus Version 9.5.', 2021). During simulation, energy and mass conservation equations are solved at each time step to quantify the evolution of temperature and humidity of room air, building surface temperatures, and fluid systems like hot/chilled water-loops. If one is modeling an existing building, the resulting simulation outputs should be benchmarked against the measured performance and the model should be updated until there is satisfactory agreement with the measured data (L. Capehart et al., 2016). If one is modeling a proposed building construction, the input data and modeling assumptions should adhere to best practices which consider interactivity between ECMs (L. Capehart et al., 2016).

Notes on units. Engineers and buildings scientists typically consider the key performance indicator for energy efficiency in buildings as the energy use index, EUI (MJ $m^{-2}$) which is the annual energy consumption per unit of building area (L. Capehart et al., 2016). However, growers and PFAL managers are more likely to consider the EIP (MJ $kg^{-1}$), i.e. energy use per kilogram of harvested crop, as the most appropriate performance indicator. Both EUI and EIP can be further parsed out into the energy contribution of electricity and fossil fuels like natural gas.

## 2.2 Existing plant factories with artificial lighting energy models

Making a reasonable comparison of predicted energy consumption in PFALs in the literature is somewhat challenging due to the variety of building configurations, target crops, growth parameters, and assumptions built into models. However, we can identify differences in modeling inputs, explain differences in simulation outputs, highlight themes, and compare EIP. The majority of PFAL modeling studies focus on lettuce production (as the most common crop grown in PFALs), and this review covers four studies which model PFALs growing lettuce or leafy greens. Table 1 summarizes assumptions and differences among modeling studies for three components: crop parameters,

**Table 1** Input summary of PFAL energy modeling studies sorted by crop, lighting, HVAC system descriptions, and assumptions

| | | **General information** | | | |
|---|---|---|---|---|---|
| | Facility type | Location(s) | Building footprint ($m^2$) | Layers | Production Area multiplier |
| Eaton et al. (2023) | PF | {NYC, LA, SEA, ATL, CHI} | 4,095 | 10 | 5x |
| Zhang and Kacira (2020) | PF | {RIY, ABU, PHX, MIA, SEA, DU} | 1,891 | {2,5,11} | {0.72 - 7.9} x |
| Graamans et al. (2018) | PF | {SWE, UAE} | 10,000 | 6 | 5x |
| Harbick and Albright (2016) | PF | {ATL, HLA, MIN, PHX} | 1,712 | 10 | 10x |
| | | **Crop information** | | | |
| | Crop type | $CO_2$ level (ppm) | Evapotranspiration rate | Cycle Time (days) | Yield Fresh Weight (kg $m^{-2}$ $year^{-1}$) |
| Eaton et al. (2023) | Lettuce | {400, 839} | {56-58} ml $head^{-1}$ $day^{-1}$ | 35 | 103.6 |
| Zhang and Kacira (2020) | Lettuce | 400 | {21.5, 43, 64.5} ml $head^{-1}$ $day^{-1}$ | 38 | {95, 81} |
| Graamans et al. (2018) | Lettuce | 1200 | Not reported | 30 | {71.4} |
| Harbick and Albright (2016) | Lettuce | 400 | 100 ml $head^{-1}$ $day^{-1}$ | Not reported | Not reported |
| | | **Lighting Systems** | | | |
| | DLI (mol $m^{-2}$ $day^{-1}$) | PAR Lighting Efficacy (umol $J^{-1}$) | Photoperiod (hrs) | PPFD (umol $s^{-1}$ $m^{-2}$) | Installed Lighting Capacity (W $m^{-2}$) |
| Eaton et al. (2023) | {17,14} | {2.0, 2.5, 3.0} | 24 | {196.8, 162} | {55-98.6} |

| | | | | | |
|---|---|---|---|---|---|
| Zhang and Kacira (2020) | {13-15} | {2.0, 2.5, 3.0} | 16 | {191, 260} | {63.6-121.5} |
| Graamans et al. (2018) | 28.8 | 2.6 | 16 | 500 | 192.3 |
| Harbick and Albright (2016) | 17 | 1.7 | 24 | 196.8 | 115.7 |
| | | **HVAC systems** | | | |
| | System type | Economizer Scenarios | Nominal Cooling/Heating COP | Heating/Cooling Setpoint (°C) | Min/Max Humidity |
| Eaton et al. (2023) | Unitary HVAC + Dehumidifier | Yes | 4.0/80% | 21/24 | None/85% |
| Zhang and Kacira (2020) | Unitary HVAC | Yes | 3.45/80% | 19/24 | None/None |
| Graamans et al. (2018) | Air-cooled Chiller, FCU | No | 5.0/5.0 | 24/30 | 60%/90 % |
| Harbick and Albright (2016) | Air-cooled Chiller + Gas boiler | Yes | 4.0/80% | 19/24 | None / Not reported |

lighting systems, and the HVAC system. Some studies investigated multiple building configurations, giving a wide range of input parameter values; curly brackets are used to indicate a range of parameter values. Figure 1 shows the comparison of EIP for these recent modeling studies.

## 2.3 Comparison of existing plant factories with artificial lighting models

We will briefly summarize the findings and discuss the connection between predicted EIP and model inputs in selected studies. Modeling inputs significantly affect the predicted energy consumption and can impact the final comparison. For example, crop productivity in each published study is dependent on environmental parameters, growth model, planting density, growth cycle time, and target weight. Differences in lighting systems across studies can result in a large range of outcomes as lighting is the largest energy end-use in PFALs. Differences in the type and configuration of HVAC equipment, control strategy, setpoints, and evapotranspiration rate will impact the HVAC energy consumption.

Harbick and Albright modeled a ten-layer PFAL growing lettuce with a total canopy area of 1 acre (4046 $m^2$) (Harbick and Albright, 2016). The authors did not report yield, and so equivalent yield to Eaton et al. is assumed due to the identical lighting target (Harbick and Albright, 2016). This study targeted a daily light integral (DLI) of 17 mol $m^{-2}$ $day^{-1}$, which was found to be optimal for some varieties of lettuce, in that it drives photosynthesis without inducing a physiological disorder known as tipburn (Both, 1995). The lighting system modeled in their work assumed a photosynthetically active radiation (PAR) efficacy of 1.7 μmol $J^{-1}$, resulting in high electrical input for lighting and more waste-heat. Harbick et al assumed a large evapotranspiration rate for lettuce at 100 ml $head^{-1}$ $day^{-1}$. Temperature and humidity controls were modeled using a cool/reheat method, an air-cooled chiller, and a boiler supplying hot water for reheating. The high evapotranspiration rate combined with the simultaneous heating and cooling resulted in a high energy input for conditioning the space. Of the studies reviewed here, Harbick and Albright estimated the highest EIP of about 23 kWh $kg^{-1}$ freshweight (FW) for PFALs growing lettuce.

Graamans et al. (2018) calculated lettuce yields associated with meeting a 30-day crop growth cycle, resulting in an annual yield rate of 71.4 kg $m^{-2}$ FW. They targeted a DLI of 28.8 mol $m^{-2}$ $day^{-1}$ while maintaining a $CO_2$ concentration of 1200 ppm, assuming a lighting efficacy of about 2.6 μmol $J^{-1}$. This high level of lighting and $CO_2$ can explain the fast 30-day crop cycle, though pushing lettuce growth with such high levels of lighting would require a high volume of airflow across the canopy to avoid tipburn (Both, 1995; Frantz et al., 2004). They modeled evapotranspiration via a method described in Graamans et al. (2017),

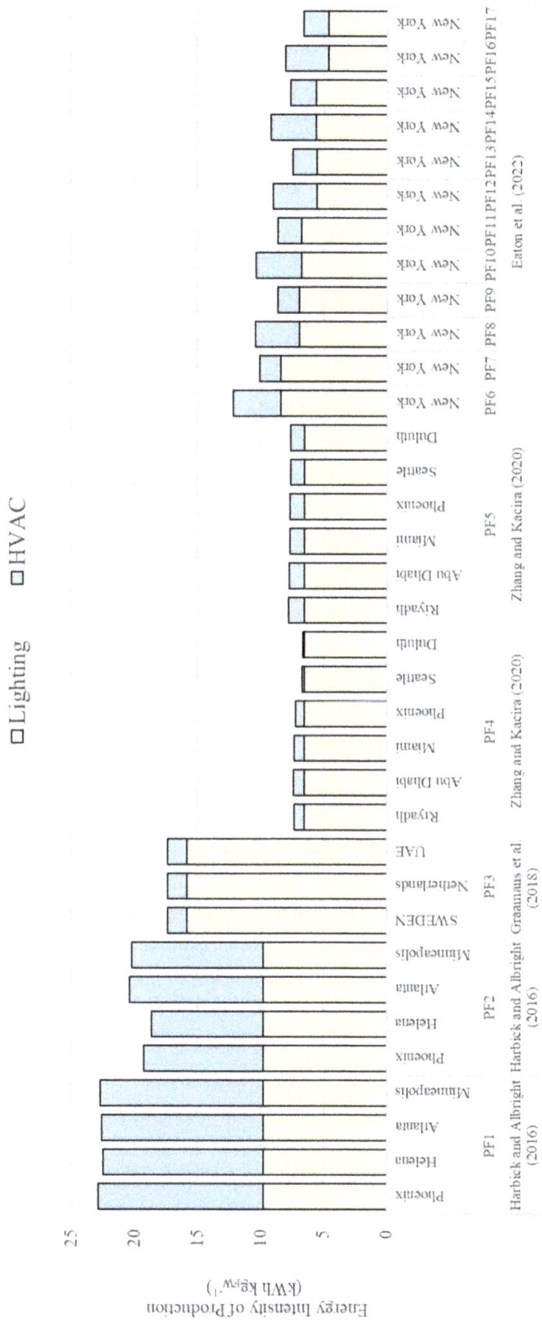


**Figure 1** EIP of PFALs in selected energy modeling studies from the literature.

though the transpiration rate was not reported directly. This work modeled a high-efficiency air-cooled chiller, fan-coil unit forced air HVAC system, with a dedicated heat pump for removing waste heat from lighting, resulting in a low proportion of facility energy used for HVAC. Their EIP for each scenario was reported to be around 17 kWh $kg^{-1}$ FW.

Zhang and Kacira generated a significant number of scenarios with parameter analysis around light efficacy {2.0, 2.5, 3.0} μmol $J^{-1}$, two DLI targets {13, 15} mol $m^{-2}$ $day^{-1}$, multiple numbers of layers, HVAC configuration, and evapotranspiration rate {21.5, 43, 64.5} mL $head^{-1}$ $day^{-1}$ (Zhang and Kacira, 2020). They calculated a 38-day growth cycle to reach a target weight of 150 g $head^{-1}$ (Zhang and Kacira, 2020). Their productivity estimates ranged between 81 and 95 kg $m^{-2}$. They modeled a packaged unitary HVAC with a coefficient of performance (COP) of 3.45, serving a production space of about ½ acre with an EIP between 6.7 and 7.8 kWh $kg^{-1}$ FW. While this study modeled scenarios with three evapotranspiration rates, humidity was not controlled in the production space, which resulted in a much lower HVAC energy consumption than if humidity was controlled. At a lower transpiration rate, the cooling load increased slightly due to less sensible heat being dissipated by moisture emitted through transpiration. At a higher transpiration rate, the sensible cooling load dropped dramatically due to more heat being converted to latent heat, while heating demand increased. Without humidity controls, the energy balance of the zone shifted from requiring cooling to requiring heat due to an increase in energy dissipated by evapotranspiration. Notably, this study compared predicted energy estimates to measured data in a growth chamber, finding good agreement and reporting the experimental energy consumption data.

Eaton et al. (2023) modeled PFALs growing lettuce in five U.S. locations, varying lighting efficacy {2.0, 2.5, 3.0} μmol $J^{-1}$, HVAC configuration, and $CO_2$ concentration to quantify resource consumption and carbon footprint associated with production (Eaton et al., 2023). They calculated a 35-day growth cycle driven by a DLI of 17 mol $m^{-2}$ $day^{-1}$ (Eaton et al., 2023) while respacing plants according to the Cornell University spacing strategy (Brechner et al., 1996), calculating an annual yield rate of 103.6 kg $m^{-2}$ of canopy area. This work calculated the evapotranspiration of lettuce via the method described in Graamans et al. (2017) as 58 mL $head^{-1}$ $day^{-1}$ which matched the observed data in an unpublished study. The HVAC system contained an air-handling unit (AHU) in combination with a zone dehumidifier. The EIP in this study ranged between 6.6 and 12 kWh $kg^{-1}$ FW depending on the scenario inputs.

## 2.4 Notable publications not listed in Table 1

The researchers in Wang and Iddio (2022) audited and modeled the energy usage of an existing PFAL in Wyoming growing lettuce, tomato, and cucumber and recommended ECMs. Energy saving calculations were supported by the data measured at the facility. Interestingly, this PFAL uses a combination of crossflow ventilation, evaporative cooling, and heating to maintain zone setpoints. This article published energy consumption data which serves as a useful benchmark for one facility in the Wyoming location. This article was not included in the summary table due to not reporting crop yield or productivity.

The researchers in Weidner et al. (2021) modeled and compared the EIP of a mixed variety of crops grown in open and closed greenhouses and a PFAL with growth parameters individually optimized for each growing facility and location. This article was not included in the summary table due to modeling a mixed crop rather than only lettuce and due to utilizing custom software rather than BEM software EnergyPlus.

The researchers in Benis et al. (2017) created a framework for modeling multiple types of urban agriculture facilities including rooftop greenhouses, and PFAL-type production systems, providing case studies to show the utility of the tool for modeling building integrated agriculture for growing tomatoes in Portugal. Though the software package is quite powerful, this article was not included in the summary table because there were flawed assumptions used in the case studies which resulted in erroneous calculations for energy consumption. The error occurred due to assuming a DLI target of 20 mol $m^{-2}$ $day^{-1}$ which could be supplied over 13 h, using only 47 W $m^{-2}$ of installed lighting power, resulting in a lighting efficacy > 9.0 μmol $J^{-1}$, which is beyond physical limits (Pattison et al., 2018).

The researchers in Liebman-Pelaez et al. (2021) modeled the energy consumption of a shipping container farm growing lettuce, using data measured in a test facility. The aim of this study was to test a dynamic model for an installed heat pump COP and to calibrate the model with direct measurements. The researchers found good agreement between the dynamic COP model and measured energy data over a crop cycle. This study was omitted from Table 1 because the annual energy consumption and total yield were not reported.

## 3 Plant factories with artificial lighting system components and efficiency case studies

The schematic for a PFAL is shown in Fig. 2, with the six essential building components to production as described in Kozai and Niu (2016) which include: (1) thermally insulated barrier to the external environment, (2) an air-conditioning and ventilation system, (3) lighting devices which can be integrated with tiered growing shelves, (4) a unit for supplying $CO_2$ to the production space, (5) a system to supply water and nutrients to crops, and (6) an environmental control unit. While each component in Fig. 2 is essential for

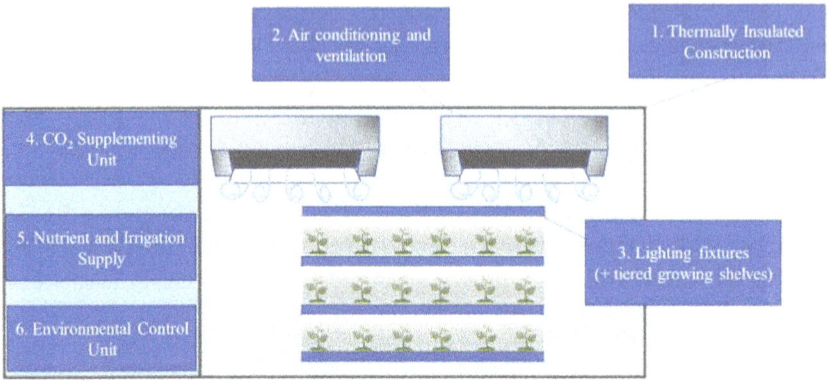


**Figure 2** PFAL with six essential components. This figure was adapted from Kozai's conception of closed plant production system (CPPS) elements. *Source*: Adapted from: Kozai and Niu (2016).

PFALs, the lighting devices, air-conditioning system, and environmental control units are most consequential for energy consumption.

### *3.1 Building thermal envelope*

The building thermal envelope refers to the total insulative value of construction layers in the building; improving the thermal envelope can reduce heat transfer between internal zones of the building and the external environment. The heat flow through the building envelope is governed by conduction, convection, and radiation heat transfer. Researchers have found that through intelligent building façade design (i.e. optimizing architectural design of the external faces of the building), one can help reduce the energy consumption of PFALs by around 10% (Graamans et al., 2020). Heating and cooling loads can also be affected the by the infiltration rate of unconditioned outdoor air into the production space. While reducing heat transfer with the environment is important, the primary concern for PFAL HVAC systems is the large internal heat loads from lighting devices and its partial conversion to latent heat by crop evapotranspiration (Graamans et al., 2018).

#### *3.1.1 Case study: thermal envelope*

To illustrate relative magnitudes of internal heat gains and environmental heat gains, we present an example considering a PFAL of rectangular building construction with a length, width, and height of 20, 10, and 5 m, respectively. We assume there is a temperature difference of 20°C between the indoor and outdoor environment for these calculations and present two scenarios: scenarios A and B which correspond to better and worse building thermal envelopes. Scenario A contains double the insulative value of B and reduces the air infiltration by 50%.

### 3.1.1.1 Conduction and convection gains

Though the total heat transfer to a building space is a complex function of conduction, convection, and incoming radiation, for the purpose of this illustration, the overall heat transfer rate for the building can be approximated using Equation (1).

$$\dot{Q}_{total} = \sum_{i}^{i=n} \dot{Q}_{surface_i} = \sum_{i}^{i=n} U_i A_i \Delta T \quad (1)$$

where $\dot{Q}_{total}$ is the total heat loss or gain rate to the indoor environment (W), the subscript *i* denotes the association with the '$i_{th}$ surface' for *n* total surfaces, $\dot{Q}_{surface}$ is the heat transfer rate through the building surface (W), *U* is the heat transfer coefficient (W $m^{-2}$ $K^{-1}$), *A* is the surface area ($m^2$), and $\Delta T$ is the temperature difference between indoor environment and exterior (K or °C). For the surfaces considered in our rectangular PFAL, the heat transfer rate calculations are shown in Table 2 for scenario A.

One should note that the temperature difference across the floor is negative due to the assumption that the ground temperature is lower than the temperature of the production space.

### 3.1.1.2 Infiltration

The heat gain rate from the infiltration of outside air into the building can be calculated using Equation (2).

$$\dot{Q}_{infiltration} = ACH \bullet V \bullet C_1 \bullet \rho \bullet c \bullet \Delta T \quad (2)$$

where $\dot{Q}_{infiltration}$ is the heat gain rate from infiltration air (W), ACH is the air changes per hour, *V* is the building volume ($m^3$), $C_1$ is the conversion factor from $h^{-1}$ to $s^{-1}$ ($3600^{-1}$ s $h^{-1}$), $\rho$ is the density of air = 1.225 (kg $m^{-3}$), and *c* is the heat capacity of air = 1005 (J $kg^{-1}$). Table 3 summarizes the input and calculated values for infiltration heat gain rate. The values used for calculations are shown in Table 3.

**Table 2** Summary of case study calculations for heat loss through walls. UA is the product of the heat transfer coefficient 'U' and surface area 'A'.

| | Area ($m^2$) | $U$ ($W\ m^{-2}\ K^{-1}$) (A, B) |
|---|---|---|
| Wall 1 | 50 | 0.43, 0.215 |
| Wall 2 | 100 | 0.43, 0.215 |
| Wall 3 | 50 | 0.43, 0.215 |
| Wall 4 | 100 | 0.43, 0.215 |
| Roof | 200 | 0.43, 0.215 |
| Floor | 200 | 0.18, 0.09 |
| Overall UA except floor ($W\ K^{-1}$) | | 215, 107.5 |
| Overall UA floor ($W\ K^{-1}$) | | 36, 18 |
| Temperature difference (K) | | 20 |
| Temperature difference floor (K) | | -15 |
| Total heat gain rate through building (kW) | | 3.76, 1.68 |

Calculations shown for scenarios A and B.

#### 3.1.1.3 Internal heat gains

We can calculate the lighting heat gains by assuming 5 tiers of growing shelves, occupying 70% of the building floor area, with an installed lighting power density (LPD) of 98 $W\ m^{-2}$ of the canopy area. Multiplying these values by the building area, the total heat gain rate from lighting amounts to 68 600 W. Figure 3 shows the relative magnitudes of conduction/convection and infiltration loads for scenarios A and B compared to the internal lighting load. Though heat transfer from the external environment in scenario A is twice that in scenario B, the internal lighting load is still six and a half times greater than the sum of the other heat gain rates in scenario A and 13 times greater than those in scenario B.

It is useful to note that if outdoor air is cooler than the indoor environment and cooling is required, higher infiltration and heat loss through the walls could be beneficial for meeting the cooling load; though, this would result in $CO_2$ loss, increased risk of dust and pests, and larger external disturbances for the control systems to mitigate (Kozai, 2022). While a tight thermal envelope is the first defense against external disturbances, internal loads such as lighting and crop evapotranspiration will most likely have the biggest impact on the overall energy balance.

### 3.2 Heating, ventilation and air-conditioning

After lighting, the largest use of energy is associated with providing HVAC to the crop production space. Lighting fixtures introduce large quantities of waste

**Table 3** Parameter values for infiltration calculations in the case study.

| Property | Value |
|---|---|
| Air changes per hour | 1, 0.5 |
| Volume of building ($m^3$) | 1000 |
| Mass flow rate (kg $s^{-1}$) | 0.278 |
| Density of air (kg $m^{-3}$) | 1.225 |
| Heat capacity of air (J $kg^{-1}$ $K^{-1}$) | 1005 |
| Temperature Difference (K) | 20 |
| Infiltration heat gain rate (kW) | 6.84, 3.42 |

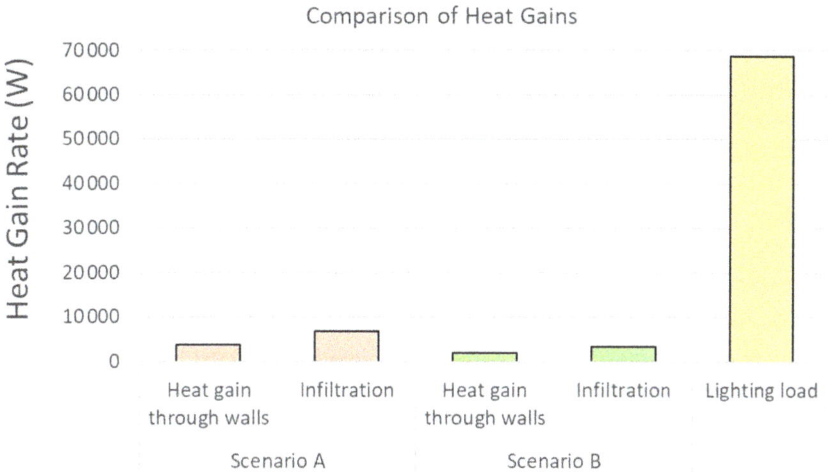


**Figure 3** Heat gain rates (W) of internal versus external PFAL loads under two scenarios.

heat which must be removed by the HVAC system to maintain acceptable temperatures, though crop transpiration can introduce some degree of evaporative cooling lowering the sensible cooling load. Dehumidification is another large component of the HVAC energy demand in PFALs and can make up more than 50% of the cooling load (Graamans et al., 2018; Harbick and Albright, 2016). The performance of air-conditioning equipment can be quantified using the COP, defined as the ratio of the total heat transferred to the total input energy [Equation (3)].

$$COP = Q_{transferred} / E_{input} \tag{3}$$

The COP of equipment in a PFAL environment will differ from designed performance based on operating conditions and compressor part-load ratio. At favorable outdoor conditions, the COP can range between 8 and 12 according to Kozai and Niu (2016), though at warmer outdoor conditions, the COP can be considerably lower. Purchasing high-performance equipment that is optimized for conditions in a PFAL environment will reduce energy consumption and costs throughout the useful life of the equipment. An AHU (Fig. 4a) will contain equipment to move air, dehumidify, and add and remove sensible heat before resupplying to the production space.

### 3.2.1 Case study: air-side economizing

Dehumidification can be accomplished using mechanical cooling, a desiccant dehumidifier, or simply replenishing air in the production space with fresh outdoor air if conditions allow. Air-side economizing refers to the practice of introducing outdoor air when it is advantageous (i.e. outdoor air is dry/cool) rather than recirculating air from the production zone through the central air-handling system. This practice can reduce the amount of conditioning required to meet the cooling loads. A case study is presented to show the air-handling processes with and without air-side economizing for a given outdoor air condition. Figure 4 shows a process diagram (4i) and accompanying psychrometric diagram (4ii) for the two cases: no economizing (4a) and with economizing (4b). For this example, we take outdoor air temperature to be 16°C and relative humidity as 45% (Figure 4ii, point 0) and air returned from the production space to be at 24°C, with 80% relative humidity. To balance the heat and moisture gain of the room, we supply air at 22°C, with 50% relative humidity.

In a closed, constant air-volume air handling system (Fig. 4a), air returned from the production space is passed over cooling coils (Fig. 4a, 4→2). Only part of the air stream is near enough to the cooling coil to reach the point of condensation (Fig. 4a, 4→1), whereas part of the airstream bypasses the coil, remaining unaffected until the two airstreams recombine at point 2 (Fig. 4a, 1→2). To achieve a target humidity ratio, the air temperature may need to drop significantly, which would affect the building energy balance if resupplied directly back to the space. To prevent temperature in the production space from dropping below a setpoint, the air stream is reheated to appropriate supply conditions using heating coils (Fig. 4a, 2→3). The supplied air then picks up heat and moisture from lighting devices and crop evapotranspiration and finally restarts the conditioning cycle (Fig. 4a, 3→4).

Using an air-side economizer, air returned from the production space (Fig. 4b, point 3) can be blended with outdoor air (Fig. 4b, point 0) to obtain mixed air conditions (Fig. 4b, point 1). The mixing ratio can be controlled with an air damper to minimize the temperature difference or enthalpy difference between mixed air and the required supply air conditions. The mixed air is then heated to meet the required supply air conditions (Fig. 4b, 2→3).

The main difference between the two processes is the energy required to condition the return air to supply conditions. Without the economizer, the cooling process (Fig. 4a, points 4→2) is achieved through the use of a vapor-compression cycle using a compressor and fans for the evaporator and condenser. Whereas an air-side economizer allows the air to reach dry and cool conditions (Fig. 4b, point 1) simply by mixing. After introducing air-side economizing, previous studies have found between 20% and 70% HVAC energy savings (Eaton et al., 2023; Harbick and Albright, 2016; Zhang and Kacira, 2020). Though air-side economizing can be very useful in areas with favorable outdoor conditions, there is a trade-off between energy and water efficiency when introducing outdoor air according to Eaton et al. (2023).

### 3.3 Supplementing carbon dioxide

Supplementing carbon dioxide in the production space increases the carbon dioxide concentration in leaves, ultimately increasing the photosynthetic efficiency and the amount of carbon fixed (Drake et al., 1997). A PFAL may take advantage of carbon dioxide enrichment by reducing the amount of lighting required to achieve a desired growth rate. Figure 5 shows combinations of DLI and $CO_2$ concentration that result in equivalent growth of {13, 15, 17} mol m$^{-2}$ day$^{-1}$ at ambient $CO_2$ levels according to Both (1998).

### 3.3.1 Case study: $CO_2$ supplementation system

Maintaining high concentrations of $CO_2$ will also affect the LPD and annual energy usage (AEU) required to achieve the desired growth condition. Three baseline scenarios are considered in Table 4 showing how elevated levels of $CO_2$ will impact LPD, AEU, and annual energy savings relative to the baseline scenario, that is 400 ppm $CO_2$, with 13, 15, or 17 mol m$^{-2}$ day$^{-1}$ of light.

Relative to the baseline scenario (400 ppm, 17 mol m$^{-2}$ day$^{-1}$), increasing $CO_2$ concentration in the production area can reduce the total energy consumption associated with lighting throughout the year significantly. In addition to energy and operational cost savings, boosting $CO_2$ concentration can also reduce the amount of installed lighting capacity required to obtain a desired growth rate, saving in upfront investment costs.

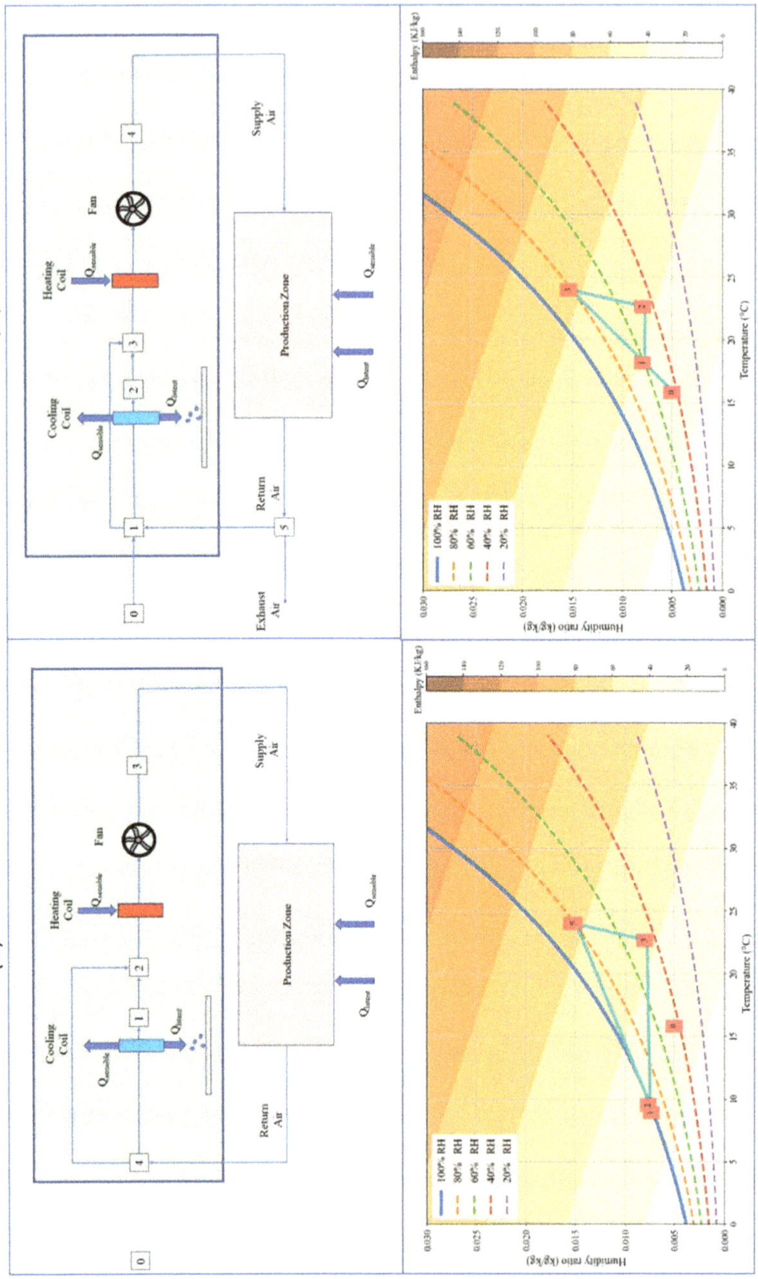


**Figure 4** (i) PFAL HVAC system process diagram. (ii) Psychrometric chart showing air-conditioning process steps of the HVAC system. (a) Scenario without economizing. (b) Scenario with economizing. For a detailed description of the process, see the text.

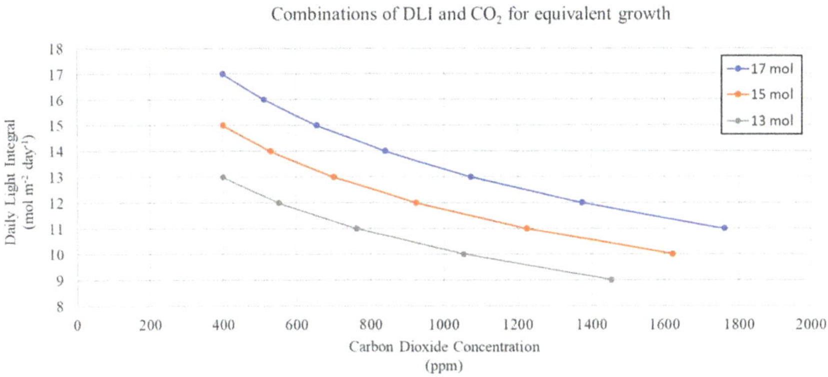


**Figure 5** Combinations of daily light integral (DLI) and $CO_2$ concentration which result in equivalent growth rates.

### *3.4 Lighting systems*

Lighting is the most important component of PFALs due to driving photosynthesis, influencing crop appearance and quality, and being the largest energy end-use in PFALs (Graamans et al., 2018; Harbick and Albright, 2016; Zhang and Kacira, 2020). Investing in high-efficacy lighting will reduce lighting energy consumption and lower the associated cooling load (Zhang and Kacira, 2020). Though, beyond the efficacy of lighting fixtures themselves, overall light use efficiency can be improved at multiple points between input electrical energy and useful light absorbed by plants (Kozai and Niu, 2016). This section provides a brief discussion on improving the sustainability of artificial lighting systems, and for a more in-depth analysis, please read Elliott et al. (2020), Kusuma et al. (2020), and Pattison et al. (2016, 2018). Though this section focuses on operational energy consumption, the indirect energy costs and environmental impacts (carbon emissions, etc.) associated with production should be considered when accounting for overall lifecycle sustainability (Chen et al., 2020; Zhang et al., 2017). Figure 6 shows example values for energy flow and losses between lighting fixtures and crops, noting three main areas for improvement: increasing the PAR output of lighting devices, increasing light utilization, and improving light use efficiency of plants through optimizing environmental conditions and breeding. Apart from the light used by photosynthesis, the remaining 99% is ultimately turned into waste heat.

**Table 4** Summary of lighting impacts from $CO_2$ supplementation in terms of minimum lighting power density (LPD), annual lighting energy usage (ALEU), annual lighting energy savings (ALES) as compared to the baseline lighting efficacy, and 400 ppm $CO_2$.

| Equivalent Growth Target | Minimum PPFD (μmol m$^{-2}$day$^{-1}$) | DLI (mol m$^{-2}$ day$^{-1}$) | $CO_2$ Concentration (ppm) | Minimum Lighting Power Density (W m$^{-2}$) | | | | Annual Lighting Energy Consumption (kWh m$^{-2}$year$^{-1}$) | | | | Annual Lighting Energy Savings (kWh m$^{-2}$year$^{-1}$) | | | |
|---|---|---|---|---|---|---|---|---|---|---|---|---|---|---|---|
| | | | | Photosynthetic Photon Efficacy (μmol J$^{-1}$) | | | | Photosynthetic Photon Efficacy (μmol J$^{-1}$) | | | | Photosynthetic Photon Efficacy (μmol J$^{-1}$) | | | |
| | | | | 2.0 | 2.5 | 3.0 | 3.5 | 2.0 | 2.5 | 3.0 | 3.5 | 2.0 | 2.5 | 3.0 | 3.5 |
| 13 mol m$^{-2}$ day$^{-1}$ | 104 | 9 | 1455 | 52 | 42 | 35 | 30 | 456 | 365 | 304 | 261 | 203 | 294 | 355 | 398 |
| | 116 | 10 | 1053 | 58 | 46 | 39 | 33 | 507 | 406 | 338 | 290 | 152 | 253 | 321 | 369 |
| | 127 | 11 | 762 | 64 | 51 | 42 | 36 | 558 | 446 | 372 | 319 | 101 | 213 | 287 | 340 |
| | 139 | 12 | 552 | 69 | 56 | 46 | 40 | 608 | 487 | 406 | 348 | 51 | 172 | 253 | 311 |
| | 150 | 13 | 400 | 75 | 60 | 50 | 43 | 659 | 527 | 439 | 377 | 0 | 132 | 220 | 282 |
| 15 mol m$^{-2}$ day$^{-1}$ | 116 | 10 | 1621 | 58 | 46 | 39 | 33 | 507 | 406 | 338 | 290 | 253 | 355 | 422 | 471 |
| | 127 | 11 | 1225 | 64 | 51 | 42 | 36 | 558 | 446 | 372 | 319 | 203 | 314 | 389 | 442 |
| | 139 | 12 | 925 | 69 | 56 | 46 | 40 | 608 | 487 | 406 | 348 | 152 | 274 | 355 | 413 |
| | 150 | 13 | 700 | 75 | 60 | 50 | 43 | 659 | 527 | 439 | 377 | 101 | 233 | 321 | 384 |
| | 162 | 14 | 529 | 81 | 65 | 54 | 46 | 710 | 568 | 473 | 406 | 51 | 193 | 287 | 355 |
| | 174 | 15 | 400 | 87 | 69 | 58 | 50 | 760 | 608 | 507 | 435 | 0 | 152 | 253 | 326 |
| 17 mol m$^{-2}$ day$^{-1}$ | 127 | 11 | 1760 | 64 | 51 | 42 | 36 | 558 | 446 | 372 | 319 | 304 | 416 | 490 | 543 |
| | 139 | 12 | 1375 | 69 | 56 | 46 | 40 | 608 | 487 | 406 | 348 | 254 | 375 | 456 | 514 |
| | 150 | 13 | 1073 | 75 | 60 | 50 | 43 | 659 | 527 | 439 | 377 | 203 | 335 | 423 | 485 |
| | 162 | 14 | 839 | 81 | 65 | 54 | 46 | 710 | 568 | 473 | 406 | 152 | 294 | 389 | 456 |
| | 174 | 15 | 655 | 87 | 69 | 58 | 50 | 761 | 608 | 507 | 435 | 101 | 254 | 355 | 427 |
| | 185 | 16 | 512 | 93 | 74 | 62 | 53 | 811 | 649 | 541 | 464 | 51 | 213 | 321 | 398 |
| | 197 | 17 | 400 | 98 | 79 | 66 | 56 | 862 | 690 | 575 | 493 | 0 | 172 | 287 | 369 |

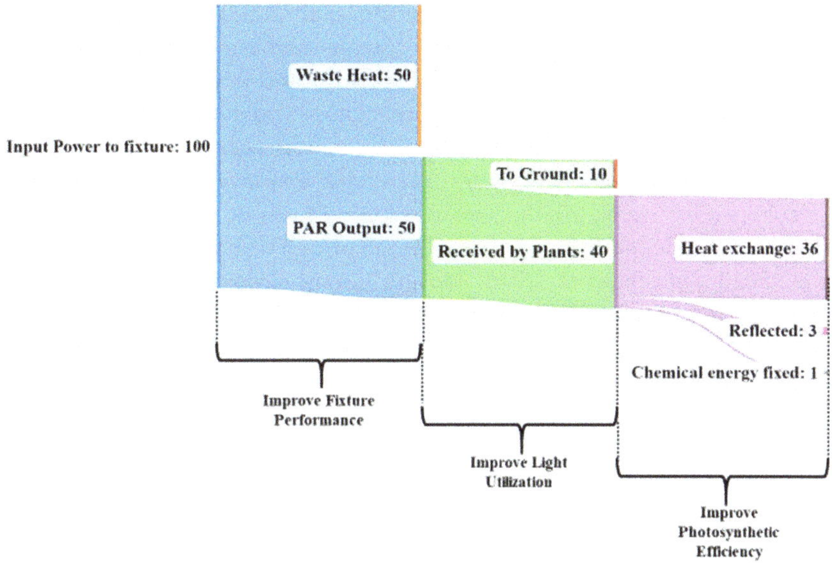


**Figure 6** Sankey diagram of light energy flow. Proportions use example values to illustrate the losses. Generated with SankeyMATIC.

### 3.4.1 Improving light fixture performance

The performance of horticultural fixtures can be characterized in terms of PAR which is the portion of the light spectrum used in photosynthesis, ranging between 400 nm and 700 nm (McCree, 1971). Some scientists point out that the definition of PAR could be extended to include wavelengths up to 750 nm, citing synergistic benefits of far-red light in photosynthesis (Zhen et al., 2021; Zhen and Bugbee, 2020). The energy efficiency of horticultural lighting fixtures can be defined by Equation (4), which relates the total electric energy input, $E_L$, to the PAR energy output of lamps, $PAR_L$.

$$\eta = PAR_L / E_L \tag{4}$$

The value of $\eta$ is around 0.25 for fluorescent lamps (Kozai and Niu, 2016), 0.38 for high-pressure sodium lamps, and 0.3–0.6 for high-efficiency LEDs (DLC Horticultural Lighting Database, 2021; Katzin et al., 2020; Kozai and Niu, 2016; Shelford and Both, 2021). While fixture energy efficiency is important, photosynthetic photon efficacy (PPE, units is μmol $J^{-1}$) is seen as a more suitable performance metric for horticultural lighting because it is defined by the ratio of useful output (number of photons) to the costly input (electrical energy) (Shelford and Both, 2021). The number of output photons is a good marker for light utility because biomass production is proportional to the accumulation of PAR photons delivered to a crop (Marcelis et al., 2006). In recent years, the horticultural lighting industry has seen a large increase in efficacy with the rise of the light-emitting diode (LED) technology and increasing demand for electric lighting in agriculture (DLC Horticultural Lighting Database, 2021; Pattison et al., 2016). While there have been gains in light fixture efficacy and large estimated energy savings for operations employing LEDs for indoor agriculture (Elliott et al., 2020), there are physical limits bounding the potential efficacy of LED depending on the wavelength of light being produced (Kusuma et al., 2020; Pattison et al., 2018). Moving toward sustainability in PFALs will largely depend on the improvement of future lighting devices and increased adoption of efficacious fixtures. This section provides an overview of horticulture lighting and is not comprehensive; for further reading on fixture efficacy and the potential of LED fixtures in horticultural applications, refer to Dutta Gupta (2017), Kusuma et al. (2020), and Pattison et al. (2018).

### 3.4.2 Improving light utilization

Not all light produced by fixtures reaches the plant canopy as it can be lost to the floor, walls, and spaces not occupied by crops. Increasing the amount of light intercepted by a plant canopy will ensure that the input energy is not wasted. This efficiency can be characterized by the utilization factor, $U_f$, which is the ratio of PAR received at the plant canopy, $PAR_P$, to PAR emitted by lamps, $PAR_L$. $U_f$ can be maximized through well-planned crop spacing regiments that reduce the empty space between plant canopy, using light reflectors appropriate to the application, having reflective walls in growth chambers, and reducing the distance between lights and plants (Kozai and Niu, 2016; van Delden et al., 2021). Additionally, adjusting light spectrum may improve light utilization through influencing the rate of leaf expansion and thereby photon interception (Legendre and van Iersel, 2021).

### 3.4.3 Improving light use efficiency of plants

The light use efficiency at the plant canopy, $LUE_P$, is defined using Equation (5) (Kozai and Niu, 2016).

$$LUE_P = f \times D / PAR_P \tag{5}$$

where $f$ is the conversion factor from dry mass to chemical energy fixed in dry mass, $D$ is the dry mass increase of plants in the PFAL, and $PAR_P$ is the PAR received at the plant canopy. Discovering the conditions which maximize $LUE_P$, photosynthetic rate, and yield has been the aim of many scientific studies. Factors such as light quantity, light quality (spectrum), photoperiod, temperature, $CO_2$ concentration, and the nutrient supply makeup can all impact photosynthetic efficiency (Caplan, 2018; Carotti et al., 2020; Legendre and van Iersel, 2021). PFALs growing lettuce were found to achieve high $U_f$ and $LUE_p$ compared to greenhouses and open field, with the highest measured value approaching the maximum theoretical limit (1.63 mol/g) (Jin et al., 2023). Improvements in these three key areas will allow PFALs to reduce the installed lighting power required to meet lighting targets and consequently reduce the HVAC system sensible cooling load.

### 3.4.4 Case study: lighting systems

Case studies are presented here calculating the energy consumption and carbon footprints of PFALs growing lettuce in New York City, New York, under 18 scenarios combining the following: two lighting efficacies {2.0, 3.0 μmol $J^{-1}$}, three light utilization factors, $U_f$, {70%, 90%, 100%} (100% is the theoretical best), and three levels of $CO_2$ {400, 655, 1073 ppm} which correspond to equivalent growth with daily lighting targets {17, 15, 13 mol $m^{-2}$ $day^{-1}$}. The PFAL is growing ten layers of lettuce in tiered shelves with 50% space utilization. A simplified HVAC system model was adopted here utilizing the EnergyPlus *IdealAirLoads* component to calculate HVAC system loads, and purchased energy is calculated post-runtime. Purchased energy is calculated using Equation (3), assuming an annual average COP of 4.0 for sensible cooling and a dehumidification efficiency of 3.412 L $kWh^{-1}$. Facility annual energy consumption under each scenario is shown in Fig. 7.

There is a great range of values in energy performance for PFALs producing lettuce; the most efficient (DLI: 13, $U_f$: 100%, PPE: 3.0) case used 58% less energy than the least efficient case (DLI: 17, $U_f$: 70%, PPE: 2.0). Combinations of lighting and $CO_2$ were chosen to have equivalent growth such that the energy consumption per unit area is proportional to the EIP. Improvements in fixture efficacy, light utilization, and photosynthetic efficiency (via $CO_2$ supplementation) all gave significant reductions in the facility energy consumption. Increasing PPE from 2.0 to 3.0 alone can result in around 30% energy savings. Improving light utilization from 70% to 90% can reduce the required installed lighting capacity, lighting energy consumption, and sensible cooling, resulting in around 20% energy savings. Supplementing $CO_2$ such that one can reduce the lighting target results in about 11% facility energy savings from baseline when increasing from 400 ppm to 655 ppm and 21% savings when increasing from 400 ppm to 1073 ppm $CO_2$.

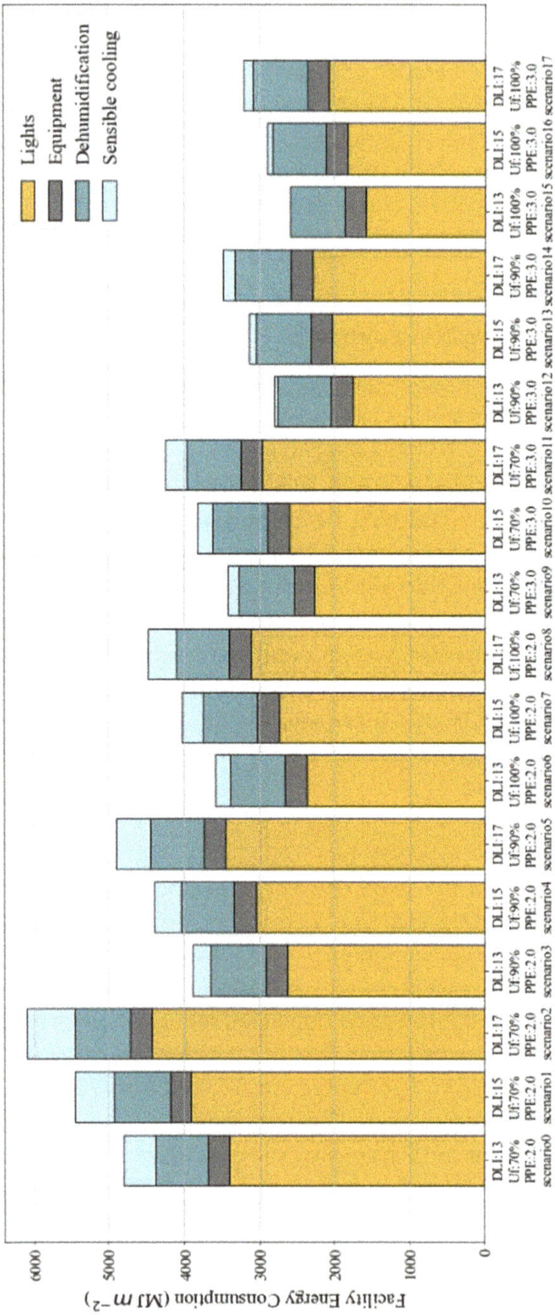


**Figure 7** Energy consumption per unit canopy area of PFALs under 18 scenarios varying PPE, CO2 level, and light Uf.

## 4 Concepts and pathways toward sustainability

Improvements of sustainability in PFALs can be made by establishing efficient practices to minimize energy consumption and waste, acquiring clean primary sources of energy through investment in renewable energy, and utilizing sustainable input materials like growing media and packaging. Concerning operational energy consumption, there are a variety of promising energy-efficiency measures related to lighting include:

- Increasing light utilization for example by using reflective walls or decreasing distance between lights and crop;
- Increasing fixture PAR efficacy by investing in high-efficacy fixtures (Both et al., 2017; Kusuma et al., 2020; Pattison et al., 2018);
- Supplementing $CO_2$ to the production space to enhance the photosynthetic rate (Both, 1998);
- Tuning light spectrum to enhance leaf expansion and consequently photon capture; and
- Optimizing environmental conditions (light, temperature, $CO_2$, and airflow) to maximize $LUE_p$ (Carotti et al., 2020).

Promising energy-efficiency measures related to HVAC systems include:

- Optimizing façade design and increasing insulation as done in Graamans et al. (2020);
- Employing an air-side economizer for cooling (Eaton et al., 2023; Harbick and Albright, 2016; Zhang and Kacira, 2020);
- Employing hot gas reheat or heat recovery in HVAC systems or utilizing passive cooling strategies (Singh et al., 2018);
- Purchasing high-efficiency air-cooled chiller systems and high-efficiency condensing boilers for heating;
- Installing variable speed drives on pumps and fans to modulate flow, minimizing the use of dampers and valves;
- Improving air distribution in production space to increase the effectiveness of ventilation and reduce excessive flow, thereby reducing fan energy consumption (Zhang and Kacira, 2022); and
- Using advanced control strategies and artificial intelligence (Cao et al., 2021; Hemming et al., 2020).

Energy supply measures are as follows:

- Purchasing renewable energy directly or renewable energy credits;
- Utilizing waste heat from power plants and employing absorption chillers to meet cooling load;
- Locating PFALs near clean sources of primary energy (Eaton et al., 2023);

- Cogeneration of heat and power utilizing waste heat and $CO_2$;
- Colocation of PFALs and greenhouses for complementary production methods; and
- Colocation of PFALs with district energy systems for utilizing waste heat to supply heat to district heating systems.

Beyond efficiency improvements at the facility level, there is saving potential through improving crop interaction with the environment through:

- reducing evapotranspiration if it can be done without sacrificing nutrient transport;
- breeding crops to achieve higher photosynthetic efficiency under low light conditions; and
- breeding crops to tolerate and thrive under expanded ranges of temperature and humidity.

## 5 Conclusion

This chapter provided a summary of PFAL energy modeling considerations, a review of models and findings in literature, and a case study to illustrate the potential for ECMs to reduce EIP. Energy modeling is a good tool to explore the increasing energy efficiency in PFALs. There is more need for real-world data to validate these models. Beyond light efficacy, there are many other opportunities to improve energy efficiency, including improving crop interception of light and a variety of HVAC system improvements. This article documents a greater need for real-world data; we also recommend improvements in HVAC efficiency and connecting renewable energy or waste energy sources that could be used.

## 6 Where to look for further information

A standard introduction to the subject is: Kozai, T. and Niu, G. 2016. *Plant Factory*. Academic Press. In addition to the publications listed in the review section, some recommended materials are as follows:

Lighting:

- Pattison et al. (2018).
- Kusuma et al. (2020).
- Elliott et al. (2020).
- Dutta Gupta (2017).

Environment:

- Niu et al. (2020).
- Kozai (2022).

Other publications:

- van Delden et al. (2021).
- Zhang and Kacira (2020).
- Gómez et al. (2019).
- Nicholson et al. (2020).

Professional associations:

- Japan Plant Factory Association (https://npoplantfactory.org/en/).
- American Society of Agricultural and Biological Engineers (www.asabe.org).

Conferences:

- NCERA-101 Committee on Controlled Environment Technology and Use https://www.controlledenvironments.org/.

Full further information citations:


- Dutta Gupta, S. (Ed.). 2017. *Light Emitting Diodes for Agriculture*. Springer Singapore, Singapore. https://doi.org/10.1007/978-981-10-5807-3.
- Elliott, C., Lee, K. and Pattison, M. 2020. Energy savings potential of SSL in agricultural applications (No. DOE/EE--2077, 1644398, 8212). https://doi.org/10.2172/1644398.
- Gómez, C., Currey, C. J., Dickson, R. W., Kim, H.-J., Hernández, R., Sabeh, N. C., Raudales, R. E., Brumfield, R. G., Laury-Shaw, A., Wilke, A. K., Lopez, R. G. and Burnett, S. E. 2019. Controlled environment food production for urban agriculture. *HortScience horts* 54(9), 1448–1458. https://doi.org/10.21273/HORTSCI14073-19.
- Kozai, T. 2022. Balances and use efficiencies of CO2, water, and energy. In: *Plant Factory Basics, Applications and Advances*. Elsevier, pp. 129–151. https://doi.org/10.1016/B978-0-323-85152-7.00003-3.
- Kusuma, P., Pattison, P. M. and Bugbee, B. 2020. From physics to fixtures to food: current and potential LED efficacy. *Horticulture Research* 7, 56. https://doi.org/10.1038/s41438-020-0283-7.
- NCERA-101 Committee on Controlled Environment Technology and Use https://www.controlledenvironments.org/.
- Nicholson, C. F., Harbick, K., Gómez, M. I. and Mattson, N. S. 2020. An economic and environmental comparison of conventional and controlled environment agriculture (CEA) supply chains for leaf lettuce to US cities. In: Aktas, E. and Bourlakis, M. (Eds). *Food Supply Chains in Cities*. Palgrave Macmillan.
- Niu, G., Kozai, T. and Sabeh, N. 2020. Physical environmental factors and their properties. In: *Plant Factory*. Elsevier, pp. 185–195. https://doi.org/10.1016/B978-0-12-816691-8.00011-X.

- Pattison, P. M., Tsao, J. Y., Brainard, G. C. and Bugbee, B. 2018. LEDs for photons, physiology and food. *Nature* 563(7732), 493–500. https://doi.org/10.1038/s41586-018-0706-x.
- van Delden, S. H., SharathKumar, M., Butturini, M., Graamans, L. J. A., Heuvelink, E., Kacira, M., Kaiser, E., Klamer, R. S., Klerkx, L., Kootstra, G., Loeber, A., Schouten, R. E., Stanghellini, C., van Ieperen, W., Verdonk, J. C., Vialet-Chabrand, S., Woltering, E. J., van de Zedde, R., Zhang, Y. and Marcelis, L. F. M. 2021. Current status and future challenges in implementing and upscaling vertical farming systems. *Nature Food* 2(12), 944–956. https://doi.org/10.1038/s43016-021-00402-w.
- Zhang, Y. and Kacira, M. 2020. Enhancing resource use efficiency in plant factory. *Acta Horticulturae* (1271), 307–314. https://doi.org/10.17660/ActaHortic.2020.1271.42.

## 7 References

Avgoustaki, D. D. and Xydis, G. 2020. How energy innovation in indoor vertical farming can improve food security, sustainability, and food safety? *Advances in Food Security and Sustainability* 5, 1–51. https://doi.org/10.1016/bs.af2s.2020.08.002.

Benis, K., Reinhart, C. and Ferrão, P. 2017. Development of a simulation-based decision support workflow for the implementation of Building-Integrated Agriculture (BIA) in urban contexts. *Journal of Cleaner Production* 147, 589–602.

Both, A.-J. 1995. *Dynamic Simulation of Supplemental Lighting for Greenhouse Hydroponic Lettuce Production*. Cornell University.

Both, A. J. 1998. *Coordinated Management of Daily PAR Integral and Carbon Dioxide for Hydropnic Lettuce Production. Crop Models in Protected Cultivation*.

Both, A.-J., Bugbee, B., Kubota, C., Lopez, R. G., Mitchell, C., Runkle, E. S. and Wallace, C. 2017. Proposed product label for electric lamps used in the plant sciences. *hortTechnology* 27(4), 544–549. https://doi.org/10.21273/HORTTECH03648-16.

Brechner, M., Both, A. J. and Staff, C. E. A. 1996. Hydroponic lettuce handbook. *Cornell Controlled Environment Agriculture*, 504–509. https://cea.cals.cornell.edu/files/2019/06/Cornell-CEA-Lettuce-Handbook-.pdf.

Cao, X., Yao, Y., Li, L., Zhang, W., An, Z., Zhang, Z., Guo, S., Xiao, L., Cao, X. and Luo, D. 2021. iGrow: A Smart Agriculture Solution to Autonomous Greenhouse Control. *Proceedings of the AAAI Conference on Artificial Intelligence*, 36(11), 11837-11845. https://doi.org/10.1609/aaai.v36i11.21440.

Capehart, L. B., Turner, C. W. and Kennedy, J. W. 2016. *Guide to Energy Management* (8th edn.). River Publishers.

Caplan, B. 2018. *Optimizing Carbon Dioxide Concentration and Daily Light Integral Combination in a Multi-level Electrically Lighted Lettuce Production System*. University of Arizona.

Carotti, L., Graamans, L., Puksic, F., Butturini, M., Meinen, E., Heuvelink, E. and Stanghellini, C. 2020. Plant factories are heating up: hunting for the best combination of light intensity, air temperature and root-zone temperature in lettuce production. *Frontiers in Plant Science* 11, 592171. https://doi.org/10.3389/fpls.2020.592171.

Chen, P., Zhu, G., Kim, H.-J., Brown, P. B. and Huang, J.-Y. 2020. Comparative life cycle assessment of aquaponics and hydroponics in the Midwestern United States. *Journal of Cleaner Production* 275, 122888. https://doi.org/10.1016/j.jclepro.2020.122888.

Crawley, D. B., Lawrie, L., Winkelmann, F. C., Buhl, W. F., Huang, Y. J., Pedersen, C. O., Strand, R. K., Liesen, R. J., Fisher, D. E., Witte, M. J. and Glazer, J. 2001. EnergyPlus: creating a new-generation building energy simulation program. *Energy and buildings* 33(4): 319-331.

*DesignBuilder.* https://www.designbuilder.co.uk/software/product-overview.

DLC Horticultural Lighting Database, D. 2021. *DLC Horticultural Lighting Database.*

Drake, B. G., Gonzàlez-Meler, M. A. and Long, S. P. 1997. MORE EFFICIENT PLANTS: A consequence of rising atmospheric $CO_2$? *Annual Review of Plant Physiology and Plant Molecular Biology* 48, 609–639. https://doi.org/10.1146/annurev.arplant.48.1.609.

Dutta Gupta, S. (Ed.) 2017. *Light Emitting Diodes for Agriculture.* Springer.

Eaton, M., Cole, M., Shelford, T. and Mattson, N. 2023. Modeling resource consumption and carbon footprint of lettuce production in plant factories. *Journal of Cleaner Production* 384, 135569.

Elliott, C., Lee, K. and Pattison, M. 2020. *Energy Savings Potential of SSL in Agricultural Applications (No. DOE/EE--2077, 1644398, 8212).* https://doi.org/10.2172/1644398.

Engler, N. and Krarti, M. 2021. Review of energy efficiency in controlled environment agriculture. *Renewable and Sustainable Energy Reviews* 141, 110786. https://doi.org/10.1016/j.rser.2021.110786.

Frantz, J. M., Ritchie, G., Cometti, N. N., Robinson, J. and Bugbee, B. 2004. Exploring the limits of crop productivity: beyond the limits of Tipburn in lettuce. *Journal of the American Society for Horticultural Science. American Society for Horticultural Science* 129(3), 331-338. https://doi.org/10.21273/JASHS.129.3.0331.

Graamans, L., Baeza, E., van den Dobbelsteen, A., Tsafaras, I. and Stanghellini, C. 2018. Plant factories versus greenhouses: comparison of resource use efficiency. *Agricultural Systems* 160, 31–43. https://doi.org/10.1016/j.agsy.2017.11.003.

Graamans, L., Tenpierik, M., van den Dobbelsteen, A. and Stanghellini, C. 2020. Plant factories: reducing energy demand at high internal heat loads through façade design. *Applied Energy* 262, 114544. https://doi.org/10.1016/j.apenergy.2020.114544.

Graamans, L., van den Dobbelsteen, A., Meinen, E. and Stanghellini, C. 2017. Plant factories; crop transpiration and energy balance. *Agricultural Systems* 153, 138–147. https://doi.org/10.1016/j.agsy.2017.01.003.

Guglielmetti, R., Macumber, D. and Long, N. 2011. *OpenStudio: An Open Source Integrated Analysis Platform*; In: Proceedings of the building simulation. Sydney, Australia; doi:Preprint NREL/CP-5500-51836.11.

Harbick, K. and Albright, L. D. 2016. Comparison of energy consumption: greenhouses and plant factories. *Acta Horticulturae* (1134), 285–292. https://doi.org/10.17660/ActaHortic.2016.1134.38.

Hemming, S., Righini, I., de Zwart, F. and Elings, A. 2020. Remote control of greenhouse cucumber production with artificial intelligence – results from the first international autonomous greenhouse challenge. *Acta Horticulturae* (1296), 987–996. https://doi.org/10.17660/ActaHortic.2020.1296.125.

Hensen, J. and Lamberts, R. (Eds.) 2019. *Building Performance Simulation for Design and Operation* (2nd edn.). Routledge.

*IES Whole Building Simulation Virtual Environment* www.iesve.com.
Jin, W., Formiga Lopez, D., Heuvelink, E. and Marcelis, L. F. M. 2023. Light use efficiency of lettuce cultivation in vertical farms compared with greenhouse and field. *Food and Energy Security* 12(1). https://doi.org/10.1002/fes3.391.
Katzin, D., van Mourik, S., Kempkes, F. and van Henten, E. J. 2020. GreenLight: an open source model for greenhouses with supplemental lighting: evaluation of heat requirements under LED and HPS lamps. *Biosystems Engineering* 194, 61–81. https://doi.org/10.1016/j.biosystemseng.2020.03.010.
Kozai, T. 2022. Balances and use efficiencies of CO2, water, and energy. In: *Plant Factory Basics, Applications and Advances*. Elsevier, pp. 129–151.
Kozai, T. 2018. *Smart Plant Factory: The Next Generation of Indoor Farms*. Springer.
Kozai, T. and Niu, G. 2016. Chapter 4. Plant factory as a resource-efficient closed plant production system. In: Kozai, T., Niu, G. and Takagaki, M. (Eds). *Plant Factory*. Academic Press, pp. 69–90.
Kusuma, P., Pattison, P. M. and Bugbee, B. 2020. From physics to fixtures to food: current and potential LED efficacy. *Horticulture Research* 7, 56. https://doi.org/10.1038/s41438-020-0283-7.
Legendre, R. and van Iersel, M. W. 2021. Supplemental Far-red light stimulates lettuce growth: disentangling morphological and physiological effects. *Plants* 10(1), 166. https://doi.org/10.3390/plants10010166.
Liebman-Pelaez, M., Kongoletos, J., Norford, L. K. and Reinhart, C. 2021. Validation of a building energy model of a hydroponic container farm and its application in urban design. *Energy and Buildings* 250, 111192. https://doi.org/10.1016/j.enbuild.2021.111192.
Lubna, F. A., Lewus, D. C., Shelford, T. J. and Both, A.-J. 2022. What you may not realize about vertical farming. *Horticulturae* 8(4), 322. https://doi.org/10.3390/horticulturae8040322.
Marcelis, L. F. M., Broekhuijsen, A. G. M., Meinen, E., Nijs, E. M. F. M. and Raaphorst, M. G. M. 2006. Quantification of the growth response to light quantity of greenhouse grown crops. *Acta Horticulturae* (711), 97–104. https://doi.org/10.17660/ActaHortic.2006.711.9.
McCree, K. J. 1971. The action spectrum, absorptance and quantum yield of photosynthesis in crop plants. *Agricultural Meteorology* 9, 191–216. https://doi.org/10.1016/0002-1571(71)90022-7.
Nicholson, C. F., Harbick, K., Gómez, M. I. and Mattson, N. S. 2020. An economic and environmental comparison of conventional and controlled environment agriculture (CEA) supply chains for leaf lettuce to US cities. In: Aktas, E. and Bourlakis, M. (Eds). *Food Supply Chains in Cities*. Springer International Publishing, pp. 33–68.
Pattison, P. M., Tsao, J. Y. and Krames, M. R. 2016. Light-emitting diode technology status and directions: opportunities for horticultural lighting. *Acta Horticulturae* (1134), 413–426. https://doi.org/10.17660/ActaHortic.2016.1134.53.
Pattison, P. M., Tsao, J. Y., Brainard, G. C. and Bugbee, B. 2018. LEDs for photons, physiology and food. *Nature* 563(7732), 493–500. https://doi.org/10.1038/s41586-018-0706-x.
Shelford, T. J. and Both, A.-J. 2021. On the technical performance characteristics of horticultural lamps. *AgriEngineering* 3(4), 716–727. https://doi.org/10.3390/agriengineering3040046.
Singh, M. C., Singh, J. P. and Singh, K. G. 2018. Development of a microclimate model for prediction of temperatures inside a naturally ventilated greenhouse under

cucumber crop in soilless media. *Computers and Electronics in Agriculture* 154, 227–238. https://doi.org/10.1016/j.compag.2018.08.044.

van Delden, S. H., SharathKumar, M., Butturini, M., Graamans, L. J. A., Heuvelink, E., Kacira, M., Kaiser, E., Klamer, R. S., Klerkx, L., Kootstra, G., Loeber, A., Schouten, R. E., Stanghellini, C., van Ieperen, W., Verdonk, J. C., Vialet-Chabrand, S., Woltering, E. J., van de Zedde, R., Zhang, Y. and Marcelis, L. F. M. 2021. Current status and future challenges in implementing and upscaling vertical farming systems. *Nature Food* 2(12), 944–956. https://doi.org/10.1038/s43016-021-00402-w.

Wang, L. and Iddio, E. 2022. Energy performance evaluation and modeling for an indoor farming facility. *Sustainable Energy Technologies and Assessments* 52, 102240. https://doi.org/10.1016/j.seta.2022.102240.

Weidner, T., Yang, A. and Hamm, M. W. 2021. Energy optimisation of plant factories and greenhouses for different climatic conditions. *Energy Conversion and Management* 243, 114336. https://doi.org/10.1016/j.enconman.2021.114336.

Zhang, H., Burr, J. and Zhao, F. 2017. A comparative life cycle assessment (LCA) of lighting technologies for greenhouse crop production. *Journal of Cleaner Production* 140, 705–713. https://doi.org/10.1016/j.jclepro.2016.01.014.

Zhang, Y. and Kacira, M. 2022. Analysis of climate uniformity in indoor plant factory system with computational fluid dynamics (CFD). *Biosystems Engineering* 220, 73–86. https://doi.org/10.1016/j.biosystemseng.2022.05.009.

Zhang, Y. and Kacira, M. 2020. Comparison of energy use efficiency of greenhouse and indoor plant factory system. *European Journal of Horticultural Science* 85(5), 310–320. https://doi.org/10.17660/eJHS.2020/85.5.2.

Zhen, S. and Bugbee, B. 2020. Far-red photons have equivalent efficiency to traditional photosynthetic photons: implications for redefining photosynthetically active radiation. *Plant, Cell and Environment* 43(5), 1259–1272. https://doi.org/10.1111/pce.13730.

Zhen, S., van Iersel, M. and Bugbee, B. 2021. Why far-red photons should be included in the definition of photosynthetic photons and the measurement of horticultural fixture efficacy. *Frontiers in Plant Science* 12, 693445. https://doi.org/10.3389/fpls.2021.693445.

# Chapter 8

# Closed plant production systems in vertical farms for a circular economy

*Yoshiaki Kitaya, Osaka Metropolitan University, Japan*



## 1 Introduction

Reducing waste generation and promoting recycling in order to reduce environmental impact has become a priority in many industrial sectors. The establishment of a resource-recycling production system with reduced waste has also become an important issue in agriculture. Conventionally, a wide range of recycling ecosystems incorporating organic substances (such as leaf and other residues from forests and livestock excrement) have been created over the years in open-field plant production, such as in grain and vegetable crops. However, the focus in current agricultural production systems is on improving productivity, and thus the environmental impact of plant production is not always considered. Consequently, eutrophication of rivers and lakes occurs due to leaching of excess fertilizer together with groundwater contamination from percolation of fertilizers through soil. The accumulation of salt in the soil due to improper irrigation and desertification from overgrazing of livestock occurs in areas where groundwater drainage is caused by excessive pumping of groundwater. To address these issues, especially in urban areas, the use of water, raw materials, and energy should be reduced, while waste generated during the production process should be recycled to minimize the use of resources and waste generation. In addition, it is necessary to move from fossil-fuel-derived materials to biologically derived biomaterials from a carbon-neutral perspective.

http://dx.doi.org/10.19103/AS.2023.0126.09

In the context of a shortage of food, biomaterials, and energy as well as rampant environmental damage, plants can play an essential role by providing food, biomaterials, and energy sources, and contributing towards environmental conservation. In this chapter, the author focuses on the material cycle within systems to achieve a sustainable balance between human activities and ecological protection. The chapter proposes the following techniques:

(1) Mushroom and plant co-culture;
(2) Co-culture of fish and plants; and
(3) Combined systems for plant production and methane fermentation in urban agriculture.

These concepts can be extended to urban ecosystems that incorporate agriculture. To achieve a closed-loop, circular material system, it is necessary to make the best use of plant functions in urban ecosystems.

## 2 Environmental control in facility-based plant production systems as semi-closed systems

Developing efficient food production systems with a lower environmental impact is important in enclosed plant production systems such as plant factories and vertical farms. Therefore, it is necessary to add more value to plant production from the perspective of the environment. Plant factories are semi-closed ecosystems that play an important role in human health and welfare, especially in urban areas. To produce environment-friendly food efficiently, semi-closed ecosystem farming is necessary to promote a material-cycling-oriented society.

In recent years, there has been a push for emission reduction, reuse, and recycling of large-scale food waste in urban areas. Developing environment-friendly food production systems is also a priority in facility-based plant production systems, such as plant factories and vertical farms. Accordingly, the value of plant production must be enhanced in the face of environmental concerns. Plant factories are semi-closed ecosystems that play a significant role in human health and welfare, particularly in urban areas. To produce food efficiently while also protecting the environment, a material-cycling society must achieve semi-closed ecosystem farming.

It is critical to have an environmental control system that optimizes plant production by minimizing energy and material input rates. This section summarizes the effects of major physical environmental factors on photosynthesis, transpiration, and growth and discusses the optimal control of environmental factors in plant production systems based on facilities.

### 2.1 Controlling light, temperature, moisture and atmospheric $CO_2$ environment

Generally, plants release carbon dioxide ($CO_2$) through respiration and absorb it via photosynthesis. The net photosynthetic rate is calculated by subtracting the $CO_2$ release rate from respiration (respiration rate) and the $CO_2$ absorption rate from photosynthesis (gross photosynthetic rate). This correlates positively with the growth rate and rises with increasing light intensity up to light saturation. Increasing plant density while improving lighting efficiency is necessary when space and energy consumption for plant cultivation are limited. The properties of a controllable light environment include light intensity, quality, and duration. These elements have significant effects on plant growth. The photosynthesis and morphology of plants are primarily affected by light intensity. Light quality considerably affects morphology whilst the photoperiod significantly affects growth rate, dormancy, and flower bud formation.

When discussing the effect of temperature on plant photosynthesis or respiration, the term 'air temperature' is frequently used. However, the body temperature of the plants (or the temperature of their leaves) directly affects their biochemical reactions and physiological functions. For instance, a photosynthetic reaction reaches its maximum at a particular temperature. In general, the optimal leaf temperature for the maximum net photosynthetic rate is between 20°C and 30°C. Leaf temperature increases as the leaves absorb more radiant energy and decreases as latent heat energy is lost to the surrounding air. Therefore, environmental factors determining energy balance (light intensity, temperature, humidity, and airflow velocity) affect leaf temperature, thereby affecting photosynthesis and growth. Accordingly, understanding the concept of combined environmental control is critical.

Although the relationship between relative humidity and net photosynthetic rate varies depending on other environmental conditions (e.g. airflow velocity), the net photosynthetic rate is generally greatest between 75% and 85% relative humidity. When the relative humidity reaches 90%, the net photosynthetic rate decreases; however, water loss due to transpiration is suppressed, while root water absorption is maintained. Therefore, leaf moisture content is considered excessive, and excessive water pressure reduces stomatal pore opening. As the rate of transpiration temporarily increases in low relative humidity and the root's water absorption rate cannot keep pace with the rate of transpiration, plants experience water stress. Subsequently, the stomatal pores close and photosynthesis is inhibited.

The net photosynthetic rate generally increases as the $CO_2$ concentration rises, up to the $CO_2$ saturation point. $CO_2$ concentration has a significant effect on photosynthesis in plants and is highly variable in response to other environmental factors, such as light intensity and temperature. The $CO_2$ saturation point rises with increasing light intensity, whilst the $CO_2$ compensation point (the $CO_2$ concentration at which the net photosynthetic rate becomes zero) decreases with increasing light intensity. As $CO_2$ concentration increases, the effect of leaf temperature on the net photosynthetic rate becomes significant, and the optimal leaf temperature increases (Yabuki, 2004). Even if the leaf temperature is raised to increase the activity of the enzymes involved in the photosynthetic reaction, the carbon supply becomes the limiting factor at low $CO_2$ concentrations. Therefore, a higher leaf temperature with an increase in the photosynthetic rate is negligible.

## 2.2 Controlling air movement as an important factor combined with other environmental elements

The exchange of $CO_2$ and water vapor between plants and the atmosphere is controlled by resistance to gas diffusion from the atmosphere to the chloroplast for $CO_2$ diffusion (photosynthesis), and from the stomatal cavity to the atmosphere for water vapor diffusion (transpiration). In general, photosynthesis and transpiration are controlled by stomatal and boundary layer resistance, as described subsequently. Several environmental factors, including air temperature, humidity, atmospheric $CO_2$, light intensity, soil moisture, soil temperature, and soil salinity have been shown to affect photosynthesis and transpiration mainly through the stomatal aperture. However, there have been fewer studies on the effects of environmental factors on photosynthesis and transpiration through the leaf boundary layer than through the stomata. The leaf boundary layer resistance is typically controlled predominantly by the airflow velocity.

In agriculture, semi-closed plant culture facilities such as environmentally controlled greenhouses and plant factories with artificial lighting are becoming increasingly popular. In facilities without adequate air circulation systems, air movement is highly restricted compared with that under field conditions. Insufficient air movement around plants increases resistance to gas diffusion in the leaf boundary layer, thus limiting photosynthesis and transpiration (Monteith and Unsworth, 1990; Jones, 1998; Yabuki, 2004), resulting in the suppression of plant growth and development. Therefore, the enhancement of gas exchange in leaves and the growth of plants depend on the appropriate control of air movement. Accordingly, this section emphasizes the importance of air movement for efficient photosynthesis and transpiration, and hence plant growth.

Restricted air movement around plants generally increases the resistance of leaf boundary layers to gas diffusion. Therefore, it inhibits photosynthesis and transpiration, thereby impairing plant growth and development. The leaf boundary layer acts as a significant barrier to $CO_2$ and water vapor transport between the atmosphere and leaf surface. This layer becomes thicker as the airflow velocity decreases, increasing resistance to gas diffusion.

When a flat leaf is placed parallel to the direction of airflow with a uniform velocity distribution, the airflow velocity on the flat leaf decreases to approximately 0 m $s^{-1}$ at the surface of the flat leaf. The area in which the airflow velocity decreases near the surface is called the surface boundary layer, and specifically on the leaves, the leaf surface boundary layer. The surface boundary layer becomes thicker rearwards from the front edge of the flat leaf. The vicinity of the leading edge is called the laminar flow area, where air flows parallel to the flat leaf surface. Subsequently, the air flows to the rear through the transition region known as the turbulent flow region, but the layer in contact with the flat leaf surface is in the laminar flow state with a remarkably low flow rate. In the case of a horizontal laminar flow with less turbulence, the turbulent diffusion velocity in the direction perpendicular to the laminar flow is limited, and the substance is hardly diffused in that direction. Specifically, the leaf boundary layer provides a relatively strong resistance against the transport of $CO_2$ and water vapor between the atmosphere and the leaf surface. This layer becomes thicker with decreased airflow velocity and increases resistance against gas diffusion.

Photosynthesis and transpiration are commonly controlled by stomatal and boundary-layer resistance. The effects of airflow velocity lower than 1 m $s^{-1}$ on the net photosynthetic (Pn) and transpiration (Tr) rates of single leaves and plant seedling canopies through the boundary layer are more significant. The Pn and Tr of leaves double as airflow velocity increases from 0.01 m $s^{-1}$ to 0.3 m $s^{-1}$ and are almost constant at an airflow velocity of 0.4–1.0 m $s^{-1}$ (Fig. 1). Increases in Pn and Tr are more significant in the plant canopy of seedlings than in the single leaf. The Pn of the plant canopy rises with increasing airflow velocities to 1.0 m $s^{-1}$ above the plant canopy (Fig. 2). Pn in the plant canopy at the airflow velocity of 1.0 m $s^{-1}$ is two times higher than at an airflow velocity of 0.1 m $s^{-1}$ above the plant canopy. The airflow velocity inside the plant canopy decreases to 30% of that above the plant canopy (Fig. 3).

The effect of the airflow velocity inside the plant canopy on the net photosynthetic rate based on leaf area varies under different leaf area indices (LAIs) and $CO_2$ conditions (Fig. 4). The Pn of the plant canopy increases with increasing airflow velocity inside the plant canopy and is saturated at 0.2 m $s^{-1}$ (0.5 m $s^{-1}$ above the plant canopy). Pn at the airflow velocity of 0.2 m $s^{-1}$ is 1.3 times that at 0.05 m $s^{-1}$ at 400 μmol $mol^{-1}$ and 800 μmol $mol^{-1}$ $CO_2$. The Pn at 800 μmol $mol^{-1}$ $CO_2$ is 1.2 times that under 400 μmol $mol^{-1}$ $CO_2$ at airflow velocity ranging from 0.1 m $s^{-1}$ to 0.8 m $s^{-1}$.

Control of air movement is therefore essential to enhance the gas exchange between plants and the ambient air, and important for promoting plant growth.

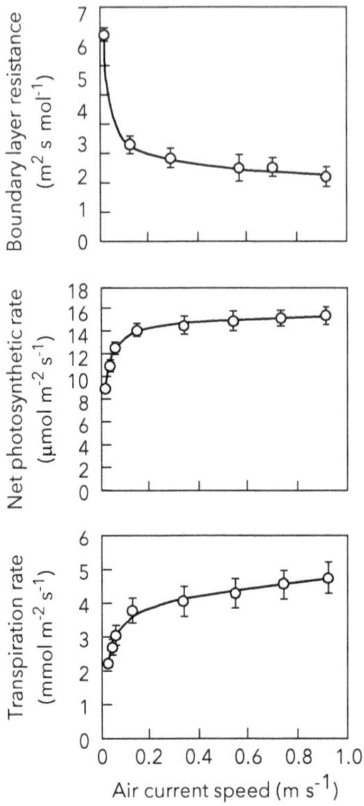


**Figure 1** Effect of airflow velocity on leaf boundary layer resistance, net photosynthetic, and transpiration rates in sweet potato leaves (from Kitaya et al., 2003). Measurements were conducted via a leaf chamber method. PPFD: 1000 µmol $m^2$ $s^{-1}$, air temperature: 28°C, relative humidity: 65%, $CO_2$ concentration: 400 µmol $mol^{-1}$. Bars indicate standard deviations.

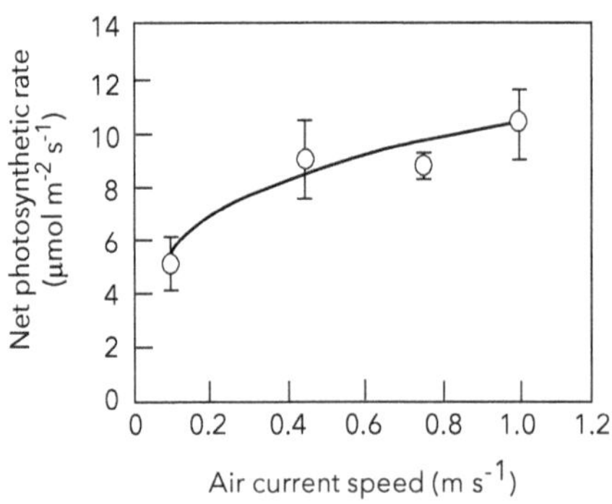


**Figure 2** Effect of the airflow velocity on the net photosynthetic rate based on the rooting bed area of a tomato plant canopy. The canopy height was 0.14 m and the LAI 2.1 $m^2$ $m^{-2}$. The airflow velocity was measured at 0.1 m above the canopy. PPFD: 250 µmol $m^2$ $s^{-1}$; air temperature: 28°C; relative humidity: 65%; $CO_2$ concentration: 400 µmol $mol^{-1}$. Bars indicate standard deviations. Source: Adapted from: Kitaya et al. (2003).

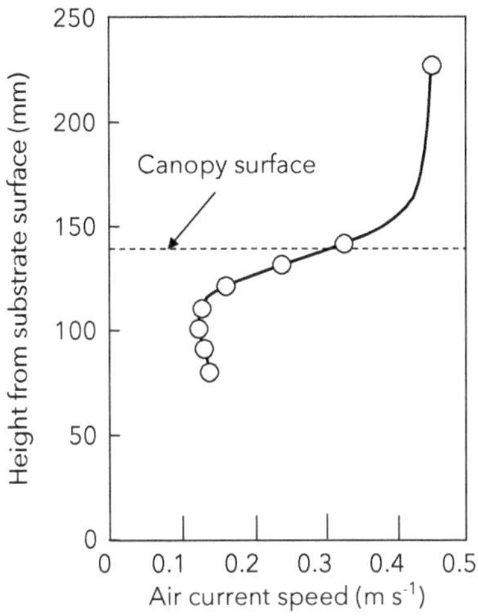


**Figure 3** Profile of the airflow velocity inside and outside a tomato seedling canopy. Source: Adapted from: Kitaya et al. (2003).

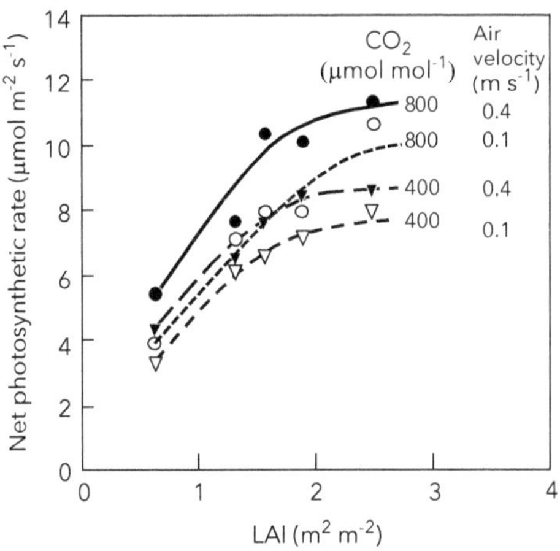


**Figure 4** Effects of airflow velocity, LAI, and $CO_2$ concentrations on the net photosynthetic rate of a tomato plant canopy. The canopy height was 0.14 m. The airflow velocity was measured at 0.1 m above the canopy. The PPFD measured on the canopy was 250 μmol $m^2$ $s^{-1}$, air temperature was 28°C, and the relative humidity was 65%. Bars indicate standard deviations. Source: Adapted from: Kitaya et al. (2004).

Environmental variables surrounding plants must be precisely controlled under adequate air currents to schedule plant production associated with rapid growth and a high turnover rate. Effective air movement is also essential for promoting plant growth and development during vegetative and reproductive growth stages. Suitable air movement ensures that environmental variables remain constant inside the plant canopy.

## 2.3 Dynamics of environmental variables inside the plant canopy as a semi-closed system

Precisely controlling the environment for optimal plant growth within plant cultivation facilities is challenging, especially when a large number of plants are grown in a limited space. Each environmental variable within the canopy is spatially uneven and varies in response to its surroundings. In general, the values of the variables inside the canopy differ from those outside. These environmental characteristics inside the plant canopy are carefully considered to avoid poor or uneven growth.

Understanding the impact of light intensity on air temperature, water vapor pressure, and $CO_2$ concentration inside a plant canopy is essential for precise control.

Each environmental element inside the plant canopy (Kitaya et al., 1998) is spatially uneven and varies according to light intensity. Effectively, the values differ from the values set for mechanically controlling the environment outside the plant canopy. These characteristics often result in poor or uneven plant growth. The effects of light intensity on air temperature, water vapor pressure deficit, and $CO_2$ concentration inside the plant canopy and the potential to approach the environmental elements inside the plant canopy to those outside the plant canopy by controlling airflow velocity above the plant canopy have been reported by Kitaya et al. (1998).

Restricted air movement induces spatial variations in environmental variables such as air temperature, water vapor pressure, and $CO_2$ concentration inside seedling canopies (Kim et al., 1996; Kitaya et al., 1998). Under a photosynthetic photon flux density (PPFD) of 500 μmol $m^{-2}$ $s^{-1}$, 2–3°C higher air temperatures, 0.6 kPa higher water vapor pressures, and 25–35 μmol $mol^{-1}$ lower $CO_2$ concentrations were observed around the canopy height than at a height 60 mm above the canopy (Fig. 5). Air temperature and water vapor pressure increased and $CO_2$ concentrations decreased inside the canopy with increasing PPFD. Lower air temperature and higher $CO_2$ concentration inside the canopy were observed at an airflow velocity of 0.3 m $s^{-1}$ than at 0.1 m $s^{-1}$ (Fig. 6). Therefore, air movement is also important to ensure that environmental variables remain uniform inside the plant canopy.

The control of environmental variables surrounding plants is important for scheduling crop production and obtaining high yields with a rapid turnover rate. Precise control of environmental variables inside plant canopies with sufficient air movement is necessary to enable controlled plant growth.

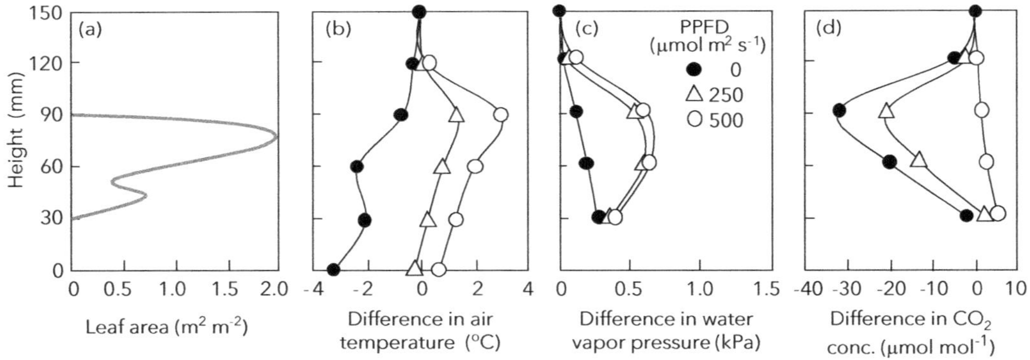


**Figure 5** Profiles of leaf area (a), air temperature (b), water vapor pressure (c), and $CO_2$ concentration (d), inside and outside the canopy of eggplant seedlings as affected by PPFD at an airflow velocity of 0.1 m $s^{-1}$. Source: Adapted from: Kitaya et al. (1998).

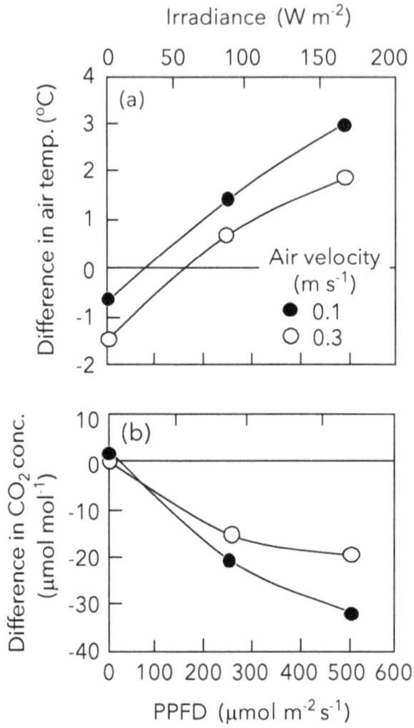


**Figure 6** Effects of PPFD and airflow velocity on differences in air temperature (a) and $CO_2$ concentration (b) between two heights around the canopy height and 60 mm above the canopy. Source: Adapted from: Kitaya et al. (1998).

# 3 Facility-based plant production systems with efficient circulation of materials

The following three examples of feasibility studies demonstrate the potential of plant factories with efficient material circulation in a closed-loop, circular material production system.

## 3.1 Co-culture of mushrooms and plants

Photosynthetic plants absorb $CO_2$ and release $O_2$ during periods of light, whereas mushrooms absorb $O_2$ and release $CO_2$ at all times. Plant residues discharged from the plant production system can also be used as a medium for mushroom mycelial culture.

### 3.1.1 Growth of plants with $CO_2$ produced by the respiration of shiitake mushroom mycelia

Growth characteristics were compared for plants with and without $CO_2$ produced by the respiration of mycelia of shiitake mushrooms (*Lentinus edodes* Sing.) as the fundamental experiment (Kitaya et al., 1995). Plants were cultured in plant culture vessels at a PPFD of 90 μmol $m^{-2}$ $s^{-1}$ and an air temperature of 23–25°C. In the treatment with $CO_2$ supplied by mushroom mycelia, each plant culture vessel was tightly combined with the mushroom culture vessel. The $CO_2$ concentration inside the plant culture vessels during the photoperiod was 300-600 μmol $mol^{-1}$ with $CO_2$ supplied by mushroom mycelia and 100 μmol $mol^{-1}$ without $CO_2$ from mushrooms. The dry weights of mint (*Mentha rotundifolia* L. Huds. cv. apple mint) and potato (*Solanum tuberosum* L. cv. Benimaru) plants cultured with $CO_2$ from mushrooms for 17 days were 1.5 and 1.3 times heavier, respectively, than those without $CO_2$ supply from mushroom mycelia.

### 3.1.2 Plant growth and gas balance in a plant and mushrooms co-culture system

In order to obtain primary data for the construction of a plant culture system incorporating a mushroom culture subsystem (Fig. 7), the relationship between plant growth and the atmospheric $CO_2$ balance in the system was investigated (Kitaya et al., 1994). Plant growth was promoted by the high level of $CO_2$ generated by the respiration of mushroom mycelia in a semi-closed system. The atmospheric $CO_2$ concentration inside the system changed significantly due to a marginal change in the net photosynthetic rate of the plants and/or the respiration rate of the mushroom mycelia when the plant culture system was combined directly with the mushroom culture subsystem. Plant residues discharged from the plant production system can also be used as a nutritional medium for mushroom mycelial culture.

The dry weight of lettuce shoots increased with rising $CO_2$ supply from the mushroom mycelia. However, as the $CO_2$ supply from the mushroom mycelia increased, the rate of dry weight increase of lettuce shoots slowed, and fresh

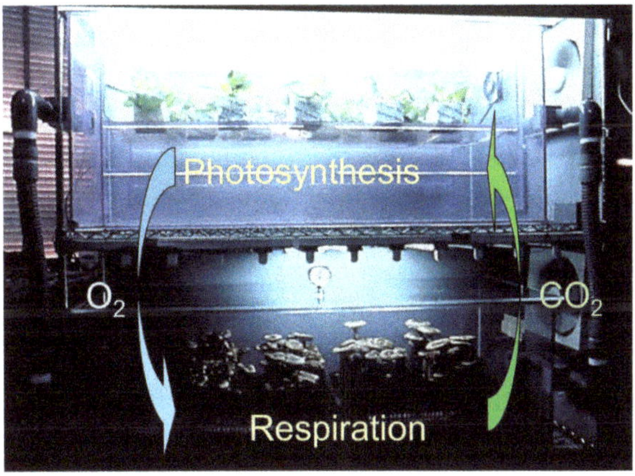


**Figure 7** Lettuce and shiitake-mushroom co-culture system. Source: Adapted from: Kitaya et al. (1994).

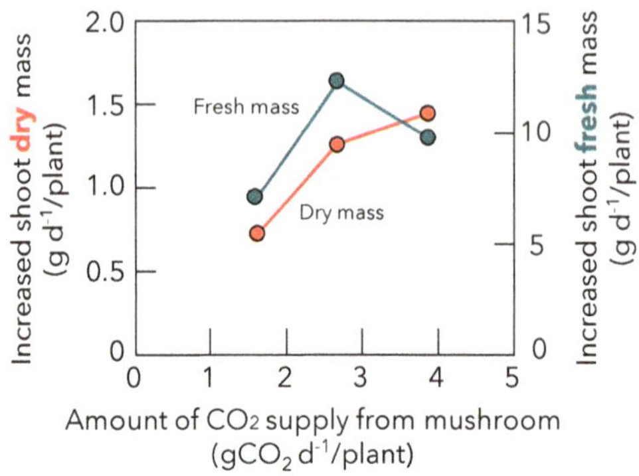


**Figure 8** Relationship between supply rates of $CO_2$ from mushroom mycelia to lettuce plants and increased rates of dry and fresh masses of lettuce shoots during the experiments with three combinations of various quantities of mushroom mycelia and lettuce plants. Source: Adapted from: Kitaya et al. (1994).

weight decreased at the highest amount of $CO_2$ supply (Fig. 8). This can be attributed to the accumulation of toxic substances such as ethylene generated in the system (Tani et al., 1993). Ethylene was released from both lettuce plants and shiitake mushroom cultures. The release rates were 1.1 μg per day/plant for lettuce and 2.8 μg per day/substrate block for shiitake mushroom cultures. Other organic solvents, such as xylene, toluene, and organic siloxane, have also been detected (Tani et al., 1995).

The net photosynthetic rate per plant can alter considerably with changes in environmental conditions and plant age. Accordingly, it may be difficult to control the atmospheric gaseous composition in such a co-culture plant and mushroom system. To establish a plant and mushroom culture system, it is necessary to prevent the accumulation of toxic substances while enhancing the $CO_2$ exchange between the plant and the mushroom culture chambers.

### 3.2 Co-culture of fish and plants as aquaponics

The integration of hydroponics with aquaculture, also known as aquaponics, is gaining increasing attention as a bio-integrated food production system. In aquaponics, water reuse and the balance of nutrient elements are sustained between both cultures, using dissolved elements in the feces of fish for plant growth.

#### 3.2.1 Integrated culture of lettuce and loach fish

To evaluate the potential of aquaponics in combining lettuce hydroponics and loach aquaculture, the growth performance of lettuce and loach, and the balance of nutrient elements, especially nitrogen, between the input and output of the entire system and nitrogen flow in the system were analyzed (Kitaya et al., 2020). Consequently, lettuce grew in aquaponics with 50% of the standard nutrient solution as well as hydroponics with the standard nutrient solution using the optimal combination of the number of plants and fish (Fig. 9). In the nitrogen balance, almost 70% of nitrogen from the feed was used by the growing lettuce, about 25% nitrogen from the feed was accumulated in loach, about 80% of the supplied nitrogen from the whole system from outside was accumulated in plants, and about 20% in fish (Fig. 10).

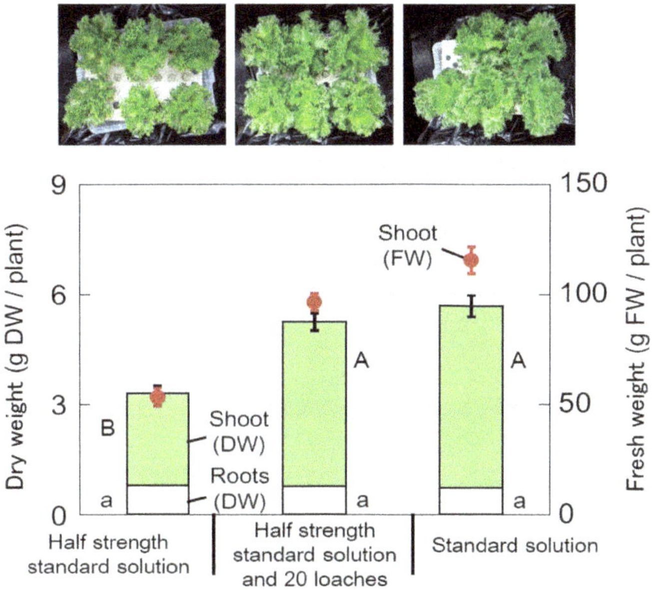


**Figure 9** Growth of lettuce plants in half-strength standard solution, half-strength standard solution with aquaponics, and full-strength standard solution at day 21.

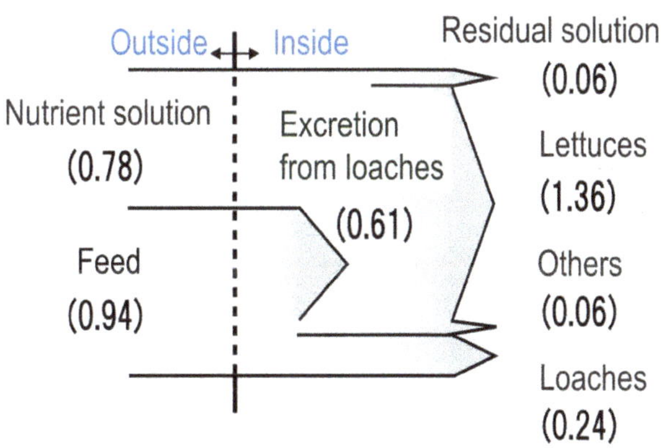


**Figure 10** Nitrogen balance in experimental aquaponics with six lettuce plants and 20 loaches. Nitrogen content (g/system) in each compartment is shown in parentheses.

**Figure 11** Growth of garlic plants and tilapia fish in aquaponics for 2 months. Source: Adapted from: Naznin et al. (2015).

### 3.3.2 Integrated culture of garlic and tilapia fish

To determine the mutual relationship between hydroponic garlic culture and tilapia fish production, an experiment was conducted with three treatments – garlic culture without fish, garlic culture with tilapia fish, and tilapia fish culture without garlic (Naznin et al., 2015).

The fresh and dry mass of the plants was greater when cultured with tilapia (Fig. 11). The fresh mass of the bulbs and leaves was 1.7 and 1.5 times greater, respectively, while the dry mass of bulbs and leaves was 1.4 and 1.6 times greater, respectively, with tilapia than garlic alone. The overall levels of medicinal compound ajoene in garlic with tilapia were 3.1 times greater than that of garlic alone. The total levels of ajoene content in bulbs and leaves were 2.6 and 2.2 times greater, respectively, with tilapia than garlic alone. The growth rate of tilapia in the combined culture was 1.3 times greater than individually. This research demonstrates that garlic plants and tilapia can be successfully co-cultured in an integrated system, resulting in increased production of both fish and plants by water purification for fish and by adding nutrients to plants.

## 4 Sustainable balance of human activities and environmental conservation in urban ecosystems

Increasing global concern about environmental issues, depletion of natural resources such as fossil fuels and minerals, and a decrease in landfill sites for waste treatment have led to growing interest in reducing, reusing, and recycling waste.

The purpose of urban agriculture is to provide fresh and safe agricultural products produced and consumed locally, conserve the local environment, educate people about agriculture through familiar agricultural experiences, support food and environmental education, and provide a venue for community activities. The provision of amenity spaces in green areas and disaster prevention spaces is an additional concern. To maximize the effectiveness of these diverse values in urban agriculture, facility-based plant production systems, such as semi-closed systems, critical to urban agriculture, are necessary.

### 4.1 Urban agriculture to realize local food production for local consumption and a healthy living environment

Developing efficient food production systems without damaging the environment is an important issue in facility-based plant production systems. Therefore, environmental value needs to be incorporated into plant production in order to implement a sustainable balance between human activities and ecological conservation in urban areas.

This section argues that urban ecosystems, including human activities, are semi-closed systems and discusses the development of agricultural production systems based on material circulation within semi-closed systems.

According to the material cycle, the ideal circulation of materials can be achieved in suburban ecosystems, including farmlands, secondary forests, grasslands, and marine and coastal areas, along with human settlements, as well as in urban areas, at the point at which materials and energy inflow and outflow. In addition to reducing the material input, energy input must be kept to a minimum. Consequently, system materials and energy consumption are minimal.

The Japanese Ministry of Agriculture, Forestry, and Fisheries has been promoting environment-friendly agriculture since the 1990s to solve environmental issues, strengthen regional resource management, and resolve regional socio-economic issues in agricultural areas. In addition, Article 36 of the Food, Agriculture, and Rural Area Basic Law, revised in 2015, describes the importance of promoting agriculture in and around cities as urban agriculture to meet the demands of urban residents. Facility-based plant production systems play an essential role in urban agriculture to make more effective use of these diverse meanings.

## 4.2 Combined systems with plant production and methane fermentation in urban agriculture

The reuse, recycling, and control of emissions from large-scale food waste in urban areas has increased in significance in recent years. Consequently, developing efficient food production systems with low environmental impact is critical in agroecosystems, particularly facility-based plant production, to enhance food production from an ecological standpoint. Plant production plays a crucial role in human health and welfare, especially in urban areas. Therefore, to produce food efficiently while protecting the environment, semi-closed and material-cycling farming is required for a material-cycling-oriented society.

To apply resource cycling and energy conversion technology to a material-cycling society, facility-based plant production systems such as plant factories should be established to reduce environmental negative impact. The author proposes an approach that combines plant cultivation and methane fermentation. Organic waste is anaerobically biodegraded in this system and converted to biogas as a renewable energy source. Biogas is composed of methane and $CO_2$. Methane can be burned or converted into hydrogen and used as a source of heat and electricity. $CO_2$ can be added to the atmosphere of a plant production facility to increase the photosynthetic rate of the plants. Digestive juice is a nutrient solution that can be used for water supply and fertilization. A semi-closed hydroponic system, coupled with methane fermentation, can use resources (electricity, heat, $CO_2$, and mineral nutrients) converted from organic waste generated in urban and suburban agricultural areas (Fig. 12).

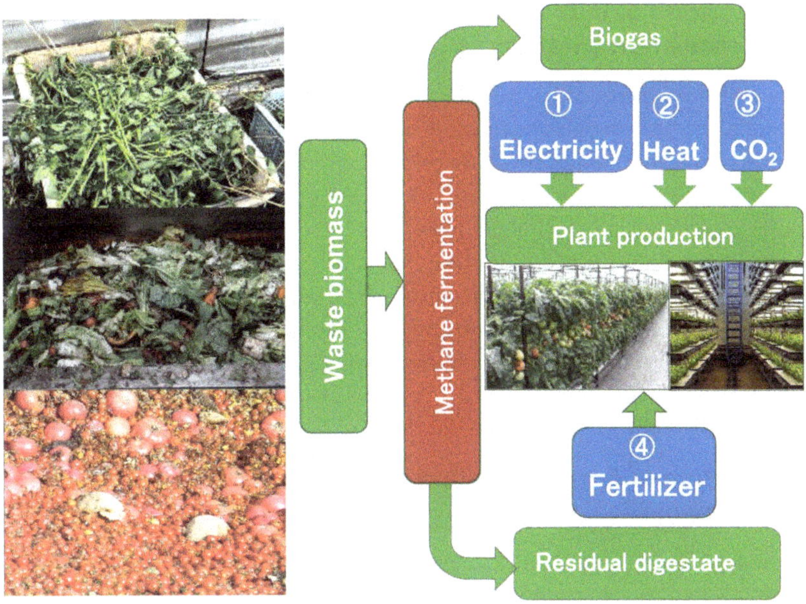


**Figure 12** Plant production with material recycling and quad generation system with methane fermentation.

In recent years, methane fermentation has been promoted to effectively utilize waste biomass as an energy resource. The main purpose of methane fermentation is to convert primary organic waste into methane ($CH_4$) gas as an energy source. Most of the digestate produced as a by-product is discharged or buried as secondary waste. Since methane fermentation digestate contains many fertilizer components, effective utilization as a liquid fertilizer is being trialed in agriculture, but the usable amount is nevertheless limited.

The author proposes a combined system of facility-based plant production and methane fermentation as an application of resource recycling and energy conversion technology for food production. In this system, organic waste is anaerobically biodegraded during fermentation and converted into biogas as renewable energy. The residue left over after fermentation can be used as a nutrient solution for plants.

The digestate is rich in nitrogen and potassium, but insufficient in phosphorus compared to chemical fertilizers (Takemura et al., 2016). In addition, most of the nitrogen components consist of ammonium, which plants tend not to absorb (Endo et al., 2016). If these problems can be resolved and methane fermentation digestate is effectively used as a water and nutrient supply for plant production, the system can be placed in the center of the material cycle.

## 4.3 Co-culture of microalgae and plants with digestate from methane fermentation

To establish a system of co-culturing microalgae and crop plants with biogas digestate, the growth performance of microalgae, *Euglena gracilis*, cultured with digestate from methane fermentation (biogas digestate) has been analyzed. The growth performance of hydroponically cultured lettuce and co-cultured *E. gracilis* in the root zone of lettuce plants has been evaluated.

As a result, it was found that specific growth rates of lettuce and microalgae cultured in the same nutrient solution were identical to those of the lettuce cultured independently. In addition, *E. gracilis* reduced the levels of nutrient components in the soil, making it more suitable for lettuce growth.

### 4.3.1 Culture of E. gracilis with digestate from methane fermentation

To investigate the potential for *E. gracilis* to convert digestate from methane fermentation of organic wastes into a medium for hydroponics, the growth of *E. gracilis* cultured in aqueous solutions containing various levels of filtrate of raw digestate and nitrified digestate was examined (Takemura et al., 2017). Concentrations of plant macronutrient ions in the nitrified digestate before and after culturing *E. gracilis* were also examined. Specific growth rates in aqueous solutions containing 1–10% (v/v) filtrate of raw digestate and 10–100% (v/v) nitrified digestate showed no significant differences. The rates in the filtrate of nitrified digestate were significantly higher than those in the filtrate of raw digestate. Moreover, there were no significant differences between concentrations of plant macronutrient ions other than ammonium ions ($NH_4^+$) in the filtrate of the nitrified digestate before and after culturing *E. gracilis*. The concentration of $NH_4^+$ decreased by 10.5% of the initial concentration. The constituent ratio of plant macronutrient ions other than magnesium in the solution after culturing *E. gracilis* was similar to that in a standard nutrient solution for hydroponics.

### 4.3.2 Co-culture of E. gracilis and lettuce plants

To establish a system of co-culturing microalgae and plants with biogas digestate from organic wastes, the growth performance of *L. sativa* 'frill-ice lettuce' co-culturing with *E. gracilis* was investigated using a simulated solution of nitrified digestate derived from organic waste supplemented with insufficient plant nutrients (Mg, Fe, and Mn) as shown in Fig. 13 (Takemura et al., 2018). The specific growth rate of the co-cultured *E. gracilis* was 0.761 per day, the same rate that *E. gracilis* achieved when grown as a sole culture.

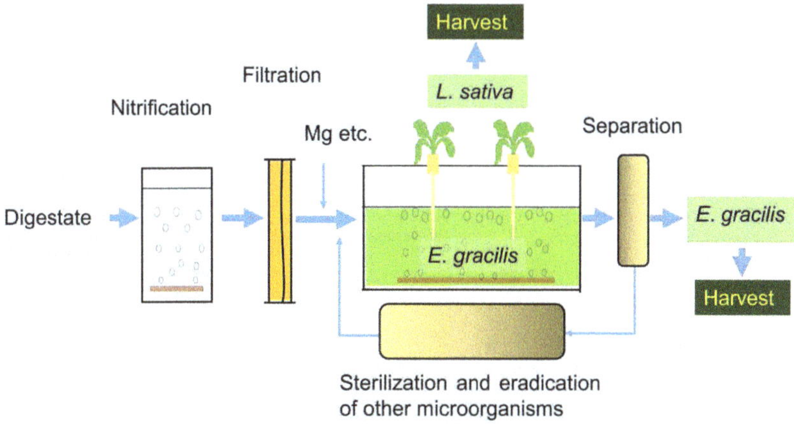


**Figure 13** Co-culture of *E. gracilis* microalgae and lettuce plants with digestate from methane fermentation. Source: Adapted from: Takemura et al. (2018).

There were no significant differences between the relative growth rates of lettuce plants cultured with and without *E. gracilis*. The excess ammonia contained in the biogas digestate was absorbed by *E. gracilis* at a sufficiently low level suitable for plants. These results suggest that it is technically feasible to establish a co-culture system of *E. gracilis* and lettuce plants with biogas digestate after nitrifying and filtering and by adding insufficient minerals such as Mg, Fe, and Mn.

### 4.4 Urban ecosystems incorporated with facility-based plant production systems

This concept can be extended to urban ecosystems that incorporate agriculture. Figure 14 shows the inflow and outflow of energy and the ideal material cycle in urban areas, including suburban agricultural land and water. In urban ecosystems, where a sound material-cycle society has been achieved, the amount of substances input into the system and the amount of energy input to the system other than solar energy become limited. Therefore, the total amount of substances and energy emitted outside the system also decreases. To achieve a functioning material-cycle society, it is necessary to make optimal use of plant functions. Urban ecosystems, including human activities within them, are semi-closed systems. If we can build a sound material-cycle system and minimize the input of substances and energy derived from fossil fuels from outside the system, we can create a society with minimal impact on the environment.

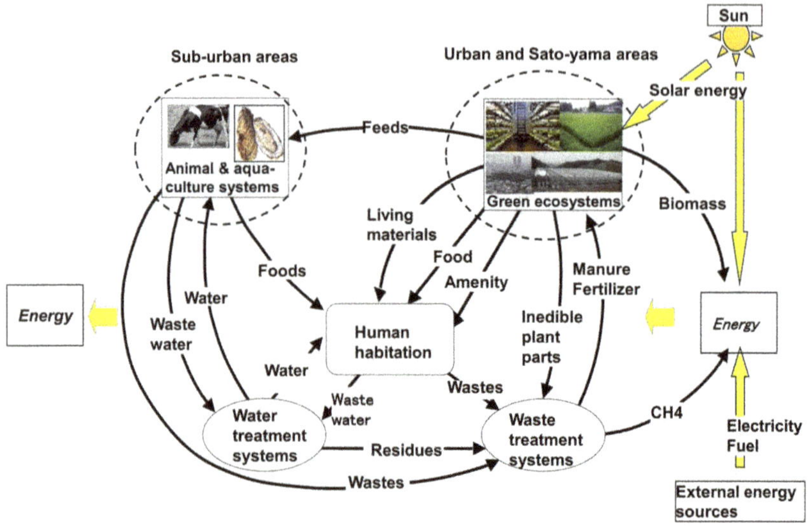


**Figure 14** Material cycle centered on agriculture in the ecosystem, including human activities in urban and suburban areas.

## 5 Conclusion

Traditionally, a wide range of semi-closed circulating ecosystems that incorporate organic substances (such as undergrowth and litter leaves in forests and livestock excrement) has been used as nutrients for crops in plant production in open fields, such as paddy fields and arable fields. However, current agricultural production systems have focused solely on increasing productivity, ignoring the burden on the environment. To achieve a material-cycling society, it is necessary to maximize the plant functions of suburban and urban green spaces and the functions of facility-based plant production systems such as plant factories in order to guarantee environmental conservation as well as efficient food production.

## 6 References


Endo, R., Yamashita, K., Shibuya, T. and Kitaya, Y. Use of methane fermentation digestate for hydroponic culture: analysis of potential inhibitors in digestate to cucumber seedling, *Eco-Engineering* 28(3), 67–72, 2016.

Jones, H. G. *Plants and Microclimate*, Cambridge University Press, Cambridge, UK (vol. 428), 1998.

Kim, Y. H., Kozai, T. and Kubota, C. Effects of air velocities on the microclimate of plug stand under artificial lighting, *Acta Horticulturae* 440, 354–359, 1996.

Kitaya, Y., Sakami, K. and Kozai, T. Development of photoautotrophic plant tissue culture system using $CO_2$ from shiitake mushrooms, *Acta Horticulturae* (393), 195–202, 1995.

Kitaya, Y., Shibuya, T. and Kozai, T. Effects of light intensity and air current speed on air temperature, water vapor pressure, and $CO_2$ concentration inside a plants stand under an artificial lighting condition, *Life Support and Biosphere Science* 5, 199-203, 1998.

Kitaya, Y., Shibuya, T., Yoshida, M. and Kiyota, M. Effects of air velocity on photosynthesis of plant canopies under elevated $CO_2$ levels in a plant culture system, *Advances in Space Research* 34(7), 1466-1469, 2004.

Kitaya, Y., Shimakawa, Y., Wada, T., Shibuya, T., Endo, R., Kikuchi, M. and Nakamura, K. *Food Production and Nitrogen Flow in Aquaponics for Application to CELSS, Abstract of 2020 Committee on Space Research (COSPAR)*, 2020.

Kitaya, Y., Tani, A., Kiyota, M. and Aiga, I. Plant growth and gas balance in a plant and mushrooms culture system, *Advances in Space Research* 14(11), 281-284, 1994.

Kitaya, Y., Tsuruyama, J., Shibuya, T., Kiyota, M. and Kiyota, M. Effects of air current speed on gas exchange in plant leaves and plant canopies, *Advances in Space Research* 31(1), 177-182, 2003.

Monteith, J. L. and Unsworth, H. M. *Principles of Environmental Physics*, Edward and Arnold Publishing Co London, p. 291, 1990.

Naznin, M. T., Kitaya, Y., Shibuya, T., Hirai, H., Endo, R. and Lefsrud, M. G. Integrated culture of garlic and tilapia fish as a source of protein and medicinal compound ajoene, *Acta Horticulturae* (1098), 95-101 2015.

Takemura, K., Endo, R. and Kitaya, Y. Possibility of co-culturing Euglena gracilis and Lactuca sativa L. with biogas digestate, *Environmental Technology* 38(18), 1007-1014, 2018.

Takemura, K., Endo, R., Shibuya, T. and Kitaya, Y. Modifications of concentrations of plant macronutrient ions in digestate from anaerobic digestion during nitrification processes, *Journal of Residuals Science and Technology* 13(3), 207-214, 2016.

Takemura, K., Endo, R., Shibuya, T. and Kitaya, Y. Growth performances and changes of macronutrient ion concentrations in the culture medium when Euglena gracilis was cultured with nitrified digestate, *Environmental Technology* 38(18), 2273-2279, 2017.

Tani, A., Kiyota, M. and Aiga, I. Study of trace contaminant accumulated in a closed system - Measurement of trace contaminant generated in the closed chamber in which lettuce and shiitake mushrooms are cultivated, *Environment Control in Biology* 33(2), 151-154, 1995. (in Japanese).

Tani, A., Yamanami, A., Kiatya, Y., Kiyota, M. and Aiga, I. Changes in carbon dioxide and ethylene concentrations in the closed ecological chamber in which lettuce and shiitake mushrooms are cultivated. *CELSS Journal* 5(2), 11-16, 1993. (in Japanese).

Yabuki, K. *Photosynthetic Rate and Dynamic Environment*, Kluwer Academic Publishers, Dordrecht, Netherlands, p. 121, 2004.

# Part 3

## Phenotyping

# Chapter 9

# Application of machine vision in plant factories

*Wei Ma and Zhiwei Tian, Institute of Urban Agriculture, Chinese Academy of Agricultural Sciences, China*



## 1 Introduction

The world's population continues to grow and is expected to exceed 9 billion by 2050. Per capita arable land will be less than 0.1 hectares at that time (Fatemeh et al., 2017) and food scarcity will become the primary factor threatening human survival, so 'growing more with less' is a huge priority for scientists (Franchetti et al., 2019). A plant factory is a promising option to solve these problems. Compared with traditional agriculture, plant factories can create a growth environment that is controllable, and crops can be planted in a vertical space so that less land is required. Factories can produce more food due to year-round production capacity and are often deployed in the vicinity of consumers to save transportation costs (Yang and Sun, 2019; Story et al., 2010; Asseng et al., 2020; Redmond et al., 2018; Aya et al., 2015; Story and Kacira, 2015; Aznar-Sánchez et al., 2020). However, building and labor costs are the main factors constraining plant factory development, and in this context, automatic production becomes promising and sustainable.

http://dx.doi.org/10.19103/AS.2023.0126.10

Machine vision has recently been widely applied in the intelligent systems of plant factories (Dewi et al., 2020; Guanghui and Changyin, 2002; Scho et al., 2017; Ikhlef and Sabine, 2010; Pereira Valente and Guiffrida, 2019). This technology covers virtually all production stages, from raising seedlings, transplanting, management and harvesting to product packaging. It uses computers to simulate human vision and extract information from images and then analyzes them and guides practical production (Ren, 2007; Yang Si, 2019; He Zili, 2019). Compared with chemical and physical methods, machine vision conducts detection in a non-invasive way and causes no damage to plants. It works continuously at high efficiency and low cost.

The classic machine vision system in plant factories is shown in Fig. 1, which consists of the image acquisition layer, processing and decision-making layer, and task execution layer. The image acquisition layer is used to capture images and transmit them to the processing and decision-making layer. The components of this part consist of light sources, optical systems and image acquisition cards. The processing and decision-making layer extracts and analyzes information from images, then makes decisions based on its learning results, before finally giving order to the execution layer. The main component in this layer is a computer system. Generally, the task execution layer consists of mechanical modules, which perform tasks such as irrigation and fruit picking. In this layer, the robot system, environmental controlling system and nutrition supply system are included.

Many reviews have reported machine vision applications in agriculture, but most are about feild crops (Kaiyan et al., 2013; Hongkun et al., 2019) and few refer to plant factories. The internal environment of a plant factory is complex and completely different from the outdoors. In addition to the plants, there are irrigation pipes, hanging ropes, mechanical equipment and other supporting facilities. Lighting conditions also change periodically according to the plants' needs, which bring with it various challenges to machine vision applications. In this chapter, we attempt to present a comprehensive survey on the subject, as a starting point for research on this issue. It is presented from

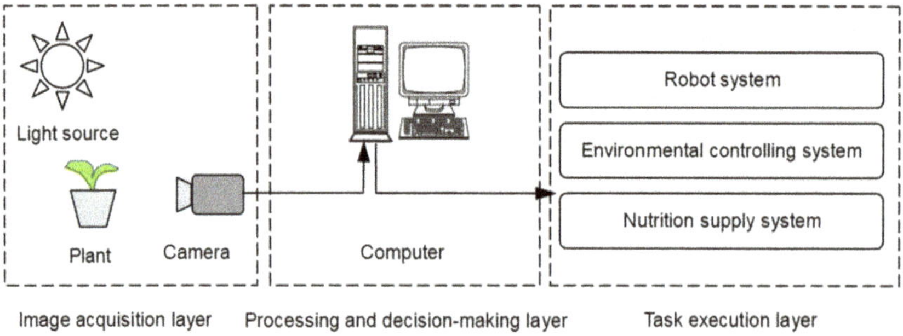


**Figure 1** Composition of the machine vision system in a plant factory.

three aspects - plant growth monitoring, robot operation assistance and fruit grading. The advantages and disadvantages of various methods are compared and the challenges that machine vision faces in plant factories are analyzed.

## 2 Plant growth monitoring

Information regarding plant health levels can support management decision-making. Rapid information acquisition and decision-making can often avoid risks and reduce economic losses. In a traditional greenhouse, environmental parameters such as air temperature, relative humidity and lighting intensity can be adjusted based on the grower's observations or an unchanged threshold, rather than the needs of plants (Lakhiar et al., 2018; Kim, 2010). When machine vision is introduced, plants will be fed based on growth information collected continuously and dynamically. At present, this technique is mainly used in plant phenotype, pest and disease warning, and nutrient stress detection.

### *2.1 Plant morphological characterization*

Plant phenotype parameters include color, size, texture and shape for different plant organs such as flowers, leaf, stem, fruit, canopy and roots (estimated indirectly) (Story et al., 2011). This information reflects the growth state and health of plants and is important for decision-making and management of horticultural systems. As shown in Fig. 2, leaf area and morphology can be used to identify plant species. Plant height is a quantitative morphological parameter used to ascertain the growth rate or annual growth cycle of a plant grown in different/same weather or nutrient conditions (Chaudhury et al., 2019; Gupta and Pahuja, 2019). Fall foliage is a qualitative feature to indicate the health levels of a plant (Kumar et al., 2019). The top canopy area can also be related to the plant's fresh and dry weight (Shimizu and Heins, 1995; Lee et al., 2018).

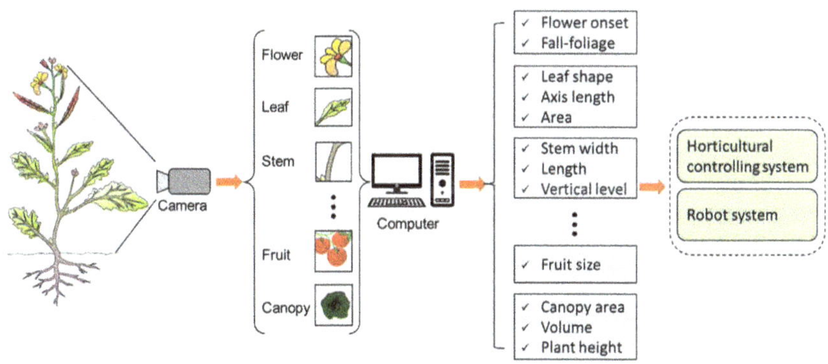


**Figure 2** Different parts of the plant for morphology information acquisition.

Historically, this information was obtained through physical measurement methods or observation by the naked eye, which is labor-intensive and requires a high level of grower expertise. It is also impossible to collect this information in real time. To solve these problems, machine vision has been used in many researches, the results of which are listed in Table 1.

Plant size can be used to evaluate its growth rate and estimate yield. Prior to transplanting, it is necessary to measure the seedling size and judge whether it has grown to a suitable level. Yang Zhenyu (2014) combined the S component in the HSI (Hue Saturation and Intensity) color model and the Otsu algorithm to segment seedlings from the background; the Vertex Chain Code (VCC) and the Minimum Bounding Rectangle (MBR) were used to calculate the seedling height and the seedling skeleton was extracted to calculate the vertical degree. These two parameters were combined to select the available seedlings. However, images in this study were taken manually. He Zili (2019) designed an inspection robot that can capture images of a single plant automatically and calculate its growth parameters. In addition to providing supporting information for the robot's application, plant growth parameters can also be input into the environmental control system. Sun Guoxiang (2016) extracted cucumber population canopy coverage and width through the Excess Green Minus Excess Red (MExg-ExR) algorithm, and this information was used as a decision-making input along with air temperature, relative humidity and soil moisture. An integrated irrigation system for water and fertilizer was then designed.

The research mentioned earlier was conducted in a sunlit greenhouse. Indeed, supplemental lighting systems are used in most plant factories to improve light conditions indoors, to help plants grow healthily. To meet a plant's particular needs at different stages and to save energy, artificial light only radiates a narrow band of wavelength, which produces a monochromatic light. Some algorithms perform well when dealing with images taken in sunlight, but often fail to segment the feature area from the images taken in artificial light. It is sometimes difficult to separate the plant and its organs from the background. The study by Sisi et al. (2014) attempted to use the Kubelka-Musk theory to solve this problem but achieved an imperfect result. Image segmentation is the first step for image analysis, although artificial light adds challenges during this process.

Plant phenotype accuracy is an issue worth noting, but few reports mention it. Larger deviations will cause the control system to make wrong decisions, which could have serious consequences. Plant parameters are variable even under the same growth conditions, so individual plant growth parameters are not representative and should not be taken as a decision-making basis for control systems. In future studies, the measurement accuracy of machine vision should be validated through physical and other methods, and the accuracy of phenotype modeling should be revised and improved. Statistical methods

**Table 1** Summary of machine vision used for plant phenotype

| Ref. | Task | Color model | Method | Advantages | Disadvantages |
|---|---|---|---|---|---|
| Story et al. (2011) | Extract plant outlines, stem width and lengths | RGB | Least-squares threshold selection, thinning algorithms | High accuracy of 0.025 mm | Analysis was not possible when leaves shaded the stem |
| Chaudhury et al. (2019) | Assess variations in plant parameters (such as area) over time | CIE, L*a*b* | Super pixel-based random forest, Simple linear iterative clustering (SLIC) algorithm, K-means | An automated, high-throughput pipeline to analyze large-scale image data | Poor performance when different color distribution, leaves overlapped |
| Gupta and Pahuja (2019) | Extract leaf centroid, major axis length, minor axis length | RGB | K-NN, Decision Tree, multilayer perceptron, AdaBoost methodology | High precision rate of 95.42% | Any inclination or curling of the leaves will cause the morphological features extraction to fail |
| Kumar et al. (2019) | Extract leaf area, plant height, detection of onset of flowers, fall-foliage condition | RBG | PIPME, binary morphological operation, threshold method, linear spatial filtering | High level of accuracy, display the results on the user interface | The device is fixed and cannot capture more images automatically |
| Shimizu and Heins (1995) | Extract plant top projected canopy area, plant perimeter and weighted radius | RGB, HSL | Co-occurrence matrix and gray tone spatial dependence matrix, Bresenhams line algorithm | The user interface enables the operator to visualize dynamic camera motion | The accuracy of plant parameters detected is unknown |
| Lee et al. (2018) | Detect surface area and volume computations | None | Clustering algorithm, Expectation maximization algorithm, α-Shape triangulation | The system works continuously for the life cycle of the plant | Can only scan a single plant at a time |

*(Continued)*

**Table 1** (*Continued*)

| Ref. | Task | Color model | Method | Advantages | Disadvantages |
|---|---|---|---|---|---|
| Yang (2014) | Extract the vertical degree and height of pot seedlings | HSI | Otsu algorithm, vertex chain code method and MBR (Minimum Bounding Rectangle) | Improve the efficiency and robustness of the image segmentation | The image background needs to be processed manually |
| He (2019) | Estimate the leaf area | RGB | Threshold segmentation, Gaussian smoothing | Robot mobile platform makes image acquisition flexible | Only classify three different plants |
| Sun (2016) | Extract canopy coverage, width and length | HIS, YcbCr | Retinex, MExg-ExR (excess green minus excess red) | With high accuracy of 98.31% under different light conditions | It is only suitable for vertical viewing angle conditions |

need to be applied for the deployment of sampling points, ensuring the information collected can reflect all plants.

### 2.2 Pest and disease detection

High levels of humidity and air temperature in plant factories provide a suitable living environment for crop pests and diseases, which cause irreversible economic loss if not found and controlled in time. Traditionally, pests and diseases were observed and alerted manually, which was not only time-consuming but also labor-intensive. Observation results were easily affected by human subjectivity, as each person has inconsistent definitions of levels of severity, often leading to optimal controlling opportunities being missed (Al and LNO, 2017; Wudneh and Vanitha, 2019). The initial symptoms for various pests and diseases are not obvious to the human eye, and so large areas of plants become infected before they can be seen by growers. Machine vision has been suggested by many studies to monitor plant health conditions and research results are listed in Table 2.

#### 2.2.1 Disease detection

The process of disease diagnosis through machine vision includes (i) image pre-processing, (ii) lesion area segmentation, (iii) feature-searching (color, shape and texture) of the lesion area, (iv) model training and (v) disease recognition.

Mi Yating (2016) analyzed the foreground and background of tomato leaves by a Sobel operator and used the Watershed algorithm to segment disease spots. A disease recognition model was built based on BP and GA-BP neural networks. The recognition rate of early blight, late blight and leaf mold reached 100%, 98% and 96%, respectively, although fewer than 100 images were tested in this research. Juncheng et al. (2017) used Conditional Random Fields (CRF) and the Decision Tree model to segment cucumber downy mildew. Rough Sets were adopted to select classification features. Finally, an support vector machine (SVM) was built to diagnose cucumber downy mildew. Significantly more images have been tested this way and the accuracy rate was 90%. To improve recognition accuracy, (Juncheng et al., 2017) and (Ding Lian, 2020) trained an SVM model based on the color and shape features of powdery mildew, brown spot and anthracnose spots of cucumber leaves. This SVM model combined several features and improved the average accuracy of lesion identification to 96.39%. However, these methods are time-consuming and cannot meet real-time work requirements. The Hough Transform and Random Forest algorithm were used in Kiani and Mamedov's (2017) research, in which detection and segmentation accuracy of the plant diseases reached 97% and processing time totaled 1.2 s. A system based on color processing detection algorithm (CPDA) and fuzzy logic classification algorithm (FLCA) shortened the

**Table 2** Summary of machine vision used for plant disease detection

| Ref. | Plant type | Task | Color model | Method | Advantages | Disadvantages |
|---|---|---|---|---|---|---|
| Mi Yating (2016) | Tomato | Identify and grade tomato early blight, late blight, leaf mold | RGB, HIS, LAB | Divide and conquer median filter, Sobel-Watershed, GA-BP | High denoising efficiency, BP neural network optimized by genetic algorithm converges faster | A small number of images leads to incomplete test results |
| Ma Juncheng et al. (2017) | Cucumber | Diagnosing cucumber downy mildew | HSV | Rough set method, SVM-based RBF, image segmentation based on conditional random fields | The algorithm has high computational efficiency and high accuracy in lesion segmentation | Different lighting conditions will affect the results |
| Liu Shuai (2018) | Cucumber | Detection of powdery mildew on cucumber leaves | HIS | Threshold segmentation, SVM classifier based on the radial basis kernel function | The system is mounted on the disease warning robot, so it can collect images autonomously | Time-consuming |
| Ding Lian (2020) | Cucumber | Detection of brown spot, downy mildew and anthracnose of cucumber | HIS | Histogram threshold segmentation, Gray-level co-occurrence matrix | The average recognition rate of the three diseases reached 96.39% | Algorithm running is time consuming |
| Kiani and Mamedov (2017) | Greenhouse plant | Powdery mildew recognition | RGB | Hough transform, random forest | High speed and detection accuracy of 97% | The result will be affected by shadows |
| Scho et al. (2017) | Strawberry | Detect healthy and disease-infected strawberry leaves | RGB | Color processing detection algorithm (CPDA), Fuzzy logic classification algorithm (FLCA) | Fast running speed (0.06 s) and has minimal computational load | It is only suitable for images under outdoor sun illumination conditions |

| | | | | | | |
|---|---|---|---|---|---|---|
| Ali and Hassanein (2019) | Tomato, bell pepper | Powdery mildew of bell pepper plants and spotted wilt virus of tomato | RGB | Random forest and Hough transform algorithms | High accuracy of mold recognition | Low running speed |
| Sun Wenjuan (2017) | Cucumber | Detect LWD (leaf wetness duration) to warn of the occurrence of diseases | L*a*b* | K-means | Innovatively use image processing to estimate the wetting time of cucumber leaves | Compared with manual detection, monitoring error within 1 h |

working time to 0.06 s (Scho et al., 2017). In plant factories, lighting conditions and complex backgrounds typically lead to a poor image segmentation effect and disease recognition accuracy. Ali et al. (2019) used Random Forest and Hough Transform algorithms to extract the area of the lesion, in which the learning characteristics of the Random Forest and the detection characteristics of the Hough Transform are well combined, so recognition accuracy is improved. These studies showed that machine vision has great potential for the detection of diseases. Although SVM was adopted to classify lesions widely in these reports, it consumed a large amount of machine memory and time.

Existing studies have focused on obvious disease lesion areas. In this case, the disease has caused irreversible damage to the quality and yield of the crop, so predicting disease in advance is a promising field of study. Leaf Wetness Duration (00) is regarded as the main factor that can be correlated with infection and the prevalence of many leaf diseases. In plant factories, there is high relative humidity due to plant transpiration and evaporation of nutrient fluid, which causes plant leaves to be moist for a long time, and eventually induces the occurrence of downy mildew and other diseases. However, the leaf wetness level is a changing process and is difficult to observe by the naked eye. Sun Wenjuan (2017) tracked the ratio of the wetness area to the whole cucumber leaf area dynamically based on the L*a*b color model and K-means algorithm, after which the LWD value was calculated. Compared with the manual observation, the monitoring error was within 1 h.

### 2.2.2 Pest detection

Pests present in plant factories include aphids, whiteflies and thrips. They are small, migratory and concealed (often located on the underside of leaves). Checking plants and searching for pests periodically is challenging for greenhouse staff. To improve intelligent management, various studies have proposed an automatic monitoring system based on machine vision (listed in Table 3). In these systems, the shape, color, texture and numerical features of pests were extracted, and models were built for pest species identification based on artificial neural networks (ANN) and SVM (Lima et al., 2020).

Li et al. identified and located the pest on the plant directly by image segmentation and binocular stereo vision (BSV) technique (Li et al., 2009). To measure the infected level of the whitefly, S. R. Pokharkar and Thool (2012) calculated the percentage of the infected area of the leaf and the size of each pest, focusing only on adult whitefly detection. Rupesh (2013) proposed a system for early whitefly detection and achieved an accuracy of 100%, although pest eggs were ignored in these studies. Wang et al. (2018) presented a cognitive segmentation approach, the results of which showed that the whitefly and its eggs can be segmented accurately from leaf images.

**Table 3** Summary of machine vision used in plant pest detection

| Ref. | Task | Color Model | Method | Result | Advantages | Disadvantages |
|---|---|---|---|---|---|---|
| Li et al. (2009) | Identify and locate pests on the plant | HSV | Centroid-matching, Sum of Squared Differences (SSD) | Pest error focused on the range of 20 mm when the depth measurement is less than a range of 500 mm | This system can locate pests | The illumination conditions, size and color of pests lead to measurement errors |
| Pokharkar and Thool (2012) | Detection and calculating the area of infection on leaves of a whitefly | RGB | Linear spatial filtering, Sobel operator | The percentage of the infected area and the size of each pest was calculated | High calculation accuracy of whitefly size and leaf infection area | Only for whitefly adult detention |
| Rupesh (2013) | Early whitefly detection | RGB | Smoothening filter, SVM | SVM is done with 100% accuracy | High detection accuracy for early pests | High image acquisition requirements |
| Wang et al. (2018) | Whitefly detection | CIE L*a*b* | K-means Clustering Algorithm, Initial cluster centers self-learning algorithm | The new method can accurately segment whiteflies from crop leaf images | High accuracy and good robustness for both egg and pest detection | Results are easily affected by light conditions |
| Liu Mengmeng (2018) | Detect whitefly and thrips on cucumber | HIS L*a*b* | Prewitt and Canny operator, SVM, BP | Recognition rates of whiteflies and thrips are 91% and 96% | SVM classification performance based on linear kernel function is good and stable | The method was greatly affected by light and has high requirements for image acquisition |
| Cheng Hong (2015) | Count whiteflies and thrips | RGB | Two-dimensional Fourier transform (2DFT) | The correlation coefficient $R^2$ was above 0.99 compared with human counting result | High counting accuracy and little influence by light | Only two pests were detected and tested |

Pests have migratory characteristics. If they can be captured by camera easily on plant leaves, this implies severe infestation. Counting pests on the trap board is a common statistical method (Meixiang et al., 2017). Trap boards attract pests and catch them based on their biological habits. It is believed that pest density on trap boards is larger than on the leaf. Mengmeng (2018) used Prewitt and Canny operators to detect a single pest edge in the HSI and L*a*b color model, then the two images were fused to extract the pest area. Finally, nine color characteristics and five morphological characteristics of pests were selected to train SVM, and the average recognition accuracy of the model reached 93.5%. This method was easily affected by lighting and had a higher requirement for image acquisition. To solve this problem, Cheng Hong (2015) proposed a method based on Two-Dimensional Fourier Transform (2DFT), in which pests on the trap board were regarded as noise in the image, and the 2DFT was used as a noise collector. Finally, spectral analysis was used to count the number of pests. The correlation coefficient of results between the system and manual counting reached 0.99, and the relative error was within ±4%. In practical application, trap boards need to be replaced manually after they are scrapped, which is inefficient and prevents automation of the entire monitoring system. Meixiang et al. (2014) has designed a pest monitoring system in which the trap boards can be replaced automatically.

Pesticide application is not the best strategy to control pests. The most environmentally friendly approach is prevention and comprehensive management before pests occur. Their occurrence is closely related to environmental situations, and an early warning model can be created through collecting, analyzing and excavating information such as the plant's nutritional status, environmental data and pest levels.

## *2.3 Nutrient stress detection*

In plant factories, plants are supplied with nutrients by hydroponic technology. When nutrient supply is insufficient or excessive, it will inhibit plant growth, a phenomenon known as nutrient stress. Nutrient stress is the most important information used to assess nutritional balance. However, there is no commercially available sensor for real-time detection of plant stress conditions (Kirongo, 2016). Machine vision may bridge the gap as it can 'let the plant speak' and help growers administer nutrients according to the plant's needs. The latest research results in this field have been listed in Table 4.

Yane (2005) extracted the color and texture feature of cucumber leaves and growing points to build a model describing the relationship between them and plant nutrient levels. Hendrawan Yusuf (2014) combined the N-DHRIO algorithm and ANN to detect the moisture of moss and designed an automatic irrigation

**Table 4** Summary of machine vision used in plant stress detection

| Ref. | Plant type | Task | Color Model | Method | Result | Advantages | Disadvantages |
|---|---|---|---|---|---|---|---|
| Yane (2005) | Cucumber | Nutrient stress | RGB | Threshold segmentation, Gray level co-occurrence matrix | The correlation coefficient between G component and leaf nitrogen content and chlorophyll reached 0.88. | High prediction accuracy of the model | Only for cucumbers at a certain growth stage |
| Hendrawan Yusuf (2014) | Lettuce | The relationship between light intensity and color characteristics | RGB | Neural-Discrete Hungry Roach Infestation Optimization (N-DHRIO), ANN | Adjust light intensity automatically according to the needs of different parts of the lettuce | High detection accuracy, accurate and stable controlling result | Light control only for individual lettuce |
| Story et al. (2010) | Lettuce | Calcium deficiency | RGB, SHL | Gray-level co-occurrence matrix, Dual-segmented regression analysis | Identify calcium-deficient one day earlier than human vision | High detection sensitivity | The variable nature of lighting is problematic |
| Stephen andHaruhiko (2014) | Lettuce | Water stress detection | RGB, SHL | HSL filtering | Able to detect water stress 2 h before human vision | Combine RGB image, NIR image and thermal image | Slow running speed, the whole process took about 11.2 min |
| Stephen and Haruhiko (2008) | Moss | Water stress detection | RGB, HIS, CIE, LAB, XYZ | Color Co-occurrence Matrix (CCM) | The HSI texture features achieved 99.45% water stress classification efficiency | Good detection result under varying light intensities | Some errors still need to be corrected |

*(Continued)*

**Table 4** (*Continued*)

| Ref. | Plant type | Task | Color Model | Method | Result | Advantages | Disadvantages |
|---|---|---|---|---|---|---|---|
| Kacira et al. (2002) | Impatiens | Identify plant water stress | RGB | Typical gray-level histogram segmentation | Five to 45 h earlier stress detection compared to human vision | Establish a water stress index using plant movement information | Only one type of plant has been tested |
| Gong Zhen (2018) | Tomato, pepper | Nutrient stress | None | Median filter | Established a spectral prediction model for crop nutrition | Fusion of hyperspectral, image, and chemical detection results | The accuracy of nutrient stress prediction has not been verified |

system. In this study, the color and texture of lettuce canopy were also extracted to analyze the light requirements in different parts of lettuce and to provide suitable light intensity. In David's (2010) research, the top canopy area of lettuce was taken to deduce its morphological characteristics. Hue, saturation and brightness were used as color characteristics. Entropy, energy, contrast and homogeneity were used as texture characteristics, and these features were combined to evaluate the calcium deficiency of early lettuce. The method can identify whether plants are calcium-deficient one day earlier than human vision. The studies mentioned above used RGB cameras to capture plant images, although the information contained in the RGB image was limited. Three different cameras were used in Story and Kacira (2014). The image from an RGB camera includes color, shape and texture of lettuce, whilst that from a NIR camera infers the vegetation index whilst the thermal image can be used to measure plant temperature. Three images were fused to diagnose the health and growth of the plant in real time. To improve water stress detection accuracy, Stephen and Haruhiko (2008) tested different color models and found that the texture classification efficiency in the HIS color model achieved 99.45%. Experiments also showed that the system can recognize plant water stress 2 h before the human visual inspection. Changes in plant morphology are also a sign of water stress. Plants under water stress generally have drooping leaves and are wilted due to the imbalance between water demand and water supply (Kacira et al., 2002). Therefore, detecting water stress by observing plant morphology is an interesting area of study.

Although machine vision can be used to monitor plant nutrient stress more effectively than human vision, the accuracy of the method needs to be improved. Gong Zhen (2018) took tomatoes and sweet peppers, collected the characteristic spectra of nitrogen, phosphorus and water, and extracted the canopy width, height and stalk thickness of the plant through machine vision. These data were combined with chemical analysis results and plant growth data. A prediction model showing nutritional stress and growth was built. Compared with the model based on single data, this model performed better.

## 3 Robot operation assistance

To reduce labor costs and improve production efficiency, robots are in widespread use in plant factories. They are mainly responsible for the production (seedling, transplanting, ad picking), transportation and management. Machine vision is as important to robots as eyes are to humans and is essential to navigate, recognize and locate the target.

### *3.1 Visual navigation*

Navigation and localization systems indoors include (i) map-based navigation systems, (ii) map-building-based navigation systems and (iii) mapless navigation systems. The first method needs to build a robust and reliable map with sensors ahead of time and is time consuming and labor intensive. On the other hand, the map built before is easily broken due to the environment changing in plant factories, because other robots and equipment need to be introduced according to crop growth stages and thus change space distributions. The second system requires more computational resources, time and storage capability. A mapless navigation system based on machine vision is a potential technique (Gao et al., 2019; Bonin-Font et al., 2008), which has the advantages of low cost, high accuracy and rich information, and is suitable for robot navigation indoors.

Many researchers have proposed a path detection algorithm based on machine vision (Table 5), of which the most famous are Hough Transform, Least-Square Method (LSM) and BSV. Hough Transform has good anti-interference ability as it reduces the influence of navigation point errors and improves the fitting accuracy of the navigation route, but the algorithms run slowly and cannot meet real-time requirements. LSM is a mathematical optimization technique and selects the best function to match the path by minimizing the sum of squares of deviations. This is a common method for crop row extraction, but the returned result is inaccurate when there are many error points. BSV is a technology that obtains the parallax of two images through two cameras and generates 3D information of the target image through a series of inverse projection transformations. It is difficult to synchronize two cameras, so the stereo vision algorithm is usually more complicated and has poor real-time performance (Xie Zhonghong, 2013; Lin Rui, 2011; Yongxin et al., 2009). Realizing these existing problems, Jiqing et al. (2020) proposed a midpoint Hough Transform algorithm based on an optical system. This method used an improved gray factor and Otsu algorithm to segment soil and plants, then extracted navigation points by relative coordinate centers in the binary image. Finally, the navigation path was fitted through the midpoint Hough Transform. Compared with the traditional Hough Transform (52.37 ms), the improved method requires only 7.13 ms, and its average accuracy is 5.1° higher than that of the LSM.

Different lighting conditions and shadows often present challenges to robot navigation. In studies by Guoqin and Ming (2014) and Bai Ruyue (2018), the K-means algorithm was used to cluster and segment images, after which redundant and interfering information was removed by morphological corrosion method and the Canny operator was used to detect the edge of the road. This method performed better under different light conditions and has a higher road extraction rate in less time.

However, when the robot turns at the end of the road, the visual navigation suffers from low efficiency and large deviation (Younse and Burks, 2007; Younse and Burks, 2005). Wang Peng (2019) designed a QR code navigation system, in which a QR code was situated at the end of the road with the navigation

**Table 5** Summary of machine vision used in navigation systems

| Ref. | Color Model | Method | Result | Advantages | Disadvantages |
|---|---|---|---|---|---|
| Ren Yongxin et al. (2009) | RGB | Chromatic aberration operator (2G-R-B), Improved Hough transform | The average time of this method is less than 150 ms, and the maximum error is less than 5 cm when the path is tracked at 0.3 m $s^{-1}$. | The method of path extraction is more reliable | The method is time-consuming and the robot moves slowly |
| Jiqing et al. (2020) | RGB | Improved grayscale factor, Otsu, Median point Hough transformation | The maximum error of the navigation path was 0.5° and the time taken was 7.13 ms, which is 45. 24 ms faster than the traditional Hough transform and on average 5.1° more accurate than the least-squares method. | Faster running speed and more accurate | Easily affected by shadows and light |
| Bai Ruyue (2018) | HIS | K-means, Morphological corrosion, Canny | Compared with common threshold segmentation methods, the road extraction rate can reach 95%, and the average single image processing time is reduced by 53.26%. | The algorithm is not interfered with by light High road recognition rate faster calculation speed | The shelter will affect path recognition result |
| Bai Ruyue (2018) | HIS | K-means, Improved Hough transform, Canny | It takes an average of 12.36 ms to extract the centerline of the crop row and the maximum lateral deviation within 5 cm. | Can effectively adapt to different light conditions | Low running speed |
| Younse and Burks (2007) | RGB | KLT algorithm | The vehicle demonstrated the ability to follow the center of a 61-cm wide path. | The vehicle demonstrated the ability to turn at intersections, as well as drive through them | Low speed at intersections |
| Wang Peng (2019) | RGB | Multi-scale retinex, Improved Hough Transform, QR code | The maximum lateral position deviation is 3.9 cm, the average heading angle deviation is 2.1°, and the average steering angle deviation is 5.6°. | High accurate navigation by QR code | Robot speed is slower at corners |

information stored in it ahead of time. Robots can accurately turn a corner by recognizing the QR code. The test results showed that the maximum navigation deviation was 3.9 cm when the robot traveled at 2 $m \cdot s^{-1}$.

### 3.2 Target recognition and location

Target recognition is the process of separating the plant from the background or fruit from the rest of the plant. It is a prerequisite for robots to complete tasks such as fruit picking and seedling transplantation automatically. The calculation is conducted simultaneously between the global, camera and image coordinates to obtain the spatial coordinate of the target, after which the robot's manipulator moves to that location and performs an action (Mavridou et al., 2019).

To ensure the transplanting robot works accurately and reliably, Ren (2007) used the Single-Connected Domain Analysis algorithm to extract the leaf area of seedlings and located them, then judged the seedling as to whether it was suitable for transplanting according to leaf area. The recognition accuracy rate reached more than 98%. Often, there are empty holes in the seedling tray due to seed quality and mechanical damage. To replant the holes automatically, Zhang et al. (2019) segmented the seedling tray into grids and then calculated the leaf area of the seedling and set a threshold to detect and locate the empty holes. The system has high recognition accuracy and a good replenishing effect. To enable the picking robot to work stably, a Pulse Coupled Neural Network (PCNN) was used by Haiqing et al. (2012) to segment and recognize cucumber based on Least-Squares Support Vector Machine (LS-SVM). Seventy images were tested and the correct recognition rate was 82.9%.

In practical applications, not all targets can be easily separated. The process becomes more difficult when target colors change due to lighting conditions, or are similar to the background. Some researchers have tried to segment the target based on its texture and shape features, but a threshold defining the feature needs to be given and the calculation results are greatly affected by the threshold. Using a statistical method to extract image features is a practical technique. First, the image is modeled from a statistical perspective. Second, it finds the set of pixels with the greatest probability. Ping and Lihong, 2018 proposed an Unsupervised CRF Image segmentation algorithm (ULCRF) by combining the CRF and Latent Dirichlet Allocation (LDA) algorithms. The LDA's segmentation result was used as the initial mark of CRF, and then CRF was used to reflect the difference in pixels. This algorithm can segment the plant from the background or the organ (fruit, leaf and stem) from the plant quickly with a higher level of accuracy. Many studies focused purely on the daytime scenario, but operating at nighttime can overcome the problem of low work efficiency. Long et al. (2015) recognized kiwifruit at nighttime based on machine vision with the help of artificial lighting. First, an RGB camera was placed underneath the canopy so that clusters of kiwifruits could be included in the images. Next, the images were segmented into an R-G color model. Finally, a group of image-processing conventional methods such as the Canny operator were applied to detect the fruits. The developed algorithm recognized 88.3% of the fruits successfully.

# 4 Fruit grading

Machine vision is also applied widely in the field of fruit and vegetable classification, which has promoted the development of the food processing industry greatly. In these studies, the SVM and ANN were commonly used to achieve fruit maturity detection or quality grading. The accuracy of these methods depends on whether there are a large number of reliable data sets to train it. Table 6 compares various methods and results. It can be seen that the sample for model training in most existing studies does not exceed 100 images.

## *4.1 Maturity grading*

Vegetables such as tomatoes and cucumbers have different flowering and fruiting times - the maturity degree varies with the individual. One-time harvesting cannot meet the requirements of maturity, quality and marketing standards, so it needs to detect and select ripe fruits before picking.

Zhang Jingqi (2019) adopted the GrabCut algorithm to segment the tomato image, and the K-means clustering operation was carried out to divide the tomato into green, half-ripe and mature in the HSV color model. Finally, the SVM classifier was designed to classify different fruits. The accuracy rate of classification reached more than 92%. To solve the problem of plant stems and hanging ropes sheltering tomato fruits and interfering with machine recognition, Wang Xinzhong (2012) used color features to segment fruit-sensitive areas and removed background noises such as leaves, stems and facilities in plant factories through morphological calculation. The regional hole-filling algorithm was used to eliminate the bright holes formed by sunlight, and, finally, the sequential method was used to track and segment the boundary of the multi-fruit area. In this study, the tomato was divided into green ripening, ripening, semi-mature, mature and perfect mature. To improve classification efficiency, Iraji (2018) combined neural network, regression model and extreme learning machine together and proposed a Deep Stacking Sparse Auto-Encoder to grade tomato maturity by image data directly. This method was fast as it did not need to extract features from the image.

**Table 6** Comparison of grading methods and results in different studies

| Ref. | Plant | Task | Color model | Method | Sample number | Results |
|---|---|---|---|---|---|---|
| Jinqi (2019) | Tomato | Maturity classification, divided into green ripe, half ripe and mature | HSV | K-means, SVM | 65 | The maturity classification accuracy rate in 100 images can reach more than 92% |
| Wang et al. (2012) | Tomato | Maturity classification is divided into green ripeness, ripening, half ripening, ripe and perfect ripe | HIS, RGB | BPNN | 70 | The correct grading rate in 100 images reaches 93% |
| Iraji (2018) | Tomato | The quality is divided into five levels from good to bad | RGB | ANN, ELM | 840 | The grading accuracy rate in 126 images is 95.5% |
| Li Han et al. (2017) | Tomato | Detect the potential area of the fruit | HSV, RGB | FNCC, Hough Transform circle | Not mentioned | Recognition accuracy for 83 images was 89.2% |
| Hongfei et al. (2018) | Tomato | Cracking fruit detection | RGB | SVM | 50 | The correct recognition rate of 35 images is 97.14% |
| Xie et al. (2019) | Carrot | Classification based on the green shoulder, bending, root, cracked, broken | HSV, RGB | Feature extraction in different color model components | Not mentioned | The total recognition rate among 720 images is 90.9% |

Most reports regard red as indicative of a ripe tomato. However, to reduce damage during transportation and ensure fruits are perfectly ripe when they are shipped to retailers, the grower needs to pick ripe fruit with a green color and hard appearance. These kinds of fruits remain fresh and good-looking when delivered to retailers. These particular demands cause a new problem as the green tomato is similar to the leaves and stems in color, which makes it challenging to separate fruits from the background. Han et al. (2017) used the Fast Normalized Cross-Correlation function (FNCC) to detect the potential area of the fruit, then separated the potential fruit region and removed the non-fruit region based on the histogram information. The Hough Transform circle was used to extract the position of green fruit. The two results were fused to recognize green tomatoes. The recognition rate for 83 images was 89.2%. Different varieties of tomatoes have different colors when they are mature. Judging harvesting time based on color is limited. Fujinaga et al. (2018) outlined a method of generating a map of the tomato growth states to monitor the various stages of tomatoes and to decide on harvesting timing for robots. The tomato growth state map visualized the relationship between the maturity stage, harvest time and yield. The current tomato picking system uses multi-spectral sensors, color and other passive sensors for tomato detection and recognition. There is a narrow detection range and poor anti-interference ability, and detection cannot be performed accurately in real time. Liu (2020) proposed an improved DenseNet Deep Neural Network architecture and the Focal Loss function to enhance the accuracy of feature propagation and reduce the amount of stored data. Experiments showed that the improved network was superior to other existing deep models in terms of detection rate and FPPI, and its computational complexity lower than that of DenseNet algorithm 18% under the same hardware and software configuration.

Additionally, various researchers suggested maturity grading of other fruits based on machine vision, the results of which proved to be pretty good. Dhakshina Kumar et al. (2020) suggested two approaches based on machine learning and transfer learning for the classification of papaya maturity status. Among the transfer learning approaches, VGG19 performed better with 100% accuracy and 112 s training time, which is 6% more than the existing method. In Villaseñor et al.'s (2020) investigation, the maturity stages of bell peppers were referenced by measuring total soluble solids, brix and refractometry. The maturity classifier models using neural networks (RBF-ANN) and fuzzy logic models (ANFIS) were tested and accuracy rates were 100% and 88%, respectively. Absorbance data were used in Yue Xiaoqin's research to establish three maturity recognition models (multivariate linear, multivariate nonlinear and SoftMax regression classifier). Strawberry maturity was divided into three levels with the results showing that the multivariate nonlinear model gives the highest identification rate (over 94%) in the greenhouse (Yue et al., 2020).

These research results confirm the great advantages of machine vision in fruit maturity grading.

### *4.2 Quality classification*

Other studies have attempted to grade fruit by detecting quality. This is important as if one tomato gets squashed, the quality of all tomatoes in the box decreases during transportation, causing economic loss. Removing squashed fruit early on can avoid such losses from happening.

Hongfei et al. (2018) fitted the contour of a single tomato through the improved Hough Transform based on the LSM, and the 2D Gabor wavelet operator was used to extract the texture of the fitted image and judge the fruit. The correct recognition rate of fruit was 91.41%, while the correct recognition rate of squashed fruit was 97.14%. Xie et al. (2019) extracted features of the green shoulder, bending, fibrous root, surface cracking and fracture to grade carrots. In this study, the green shoulder was identified by binarizing the image under the H component of the color model, bending characteristics were obtained by extracting the carrot skeleton, fibrous roots were determined by the slope of the carrot edge under the S component, surface cracking was obtained by binarizing the G and B components, and the fracture was judged by the slope of the carrot tip under the H component. Fu et al. (2016) demonstrated the feasibility of classifying kiwifruit into different grades by calculating fruit length, the maximum diameter of equatorial section and the projected area. Results showed that fruit classification based on the estimated minimum diameter of the equatorial section achieved a success rate of 84.6%. This number increased to 98.3% when a linear combination of the length/maximum diameter of the equatorial section and projected area/length ratios were used.

## 5 The application of deep learning in the plant factory

Machine vision application in plant factories includes but is not limited to the aspects mentioned above. Image analysis is an important research area in these applications. Most methods for image analysis in our survey consist of K-means, SVM, ANN and other basic image processing algorithms, such as Ostu, Decision Tree, Random Forest, Threshold Segmentation and Hough Transform. Features must be handcrafted to suit the particular object in these classical machine-vision approaches. A decision is then made by a hand-crafted approach or learning-based classifiers such as SVM and Decision Tree. Hand-crafted features play a very important role (Kamilaris and Prenafeta-Boldú, 2018). The uncertainty of factors such as lighting conditions, crop growth conditions and viewing distance increases the difficulty of feature selection and design, which affects the robustness and generalization of algorithms.

In recent years, computer vision-based deep learning has proven promising for real-time applications. Based on massive original data and the neural network structure with multiple hidden layers, this technique realizes autonomous learning and is faster, which avoids interference caused by hand-crafted features (Xiaofen Yang, 2019; Nilay and Atul, 2020). Their highly hierarchical structure and large learning capacity make them flexible and adaptable to extremely complex challenges when performing classification and predictions. Deep learning has now been used in the precision manufacturing industry. It is helping manufacturing transform into highly optimized smart facilities by finishing descriptive-analytic, diagnostic-analytic, predictive-analytic and prescriptive-analytic tasks. The application reduces operating costs, keeps up with changing consumer demand, improves productivity and reduces downtime (Wang et al., 2018). In studies by Aditya (2020) and Tabernik et al. (2020), the technique was used to carry out one-shot recognition for defects on the steel surface and to detect surface anomalies of products. The results showed that deep learning can significantly reduce data training requirements and have a good performance.

Methods used in manufacturing are also being used in precision agriculture and plant factories. However, the image acquisition process in agriculture is difficult, as there are complex crop features and backgrounds. Additionally, image acquisition must be timed and synchronized to the crop growth cycle (just once a year for many crops). Zhao Qihui proposed a kind of leaf water stress degrees classification method based on R-CNN and Dense Net169, in which the classification accuracy of the model on the test set was 94.68% (Qihui et al., 2020). Wang Tiewei proposed an approach based on data-balanced Faster R CNN for ripe degree recognition of winter jujube, in which the average recognition accuracy was 98.50%, higher than YOLOv3 (Tiewei et al., 2020). In some studies, deep learning was used to classify the different fruits (Lei et al., 2020), organs of plants (Shi et al., 2019) and small sample data (Wang et al., 2020) and proved to be faster and had good real-time performance.

Nevertheless, while visual pattern recognition based on deep learning has enormous economic value in agriculture, little progress has been made to combine computer vision and crop sciences due to a lack of suitable agricultural image datasets. Most existing datasets primarily focus on common objects or street views, such as Pascal VOC, MS-COCO and ADE20K segmentation datasets (Chiu et al., 2020). In the study by Sebastian and Jörn (2015), a benchmark dataset for crop/weed discrimination, a single plant phenotype was proposed. There were 60 images with annotations in the datasets. For each image, a ground truth vegetation segmentation mask and manual annotation of the plant type (crop vs. weed) were supplied. To improve the detection accuracy and efficiency of crop pests, Liu Liu constructed the standard pest detection datasets of MPD2018 and AgriPest based on accumulated data from the past 5 years. 88K pest images and 582K pest annotations were contained in MPD2018, while 49K pest images and 264K pest annotations were contained in AgriPest (Liu, 2020). However, these datasets are not applicable to plant factories. There is an urgent need to collect and construct plant factory datasets for deep learning techniques in the future.

## 6 Challenges faced by machine vision in plant factories

The application of machine vision in plant factories covers virtually all production stages, from plant seedling, transplanting, cultivation, harvesting and quality inspection to product packaging (Daoliang and Zhen, 2020). This technology has great potential for promoting the intelligent development of plant factories. Although existing studies have solved various problems in practical application scenarios and realized some achievements, there remain challenges.

### *6.1 Failure of image processing algorithms*

In a plant factory, artificial light (LED) is used to provide plants with suitable lighting (Hans et al., 2012; Mehdi and Naruemon, 2017). The two light sources used most are red light (660 nm) and blue light (460 nm). They are turned on at the same time to improve the photosynthetic rate of plants, casting what appears to be a purple hue after the lights are combined (Olle and Viršilė, 2013). Images captured under these lighting conditions present various challenges for segmentation algorithms. We took a healthy green plant (Epipremnum aureum) from an office as an object on which to test. Images of E. aureum under sunlight and artificial light (red and blue) were taken. The background color of both images was white and only the lighting conditions differed. Common segmentation algorithms were used to extract the canopy of the E. aureum. The results are shown in Fig. 3. For the image under sunlight, almost every algorithm could separate the canopy from the background, although there was

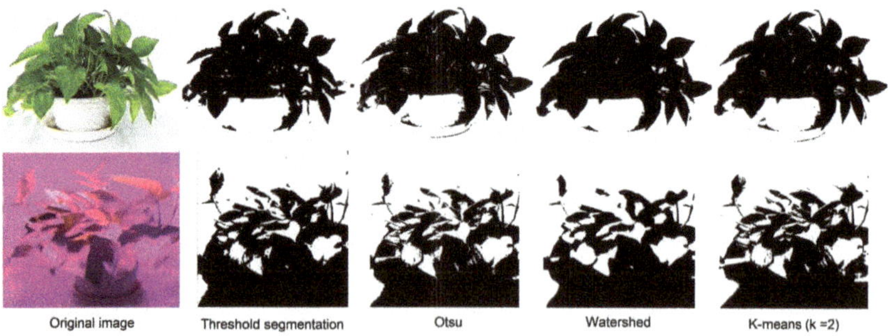


**Figure 3** Comparison of plant segmentation results of different algorithms under sunlight and artificial light.

noise for some algorithms. The watershed algorithm performed best. For the image taken under LED lights, all algorithms failed to separate the canopy area of the plant, even the watershed.

Segmentation is the first step for image processing and is the premise behind feature extraction and image analysis (Umaa Mageswari et al., 2013; Renqian et al., 2016). However, under the different lighting conditions of the plant factory, common algorithms often fail. The indoor environment is complex due to many factors, including irrigation pipes, hanging ropes, monitoring equipment and mechanical devices. The plant itself contains fruits, leaves and stems. The color of the fruit is similar to that of the leaves for some plants such as cucumber. The targets are also easily hidden by leaves. All these factors pose challenges to existing segmentation algorithms.

In the above-mentioned studies, the SVM and ANNs have been used with a supervised learning process. The results of these methods are entirely dependent on whether there are a large number of standard datasets to train it. Indeed, images used for classifier training in most of the references do not exceed 100, so the classification accuracy varies with studies. On the other hand, it takes a lot of time and computer memory from model training to model use. Unsupervised grading algorithms can support classification tasks without training samples, and these methods are the best choice when dealing with large unstructured image repositories (Abbas et al., 2014). They are also more efficient (Zhang et al., 2019). Recently, deep learning has proven promising in precision agriculture, but the progress of this technique in the plant factory is relatively slow. This is partially due to a lack of relevant datasets that encourage the study of agricultural imagery and visual patterns. Therefore, developing unsupervised grading algorithms and constructing standard datasets for agricultural image analysis is an important task for the future. Developing robust algorithms for scenarios in greenhouses and plant factories is a current study trend.

## *6.2 Gaps in basic theoretical knowledge*

Undoubtedly, machine vision technology has significant advantages when applied in a plant factory, as it works in a non-invasive manner and is more timely, accurate and efficient (Ghosal et al., 2018). When collecting information, different plants have different parameters even under the same growth conditions, so information from an individual plant is not representative and cannot be used as a basis for decision-making. However, it is not realistic to collect all plant information in a limited time. Planting architectures (vertical or plane planting) and environmental unevenness (air temperature, humidity, flow speed) will also lead to different plant growths in different areas. How to take all of these factors into account and deploy machine vision inspection points scientifically, whilst reflecting true plant growth status, is still unknown. Additionally, the detection accuracy of plant nutrition stress lacks verification (Singh et al., 2018). Much deviation would cause an excessive or insufficient response of control systems. What is the relationship between environmental parameters, nutritional conditions, plant growth and the diseases and pests in plant factories? Can they be predicted accurately to reduce economic loss? They are not mentioned in recent research.

The gap in theoretical knowledge will seriously affect the application effect of machine vision in plant factories. Therefore, plant growth information and environmental parameters need to be collected to build reliable plant growth models, phenotype models and pests and diseases alert models through deep learning technology. With the development of information technology, big data has become ubiquitous. There is potential to combine big data with machine vision to promote plant factory development in the future.

### *6.3 Lack of image acquisition hardware*

Image quality has a significant effect on image processing. Many images in the surveyed studies were taken manually by digital cameras, which inevitably causes uneven image quality. A few studies have used industrial cameras mounted on a robot to take images, but there was a poor anti-shake effect. In addition, the camera's scope is limited and cannot take images of plants growing in a vertical space. In the long term, as a vision sensor, image acquisition equipment should be the same as other sensors, meaning products with high accuracy and strong applicability at low-cost need to be developed. It is also necessary to develop supporting equipment in the plant factory to meet its application scenarios, such as slide rails, short-range or long-range shooting cameras, mounting bases, etc.

## 7 Conclusion

Machine vision plays an important role in promoting plant factory development. The recent research results of this technique have been reviewed and analyzed in this paper. Specifically, we focused on the aspects of plant growth monitoring, robot operation assistance and fruit grading. Based on the latest references, the challenges that machine vision faces in plant factories have been analyzed. We have reviewed the changing lighting conditions and the complicated indoor environment that cause existing algorithms to fail to segment feature areas from images. Datasets for deep learning are not available in plant factories. Second, there are theoretical knowledge gaps concerning machine vision's application in specific scenarios indoors, which affect the application of the technique. Third, the camera's anti-shake effect is not good and capture scope is limited, and it is incapable of taking images of plants growing in a vertical space. Moreover, the supporting equipment for image acquisition remains insufficient. These problems hinder the further application of machine vision in plant factories. Nevertheless, machine vision is a powerful tool to promote plant factory development, and it has irreplaceable advantages in precision agriculture. We believe that this technology will become more robust, efficient and reliable with the development of information technology.

## 8 Declaration of competing interest

The authors declare that they have no known competing financial interests or personal relationships that could influence the work reported in this chapter.

## 9 Where to look for further information

The latest research progress on vertical farming development and protected horticulture can be accessed on the website of the International Horticultural Society (https://www.ishs.org/) and the website of the Institute of Urban Agriculture, Chinese Academy of Agricultural Sciences (https://iua.caas.cn/).

The latest research progress and the latest algorithms and codes developed for the application of machine vision in agriculture and other fields can be found at the following websites:

i) https://paperswithcode.com/.
ii) ReadPaper (https://readpaper.com/).

Our team is committed to the research of intelligent agricultural robots and facility gardening automation equipment technology. We will publish some latest developments on ResearchGate, which can be accessed through: https://www.researchgate.net/profile/Tian-Zhiwei-2.

## 10 Acknowledgements

The authors thank the experts for editing our paper. We gratefully acknowledge the Local Finance Special Fund of Chengdu Agricultural Science and Technology Center (NASC2020KR05), the Science and Technology Innovation Project Fund of the Chinese Academy of Agricultural Sciences (ASTIP-2020-003) and the Science and Technology Innovation Project Fund of the Chinese Academy of Agricultural Sciences (ASTIP-2020-001). We also thank the anonymous reviewers for their critical comments and suggestions to improve this manuscript.

## 11 References

Abass, O., Golshah, N. and Catherine, T. (2014). Unsupervised classification of images: A review, *International Journal of Image Processing* 8, 2014–2325.

AI and LNO. (2017). Digital image processing for detecting and classifying plant diseases, *Circulation in Computer Science* 2(11), 1–7.

Ali, A. and Hassanein, H. S. (2019). A fungus detection system for greenhouses using wireless visual sensor networks and machine learning. In: *2019 IEEE Globecom Workshops (GC Wkshps)*. Handicap International, Waikoloa, pp. 1–6.

Asseng, S., Guarin, J. R. and Raman, M. (2020). Wheat yield potential in controlled-environment vertical farms, *Proceedings of the National Academy of Sciences of the United States of America* 10, 1–5.

Aya, B., Brett, A. and Shae, F. (2015). A decision support tool for greenhouse farmers in low-resource settings (Vol. 2015). In: *IEEE Global Humanitarian Technology Conference (GHTC)*, Seattle, WA, pp. 292–297.

Aznar-Sánchez, J. A., Velasco-Muñoz, J. F., López-Felices, B. and Román-Sánchez, I. M. (2020). An analysis of global research trends on greenhouse technology: Towards a sustainable agriculture, *International Journal of Environmental Research and Public Health* 17(2), 664–686.

Bonin-Font, F., Ortiz, A. and Oliver, G. (2008). Visual navigation for mobile robots: A survey, *Journal of Intelligent and Robotic Systems* 53(3), 263–296.

Chaudhury, A., Ward, C., Talasaz, A., Ivanov, A. G., Brophy, M., Grodzinski, B., Huner, N. P. A., Patel, R. V. and Barron, J. L. (2019). Machine vision system for 3D plant phenotyping, *IEEE/ACM Transactions on Computational Biology and Bioinformatics* 16(6), 2009–2022.

Chiu, M. T., Xu, X., Wei, Y., Huang, Z., Schwing, A. G., Brunner, R., Khachatrian, H., Karapetyan, H., Dozier, I., Rose, G., Wilson, D., Tudor, A., Hovakimyan, N., Huang, T. S. and Shi, H. (2020). Agriculture-vision: A large aerial image database for agricultural pattern analysis, *Computer Vision and Pattern Recognition* 1, 2825–2835.

Daoliang, L. and Zhen, L. (2020). System analysis and development prospect of unmanned farming, *Transactions of the Chinese Society for Agricultural Machinery* 51(7), 1–12.

Dewi, T., Risma, P. and Oktarina, Y. (2020). Fruit sorting robot based on color and size for an agricultural product packaging system, *Bulletin of Electrical Engineering and Informatics* 9(4), 1438–1445.

Dhakshina Kumar, S., Esakkirajan, S., Bama, S. and Keerthiveena, B. (2020). A microcontroller based on machine vision approach for tomato grading and sorting using SVM classifier, *Microprocessors and Microsystems* 76, 1–12.

Fatemeh, K., Osman, M. T. and Joni, A. (2017). Opportunities and challenges in sustainability of vertical farming: A review, *Journal of Landscape Ecology* 11(1), 98–105.

Franchetti, B., Ntouskos, V., Giuliani, P., Herman, T., Barnes, L. and Pirri, F. (2019). Vision based modeling of plants phenotyping in vertical farming under artificial lighting, *Sensors* 19(20), 43–78.

Fu, L., Sun, Shipeng, Li, R. and Wang, S. (2016). Classification of kiwifruit grades based on fruit shape using a single camera, *Sensors* 16(7), 1012–1024.

Fujinaga, T., Yasukawa, S., Li, B. and Ishii, K. (2018). Image mosaicing using multi-modal images for generation of tomato growth state map, *Journal of Robotics and Mechatronics* 30(2), 187–197.

Gao, M., Yu, M., Guo, H. and Xu, Y. (2019). Mobile robot indoor positioning based on a combination of visual and inertial sensors, *Sensors (Basel)* 19(8), 1773–1779.

Ghosal, S., Blystone, D., Singh, A. K., Ganapathysubramanian, B., Singh, A. and Sarkar, S. (2018). An explainable deep machine vision framework for plant stress phenotyping, *Proceedings of the National Academy of Sciences of the United States of America* 115(18), 4613–4618.

Guanghui, T. and Changyin, L. (2002). Application of computer vision technology in factory agriculture, *Journal of China Agricultural University* 11(2), 62–67.

Guoqin, G. and Ming, L. (2014). Navigating path recognition for greenhouse mobile robot based on K-means algorithm, *Transactions of the Chinese Society of Agricultural Engineering* 30(7), 25–33.

Guoxiang, S. (2016). *Research on fertigation system for greenhouse cucumber based on multi-information fusion*, Doctor Thesis. Nanjing: Nanjing Agricultural University (in Chinese with English abstract).

Gupta, H. and Pahuja, R. (2019). Estimating morphological features of plant growth using machine vision, *International Journal of Agricultural and Environmental Information Systems* 10(3), 30–53.

Haiqing, W., Changyin, J., Baoxin, G. and An, Q. (2012). In-greenhouse cucumber recognition based on machine vision and least squares support vector machine, *Transactions of the Chinese Society for Agricultural Machinery* 43(3), 163–167+180.

Han, L., Man, Z., Gao, Y., Li, M. and Ji, Y. (2017). Green ripe tomato detection method based on machine vision in greenhouse, *Transactions of the Chinese Society of Agricultural Engineering* 33(Suppl.1), 328–334+388.

Hans, G., Mortensen, L. M., Torre, S., Pettersen, H., Dueck, T. and Sand, A. (2012). Light and energy saving in modern greenhouse production, *Acta Horticulturae* 956, 85–97.

Hong, C. (2015). *The innovation of machine vision target identification method for horticultural application*, Doctor Thesis. Beijing: China Agricultural University (in Chinese with English abstract).

Hongfei, L., Minmin, H., Xudong, Z. and Lu, W. (2018). Detection of cracking tomato based on machine vision in greenhouse, *Transactions of the Chinese Society of Agricultural Engineering* 34(16), 170–176.

Hongkun, T., Tianhai, W., Yadong, L., Qiao, X. and Li, Y. (2019). Computer vision technology in agricultural automation: A review, *Information Processing in Agriculture* 7(1), 1–19.

Ikhlef, B. and Sabine, M. (2010). On-line counting of pests in a greenhouse using computer vision, *VAIB* 5, 1–4.

Iraji, M. S. (2018). Comparison between soft computing methods for tomato quality grading using machine vision. *Journal of Food Measurement and Characterization* 13, 1–15.

Jinqi, Z. (2019). *Research on tomato maturity detection in greenhouse based on machine vision*, Master Thesis. Taian: Shandong Agricultural University (in Chinese with English abstract).

Jiqing, C., Hu, Q., Jiahua, W., Xu, G., Wang, Z. and Liu, X. (2020). Extracting the navigation path of a tomato-cucumber greenhouse robot based on a median point Hough transform, *Computers and Electronics in Agriculture* 174, 1–11.

Juncheng, M., Haojie, W. and Xinxing, L. (2017). Downy mildew diagnosis system for greenhouse cucumbers based on image processing, *Transactions of the Chinese Society for Agricultural Machinery* 48(2), 195–202.

Kacira, M., Ling, P. P. and Short, T. H. (2002). Machine vision extracted plant movement for early detection of plant water stress, *Transactions of the ASAE: American Society of Agricultural Engineers* 45(4), 1147–1153.

Kaiyan, L., Jie, C., Hui, S. and Wu, J. (2013). A review on computer vision technologies applied in plant stress detection, *Communications in Computer and Information Science* 363, 192–200.

Kamilaris, A. and Prenafeta-Boldú, F. X. (2018). Deep learning in agriculture: A survey, *Computers and Electronics in Agriculture* 147, 70–90.

Kiani, E. and Mamedov, T. (2017). Identification of plant disease infection using soft-computing: Application to modern botany, *Procedia Computer Science* 120, 893–900.

Kim, J. W. (2010). Trend and direction for plant factory system, *Journal of Plant Biotechnology* 37(4), 442–455.

Kirongo, C. A. (2016). A review of image processing software techniques for early detection of plant drought stress, *International Journal of Computer Applications Technology and Research* 5(6), 376–379.

Kumar, M., Gupta, S., Gao, X. Z. and Singh, A. (2019). Plant species recognition using morphological features and adaptive boosting methodology, *IEEE Access* 7, 163912–163918.

Lakhiar, I. A., Jianmin, G., Syed, T. N., Chandio, F. A., Buttar, N. A. and Qureshi, W. A. (2018). Monitoring and control systems in agriculture using intelligent sensor techniques: A review of the aeroponic system, *Journal of Sensors* 2018, 1–18.

Lee, U., Chang, S., Putra, G. A., Kim, H. and Kim, D. H. (2018). An automated, high-throughput plant phenotyping system using machine learning-based plant segmentation and image analysis, *PLoS ONE* 13(4), e0196615.

Lei, Z., Junsheng, J. and Xinyu, L. (2020). Experimental research on orchard fruit detection based on fast convolutional neural network, *Journal of Chinese Agricultural Mechanization* 41(10), 183–190+210.

Li, Y., Xia, C. and Lee, J. (2009). Vision-based pest detection and automatic spray of greenhouse plant (Vol. 56). In: *IEEE international symposium industrial electronics*, pp. 920–925.

Lian, D. (2020). *Research and application of image recognition method of cucumber disease in greenhouse*, Doctor Thesis. Chengdu: University of Electronic Science and Technology of China (in Chinese with English abstract).

Lima, M. C. F., Leandro, M. E. D. de A., Valero, C., Coronel, L. C. P. and Bazzo, C. O. G. (2020). Automatic detection and monitoring of insect pests–A review, *Agriculturists* 10, 564–579.

Liu, J. (2020). A novel and high precision tomato maturity recognition algorithm based on multi-level deep residual network, *Multimedia Tools and Applications* 12(5), 21–24.

Liu, L. (2020). *Research and application on agricultural crop pest detection techniques based on deep learning*, Doctor Thesis. Hefei: University of Science and Technology of China (in Chinese with English abstract).

Long, F. S., Wang, B., Cui, Y. J., Shuai, S., Gejima, Y. and Kobayashi, T. (2015). Kiwifruit recognition at nighttime using artificial lighting based on machine vision, *International Journal of Agricultural and Biological Engineering* 8(4), 52–59.

Mavridou, E., Vrochidou, E., Papakostas, G. A., Pachidis, T. and Kaburlasos, V. G. (2019). Machine vision systems in precision agriculture for crop farming, *Journal of Imaging* 5(12), 89–121 (in Chinese with English abstract).

Mehdi, M. and Naruemon, W. (2017). Optimizing greenhouse lighting for advanced agriculture based on real time electricity market price, *Mathematical Problems in Engineering* 12, 1–11.

Meixiang, C., Mengmeng, L., Li, Z., Dongmei, W., Wenyong, L., Rui, L. and Ming, L. (2017). Research progress and prospect of monitoring technology of facility agriculture pests based on machine vision, *Agricultural Engineering Technology* 37(31), 10–15.

Meixiang, C., Wenyong, L., Xiaowei, D., Chuanhen, S., Ming, L., Tingwei, N., Yong, Z., Jianpin, Q. and Xinyan, Y. (2014). Pest monitoring system and monitoring method capable of automatically replacing sticky boards, 201410403966, X.

Mengmeng, L. (2018). *Image segmentation and recognition of adult whiteflies and thrips in greenhouse for automatic monitoring devices*, Master Thesis. Shanghai: Shanghai Ocean University (in Chinese with English abstract).

Nilay, G. and Atul, P. (2020). Deep learning methods and applications for precision agriculture, *Machine Learning for Predictive Analysis: Proceedings of ICTIS* 20, 1–13.

Olle, M. and Viršilė, A. (2013). The effects of light-emitting diode lighting on greenhouse plant growth and quality, *Agricultural and Food Science* 22(2), 223–234.

Peng, W. (2019). *Research on vision integrated navigation method of facility agriculture robot*, Master Thesis. Suzhou: Soochow University (in Chinese with English abstract).

Pereira Valente, J. R. and Guiffrida, M. V. (2019). Leaf counting from uncontrolled acquired images from greenhouse workers, *CVPPP*, 1–2.

Ping, Z. and Lihong, X. (2018). Unsupervised segmentation of greenhouse plant images based on statistical method, *Scientific Reports* 8, 1.

Pokharkar, S. R. and Thool, M. V. R. (2012). Early pest identification in greenhouse crops using image processing technique, *International Journal of Computer Science Network* 5420, 2277.

Qichang, Y. (2019). *Plant Factory*. Beijing, China: Tsinghua University Press, pp. 2–14 (in Chinese).

Qihui, Z., Li, L., Miao, Z., Lan, T. and Sigrimis, N. A. (2020). Water stress diagnosis algorithm of greenhouse tomato based on fine-tuning learning, *Transactions of the Chinese Society for Agricultural Machinery* 51(Suppl.1), 340–347+356.

Redmond, S., Fatemeh, K. and Ting, K. (2018). Advances in greenhouse automation and controlled environment agriculture: A transition to plant factories and urban agriculture, *International Journal of Agricultural and Biological Engineering* 11(1), 1–22.

Ren, Y. (2007). *Development of transplanting robot in facility agriculture based on machine vision*, Doctor Thesis, Hangzhou: Zhejiang University (in Chinese).

Renqian, Z., Yuru, H. and Sirui, F. (2016). The development of Markov random field theory and applications on image segmentation algorithm. In: *International Conference on Communication and Electronics Systems* 9, 1–5.

Rui, L. (2011). *Research on mobile robot stereo vision SLAM based on image feature points*, Master Thesis. Harbin: Harbin Institute of Technology (in Chinese with English abstract).

Rupesh, G. (2013). Mundada RGM. Detection and classification of pests in Greenhouse using image processing, *IOSR Journal of Electronics and Communication Engineering* 5, 57–63.

Ruyue, B. (2018). *Design on self-propelled robot for greenhouse based on machine vision*, Doctor Thesis. Nanjing: Nanjing Agricultural University (in Chinese with English abstract).

Scho, N., Berman, S., Dombrovsky, A., Elad, Y., Ignat, T. and Bechar, A. (2017). Development of a robotic detection system for greenhouse pepper plant diseases, *Precision Agriculture* 18(3), 394–409.

Schor, N., Berman, S., Dombrovsky, A., Elad, Y., Ignat, T. and Bechar, A. (2015). A robotic monitoring system for diseases of pepper in greenhouse, *Precision Agriculture* 45, 627–634.

Sebastian, H. and Jörn, O. (2015). A crop/weed field image dataset for the evaluation of computer vision based precision agriculture tasks, In: *Conference: European Conference on Computer Vision*, pp. 105–116.

Shi, W., van de Zedde, R., Jiang, H. and Kootstra, G. (2019). Plant-part segmentation using deep learning and multi-view vision, *Biosystems Engineering* 187, 81–95.

Shimizu, H. and Heins, R. D. (1995). Computer-vision-based system for plant growth analysis, *Transactions of the A.S.A.E. American Society of Agricultural Engineers* 38, 959–964.

Shuai, L. (2018). *Research on automatic early warning robot technology for greenhouse vegetable disease*, Master Thesis. Yinchuan: Ningxia University (in Chinese with English abstract).

Singh, A. K., Ganapathysubramanian, B., Sarkar, S. and Singh, A. (2018). Deep learning for plant stress phenotyping: Trends and future perspectives, *Trends in Plant Science* 23(10), 883–898.

Sisi, Y., Hong, Z. and Xiating, Y. (2014). Research and application of machine vision in plant seedling growth monitoring, *Jiangsu Agricultural Sciences* 33(12), 88–90.

Stephen, N. O. and Haruhiko, M. (2008). Classification of water stress in Sunagoke moss using color texture and neural networks, *Environmental Control in Biology* 23, 21–29.

Story, D. and Kacira, M. (2014). Automated machine vision guided plant monitoring system for greenhouse crop diagnostics, *Acta Horticulturae* 1037, 635–641.

Story, D. and Kacira, M. (2015). Design and implementation of a computer vision-guided greenhouse crop diagnostics system, *Machine Vision and Applications* 26(4), 495–506.

Story, D., Kacira, M., Kubota, C. and Akoglu, A. (2011). Morphological and textural plant feature detection using machine vision for intelligent plant health, growth and quality monitoring, *Acta Horticulturae* 893, 299–306.

Story, D., Kacira, M., Kubota, C., Akoglu, A. and An, L. (2010). Lettuce calcium deficiency detection with machine vision computed plant features in controlled environments, *Computers and Electronics in Agriculture* 74(2), 238–243.

Tabernik, D., Sela, S., Skvarc, J. and Skočaj, D. (2020). Segmentation-based deep-learning approach for surface-defect detection, *Journal of Intelligent Manufacturing* 31(3), 759–776.

Tiewei, W., Yao, Z., Yuxin, S., Han, Z. and Li, J. (2020). Recognition approach based on data-balanced faster R-CNN for winter jujube with different levels of maturity, *Transactions of the Chinese Society for Agricultural Machinery* 51(Suppl.1), 457–463+492.

Umaa Mageswari, S., Sridevi, M. and Mala, C. (2013). An experimental study and analysis of different image segmentation techniques, *Procedia Engineering* 64, 36–45.

Villaseñor Aguilar, M., Bravo Sanchez, M. G.,Padilla-Medina, J., Vázquez-Vera, J. L., Guevara-González, R., García-Rodríguez, F. and Barranco-Gutiérrez, A. (2020). A maturity estimation of bell pepper (*Capsicum annuum* L.) by artificial vision system for quality control, *Applied Sciences* 10(15), 1-18.

Wang, J., Ma, Y., Zhang, Laibin, Gao, R. X. and Wu, D. (2018). Deep learning for smart manufacturing: Methods and applications, *Journal of Manufacturing Systems* 48, 144-156.

Wang, Z., Wang, K., Liu, Z., Wang, X. and Pan, S. (2018). A cognitive vision method for insect pest image segmentation, *IFAC-PapersOnLine* 51, 85-89.

Wang, Z., Zou, C. and Cai, W. (2020). Small sample classification of hyperspectral remote sensing images based on sequential joint Deeping learning Model, *IEEE Access* 8, 71353-71363.

Wenjuan, S. (2017). *Research on monitoring cucumber leaf wetness duration in solar greenhouse based on machine vision*, Master Thesis. Shanghai: Shanghai Ocean University (in Chinese with English abstract).

Wudneh, T. B. and Vanitha, D. V. (2019). Implementation of IoT with image processing in greenhouse monitoring system, *International Journal of Innovative Technology and Exploring Engineering* 8(9), 509-514.

Xie, W., Wang, F. and Yang, D. (2019). Research on carrot surface defect detection methods based on machine vision, *IFAC-PapersOnLine* 52(30), 24-29.

Xinzhong, W. (2012). *Studies on information acquisition and path planning of greenhouse tomato harvesting robot with selective harvesting operation*, Doctor Thesis. Zhenjiang: Jiangsu University (in Chinese with English abstract).

Yane, Z. (2005). *Study on vision-based crop growth diagnostic mechanism and method in greenhouse*, Doctor Thesis. Beijing: China Agricultural University (in Chinese with English abstract).

Yang, X. and Sun, M. (2019). A survey on deep learning in crop planting, *IOP Conference Series: Materials Science and Engineering* 490(6), 1-7.

Yating, M. (2016). *Research on greenhouse tomato disease diagnosis based on GA-BP network*, Master Thesis. Harbin: Northeast Forestry University (in Chinese with English abstract).

Yongxin, R., Yuzhi, T., Huihua, Y. and Chen, B. (2009). Visual navigation of cucumber picking robot based on fuzzy control, *Journal of Jiangsu University (Natural Science Edition)* 30(4), 343-346.

Younse, P. and Burks, T. (2005). Intersection detection and navigation for an autonomous greenhouse sprayer using machine vision (Vol. 3). In: *SAE Annual International Meeting*, pp. 34-45.

Younse, P. J. and Burks, T. F. (2007). Greenhouse robot navigation using KLT feature tracking for visual odometry, *CIGR Journal*, IX, 1-22.

Yue, X.-Q., Shang, Z. Y., Yang, J. Y., Huang, L. and Wang, Y. (2020). A smart data-driven rapid method to recognize the strawberry maturity, *Information Processing in Agriculture* 7(4), 575-584.

Yusuf, H. (2014). Applications of intelligent machine vision in plant factory. *Proceedings of the 19th World Congress* 5, 24-29.

Zhang, L., Zhang, L., Du, B., You, J. and Tao, D. (2019). Hyperspectral image unsupervised classification by robust manifold matrix factorization, *Information Sciences* 485, 154-169.

Zhang, X., Xingxing, L., Shixiong, L., Wu, G. and Tan, Y. (2019). *Greenhouse Intelligent Recognition and Replanting Control System Based on Machine Vision* (Vol. 23). American Society of Agricultural and Biological Engineers, pp. 1-7.

Zhen, G. (2018). *Research on nondestructive detection of facility crops growth information and development of the equipment*, Doctor Thesis. Zhenjiang: Jiangsu University (in Chinese with English abstract).

Zhenyu, Y. (2014). *Study on key technology of potted-seedling transplanting information acquisition using machine vision*, Doctor Thesis. Beijing: China Agricultural University, pp. 10–50 (in Chinese with English abstract).

Zhonghong, X. (2013). *Research on key algorithms in picking robot image processing system*, Doctor Thesis. Nanjing: Nanjing Agricultural University (in Chinese with English abstract).

Zili, H. (2019). *Design and implementation of control system in plant factory based on visual patrol robot*, Doctor Thesis. Anhui: University of Science and Technology of China (in Chinese).

# Chapter 10

# Plant phenotyping of individual plants towards optimal environmental control in plant factories

*Eri Hayashi, Japan Plant Factory Association, Japan*



## 1 Introduction

A plant factory with artificial lighting (PFAL) is a closed plant production system with an almost airtight and well-insulated structure. It allows for high resource utilization efficiency and produces high quantities of high-quality plants year-round; environmental factors of PFALs can be precisely controlled, entirely independent of outdoor conditions (Goto, 2012; Kozai, 2013; Kozai et al., 2019a; Kozai et al., 2020; Hayashi et al., 2020; Ohyama et al., 2020; Orsini et al., 2020; SharathKumar et al., 2020; van Delden et al., 2021; Hayashi et al., 2022). Unlike open-field or other forms of controlled environment agriculture (i.e. greenhouses), PFALs, having an almost airtight and well-insulated structure, enable simple and accurate data acquisition and environmental control, time-series analysis, and prediction related to plant production, including resource use efficiency or productivity.

In recent years, there has been an increasing global interest in PFALs as a novel plant production system that can concurrently contribute to the solution of a range of significant global challenges, such as food security, environmental

http://dx.doi.org/10.19103/AS.2023.0126.11

preservation, resource-saving, and people's quality of life. Pre-commercial growth chamber research has been actively conducted already since the 1960s and 1970s, while commercial production in PFAL was initiated in the 1970s (Mitchell, 2022). Particularly since the mid-2010s, an increasing number of people and organizations with diverse backgrounds have become involved in business, system development, and research related to PFAL (Hayashi, 2019). More PFALs are being developed and commercially operated in many regions, including Japan and other countries in Asia, North America, Europe, and the Middle East (Hayashi, 2019).

Occasionally, terminology differs among countries, organizations, and individuals, such as PFAL, vertical farm, and indoor farm (Hayashi, 2019). Details on the differences in terminologies related to PFAL are described in Kozai (2022).

## 2 Current challenges with plant factories with artificial lighting

Considering the potential of PFALs in contributing concurrently to the solution of significant global and local challenges, including food, environment, energy and resources, and people's quality of life and health, there is further room for improvement in technical and economic aspects (Hayashi et al., 2020). The unique theoretical benefits of the PFAL can be realized only when the system is properly engineered, constructed, and operated (Kozai, 2017; Kozai et al., 2019b; Kozai et al., 2022). In reality, various PFAL challenges remain, such as high initial investment, optimal cultivation systems, environmental controls, plant growth, and socio-economic impact, which can be overcome by improving productivity in PFALs (van Delden et al., 2021; Kozai et al., 2022).

The plant productivity of a PFAL can be described by the resource type applied, including electric energy, time consumed for cultivation, cultivation space invested, and materials, together with monetary terms (Kozai et al., 2022). Table 1 describes the estimated resource and monetary productivity of major commercial PFALs producing leafy lettuce in Japan as of 2019 (Kozai et al., 2022). The values in Table 1 cover the volume of salable products after trimming for packing. In Japan, most major commercial large-scale PFALs grow, harvest, and pack leafy lettuce by each head of over 100 g or 200 g per plant, not precut lettuce of baby leaf. Therefore, uniformity of plant growth and quality is critical to improving the productivity of commercial PFALs, while ensuring the absence of physiological disorders (e.g. tip-burn), consistent plant morphology, and good taste or nutritional value to maximize the product value to meet customers' needs.

**Table 1** Estimated resource and monetary productivity of commercial PFALs in Japan

| Resources applied | Range of resource productivity | Range of monetary productivity (kg/US$) | Range of production cost (US$/kg) |
|---|---|---|---|
| Electric energy | 0.11–0.14 kg kW $h^{-1}$ | 0.645–0.755 | 1.09–1.28 |
| Labor hours | 7.7–10.0 kg $h^{-1}$ | 0.591–0.770 | 1.18–1.63 |
| Cultivation area (per day) | 0.25–0.33 kg $m^{-2}$ $day^{-1}$ | 0.482–0.645 | 1.28–1.72 |
| Other resources | - | 0.609–0.827 | 1.00–1.36 |
| Total | - | 0.166–0.219 | 4.55–5.99 |

In 2019, one US$ and one Euro€ were 110 JPY and 130 JPY, respectively. Note that wages (US$/h) are around two times higher in the United States and some European countries than in Japan as of 2021. *Source*: Adapted from: Kozai et al. (2022).

However, even though environments are controlled in PFALs, variations in the growth of individual plants have been found, which adversely affect the resource and/or monetary productivity of PFAL operations (Kozai et al., 2018; Nagano et al., 2019; Hayashi et al., 2020; Hayashi et al., 2022). Even when plant growth seemed uniform at first glance, in practice, relatively large variations in the fresh weight per plant can be found (Kozai et al., 2018). Similar variations in plant height, morphology, and concentrations of secondary metabolites per plant can also be observed (Kozai et al., 2018).

Uniformity of plant growth by reducing two- or three-dimensional (2D or 3D) variations in plant phenotype or traits (e.g. growth) over the cultivation tray or space can contribute to improving productivity in PFALs (Kozai et al., 2018). Variations in the plant phenotype of individual plants are mainly attributable to three factors (Kozai et al., 2018; Hayashi et al., 2020, Hayashi et al., 2022):

- Microenvironment (in the plant canopy) or environment (over the plant canopy);
- Genotype (species and cultivar); and
- Management (human and machine intervention), including seed treatments and transplanting.

There is a continuous interaction between plant phenotype and micro-environment or environment and management, as shown in Fig. 1. Two- or three-dimensional variations in plant phenotype across the cultivation tray or space are partially attributed to 2D or 3D variations in the surrounding microenvironment of individual plants, such as photosynthetic photon flux density (PPFD) and air current speed over and within plant individuals or communities in the tray or space, among other factors (Kozai et al., 2018).

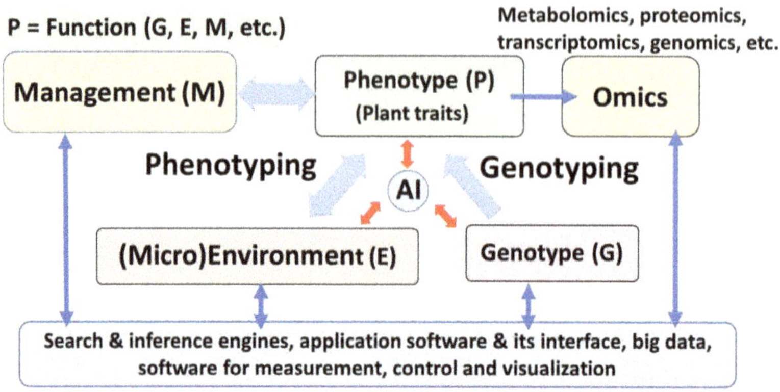


**Figure 1** Scheme of interaction between phenotype and microenvironment or environment and management, as affected by genotype. Although genotype significantly impacts phenotype, this chapter refers to the interaction between phenotype, microenvironment, and management factors. Source: Redrawn from Kozai (2018).

# 3 Plant phenotyping

Plant phenotyping plays an integral role in understanding how the microenvironment and management factors impact phenotypic variations of individual plants (Fiorani and Schurr, 2013; Kumar et al., 2015; Golbach et al., 2016; Kozai et al., 2018; Hayashi et al., 2020; Hayashi et al., 2022). Among the remaining challenges for research and development using the leading technologies to advance PFALs, plant phenotyping is crucial for improving plant productivity towards next-generation sustainable PFALs (Kozai et al., 2018).

## *3.1 Overview of plant phenotyping*

Plant phenotyping is a set of methodologies and protocols used to accurately study plant performance, growth, architecture, and composition at different organization scales, from organs to canopies; phenotyping is a nondestructive and noninvasive way of analyzing plant trait dynamics or the time course of plant phenotypes (Fiorani and Schurr, 2013; Kumar et al., 2015; Golbach et al., 2016; Kozai et al., 2018; Kozai et al., 2020; Hayashi et al., 2020; Hayashi et al., 2022).

Plant phenomics is a research field related to plant phenotypes or phenotyping and their applications. It has developed rapidly recently partly due to technological advances in sensors, geolocalization systems, image processing, data analytical techniques, and omics tools (Ninomiya et al., 2019). The number of publications related to plant phenomics has increased significantly since 2010 (Ninomiya et al., 2019). The development of plant

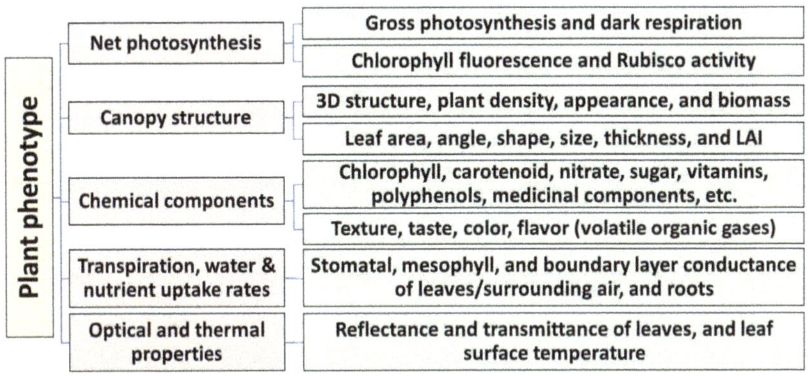


**Figure 2** Examples of plant phenotypes. Source: Redrawn from Kozai (2018).

phenomics has particularly been promoted by breeders to improve plant genetics in order to meet the mature development achieved by the high-throughput genotyping techniques (Ninomiya et al., 2019).

Figure 2 shows examples of plant phenotypes comprising morphological and physiological traits in different plant densities, including rate variables with a unit of time, such as net photosynthesis and transpiration rates and water and nutrient uptake rates.

### 3.2 Phenotyping studies in open fields and greenhouses

Phenotyping studies have recently focused on developing accurate, high-throughput, and rapid phenotyping systems and image-processing algorithms (Kumar et al., 2015). Most plant phenotyping studies have been undertaken in open fields or greenhouses. With respect to species for phenotyping studies, cereal crops (wheat, maize, rice, barley) are primarily mentioned in the published papers besides model research species such as *Arabidopsis* or others (Yang et al., 2020).

Plant phenotyping has been conducted by combining novel technologies such as advances in noninvasive imaging and spectroscopy, image analysis, robotics, and high-performance computing (Furbank and Tester, 2011; Golbach et al., 2016). Plant measurements have typically used a combination of monochromatic, red-green-blue (RGB), fluorescent, thermal (also called far-infrared thermal), multispectral, and tomographic imaging, which provide signals that are often an indirect measure of the physiological plant response or traits (e.g. chlorophyll fluorescence as an indicator of photosynthesis activity) (Golbach et al., 2016; Vialet-Chabrand and Lawson, 2019; Yang et al., 2020; Song et al., 2021).

The simplest, most cost-effective, and widely used phenotyping method involves measuring morphological traits, biomass, and shoot growth using RGB camera images (Furbank and Tester, 2011; Yang et al., 2020). By using 2D images, the projected shoot area is extracted following image preprocessing and segmentation (Furbank and Tester, 2011; Guo et al., 2017). Problems with 2D RGB imaging are typically caused by overlapping of leaves during plant development and by background noise (e.g. from soil), in which estimation of the biomass of a plant from 2D images results in high inaccuracies (Furbank and Tester, 2011; Golbach et al., 2016).

Plant measurements have also been performed using 3D models. Geometrically accurate plant models enable more accurate measurements and modeling of biological processes (Golbach et al., 2016; Song et al., 2021). Three-dimensional imaging techniques mainly use image-based and laser scanning-based techniques (Yang et al., 2020). Image-based 3D plant reconstruction can be generated using different techniques, including multi-view stereo, structure-from-motion (SfM), or other new methods (Pound et al., 2014; Golbach et al., 2016). However, it is time-consuming to create a 3D model, which is proportional to the complexity of plant morphology (Golbach et al., 2016). Golbach et al. (2016) introduced the shape-from-silhouette (SfS) method used for the 3D reconstruction of seedling phenotyping. Easy Dense Cloud Phenotyping (EasyDCP), an open-source software tool to measure phenotypic traits of potted plants using a regular digital camera and software, provides a means for affordable and high-throughput 3D phenotyping (Feldman et al., 2021).

As described earlier, chlorophyll fluorescence has also been commonly used to indicate photosynthetic activity (Furbank and Tester, 2011; Vialet-Chabrand and Lawson, 2019). Thermal imaging has been used to measure plant or leaf temperature and to assess leaf water status (e.g. evaporation) as affected by environmental conditions (Furbank and Tester, 2011; Yang et al., 2020). Vialet-Chabrand and Lawson (2019) applied thermal imaging analysis to derive temporal responses of stomatal conductance under fluctuating conditions using dynamic leaf energy balance.

Imaging spectroscopy, including multispectral and hyperspectral cameras that are capable of scanning wavebands of interest at high resolution, has also been used to provide much spectral and spatial information (Furbank and Tester, 2011; Yang et al., 2020). Notably, the near-infrared region, which includes strong water-absorbing bands, has been used to measure water content (leaf water thickness) (Furbank and Tester, 2011). Flourishing research fields include assessing plant composition using hyperspectral reflectance sensors for remote estimation of plant composition, health, disease status, and other physiological traits (Furbank and Tester, 2011; Lowe et al., 2017; Yang et al., 2020). However, processing hyperspectral imaging data can be complicated and hyperspectral cameras are costly (Yang et al., 2020).

In addition to plant shoot measurement, there has been a significant need for root phenotyping. Various methods have been explored to assess root systems dynamically, including high-resolution and dynamic 3D imaging, occasionally using transparent growth media (Furbank and Tester, 2011; Yang et al., 2020). Magnetic resonance imaging (MRI) and x-ray computed tomography have also been used for root phenotyping research (Furbank and Tester, 2011). Müllers et al. (2022) used MRI and scanning of harvested roots to measure root length in studying different water uptake rates per root length over depth. There have also been advances in seed phenotyping. Two-dimensional cameras, open-source, and user-friendly image analysis software have enabled extraction of seed size and shape (including area, perimeter, length, width, etc.) and seed color (Tanabata et al., 2012; Yang et al., 2020; Zhu et al., 2020).

### *3.3 Challenges of plant phenotyping in open fields and greenhouses*

Interpretation of raw measurement signals in plant phenotyping typically requires extra steps to convert the raw data into biologically significant findings (Fahlgren et al., 2015; Ghanem et al., 2015; Vialet-Chabrand and Lawson, 2019). Ensuring accuracy and comparability of results requires accurate measurements of environmental conditions in the laboratory or in the field at appropriate spatial and temporal scales (Furbank and Tester, 2011). The reproducibility of results in individual experiments has often been unsatisfactory due to unexplained environmental variations, along with difficulties in translating results from experimental settings into normal growing conditions in the field or greenhouse (Yang et al., 2020). Environmental and management factors have not received the same attention as phenotyping techniques. Cheaper, more accessible solutions, standardization (definition of standards), and a broad multidisciplinary approach have been needed to increase the reproducibility and applicability of plant phenotyping, which has been mainly conducted in the open fields or greenhouses under controlled (or semicontrolled) environmental laboratory settings (Furbank and Tester, 2011).

## 4 Plant phenotyping in plant factories with artificial lighting

Plant phenotyping can be applied relatively easily in PFALs which provide more control of the cultivation environment compared with a greenhouse or open field (Guo, 2018). PFAL studies using camera images have been conducted for years. The projected leaf area (using 2D camera images captured from above the plant canopy or seedling trays) of the plant canopy using camera images has been used particularly for growth prediction (Shibata et al., 1992; Giacomelli et al., 1996; Itoh and Yamamoto, 2005; Nagano et al., 2019) or for estimating electric energy consumption or efficiency (Furuyama et al., 2017; Jayalath and van Iersel, 2021) in PFALs.

There have been increasing phenotyping studies relevant to PFALs that apply machine learning or deep learning to quantify the projected leaf area or canopy structure to predict plant growth (Rizkiana et al., 2021; Hwang et al., 2022; Kim et al., 2022; Ueno et al., 2022). Besides RGB images, the depth image, which is the 2D image where each pixel represents the distance between the sensor and the in-scene object, can be captured using an RGB-depth (D) sensor (Franchetti et al., 2019; Buxbaum et al., 2022). This provides a 3D reconstruction for predicting height, weight, and leaf area across a plant community at a certain period of plant growth (Franchetti et al., 2019; Buxbaum et al., 2022). Nagano et al. (2019) focused on the time-series movement of plant leaves for growth prediction, using optical flow analysis, image processing, and machine learning in PFAL.

Plant phenotyping has also recently been conducted to identify plant physiological diseases (e.g. tip-burn) in PFALs. Shimamura et al. (2019) used binary classification of single lettuce images to identify tip-burn, using machine learning with convolutional neural networks (Shimamura et al., 2019; Gozzovelli et al., 2021). Gozzovelli et al. (2021) detected tip-burn using low-cost RGB cameras, computer vision, and deep learning, combining classification and segmentation with self-supervised masks and deep learning-based data augmentation. Current case studies on the commercial application of plant phenotyping in PFALs can be found in Part 6 of this book.

However, most previous phenotyping studies in PFALs have been limited to the method development of image analysis for monitoring or predicting plant growth, but not for analyzing the interaction between plant phenotype and microenvironment or environment and management factors in PFALs. In most previous studies, the frequency of image capture used to estimate plant growth has been limited to a few times per day or even a week only during a certain period of plant growth (Hayashi et al., 2022). To elucidate phenotypic variations in the microenvironment of individual plants and management factors, and how these plant phenotypes interact with the microenvironment and management, it is crucial to measure individual plant phenotypes as time-series data (Hayashi et al., 2022).

## 5 Plant cohort research in plant factories with artificial lighting

A recent development affecting PFALs can be cohort research (Kozai et al., 2018; Hayashi and Kozai, 2018; Kozai et al., 2019a; Kozai et al., 2020; Hayashi et al., 2020; Hayashi et al., 2022). A plant cohort refers to a relatively large group or community of plant individuals with a statistical factor (such as planting density, cultivar, and cultivation system) (Kozai et al., 2018). In plant cohort research, the life cycle phenome history of individual plants can be obtained noninvasively and continuously, and analyzed from seed sowing to harvesting with phenotyping units, together with the time-series data on the microenvironment, management, and resource inputs/outputs in PFAL (Hayashi et al., 2020).

## 5.1 Variations in the phenotype of individual plants (Hayashi et al., 2020; Hayashi et al., 2022)

The variation in plant phenotype (e.g. growth) of individual plants can be explored to analyze how a phenotype interacts with the surrounding microenvironment and management factors, using time-series 2D camera images. A modular plant phenotype measurement system for germination and seedling production has been developed that focused on practicality and scalability in commercial PFALs, achieving a low cost with small cameras and sensors and a cultivation method similar to commercial PFALs. Experiments on the germination and initial seedling stages of romaine lettuce (*Lactuca sativa* L. var. longifolia), whose growth strongly impacts plant productivity, have been conducted as part of plant life cycle research.

Seed germination and the initial seedling stage are among the most critical stages of plant growth. The eco-physiological status of young individual plants impacts subsequent plant growth processes (Kitajima, 2003; Hanley and May, 2006; Zhang et al., 2008; Zheng et al., 2011; Soltani et al., 2015). Variations in the timing of germination and individual seedling growth significantly influence the uniformity of plant growth and, consequently, monetary productivity in PFALs. It is critical to identify variations in individual germination time and initial growth of seedlings and to understand how they interact with the surrounding microenvironment and management factors.

## 5.2 Variations in the germination time of individual seeds as affected by microenvironment and management factors (Hayashi et al., 2020)

A plant phenotype measurement system was used to assess the individual germination time (GT) of romaine lettuce (*L. sativa* L. var. longifolia) in seed trays, each with 300 seeds, using time-series data captured on 2D camera images. The technique was used to analyze each GT as affected by its surrounding microenvironment [seed surface temperature of individual seed (SST), volumetric percentage of nutrient solution in a seed tray (VPNS), and air temperature] and management [uncoated seed (U) and coated seed (C)] factors of individual seeds for plant cohort research.

The results demonstrated that individual GT variations were affected by each microenvironment (SST, VPNS, and air temperature) and management factors such as seed type (U and C). Each treatment with different air temperatures or VPNS, both with U and C, reflected the variance of GT. Besides microenvironmental factors, the results showed that management factors (U and C) affected GT; GT and GT's variance varied, responding to seed type even of the same variety. The phenotyping platform revealed the impacts of microenvironment and management factors on GT of individual seeds.

## 5.3 Phenotypic variations in the growth of cotyledons and initial true leaves as affected by microenvironments (Hayashi et al., 2022)

Experiments were carried out to estimate cotyledon unfolding time from sowing (Hc) and the time-series projected leaf area of cotyledons (Ap-c) and true leaves (Ap-t) of individual seedlings of romaine lettuce (*Lactuca sativa* L. var. longifolia), using 2D camera images. The experiment analyzed how the surrounding microenvironment of PPFD and the absence or presence of nutrients in water in the seedling tray affects variations in the growth of individual seedlings. RGB images were acquired every 0.5 h in light of leaf movement (Hoshi and Takiguchi, 1995; Nagano et al., 2019).

Table 2 shows the distributions of horizontal PPFD within the trays (width 605 × length 300 mm). Eighteen different PPFDs in each tray (Tray 1 with water

**Table 2** Treatment and codes

| Code | Nutrients/water | PPFD classifications | Distribution of horizontal PPFD (μmol $m^{-2}$ $s^{-1}$) |
|---|---|---|---|
| Nutrients-H | | High | 175, 181, 187, 195, 202, 208 |
| Nutrients-M | Nutrients (Tray 2) | Medium | 140, 145, 147, 148, 151, 165 |
| Nutrients-L | | Low | 84, 102, 102, 104, 118, 126 |
| Water-H | | High | 168, 175, 181, 190, 197, 205 |
| Water-M | Water (Tray 1) | Medium | 136, 142, 144, 148, 153, 161 |
| Water-L | | Low | 84, 100, 104, 112, 121, 123 |

*Source*: Adapted from: Hayashi et al. (2022).

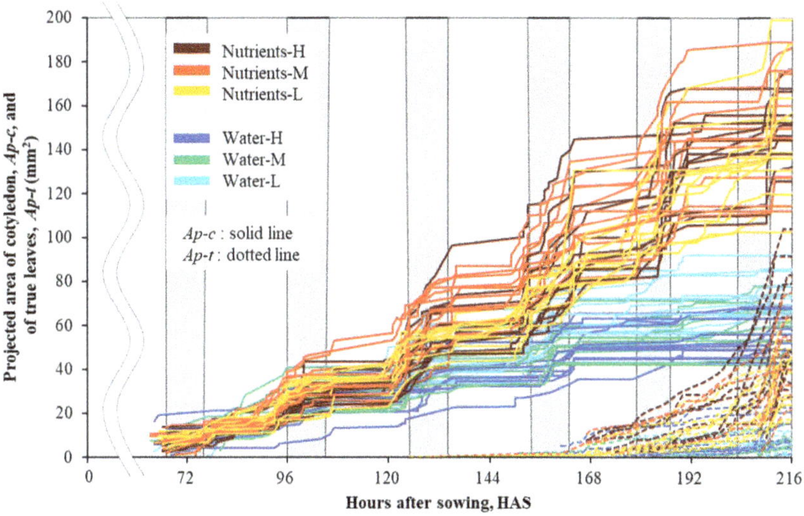


**Figure 3** Time-course increasing changes in projected areas of the cotyledon (Ap-c) and true leaves (Ap-t) of individual seedlings with nutrients and water from hours to cotyledon unfolding from sowing (Hc) to 216 h after sowing, each with distributions of horizontal PPFDs. Source: Adapted from: Hayashi (2022).

and Tray 2 with nutrients) were categorized into three classifications (6 PPFDs/classification) based on the level of PPFD: low (L), medium (M), and high (H), 84 ≤ 126, 126 < 166, 166 ≤ 208 μmol $m^{-2}$ $s^{-1}$, respectively.

Figure 3 shows the time-course increase in the Ap-c and Ap-t of individual seedlings, both with nutrients and water from Hc to 216 h after sowing, each with distributions of horizontal PPFDs. Although Ap-c and Ap-t with water remained reasonably constant over time, Ap-c and (in particular) Ap-t significantly increased with nutrients with more fluctuation over time. Ap-c with nutrients continued to increase even after the initial development of Ap-t (Fig. 3). It appeared that, particularly for Ap-c, a relative increase was caused mainly in the dark period rather than the photoperiod (Fig. 3). As clearly shown in Fig. 3, relative growth in Ap-c and Ap-t was more extensive with a higher PPFD with nutrients. However, it was found that the time-course changes of individual seedlings with nutrients showed more variations compared to water (Fig. 3). In agreement with the actual individual measurements, variations in seedling growth were identified even under similar microenvironments. Furthermore, the results demonstrated more significant variations in seedlings with higher relative growth.

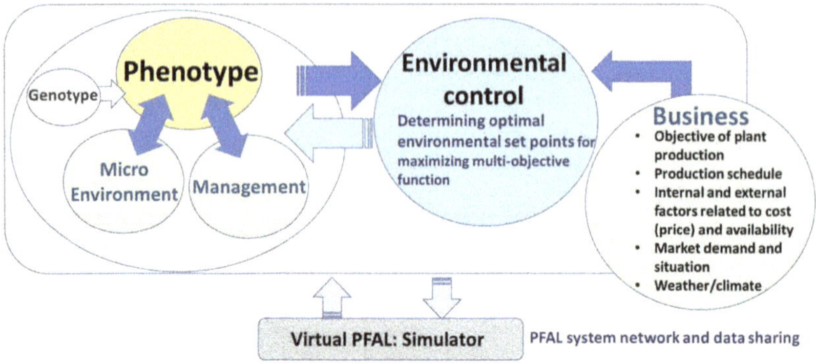


**Figure 4** Plant phenotype-based environmental control considering business-related factors.

## 6 Phenotype-based environmental control in plant factories with artificial lighting

In plant cohort research, time-series phenotype data can be acquired by noninvasive, nondestructive, and continuous plant phenotyping, together with data on the microenvironment, management, and resource inputs/outputs in PFAL (Kozai et al., 2018; Hayashi and Kozai, 2018; Kozai et al., 2019a; Kozai et al., 2020; Hayashi et al., 2020; Hayashi et al., 2022).

Figure 4 describes plant phenotype-based environmental control, considering business-related factors. The business-related factors relate to the objective of plant production and business strategy, production schedule, internal and external factors related to changing cost (price) and availability of inputs, market needs and situation, weather/climate, among others. By considering those business-related factors, phenotype-based environmental control increases the resource and monetary productivity of commercial PFALs.

Plant phenotyping in PFALs can show the interaction between plant phenotype, microenvironment, and management factors concurrently during plant production. Using time-series data sets, including rate variables, corresponding to business-related factors, plant phenotype-based environmental control makes it possible to determine optimal set points of environmental factors for maximizing multiobjective function to control plant phenotype (Kozai et al., 2018; Hayashi and Kozai, 2018; Kozai et al., 2019a; Kozai et al., 2020; Hayashi et al., 2020; Hayashi et al., 2022).

Having an almost airtight and well-insulated structure, PFALs enable simple and accurate data acquisition and environmental control, time-series analysis, and prediction related to plant production, along with improving plant productivity, selection of seedlings for grading and breeding new cultivars in PFAL (Kozai et al., 2018; Hayashi and Kozai, 2018; Kozai et al., 2019a; Kozai et al., 2020; Hayashi et al., 2020, Hayashi et al., 2022). Plant phenotyping based on cohort research in PFALs has unique benefits, such as high reproducibility, predictability, and traceability, which makes it easier for data sharing and analyzing via a PFAL system network to advance the evolutional research and business platform. These unique benefits can be further extended by integrating a real PFAL with a virtual PFAL simulator (Kozai, 2018).

## 7 Conclusion

An almost airtight and well-insulated PFAL enables simple and accurate environmental control and related data acquisition, time-series analysis, and prediction related to plant phenotypes. Plant phenotyping based on cohort research in PFALs can lead to plant phenotype-based optimal environmental control for plant production, selecting seedlings for grading and breeding in PFALs, which can simplify interactions between plant phenotype and the microenvironment and management for the life cycle of individual plants. Phenotype-based optimal environmental control taking into account business-related factors, together with virtual PFAL, can improve uniformity and productivity in PFALs and advance the evolutional PFAL research and sustainable business, playing an essential part in the well-being of society.

## 8 Acknowledgements

The author deeply appreciates Toyoki Kozai for sharing valuable ideas and providing technical advice and comments. The author also thanks Yumiko Amagai and other colleagues for experimental research related to this chapter.

## 9 References

Buxbaum, N., Lieth, J. H. and Earles, M. (2022). Nondestructive plant biomass monitoring with high spatio-temporal resolution via proximal RGB-D imagery and end-to-end deep learning. *Front. Plant Sci.* 13, 758818. doi:10.3389/fpls.2022.758818.

Fahlgren, N., Gehan, M. A. and Baxter, I. (2015). Lights, camera, action: High-throughput plant phenotyping is ready for a close-up. *Curr. Opin. Plant Biol.* 24, 93–99. doi:10.1016/j.pbi.2015.02.006.

Feldman, A., Wang, H., Fukano, Y., Kato, Y., Ninomiya, S. and Guo, W. (2021). EasyDCP: An affordable, high-throughput tool to measure plant phenotypic traits in 3D. *Methods Ecol. Evol.* 12(9), 1679–1686. doi:10.1111/2041-210X.13645.

Fiorani, F. and Schurr, U. (2013). Future scenarios for plant phenotyping. *Annu. Rev. Plant Biol.* 64, 267–291.

Franchetti, B., Ntouskos, V., Giuliani, P., Herman, T., Barnes, L. and Pirri, F. (2019). Vision based modeling of plants phenotyping in vertical farming under artificial lighting. *Sensors (Basel)* 19(20), 4378. doi:10.3390/s19204378.

Furbank, R. T. and Tester, M. (2011). Phenomics - technologies to relieve the phenotyping bottleneck. *Trends Plant Sci.* 16(12), 635-644. doi:10.1016/j.tplants.2011.09.005.

Furuyama, S., Ishigami, Y., Hikosaka, S. and Goto, E. (2017). Estimation of lighting energy consumption required for red leaf lettuce production under different blue/red ratios and light intensity conditions in a plant factory with artificial lighting. (in Japanese). *J. SHITA* 29(2), 60-67.

Ghanem, M. E., Marrou, H. and Sinclair, T. R. (2015). Physiological phenotyping of plants for crop improvement. *Trends Plant Sci.* 20(3), 139-144.

Giacomelli, G. A., Ling, P. P.. and Morden, R. E. (1996). An automated plant monitoring system using machine vision. *Acta Hortic.* 440, 377-382, doi:10.17660/ActaHortic.1996.440.66.

Golbach, F., Kootstra, G., Damjanovic, S., Otten, G. and Van de Zedde, R. (2016). Validation of plant part measurements using a 3D reconstruction method suitable for high-throughput seedling phenotyping. *Mach. Vis. Appl.* 27(5), 663-680.

Goto, E. (2012). Plant production in a closed plant factory with artificial lighting. *Acta Hortic.* (956), 37-49. doi:10.17660/ActaHortic.2012.956.2.

Gozzovelli, R., Franchetti, B., Bekmurat, M. and Pirri, F. (2021). Tip-burn stress detection of lettuce canopy grown in plant factories. In: Proceedings of the IEEE/CVF International Conference on Computer Vision (ICCV) Workshops, 1259-1268.

Guo, W., Zheng, B., Duan, T., Fukatsu, T., Chapman, S. and Ninomiya, S. (2017). EasyPCC: Benchmark datasets and tools for high-throughput measurement of the plant canopy coverage ratio under field conditions. *Sensors (Basel)* 17(4), 798. doi:10.3390/s17040798.

Guo, W. (2018). Automated characterization of plant growth and flowering dynamics using RGB images. In: *Smart Plant Factory: The Next Generation Indoor Vertical Farms*, Kozai, T. (Ed.). Springer, Berlin, 385-393.

Hanley, M. E. and May, O. C. (2006). Cotyledon damage at the seedling stage affects growth and flowering potential in mature plants. *New Phytol.* 169(2), 243-250, doi:10.1111/j.1469-8137.2005.01578.x.

Hayashi, E. (2019). Chapter 30. Selected PFALs in Japan. In: *Plant Factory: An Indoor Vertical Farming System for Efficient Quality Food Production* (2nd edn.), Kozai, T., Niu, G. and Takagaki, M. (Eds). Elsevier. 437-454.

Hayashi, E. (2022). Phenotypic variations in the growth of lettuce seedlings as affected by the surrounding microenvironment and management factors. Ph.D. thesis. Chiba University, Matsudo, Japan.

Hayashi, E. and Kozai, T. (2018). Phenotyping and AI-based environmental control and breeding for PFAL. In: *Smart Plant Factory: The Next Generation Indoor Vertical Farms*, Kozai, T. (Ed.). Springer, Berlin, 405-411.

Hayashi, E., Amagai, Y., Maruo, T. and Kozai, T. (2020). Phenotypic analysis of germination time of individual seeds affected by microenvironment and management factors for cohort research in plant factory. *Agronomy* 10(11), 1680. doi:10.3390/agronomy10111680.

Hayashi, E., Amagai, Y., Kozai, T., Maruo, T., Tsukagoshi, S., Nakano, A. and Johkan, M. (2022). Variations in the growth of cotyledons and initial true leaves as affected by photosynthetic photon flux density at individual seedlings and nutrients. *Agronomy* 12(1), 194. doi:10.3390/agronomy12010194.

Hoshi, T. and Takiguchi, T. (1995). Continuous measurement of movement in spinach leaves using video cameras. (in Japanese). *J. Agric. Meteorol.* 51(2), 123–130. doi:10.2480/agrmet.51.123.

Hwang, Y., Lee, S., Kim, T., Baik, K. and Choi, Y. (2022). Crop growth monitoring system in vertical farms based on region-of-interest prediction. *Agriculture* 12(5), 656. doi:10.3390/agriculture12050656.

Itoh, H. and Yamamoto, H. (2005). A modeling of lettuce growth by system identification (Part 1). (in Japanese). *J. Jpn. Soc. Agric. Mach.* 67(6), 71–80. doi:10.11357/jsam1937.67.6_71.

Jayalath, T. C. and van Iersel, M. W. (2021). Canopy size and light use efficiency explain growth differences between lettuce and mizuna in vertical farms. *Plants (Basel)* 10(4), 704. doi:10.3390/plants10040704.

Kim, T., Lee, S. H. and Kim, J. O. (2022). A novel shape based plant growth prediction algorithm using deep learning and spatial transformation. *IEEE Access* 10, 37731–37742. doi:10.1109/ACCESS.2022.3165211.

Kitajima, K. (2003). Impact of cotyledon and leaf removal on seedling survival in three tree species with contrasting cotyledon functions. *Biotropica* 35(3), 429–434. doi:10.1111/j.1744-7429.2003.tb00597.x.

Kozai, T. (2013). Resource use efficiency of closed plant production system with artificial light: Concept, estimation and application to plant factory. *Proc. Jpn. Acad. Ser. B Phys. Biol. Sci.* 89(10), 447–461. doi:10.2183/pjab.89.447.

Kozai, T. (Ed.) (2018). *Smart Plant Factory: The Next Generation Indoor Vertical Farms*. Springer, Berlin, 724.

Kozai, T. (2022). Terms related to PFALs. In: *Plant Factory Basics, Applications and Advances*, Kozai, T., Niu, G. and Masabni, J. (Eds). Elsevier, Amsterdam, Netherlands, 11–23.

Kozai, T., Lu, N., Hasegawa, R., Nunomura, O., Nozaki, T., Amagai, Y. and Hayashi, E. (2018). Plant cohort research and its application. In: *Smart Plant Factory: The Next Generation Indoor Vertical Farms*, Kozai, T. (Ed.). Springer, Berlin, Germany, 413–431.

Kozai, T., Amagai, Y. and Hayashi, E. (2019a). Towards sustainable plant factories with artificial lighting (PFALs): From greenhouses to vertical farms. In: *Achieving Sustainable Greenhouse Cultivation*, Marcelis, L., Heuvelink, E. (Eds). Burleigh Dodds Science Publishing, Cambridge, UK, 177–202.

Kozai, T., Uraisami, K., Kai, T. and Hayashi, E. (2019b). Indexes, definition, equation and comments on productivity of plant factory with artificial lighting. (in Japanese). *Nogyo Oyobi Engei* 94, 661–672.

Kozai, T., Hayashi, E. and Amagai, Y. (2020). Plant factories with artificial lighting (PFALs) toward sustainable plant production. *Acta Hortic.* (1273), 251–260. doi:10.17660/ActaHortic.2020.1273.34.

Kozai, T., Uraisami, K., Kai, K. and Hayashi, E. (2022). Chapter 12. Productivity: Definition and application. In: *Plant Factory: Basics, Applications and Advances*, Kozai, T., Niu, G. and Masabni, J. (Eds). Elsevier, Amsterdam, Netherlands, 197–216.

Kumar, J., Pratap, A. and Kumar, S. (2015). *Phenomics in Crop Plants: Trends, Options and Limitations*. Springer, New Delhi, India, 314.

Lowe, A., Harrison, N. and French, A. P. (2017). Hyperspectral image analysis techniques for the detection and classification of the early onset of plant disease and stress. *Plant Methods* 13, 80. doi:10.1186/s13007-017-0233-z.

Mitchell, C. A. (2022). History of controlled environment horticulture: Indoor farming and its key technologies. *Hortscience* 57(2), 247–256. doi:10.21273/HORTSCI16159-21.

Müllers, Y., Postma, J. A., Poorter, H., Kochs, J., Pflugfelder, D., Schurr, U. and van Dusschoten, D. (2022). Shallow roots of different crops have greater water uptake rates per unit length than deep roots in well-watered soil. *Plant Soil* 481(1–2), 475–493. doi:10.1007/s11104-022-05650-8.

Nagano, S., Moriyuki, S., Wakamori, K., Mineno, H. and Fukuda, H. (2019). Leaf-movement-based growth prediction model using optical flow analysis and machine learning in plant factory. *Front. Plant Sci.* 10, 227. doi:10.3389/fpls.2019.00227.

Ninomiya, S., Baret, F. and Cheng, Z. M. (2019). Plant phenomics: Emerging transdisciplinary science. *Plant Phenom.* 2019, 2765120. doi:10.34133/2019/2765120.

Ohyama, K., Yamaguchi, J. and Enjoji, A. (2020). Resource utilization efficiencies in a closed system with artificial lighting during continuous lettuce production. *Agronomy* 10(5), 723. doi:10.3390/agronomy10050723.

Orsini, F., Pennisi, G., Zulfiqar, F. and Gianquinto, G. (2020). Sustainable use of resources in plant factories with artificial lighting (PFALs). *Eur. J. Hortic. Sci.* 85(5), 297–309. doi:10.17660/eJHS.2020/85.5.1.

Pound, M. P., French, A. P., Murchie, E. H. and Pridmore, T. P. (2014). Automated recovery of three-dimensional models of plant shoots from multiple color images. *Plant Physiol.* 166(4), 1688–1698. doi:10.1104/pp.114.248971.

Rizkiana, A ., Nugroho, A. P., Salma, N. M., Afif, S., Masithoh, R. E., Sutiarso, L. and Okayasu, T. (2021). Plant growth prediction model for lettuce (Lactuca sativa.) in plant factories using artificial neural network. *IOP Conf. Ser. Earth Environ. Sci.* 733(1), 012027. doi:10.1088/1755-1315/733/1/012027.

SharathKumar, M., Heuvelink, E. and Marcelis, L. F. M. (2020). Vertical farming: Moving from genetic to environmental modification. *Trends Plant Sci.* 25(8), 724–727. doi:10.1016/j.tplants.2020.05.012.

Shibata, T., Iwao, K. and Takano, T. (1992). Growth prediction of lettuce plants by image processing. *Act. Hortic.* (319), 689–694. doi:10.17660/ActaHortic.1992.319.111.

Shimamura, S., Uehara, K. and Koakutsu, S. (2019). Automatic identification of plant physiological disorders in plant factories with artificial light using convolutional neural networks. *IJNCAA* 9(1), 25–30.

Soltani, E., Ghaderi-Far, F., Baskin, C. C. and Baskin, J. M. (2015). Problems with using mean germination time to calculate rate of seed germination. *Aust. J. Bot.* 63(8), 631–635. doi:10.1071/BT15133.

Song, P., Wang, J., Guo, X., Yang, W. and Zhao, C. (2021). High-throughput phenotyping: Breaking through the bottleneck in future crop breeding. *Crop J.* 9(3), 633–645. doi:10.1016/j.cj.2021.03.015.

Tanabata, T., Shibaya, T., Hori, K., Ebana, K. and Yano, M. (2012). SmartGrain: High-throughput phenotyping software for measuring seed shape through image analysis. *Plant Physiol.* 160(4), 1871–1880. doi:10.1104/pp.112.205120.

Ueno, T., Nagano, S., Moriyuki, S., Hayashi, T. and Fukuda, H. (2022). Optimized excess-green image binarization for accurate estimation of lettuce seedling Leaf-Area in a plant factory. *Environ. Control Biol.* 60(3), 153–159. doi:10.2525/ecb.60.153.

van Delden, S. H., SharathKumar, M., Butturini, M., Graamans, L. J. A., Heuvelink, E., Kacira, M., Kaiser, E., Klamer, R. S., Klerkx, L., Kootstra, G., Loeber, A., Schouten, R. E., Stanghellini, C., van Ieperen, W., Verdonk, J. C., Vialet-Chabrand, S., Woltering, E. J., van de Zedde, R., Zhang, Y. and Marcelis, L. F. M. (2021). Current status and future challenges in implementing and upscaling vertical farming systems. *Nat. Food* 2(12), 944–956. doi:10.1038/s43016-021-00402-w.

Vialet-Chabrand, S. and Lawson, T. (2019). Dynamic leaf energy balance: Deriving stomatal conductance from thermal imaging in a dynamic environment. *J. Exp. Bot.* 70(10), 2839–2855. doi:10.1093/jxb/erz068.

Yang, W., Feng, H., Zhang, X., Zhang, J., Doonan, J. H., Batchelor, W. D., Xiong, L. and Yan, J. (2020). Crop phenomics and high-throughput phenotyping: Past decades, current challenges, and future perspectives. *Mol. Plant* 13(2), 187–214. doi:10.1016/j.molp.2020.01.008.

Zhang, H., Zhou, D., Matthew, C., Wang, P. and Zheng, W. (2008). Photosynthetic contribution of cotyledons to early seedling development in Cynoglossum divaricatum and Amaranthus retroflexus. *New Zeal J. Bot.* 46(1), 39–48. doi:10.1080/00288250809509752.

Zheng, W., Wang, P., Zhang, H. and Zhou, D. (2011). Photosynthetic characteristics of the cotyledon and first true leaf of castor (Ricinus communis L.). *Aust. J. Crop Sci.* 5(6), 702–708. Available at: http://www.cropj.com/zheng_5_6_2011_702_708.pdf.

Zhu, F., Paul, P., Hussain, W., Wallman, K., Dhatt, B. K., Sandhu, J., Irvin, L., Morota, G., Yu, H. and Walia, H. (2020). SeedExtractor: An open-source GUI for seed image analysis. *Front. Plant Sci.* 11, 581546. doi:10.3389/fpls.2020.581546.

# Part 4

## Spectral manipulations for controlling the growth and quality of leafy greens

# Chapter 11

# Growth and quality of lettuce in vertical farms as affected by red:blue and red:far-red ratios

*Wenqing Jin, Wageningen University & Research and Priva B.V., The Netherlands; and Hua Li, Wageningen University & Research, The Netherlands*



## 1 Introduction

Rapid urbanization, climate change, and land degradation are putting the current food supply chain under pressure (van Delden et al., 2021). During the COVID-19 pandemic, the general public realized the importance of developing localized food production systems to improve the robustness of food supply. Any solution needs to be resource efficient in terms of land and water use, which are becoming increasingly scarce, while food safety assurance is also a necessity. Indoor vertical farming is a potentially feasible food production system to address all these challenges, by making food production independent of solar light and natural conditions (SharathKumar et al., 2020; van Delden et al., 2021).

Indoor vertical farming is a system in which crops are grown in a closed system, typically in stacks on top of one another, allowing a substantially greater production efficiency per land area (Asseng et al., 2020; Jin et al., 2023) and improving safety with barely any pesticide usage and potential residues in food (SharathKumar et al., 2020; van Delden et al., 2021). In indoor vertical farming, all growth factors are controlled. Specifically, the

http://dx.doi.org/10.19103/AS.2023.0126.15

light in indoor vertical farms is supplied by artificial lamps, and the climate is managed by heating, ventilation, and air-conditioning systems combined with well-developed soilless cultivation systems, such as aquaponic, aeroponic, and nutrient film techniques, etc. (Eldridge et al., 2020; SharathKumar et al., 2020; van Delden et al., 2021). Since indoor vertical farming is similar to the modern manufacturing factory, in which all processes are controlled and planned, plant factory with artificial light is another term that is widely used (Kozai, 2013).

The development of indoor vertical farming is hindered by its high energy consumption, especially in terms of lighting systems (Kozai, 2013). The application of light-emitting diodes (LED) in indoor vertical farming is a potential solution, albeit with some limitations (Haitz and Tsao, 2011; Mitchell et al., 2015; Kusuma et al., 2020). It is particularly promising for indoor vertical farming not only due to high energy efficacy but also due to its flexibility in light spectrum combinations which can be matched to the plant's photoreceptor characteristics (Kusuma et al., 2020). In addition, the low heat emission of LED can reduce the distance between the LED modules and the plant canopy, and its relatively long-term radiation output means longer periods without the necessity for replacement. LEDs can be switched on and off instantly without needing to warm up (Morrow, 2008). Overall, LED facilitates a fundamental revolution in plant photobiology research (Pattison et al., 2018) and crop production in controlled environment agriculture. However, LED is not the ultimate solution to reduce the high levels of energy consumption in indoor vertical farm production (Graamans et al., 2018). Under current circumstances, increasing light use efficiency is required.

Crop biomass production, or harvesting biomass, results from a series of ongoing physiological processes including light interception, photosynthesis, biomass partitioning, respiration, etc. This chapter discusses the effects of different light spectra and planting density on light interception, photosynthesis, biomass production, and quality of lettuce, one of the most consumed vegetables worldwide. Various features of lettuce make it a model crop to be exploited in indoor vertical farm research, including its fast growth, low plant height, marketable shoot organs, and sensitivity to different light spectra (Dougher and Bugbee, 2001). As lettuce is predominately consumed fresh (Mulabagal et al., 2010), its marketability depends not only on its visual appearance, but also the concentration of health-promoting phytochemicals such as vitamins, polyphenols, and fiber (Llorach et al., 2008). Indoor vertical farming is beneficial for improved and uniform product quality due to its highly controllable growth conditions (van Delden et al., 2021). This chapter also reviews the effects of different light spectra and planting density on lettuce quality in indoor vertical farms.

# 2 Light spectra: Red and blue

## 2.1 Red light

Red light, defined as 600–700 nm wavelengths (Pattison et al., 2018; Kusuma et al., 2020), is the first spectrum achieved by LED (~660 nm) in controlled environment agriculture. Red light LED was first combined with fluorescent lamps and utilized in indoor vertical farming (Bula et al., 1991; Morrow, 2008) due to its high efficiency in photosynthetic $CO_2$ assimilation in plants (McCree, 1972; Inada, 1976). However, when red light is applied alone, 'Red light syndrome' occurs, which manifests as lowered photosynthetic capacity and maximum quantum yield of chlorophyll fluorescence, and unresponsive stomatal conductance (Trouwborst et al., 2016; Miao et al., 2019). Therefore, monochromatic red light does not always lead to the highest biomass production in lettuce (Yanagi et al., 1996; Yorio et al., 2001; Meng and Runkle, 2019) compared to other spectra at the same photon flux density. Red light can be applied with other spectra in controlled environment agriculture for crop production. For instance, a minor fraction of additional blue (~7%) is sufficient to reverse red light syndrome in cucumber (Hogewoning et al., 2010).

## 2.2 Blue light

Blue light is typically defined as the radiation of 400–500 nm (Pattison et al., 2018; Kusuma et al., 2020). The photosynthesis efficiency (number of $CO_2$ molecules assimilated per number of blue photons absorbed) peak of blue light is less than that of red light, which makes blue light the second most photosynthetically efficient light spectrum (McCree, 1971, 1972). High output blue light LED was invented shortly after red light LED modules (Nakamura et al., 1994). The first commercial horticultural LED modules were made with a mixture of red and blue light (Kusuma et al., 2020). Blue light is known to affect plant architecture, leaf and photosynthetic characteristics, and antioxidant concentrations (Hogewoning et al., 2010; Kook et al., 2013; Pennisi et al., 2019a; Kong and Nemali, 2021). These effects are mediated by several blue-light receptors, such as cryptochrome 1 and 2, phototropin, and phytochrome A (Lin, 2000). However, the impact of monochromatic blue light is still debated for several species and cultivars. Liang et al. (2021) reported a taller crop, larger leaf area, and improved shoot dry weight of cucumber under monochromatic blue light than other combinations of red and blue light under the same photosynthesis photon flux density (PPFD), but these effects were not observed in tomato. In lettuce, Spalholz et al. (2020) found a higher stem length under monochromatic blue light than other combinations of red and blue light on loose-leaf and head lettuce. However, a larger leaf area was only found on head lettuce and neither lettuce type produced higher dry mass under monochromatic blue light. In addition, Wang et al. (2016) reported that lettuce grown under monochromatic blue light produced the least shoot dry weight and leaf area. In addition, blue light improves antioxidant capacity and the development of compact morphology was associated with effective control of gray mold by *Botrytis cinerea* in lettuce (Kook et al., 2013). A combination of red and blue light as the growing light for lettuce has been suggested in general, since red light LED modules not only have better efficacy but also higher energy conversion efficacy from energy to photon flux.

## 2.3 Red:blue ratio

The effect of red:blue ratio (R:B ratio, defined as the ratio of photon flux density between red and blue light, where any given R:B ratio below is the ratio at plant canopy surface) has been extensively studied in lettuce in indoor vertical farming. Son and Oh (2013) tested five R:B ratios (0.7, 1.1, 1.9, 2.8. and 6.7) under the PPFD of 177 μmol $m^{-2}$ $s^{-1}$ and found the fresh and dry weight and leaf area positively correlated with red light fraction. This study suggested a higher R:B ratio for a better yield. This is in line with Wang et al. (2016) who cultivated lettuce under four R:B ratios (1, 4, 8, and 12) in which the highest R:B ratio led to the highest shoot dry weight and leaf area. The photosynthetic capacity ($A_{max}$: measured under 90% red light and 10% blue light condition) and net photosynthesis rate (measured under growth PPFD 200 μmol $m^{-2}$ $s^{-1}$) dropped with the increment of R:B ratio. Pennisi et al. (2019b) found similar results in that a higher R:B ratio reduced stomatal conductance under a series of R:B ratios (0.5, 1, 2, 3, and 4) and the fresh and dry weight improvement by the higher fraction of red light showed saturation at the R:B = 3. However, Spalholz et al. (2020) tested three R:B ratios (0.25, 1, and 4) on two lettuce cultivars under 200 μmol $m^{-2}$ $s^{-1}$ PPFD and no fresh weight difference was found. Dry weight improvement was only observed on cultivar Red Oak Leaf from the highest R:B ratio treatment. R:B ratios not only have an impact on growth but also on metabolite accumulation in lettuce.

Growing lettuce under different R:B ratios in indoor vertical farms can stimulate the accumulation of health-promoting compounds and reduce an excessive amount of potentially harmful compounds (van Delden et al., 2021). Lettuce generally has a high concentration (>2500 mg $kg^{-1}$ fresh weight) of nitrate among leafy vegetables (Santamaria, 2006). Although nitrate is not inherently hazardous, it can be metabolized into nitrite in the human body which can be harmful (Gangolli et al., 1994; Chan, 2011). According to the Joint Expert Committee of the Food and Agriculture and the European Commission's Scientific Committee on Food, the daily intake of nitrate should be between 0 mg $kg^{-1}$ and 3.7 mg $kg^{-1}$ body weight (Santamaria, 2006). It has been reported that supplemental red light reduced the nitrate content of lettuce (Ohashi-Kaneko et al., 2007). An R:B ratio of around 8:1 is optimal for lowering the nitrate concentration in hydroponic lettuce (Urbonavičiūtė et al., 2007), while researchers found that a 4:1 ratio of R:B showed a significant drop from 3125.7 mg $kg^{-1}$ to 1064.6 mg $kg^{-1}$ in nitrate content of hydroponic lettuce (Wanlai et al., 2013).

Some recent studies have observed that the concentration of nitrate and soluble sugars showed opposite trends in lettuce with a change in R:B ratio (Chen et al., 2014, 2021). Soluble sugars, including glucose, fructose, and sucrose, are important nutrients in lettuce. As substrates for respiratory metabolism, retaining soluble sugars at a certain level may delay lettuce texture deterioration, i.e. shape and crispness (Min et al., 2021). Meanwhile, high soluble sugar concentration provides lettuce with more intense sweetness (Lin et al., 2013). Chen et al. (2014) found that combined red and blue light showed the lowest concentration of nitrate and the highest concentration of soluble sugars in lettuce compared to monochromatic red or blue light treatments. Chen et al. (2021) presented a more systematic study on nitrate and soluble sugar concentrations in lettuce using 11 R:B ratios. The authors found that 0–10% red light provided lettuce with about 670 mg $kg^{-1}$ nitrate, which dropped significantly to around 440 mg $kg^{-1}$ in lettuce grown under 80% red light. Correspondingly, a fraction of red light higher than 50% increased soluble sugar content (Chen et al., 2021). It is suggested that soluble sugars could lower the amount of nitrate by enhancing the accumulation of nitrate reductase messenger RNA (Lillo, 1994). Additionally, adding a broad spectrum to red and blue light could increase the sugar content and reduce the nitrate content, which adds market value to lettuce. Accordingly, red and blue light, together with other spectra like white light, increased soluble sugar concentration and decreased nitrate concentration in lettuce compared to the combination of red and blue light treatment (Lin et al., 2013). The authors suggested that this outcome could result from enhanced photosynthetic activity as white light penetrates the plant canopy deeper than red and blue light. Considering the metabolic interaction between soluble sugars and nitrate, both higher fractions of red and/or a fraction of broad spectrum light can be strategically applied during lettuce growth in indoor vertical farming.

Ascorbic acid (AsA, vitamin C) and dehydroascorbic acid (DHA) are major antioxidants in lettuce, preventing enzymatic pinking and browning (Min et al., 2021). Chen et al. (2011) observed that AsA concentrations in lettuce were higher under monochromatic blue light and a combination of red and blue light than under monochromatic red light. Likewise, under continuous LED light conditions, AsA concentration in lettuce increased with decreasing R:B ratios (Zha et al., 2020). The authors further concluded that a larger fraction of blue light increased AsA concentrations due to the elevated AsA regeneration instead of biosynthesis. However, another research group has recently found that monochromatic red light more effectively promoted AsA accumulation in lettuce than other R:B ratio treatments (Chen et al., 2021). Interestingly, according to the same report, lettuce grown under increasing fractions of blue light (between 30% and 60%) exhibited increased AsA content. These findings suggest that AsA accumulation is sensitive to R:B ratio.

Secondary metabolites, in low concentrations, also play an essential role in plant quality. The biosynthesis and accumulation of these compounds are mainly stimulated by light, mediated by photoreceptors. The responses of phytochemical compound accumulation in lettuce to several R:B ratios were investigated by Pennisi et al. (2019b). The authors recommended that an R:B ratio between 2 and 4 increased flavonoid concentrations compared to a ratio of 0.5, as well as fluorescent light. Another study found that a 3:1 ratio of R:B LED light applied during lettuce growth enhanced accumulation of anthocyanins and flavonoids and induced corresponding gene expression compared to lettuce grown under fluorescent light (Sng et al., 2021). Anthocyanin and flavonoids are stronger antioxidants than vitamin C in plants (Bagchi et al., 1998). Red and blue light increased flavonoid production by promoting the expression of *Phenylalanine ammonia-lyase* (*PAL*) in lettuce (Son et al., 2013). *PAL* catalyzes the first step of phenylpropanoid metabolism, producing precursors to flavonoids and lignins (Huang et al., 2010). Interestingly, in young lettuce plants, the *PAL* expression levels were similar between red and blue light and fluorescent light treatments; in mature lettuce plants, *PAL* expression level was higher from red and blue light treatment, indicating that flavonoid accumulation in lettuce is also plant age dependent (Sng et al., 2021).

Besides an increase in anthocyanins and flavonoids, red and blue light also affect carotenoid concentration in lettuce. Carotenoids are a vital pigment protecting photosynthetic components from damage caused by the active triplet state of the chlorophyll molecule (Landrum and Bone, 2001; Chen et al., 2017). Red combined with blue light increased carotenoid content in lettuce compared to white LED light (Amoozgar et al., 2017). Sng et al. (2021) also observed that lettuce grown under red and blue LED light showed a higher concentration of carotenoids than when grown under fluorescent light. In this research, flavor- and taste-related secondary compounds, such as terpenoids and alkaloids, were also increased by red and blue light.

Taken together, altering the R:B ratios precisely from high to low before harvest in indoor vertical farming can improve the cost efficiency of lettuce production, its visual quality, and nutritional value.

## 3 Light spectra: Far-red and red

### 3.1 Far-red light

Far-red light (FR light), defined as a range of 700–800 nm (or 700–750nm) (Pattison et al., 2018; Kusuma et al., 2020; Zhen and Bugbee, 2020), is a waveband of photons outside photosynthetic photons. FR light is partially absorbed and mostly reflected or transmitted by leaves (Taiz et al., 2015). Leaf absorption rate drops rapidly from 685 nm (McCree, 1971) and barely any absorption can be measured above 750 nm (McCree, 1971; Zhen and Bugbee, 2020). When applied alone, FR light barely drives any photosynthetic assimilation. However, FR light increases photosynthesis efficiency by interacting with shorter wavelengths (i.e. the red radiation from 600 nm to 685 nm). The synergistic effect is more pronounced than the sum of separately applied red and FR light (Emerson and Rabinowitch, 1960; Zhen and van Iersel, 2017) due to improved electron-transport efficiency. There are two reaction centers, namely photosystem I (PSI) and photosystem II (PSII), operating in series to transfer electrons along the electron-transport chain and carry out photochemical reactions (Baker, 2008). The peak wavelength for excitation of PSI and PSII are 700 nm and 680 nm, respectively. Applying red and FR light together can therefore balance the excitation level between the two photosystems and improve photosynthesis efficiency. It has been reported that by adding FR light on top of red and blue light, quantum yield (the number of molecules undergoing the process divided by the number of photons absorbed by the system) of PSII increased and non-photochemical quenching (the excitation energy within PSII is lost as heat) decreased (Baker, 2008; Zhen and van Iersel, 2017; Zhen et al., 2019).

The rate of photosynthesis can be increased by adding FR light over photosynthetically active radiation (PAR) since photosynthesis efficiency is improved. In a series of studies (Zhen and van Iersel, 2017; Zhen et al., 2018; Zhen, 2020; Zhen and Bugbee, 2020), researchers proposed that FR is necessary to keep photosynthesis efficient. Several experiments have been conducted with different narrow wavebands of FR light on greenhouse and indoor vertical farm-grown lettuce, as well as some other species. Results show that photochemical efficiency is enhanced by FR of 700–750 nm and thus the photosynthesis rate. Hence, a FR wavelength of between 700 nm and 750 nm was suggested to be included in the definition of PAR (Zhen and Bugbee, 2020). Zhen and Bugbee (2020) found the canopy gross photosynthetic rate (the sum of instantaneous net photosynthesis rate and dark respiration) increased in 12 C3 and two C4 species similarly when the same photon flux density of additional FR light and PAR was applied. Zhen (2020) further revealed that the canopy instantaneous net photosynthesis rate ($P_n$) in lettuce dropped when

FR light was switched off for a short period during growth. Similarly, Zou et al. (2019) reported a $P_n$ increase when adding FR light to the combination of red and blue background light. However, Ji et al. (2019) did not find any $P_n$ increase by adding FR light either to combined red and blue or white light in tomato. Jin et al. (2021) found similar $P_n$ of lettuce under FR-enriched and FR-absent growth environment with the same PPFD, with a combination of red and blue light as background light. In addition, Legendre and van Iersel (2021) and Jin et al. (2021) estimated canopy and crop daily PPFD interception through the growing period and concluded the additional FR light did not increase biomass production efficiency (harvesting dry mass per unit of cumulative intercepted PPFD) over identical growing PPFD.

Debate continues on the instantaneous effect of additional FR light on $CO_2$ assimilation rate. However, there is agreement that the photosynthesis system acclimatizes to growing light. In the long term, photosynthetic capacity is affected by growing light, as plants acclimatize to the growing environment. Chlorophyll is a pigment that absorbs photosynthetic photons to drive photosynthesis, which is positively correlated to leaf light absorption rate (Evans and Poorter, 2001). FR light acclimatization downregulates photosynthetic capacity through decreasing chlorophyll content (Heraut-Bron et al., 1999; Demotes-Mainard et al., 2016), as leaf thickness decreases for crops grown under FR-enriched conditions compared to FR-absent conditions (Vogelmann, 1993; Zou et al., 2019; Zhen and Bugbee, 2020). Reduced chlorophyll content lowers the photosynthetic capacity per unit of the leaf area. Therefore, a trade-off effect was proposed by Ji et al. (2019) and Kalaitzoglou et al. (2019) on tomato in which the positive effect on photosynthesis efficiency by additional FR light was canceled out by lowered photosynthetic capacity. This is in agreement with the lowered photosynthesis rate tendency of FR acclimated lettuce under high PPFD (~1500 μmol $m^{-2}$ $s^{-1}$) revealed by Zou et al. (2021).

### 3.2 Red:far-red ratio

Red:far-red ratio (R:FR ratio) is defined as the photon flux density ratio between red and FR light. Any given R:FR ratio below is the ratio at the plant canopy surface. This ratio is used in shade avoidance research. In natural conditions where the sun is the only light source, the ratio between red and FR light perceived by leaves decreases when vegetation proximity or shading by leaves occurs, as red photon is substantially absorbed but FR photon is mostly scattered. A lowered R:FR ratio, therefore, is an early signal to the proximate of neighboring plants (Ballaré et al., 1990; Pierik and De Wit, 2014). The plant perceives the varying R:FR ratio through photoreceptors known as phytochromes, which exist in two photo-convertible isomers, namely $P_r$ and $P_{fr}$ (Ballaré, 1999; Smith, 2000). These two isomers can transform into each other by absorbing red or FR photons, thus determining the equilibrium between $P_r$ and $P_{fr}$ in the plant (Smith, 2000; Pierik and De Wit, 2014). $P_{fr}$ is the FR absorbing state and $P_r$ is the R absorbing state (J. C. Sager et al., 1988). A rebalanced equilibrium through a lowered R:FR ratio induces shade avoidance syndrome (SAS), otherwise known as shade intolerance, which includes responses such as retarded germination, increased stem length and/or leaf elongation, leaves moving upward (hyponasty), a higher fraction of assimilate partitioning to fruit, early flowering, and/or increased specific leaf area (Smith and Whitelam, 1997; Franklin, 2008; Bongers et al., 2014; Ji et al., 2020).

A higher fraction of biomass partitioning to the shoot is one of the effects of shade avoidance (Bongers et al., 2014). In the case of lettuces with small stems, it is the leaf that benefits. An increased leaf/root weight ratio under FR suggests the relative sink strength of leaves has increased compared to that of the root (Marcelis, 1996; Heuvelink, 1997). Specific leaf area (SLA) has often been found to increase by additional FR light or a lowered R:FR ratio (Ballaré and Pierik, 2017), although not always (Jin et al., 2021). Taken together, leaf expansion is a morphological effect induced by lowered R:FR ratio, which has been reported by several researchers (Park and Runkle, 2017; Mickens et al., 2018; Meng and Runkle, 2019; Meng et al., 2019; Zou et al., 2019; Jayalath and van Iersel, 2021; Jin et al., 2021; Kong and Nemali, 2021; Legendre and van Iersel, 2021). For rosette crops like lettuce, leaf expansion enlarges the projected leaf area, which is the projection surface of a single crop or the canopy. The increment of projected leaf area and leaf area will improve light interception, as radiation in indoor vertical farming occurs mostly from an upright position with limited reflection of photosynthetic photons from neighboring crops. Several studies have explored the effect of adding FR light over PAR and reported promoted leaf expansion as FR was applied during the photoperiod (Meng and Runkle, 2019; Meng et al., 2019; Zou et al., 2019, 2021; Zhen, 2020; Zhen and Bugbee, 2020; Jin et al., 2021; Legendre and van Iersel, 2021) and at end of day, which only applied during the end of the photoperiod (Zou et al., 2019). This is a strong advantage of indoor vertical farm cultivation, as a faster build-up of the crop architecture will benefit light interception, resulting in higher biomass production and light use efficiency within the short period of indoor vertical farm cultivation duration.

Another metric widely used in shade avoidance is phytochrome photostationary state (PSS, or PPE, which is the abbreviation of phytochrome photoequilibrium), in addition to R:FR ratio. PPS is calculated as the fraction of active phytochrome ($P_{fr}$) over total phytochrome ($P_{total}$) where $P_{total} = P_{fr} + P_r$. A low R:FR ratio leads to a low PSS and a lowered PSS correlates with various shade avoidance responses such as stem elongation (Kalaitzoglou et al., 2019; Kusuma and Bugbee, 2020). In contrast to R:FR ratio, which focuses only on red and FR photon flux, PSS is affected by other light spectra such as blue and green as well.

Both the R:FR ratio and PSS have been challenged as ways to analyze shade avoidance. In natural conditions, in direct sunlight, the R:FR ratio is ~1.2, based on the photon flux density, which varies depending on the definition of the spectrum regime of red and FR (Kusuma and Bugbee, 2020), and the PSS is ~0.71 (Kalaitzoglou et al., 2019). Therefore, research from an ecology perspective considers the shade condition with a R:FR ratio below 1.2 (Poorter et al., 2019) and/or a PSS lower than 0.71. However, in indoor vertical farming where light comes from electric lamps, neither a 1.2 R:FR ratio nor a 0.71 PSS is applicable in practical production, which requires substantial FR photon flux density. A higher R:FR ratio is more commonly studied (e.g. 4, which equals a R:FR ratio of 4:1 (Zou et al., 2019; Jin et al., 2021)). The 'lowered' R:FR ratio in indoor vertical farm studies is normally understood as a drop from an infinite value that contains negligible FR photon flux (700–800 nm) to a FR light addition, typically 10–30% of PPFD. Debate remains as to whether the FR light effect can be attributed to shade avoidance response in indoor vertical farming conditions. Shade avoidance response includes elongation of hypocotyl, stem and petiole, hyponasty (leaf going upward), reduced shoot branching, early flowering, etc. (Smith and Whitelam, 1997; Pierik and De Wit, 2014; Ballaré and Pierik, 2017). Leaf expansion, specifically defined leaf surface enlargement, is not a typical shade avoidance response (Kong and Nemali, 2021). Thus, although previous research attributes the effect of FR addition to shade avoidance, it is debatable whether plant responses should be classified into shade avoidance syndrome while the incident R:FR ratio is higher than 1.2 (Kalaitzoglou et al., 2019). From an evolutionary point of view, the question remains as to why the plant reacts in the way of shade avoidance when it is not being shaded, as shade avoidance responses have a cost, much like the resistance to *Botrytis* reduced in tomato (Ji et al., 2019).

## 4 Planting density

Increasing planting density encompasses components that affect growth and development of individual plants – it increases the R:FR ratio and decreases the availability of PPFD per plant. A symptom of vegetation shading is a lowered R:FR ratio (Ballaré et al., 1990; Pierik and De Wit, 2014). Besides radiation cues, biogenic volatile organic compounds (BVOC) elicited from neighboring plants can also trigger shade avoidance responses (Kegge and Pierik, 2010). Moreover, touching of the leaflet tips can induce a typical shade avoidance symptom, hyponasty (de Wit et al., 2012; Pierik and De Wit, 2014), which is more easily caused by higher planting density. Last but not least, increased mechanical stress by wind shielding in high planting density induces stem elongation of tobacco (Anten et al., 2005), which is less likely to occur in indoor vertical farms.

Provided LAI is not too high, increasing planting density results in increased interception of light with a greater effect of planting density on the fraction of light interception in indoor vertical farming (Jin et al., 2021; Legendre and van Iersel, 2021). An initial high planting density might be advantageous for optimal use of cultivation space (Franklin, 2008). Since plants are relatively small in the early stages, gradually decreasing planting density as plants grow larger is an approach to maximize light interception (Nicole et al., 2016). High planting density can increase biomass production per unit area, although it can potentially reduce biomass per plant (Jin et al., 2021). As discussed above, a lowered R:FR ratio enlarges the leaf area and projected leaf area of lettuce. This suggests that the effect of additional FR will be limited when the planting density increases up to a certain level. Jin et al. (2021) grew lettuce at three density levels of 23, 37, and 51 plant $m^{-2}$, and reported reduced FR effects regarding PAR interception. The effects of combining high planting density and additional FR light require further research.

## 5 Conclusion and future trends

Combining FR light into growing light is based on prevailing common practice, in which LED modules applied in indoor vertical farms are concentrated on radiating photosynthetic photons. Research has suggested increasing light use efficiency by substituting fluorescent lamps or using a combination of fluorescent lamps (typically enriched blue) and red light LEDs by red and blue light LEDs where radiation better fits crop chlorophyll absorption peaks. Future research will continue to focus on improving light use efficiency by applying different spectra combinations. For example, the application of green light (typically defined within 500–600 nm) is proposed to improve biomass production (Liu et al., 2017; Li et al., 2020). While the top layer of the canopy becomes saturated by incident red and blue, green light can penetrate deeper into the canopy (Brodersen and Vogelmann, 2010; Zhen and Bugbee, 2020) and improve overall $CO_2$ assimilation of the canopy. In addition, green light has been proposed to increase photosynthesis efficiency (Terashima et al., 2009; Liu et al., 2017). In addition, white LEDs are becoming more and more popular due to their relative higher cost performance in comparison to red and blue LEDs (personal communication). Green light is being applied in practical indoor vertical farm production within white LEDs. Since green light composition varies, an optimal combination with red and blue in order to improve energy use efficiency, light use efficiency, and/or nutritional quality would benefit from further research. Ultraviolet (UV) radiation could also potentially contribute to improving light use efficiency. Additional UV-A (315–400 nm) has been reported to benefit lettuce growth, including enlarged leaf area, shoot dry weight, and increased anthocyanin and AsA content (Chen et al., 2019).

In addition to other spectra, a dynamic lighting strategy, in terms of spectra and flux density, is another potential approach to increase light use efficiency. For example, an intermittent light exposure with red and blue light, i.e. shifting red and blue in time increments within each photoperiod, resulted in increased lettuce fresh and dry weight (Kuno et al., 2017; Chen and Yang, 2018). Pulse red and blue light by different frequencies was also found to increase energy use efficiency and high frequency enhanced leaf antioxidant capacity. Since crops are relatively small in the beginning and get larger gradually, a constant PPFD throughout the whole growing period may lead to wasted light in the beginning and insufficient intensity at later stages. A gradually increasing PPFD could potentially improve biomass production and light use efficiency. Taken together, LED as a light source has different features to fluorescent and HID lamps, potentially leading to an optimal light strategy in biomass production and light use efficiency. Improved crop models could help, e.g. functional-structural plant modeling, which combines a representation of 3D plant structure, light interception and physiological functions (mainly photosynthesis) in response to environmental factors (Vos et al., 2010).

Improving the quality of lettuce should not come at the expense of reduced energy efficiency and crop yield. End-of-production practices, by altering lighting conditions a few days before harvest (Min et al., 2021), can boost the harvest quality of fresh products without a huge impact on yield and plant morphology. This is particularly true as the biosynthesis of certain secondary metabolites (e.g. anthocyanins, flavonols, isoflavonoids, and stilbenoids) is more active in the mature stage than the early stage of lettuce (Sng et al., 2021), especially under red combined blue and high PPFD lighting conditions.

## 6 Where to look for further information

Since indoor vertical farming is a popular topic both in academia and business fields, there are a number of academic groups and companies worldwide doing scientific and applied research on the interaction between light, in terms of spectrum and/or intensity, and crop performance, in terms of morphology, biomass production, and/or nutritional quality. In order to share the recent developments and knowledge of indoor vertical farming, academic researchers together with corporate representatives organize conferences and symposiums and invite people from all over the world. Below are two of the most appreciated events with free access, namely, Indoor Ag Science Café (https://www.scri-optimia.org/cafe.php) and ISHS Talks on Vertical Farming (https://www.ishs.org/news/ishs-talks-vertical-farming). Japan Plant Factory Association (JPFA: https://npoplantfactory.org/en/) is another organization coordinating online seminars on a regular basis. Most of the references, as well as relevant researchers, quoted within present chapters can be accessed in these abovementioned approaches.

## 7 References

Amoozgar, A., Mohammadi, A. and Sabzalian, M. R. (2017). Impact of light-emitting diode irradiation on photosynthesis, phytochemical composition and mineral element content of lettuce cv. grizzly. *Photosynthetica* 55(1), 85–95. doi:10.1007/s11099-016-0216-8.

Anten, N. P. R., Casado-Garcia, R. and Nagashima, H. (2005). Effects of mechanical stress and plant density on mechanical characteristics, growth, and lifetime reproduction of tobacco plants. *Am. Nat.* 166(6), 650–660. doi:10.1086/497442.

Asseng, S., Guarin, J. R., Raman, M., Monje, O., Kiss, G., Despommier, D. D., Meggers, F. M. and Gauthier, P. P. G. (2020). Wheat yield potential in controlled-environment vertical farms. *Proc. Natl. Acad. Sci. U. S. A.* 117(32), 19131–19135. doi:10.1073/pnas.2002655117.

Bagchi, D., Garg, A., Krohn, R. L., Bagchi, M., Bagchi, D. J., Balmoori, J. and Stohs, S. J. (1998). Protective effects of grape seed proanthocyanidins and selected antioxidants against TPA-induced hepatic and brain lipid peroxidation and DNA fragmentation, and peritoneal macrophage activation in mice. *Gen. Pharmacol.* 30(5), 771–776. doi:10.1016/S0306-3623(97)00332-7.

Baker, N. R. (2008). Chlorophyll fluorescence: A probe of photosynthesis in vivo. *Annu. Rev. Plant Biol.* 59, 89–113. doi:10.1146/annurev.arplant.59.032607.092759.

Ballaré, C. L. (1999). Keeping up with the neighbours: Phytochrome sensing and other signalling mechanisms. *Trends Plant Sci.* 4(3), 97–102. doi:10.1016/S1360-1385(99)01383-7.

Ballaré, C. L. and Pierik, R. (2017). The shade-avoidance syndrome: Multiple signals and ecological consequences. *Plant Cell Environ.* 40(11), 2530–2543. doi:10.1111/pce.12914.

Ballaré, C. L., Scopel, A. L. and Sánchez, R. A. (1990). Far-red radiation reflected from adjacent leaves: An early signal of competition in plant canopies. *Science* 247(4940), 329–332. doi:10.1126/science.247.4940.329.

Bongers, F. J., Evers, J. B., Anten, N. P. R. and Pierik, R. (2014). From shade avoidance responses to plant performance at vegetation level: Using virtual plant modelling as a tool. *New Phytol.* 204(2), 268–272. doi:10.1111/nph.13041.

Brodersen, C. R. and Vogelmann, T. C. (2010). Do changes in light direction affect absorption profiles in leaves? *Funct. Plant Biol.* 37(5), 403–412. doi:10.1071/FP09262.

Bula, R. J., Morrow, R. C., Tibbitts, T. W., Barta, D. J., Ignatius, R. W. and Martin, T. S. (1991). Light-emitting diodes as a radiation source for plants. *Hortscience* 26(2), 203–205. doi:10.21273/HORTSCI.26.2.203.

Chan, T. Y. K. (2011). Vegetable-borne nitrate and nitrite and the risk of methaemoglobinaemia. *Toxicol. Lett.* 200(1–2), 107–108. doi:10.1016/j.toxlet.2010.11.002.

Chen, W., Xu, Z., Liu, X., Yang, Y., Wang, Z. and Song, F. (2011). Effect of LED light source on the growth and quality of different lettuce varieties. *Acta Bot. Boreali-Occidentalia Sin.* 31, 1434–1440.

Chen, X., Guo, W., Zhong, X., Zhang, X. and Wang, L. (2014). Growth and quality responses of "Green Oak Leaf" lettuce as affected by monochromic or mixed radiation provided by fluorescent lamp (FL) and light-emitting diode (LED). *Sci. Hortic. (Amsterdam)* 172, 168–175. doi:10.1016/j.scienta.2014.04.009.

Chen, X., Li, Y. and Wang, L. (2021). Red and blue wavelengths affect the morphology, energy use efficiency and nutritional content of lettuce (*Lactuca sativa* L.). *Sci. Rep.* 11, 1–12. doi:10.1038/s41598-021-87911-7.

Chen, X. and Yang, Q. (2018). Effects of intermittent light exposure with red and blue light emitting diodes on growth and carbohydrate accumulation of lettuce. *Sci. Hortic. (Amsterdam)* 234, 220–226. doi:10.1016/j.scienta.2018.02.055.

Chen, X., Yang, Q. and Chang, S. (2017). Growth and nutritional properties of lettuce affected by different alternating intervals of red and blue LED irradiation. *Sci. Hortic. (Amsterdam)* 223, 44–52. doi:10.1016/j.scienta.2017.04.037.

Chen, Y., Li, T., Yang, Q., Zhang, Y., Zou, J., Bian, Z. and Wen, X. (2019). UVA radiation is beneficial for yield and quality of indoor cultivated lettuce. *Front. Plant Sci.* 10, 1563. doi:10.3389/fpls.2019.01563.

de Wit, M., Kegge, W., Evers, J. B., Vergeer-van Eijk, M. H., Gankema, P., Voesenek, L. A. C. J. and Pierik, R. (2012). Plant neighbor detection through touching leaf tips precedes phytochrome signals. *Proc. Natl. Acad. Sci. U. S. A.* 109(36), 14705–14710. doi:10.1073/pnas.1205437109.

Demotes-Mainard, S., Péron, T., Corot, A., Bertheloot, J., Le Gourrierec, J., Pelleschi-Travier, S., Crespel, L., Morel, P., Huché-Thélier, L., Boumaza, R., Vian, A., Guérin, V., Leduc, N. and Sakr, S. (2016). Plant responses to red and far-red lights, applications in horticulture. *Environ. Exp. Bot.* 121, 4–21. doi:10.1016/j.envexpbot.2015.05.010.

Dougher, T. A. O. and Bugbee, B. (2001). Differences in the response of wheat, soybean and lettuce to reduced blue radiation. *Photochem. Photobiol.* 73(2), 199–207. doi:10.1562/0031-8655(2001)073<0199:ditrow>2.0.co;2.

Eldridge, B. M., Manzoni, L. R., Graham, C. A., Rodgers, B., Farmer, J. R. and Dodd, A. N. (2020). Getting to the roots of aeroponic indoor farming. *New Phytol.* 228(4), 1183–1192. doi:10.1111/nph.16780.

Emerson, R. and Rabinowitch, E. (1960). Red drop and role of auxiliary pigments in photosynthesis. *Plant Physiol.* 35(4), 477–485. doi:10.1104/pp.35.4.477.

Evans, J. R. and Poorter, H. (2001). Photosynthetic acclimation of plants to growth irradiance: The relative importance of specific leaf area and nitrogen partitioning in maximizing carbon gain. *Plant Cell Environ.* 24(8), 755–767. doi:10.1046/j.1365-3040.2001.00724.x.

Franklin, K. A. (2008). Shade avoidance. *New Phytol.* 179(4), 930–944. doi:10.1111/j.1469-8137.2008.02507.x.

Gangolli, S. D., van den Brandt, P. A., Feron, V. J., Janzowsky, C., Koeman, J. H., Speijers, G. J. A., Spiegelhalder, B., Walker, R. and Wisnok, J. S. (1994). Nitrate, nitrite and N-nitroso compounds. *Eur. J. Pharmacol.* 292(1), 1–38. doi:10.1016/0926-6917(94)90022-1.

Graamans, L., Baeza, E., van den Dobbelsteen, A., Tsafaras, I. and Stanghellini, C. (2018). Plant factories versus greenhouses: Comparison of resource use efficiency. *Agric. Syst.* 160, 31–43. doi:10.1016/j.agsy.2017.11.003.

Haitz, R. and Tsao, Y. (2011). Solid-state lighting Why it will succeed, and why it won't be overtaken. *Light Technol.* 26–30.

Heraut-Bron, V., Robin, C., Varlet-Grancher, C., Afif, D. and Guckert, A. (1999). Light quality (red:far-red ratio): Does it affect photosynthetic activity, net $CO_2$ assimilation, and morphology of young white clover leaves? *Can. J. Bot.* 77(10), 1425-1431. doi:10.1139/cjb-77-10-1425.

Heuvelink, E. (1997). Effect of fruit load on dry matter partitioning in tomato. *Sci. Hortic. (Amsterdam)* 69(1–2), 51–59. doi:10.1016/S0304-4238(96)00993-4.

Hogewoning, S. W., Trouwborst, G., Maljaars, H., Poorter, H., van Ieperen, W. and Harbinson, J. (2010). Blue light dose-responses of leaf photosynthesis, morphology, and chemical composition of Cucumis sativus grown under different combinations of red and blue light. *J. Exp. Bot.* 61(11), 3107–3117. doi:10.1093/jxb/erq132.

Huang, J., Gu, M., Lai, Z., Fan, B., Shi, K., Zhou, Y. H., Yu, J. Q. and Chen, Z. (2010). Functional analysis of the Arabidopsis PAL gene family in plant growth, development, and response to environmental stress. *Plant Physiol.* 153(4), 1526–1538. doi:10.1104/pp.110.157370.

Inada, K. (1976). Action spectra for photosynthesis in higher plants. *Plant Cell Physiol.* 17, 355–365. doi:10.1093/oxfordjournals.pcp.a075288.

Jayalath, T. C. and van Iersel, M. W. (2021). Canopy size and light use efficiency explain growth differences between lettuce and mizuna in vertical farms. *Plants (Basel)* 10(4), 704. doi:10.3390/plants10040704.

Ji, Y., Nuñez Ocaña, D., Choe, D., Larsen, D. H., Marcelis, L. F. M. and Heuvelink, E. (2020). Far-red radiation stimulates dry mass partitioning to fruits by increasing fruit sink strength in tomato. *New Phytol.* 228(6), 1914–1925. doi:10.1111/nph.16805.

Ji, Y., Ouzounis, T., Courbier, S., Kaiser, E., Nguyen, P. T., Schouten, H. J., Visser, R. G. F., Pierik, R., Marcelis, L. F. M. and Heuvelink, E. (2019). Far-red radiation increases dry mass partitioning to fruits but reduces Botrytis cinerea resistance in tomato. *Environ. Exp. Bot.* 168, 103889. doi:10.1016/j.envexpbot.2019.103889.

Jin, W., Formiga Lopez, D., Heuvelink, E. and Marcelis, L. F. M. (2023). Light use efficiency of lettuce cultivation in vertical farms compared with greenhouse and field. *Food Energy Secur.* 12(1), 1–10. doi:10.1002/fes3.391.

Jin, W., Urbina, J. L., Heuvelink, E. and Marcelis, L. F. M. (2021). Adding far-red to red-blue light-emitting diode light promotes yield of lettuce at different planting densities. *Front. Plant Sci.* 11, 1–9. doi:10.3389/fpls.2020.609977.

Kalaitzoglou, P., van Ieperen, W., Harbinson, J., van der Meer, M., Martinakos, S., Weerheim, K., Nicole, C. C. S. and Marcelis, L. F. M. (2019). Effects of continuous or end-of-day Far-red light on tomato plant growth, morphology, light absorption, and fruit production. *Front. Plant Sci.* 10, 322. doi:10.3389/fpls.2019.00322.

Kegge, W. and Pierik, R. (2010). Biogenic volatile organic compounds and plant competition. *Trends Plant Sci.* 15(3), 126–132. doi:10.1016/j.tplants.2009.11.007.

Kong, Y. and Nemali, K. (2021). Blue and Far-red light affect area and number of individual leaves to influence vegetative growth and pigment synthesis in lettuce. *Front. Plant Sci.* 12, 667407. doi:10.3389/fpls.2021.667407.

Kook, H. S., Park, S. H., Jang, Y. J., Lee, G. W., Kim, J. S., Kim, H. M., Oh, B., Chae, J. and Lee, K. (2013). Blue LED (light-emitting diodes)-mediated growth promotion and control of Botrytis disease in lettuce. *Acta Agric. Scand. B* 63(3), 271–277. doi:10.1080/09064710.2012.756118.

Kozai, T. (2013). Resource use efficiency of closed plant production system with artificial light: Concept, estimation and application to plant factory. *Proc. Jpn. Acad. Ser. B Phys. Biol. Sci.* 89(10), 447–461. doi:10.2183/pjab.89.447.

Kuno, Y., Shimizu, H., Nakashima, H., Miyasaka, J. and Ohdoi, K. (2017). Effects of irradiation patterns and light quality of red and blue light-emitting diodes on growth of leaf lettuce (*Lactuca sativa* L. "Greenwave"). *Environ. Control Biol.* 55(3), 129–135. doi:10.2525/ecb.55.129.

Kusuma, P. and Bugbee, B. (2020). Far-red fraction: An improved metric for characterizing phytochrome effects on morphology. *J. Am. Soc. Hortic. Sci.* 146(1), 3–13. doi:10.21273/JASHS05002-20.

Kusuma, P., Pattison, P. M. and Bugbee, B. (2020). From physics to fixtures to food: Current and potential LED efficacy. *Hortic. Res.* 7, 56. doi:10.1038/s41438-020-0283-7.

Landrum, J. T. and Bone, R. A. (2001). Lutein, zeaxanthin, and the macular pigment. *Arch. Biochem. Biophys.* 385(1), 28–40. doi:10.1006/abbi.2000.2171.

Legendre, R. and van Iersel, M. W. (2021). Supplemental far-red light stimulates lettuce growth: Disentangling morphological and physiological effects. *Plants (Basel)* 10(1), 166. doi:10.3390/plants10010166.

Li, L., Tong, Y., Lu, J., Li, Y. and Yang, Q. (2020). Lettuce growth, nutritional quality, and energy use efficiency as affected by red–blue light combined with different monochromatic wavelengths. *Hortscience* 55(5), 613–620. doi:10.21273/HORTSCI14671-19.

Liang, Y., Kang, C., Kaiser, E., Kuang, Y., Yang, Q. and Li, T. (2021). Red/blue light ratios induce morphology and physiology alterations differently in cucumber and tomato. *Sci. Hortic. (Amsterdam)* 281, 109995. doi:10.1016/j.scienta.2021.109995.

Lillo, C. (1994). Light regulation of nitrate reductase in green leaves of higher plants. *Physiol. Plant.* 90(3), 616–620. doi:10.1111/j.1399-3054.1994.tb08822.x.

Lin, C. (2000). Plant blue-light receptors. *Trends Plant Sci.* 5(8), 337–342. doi:10.1016/S1360-1385(00)01687-3.

Lin, K. H., Huang, M. Y., Huang, W. D., Hsu, M. H., Yang, Z. W. and Yang, C. M. (2013). The effects of red, blue, and white light-emitting diodes on the growth, development, and edible quality of hydroponically grown lettuce (*Lactuca sativa* L. var. capitata). *Sci. Hortic. (Amsterdam)* 150, 86–91. doi:10.1016/j.scienta.2012.10.002.

Liu, H., Fu, Y., Wang, M. and Liu, H. (2017). Green light enhances growth, photosynthetic pigments and $CO_2$ assimilation efficiency of lettuce as revealed by 'knock out' of the 480–560 nm spectral waveband. *Photosynthetica* 55(1), 144–152. doi:10.1007/s11099-016-0233-7.

Llorach, R., Martínez-Sánchez, A., Tomás-Barberán, F. A., Gil, M. I. and Ferreres, F. (2008). Characterisation of polyphenols and antioxidant properties of five lettuce varieties and escarole. *Food Chem.* 108(3), 1028–1038. doi:10.1016/j.foodchem.2007.11.032.

Marcelis, L. F. M. (1996). Sink strength as a determinant of dry matter partitioning in the whole plant. *J. Exp. Bot.* 47(Spec No), 1281–1291. doi:10.1093/jxb/47.Special_Issue.1281.

McCree, K. J. (1971). The action spectrum, absorptance and quantum yield of photosynthesis in crop plants. *Agric. Meteorol.* 9, 191–216. doi:10.1016/0002-1571(71)90022-7.

McCree, K. J. (1972). Test of current definitions of photosynthetically active radiation against leaf photosynthesis data. *Agric. Meteorol.* 10, 443–453. doi:10.1016/0002-1571(72)90045-3.

Meng, Q., Kelly, N. and Runkle, E. S. (2019). Substituting green or far-red radiation for blue radiation induces shade avoidance and promotes growth in lettuce and kale. *Environ. Exp. Bot.* 162, 383–391. doi:10.1016/j.envexpbot.2019.03.016.

Meng, Q. and Runkle, E. S. (2019). Far-red radiation interacts with relative and absolute blue and red photon flux densities to regulate growth, morphology, and pigmentation of lettuce and basil seedlings. *Sci. Hortic. (Amsterdam)* 255, 269–280. doi:10.1016/j.scienta.2019.05.030.

Miao, Y., Chen, Q., Qu, M., Gao, L. and Hou, L. (2019). Blue light alleviates 'red light syndrome' by regulating chloroplast ultrastructure, photosynthetic traits and nutrient accumulation in cucumber plants. *Sci. Hortic. (Amsterdam)* 257, 108680. doi:10.1016/j.scienta.2019.108680.

Mickens, M. A., Skoog, E. J., Reese, L. E., Barnwell, P. L., Spencer, L. E., Massa, G. D. and Wheeler, R. M. (2018). A strategic approach for investigating light recipes for 'Outredgeous' red romaine lettuce using white and monochromatic LEDs. *Life Sci. Space Res. (Amst)* 19, 53–62. doi:10.1016/j.lssr.2018.09.003.

Min, Q., Marcelis, L. F. M., Nicole, C. C. S. and Woltering, E. J. (2021). High light intensity applied shortly before harvest improves lettuce nutritional quality and extends the shelf life. *Front. Plant Sci.* 12, 615355. doi:10.3389/fpls.2021.615355.

Mitchell, C. A., Dzakovich, M. P., Gomez, C., Lopez, R., Burr, J. F., Hernández, R., Kubota, C., Currey, C. J., Meng, Q., Runkle, E. S. and Bourget, C. M. (2015). Light-emitting diodes in horticulture. *Hortic. Rev. (Am. Soc. Hortic. Sci.)* 43, 1-87. doi:10.1002/9781119107781.ch01.

Morrow, R. C. (2008). LED lighting in horticulture. *Hortscience* 43(7), 1947–1950. doi:10.21273/HORTSCI.43.7.1947.

Mulabagal, V., Ngouajio, M., Nair, A., Zhang, Y., Gottumukkala, A. L. and Nair, M. G. (2010). In vitro evaluation of red and green lettuce (*Lactuca sativa*) for functional food properties. *Food Chem.* 118(2), 300–306. doi:10.1016/j.foodchem.2009.04.119.

Nakamura, S., Mukai, T. and Senoh, M. (1994). Candela-class high-brightness InGaN/AlGaN double-heterostructure blue-light-emitting diodes. *Appl. Phys. Lett.* 64(13), 1687–1689. doi:10.1063/1.111832.

Nicole, C. C. S., Charalambous, F., Martinakos, S., Van De Voort, S., Li, Z., Verhoog, M. and Krijn, M. (2016). Lettuce growth and quality optimization in a plant factory. *Acta Hortic.* 1134, 231-238. doi:10.17660/ActaHortic.2016.1134.31.

Ohashi-Kaneko, K., Takase, M., Kon, N., Fujiwara, K. and Kurata, K. (2007). Effect of light quality on growth and vegetable quality in leaf lettuce, spinach and komatsuna. *Environ. Control Biol.* 45(3), 189–198. doi:10.2525/ecb.45.189.

Park, Y. and Runkle, E. S. (2017). Far-red radiation promotes growth of seedlings by increasing leaf expansion and whole-plant net assimilation. *Environ. Exp. Bot.* 136, 41–49. doi:10.1016/j.envexpbot.2016.12.013.

Pattison, P. M., Tsao, J. Y., Brainard, G. C. and Bugbee, B. (2018). LEDs for photons, physiology and food. *Nature* 563(7732), 493–500. doi:10.1038/s41586-018-0706-x.

Pennisi, G., Blasioli, S., Cellini, A., Maia, L., Crepaldi, A., Braschi, I., Spinelli, F., Nicola, S., Fernandez, J. A., Stanghellini, C., Marcelis, L. F. M., Orsini, F. and Gianquinto, G. (2019a). Unraveling the role of red:blue LED lights on resource use efficiency and nutritional properties of indoor grown sweet basil. *Front. Plant Sci.* 10, 305. doi:10.3389/fpls.2019.00305.

Pennisi, G., Orsini, F., Blasioli, S., Cellini, A., Crepaldi, A., Braschi, I., Spinelli, F., Nicola, S., Fernandez, J. A., Stanghellini, C., Gianquinto, G. and Marcelis, L. F. M. (2019b). Resource use efficiency of indoor lettuce (*Lactuca sativa* L.) cultivation as affected by red:blue ratio provided by LED lighting. *Sci. Rep.* 9(1), 14127. doi:10.1038/s41598-019-50783-z.

Pierik, R. and De Wit, M. (2014). Shade avoidance: Phytochrome signalling and other aboveground neighbour detection cues. *J. Exp. Bot.* 65(11), 2815–2824. doi:10.1093/jxb/ert389.

Poorter, H., Niinemets, Ü., Ntagkas, N., Siebenkäs, A., Mäenpää, M., Matsubara, S. and Pons, T. (2019). A meta-analysis of plant responses to light intensity for 70 traits ranging from molecules to whole plant performance. *New Phytol.* 223(3), 1073–1105. doi:10.1111/nph.15754.

Sager, J. C., Smith, W. O., Edwards, J. L. and Cyr, K. L. (1988). Photosynthetic efficiency and phytochrome photoequilibria determination using spectral data. *Trans. ASAE* 31, 1882–1889. doi:10.13031/2013.30952.

Santamaria, P. (2006). Nitrate in vegetables: Toxicity, content, intake and EC regulation. *J. Sci. Food Agric.* 86(1), 10–17. doi:10.1002/jsfa.2351.

SharathKumar, M., Heuvelink, E. and Marcelis, L. F. M. (2020). Vertical farming: Moving from genetic to environmental modification. *Trends Plant Sci.* 25(8), 724–727. doi:10.1016/j.tplants.2020.05.012.

Smith, H. (2000). Phytochromes and light signal perception by plants – An emerging synthesis. *Nature* 407(6804), 585–591.

Smith, H. and Whitelam, G. C. (1997). The shade avoidance syndrome: Multiple responses mediated by multiple phytochromes. *Plant Cell Environ.* 20(6), 840–844. doi:10.1046/j.1365-3040.1997.d01-104.x.

Sng, B. J. R., Mun, B., Mohanty, B., Kim, M., Phua, Z. W., Yang, H., Lee, D. Y. and Jang, I. C. (2021). Combination of red and blue light induces anthocyanin and other secondary metabolite biosynthesis pathways in an age-dependent manner in Batavia lettuce. *Plant Sci.* 310, 110977. doi:10.1016/j.plantsci.2021.110977.

Son, K. H. and Oh, M. M. (2013). Leaf shape, growth, and antioxidant phenolic compounds of two lettuce cultivars grown under various combinations of blue and red light-emitting diodes. *Hortscience* 48(8), 988–995. doi:10.21273/HORTSCI.48.8.988.

Son, K. H., Park, J. H., Kim, D. and Oh, M. M. (2013). Leaf shape index, growth, and phytochemicals in two leaf lettuce cultivars grown under monochromatic light-emitting diodes. *Korean J. Hortic. Sci. Technol.* 31, 664–672. doi:10.7235/hort.2012.12063.

Spalholz, H., Perkins-Veazie, P. and Hernández, R. (2020). Impact of sun-simulated white light and varied blue:red spectrums on the growth, morphology, development, and phytochemical content of green- and red-leaf lettuce at different growth stages. *Sci. Hortic. (Amsterdam)* 264, 109195. doi:10.1016/j.scienta.2020.109195.

Taiz, L., Zeiger, E., Møller, I. M. and Murphy, A. (2015). *Plant Physiology and Development, 2015*. 6th Edition, Sinauer Associates, Sunderland, CT.

Terashima, I., Fujita, T., Inoue, T., Chow, W. S. and Oguchi, R. (2009). Green light drives leaf photosynthesis more efficiently than red light in strong white light: Revisiting the enigmatic question of why leaves are green. *Plant Cell Physiol.* 50(4), 684–697. doi:10.1093/pcp/pcp034.

Trouwborst, G., Hogewoning, S. W., van Kooten, O., Harbinson, J. and van Ieperen, W. (2016). Plasticity of photosynthesis after the "red light syndrome" in cucumber. *Environ. Exp. Bot.* 121, 75–82. doi:10.1016/j.envexpbot.2015.05.002.

Urbonavičiūtė, A., Pinho, P., Samuolienė, G., Duchovskis, P., Vitta, P., Stonkus, A., Tamulaitis, G., Žukauskas, A. and Halonen, L. (2007). Effect of short-wavelength light on lettuce growth and nutritional quality. *Sodininkystė Daržininkystė* 26, 157–165.

van Delden, S. H., SharathKumar, M., Butturini, M., Graamans, L. J. A., Heuvelink, E., Kacira, M., Kaiser, E., Klamer, R. S., Klerkx, L., Kootstra, G., Loeber, A., Schouten, R. E., Stanghellini, C., van Ieperen, W., Verdonk, J. C., Vialet-Chabrand, S., Woltering, E. J., van de Zedde, R., Zhang, Y. and Marcelis, L. F. M. (2021). Current status and future challenges in implementing and upscaling vertical farming systems. *Nat. Food* 2(12), 944–956. doi:10.1038/s43016-021-00402-w.

Vogelmann, T. C. (1993). Plant tissue optics. *Annu. Rev. Plant Physiol. Plant Mol. Biol.* 44(1), 231–251. doi:10.1146/annurev.pp.44.060193.001311.

Vos, J., Evers, J. B., Buck-Sorlin, G. H., Andrieu, B., Chelle, M. and De Visser, P. H. B. (2010). Functional-structural plant modelling: A new versatile tool in crop science. *J. Exp. Bot.* 61(8), 2101–2115. doi:10.1093/jxb/erp345.

Wang, J., Lu, W., Tong, Y. and Yang, Q. (2016). Leaf morphology, photosynthetic performance, chlorophyll fluorescence, stomatal development of lettuce (*Lactuca sativa* L.) exposed to different ratios of red light to blue light. *Front. Plant Sci.* 7, 250. doi:10.3389/fpls.2016.00250.

Wanlai, Z., Wenke, L. and Qichang, Y. (2013). Reducing nitrate content in lettuce by pre-harvest continuous light delivered by red and blue light-emitting diodes. *J. Plant Nutr.* 36(3), 481–490. doi:10.1080/01904167.2012.748069.

Yanagi, T., Okamoto, K. and Takita, S. (1996). Effects of blue, red, and BLUE/RED lights of two different PPF levels on growth and morphogenesis of lettuce plants. *Acta Hortic.* 440, 117–122. doi:10.17660/ActaHortic.1996.440.21.

Yorio, N. C., Goins, G. D., Kagie, H. R., Wheeler, R. M. and Sager, J. C. (2001). Improving spinach, radish, and lettuce growth under red light-emitting diodes (LEDs) with blue light supplementation. *Hortscience* 36(2), 380–383. doi:10.21273/HORTSCI.36.2.380.

Zha, L., Liu, W., Yang, Q., Zhang, Y., Zhou, C. and Shao, M. (2020). Regulation of ascorbate accumulation and metabolism in lettuce by the red:blue ratio of continuous light using LEDs. *Front. Plant Sci.* 11, 704. doi:10.3389/fpls.2020.00704.

Zhen, S. (2020). Substituting Far-red for traditionally defined photosynthetic photons results in equal canopy quantum yield for $CO_2$ fixation and increased photon capture during long-term studies: Implications for re-defining PAR. *Front. Plant Sci.* 11, 1–14. doi:10.3389/fpls.2020.581156.

Zhen, S. and Bugbee, B. (2020). Far-red photons have equivalent efficiency to traditional photosynthetic photons: Implications for redefining photosynthetically active radiation. *Plant Cell Environ.* 43(5), 1259–1272. doi:10.1111/pce.13730.

Zhen, S., Haidekker, M. and van Iersel, M. W. (2019). Far-red light enhances photochemical efficiency in a wavelength-dependent manner. *Physiol. Plant.* 167(1), 21–33 doi:10.1111/ppl.12834.

Zhen, S. and van Iersel, M. W. (2017). Far-red light is needed for efficient photochemistry and photosynthesis. *J. Plant Physiol.* 209, 115–122. doi:10.1016/j.jplph.2016.12.004.

Zou, J., Fanourakis, D., Tsaniklidis, G., Cheng, R., Yang, Q. and Li, T. (2021). Lettuce growth, morphology and critical leaf trait responses to far-red light during cultivation are low fluence and obey the reciprocity law. *Sci. Hortic. (Amsterdam)* 289, 110455. doi:10.1016/j.scienta.2021.110455.

Zou, J., Zhang, Y., Zhang, Y., Bian, Z., Fanourakis, D. and Yang, Q. (2019). Morphological and physiological properties of indoor cultivated lettuce in response to additional far-red light. *Sci. Hortic. (Amsterdam)* 257, 108725. doi:10.1016/j.scienta.2019.108725.

# Chapter 12

# Spectral manipulations to control growth and quality of lettuce and other leafy greens in vertical farms

*Yuxin Tong, Institute of Environment and Sustainable Development in Agriculture, Chinese Academy of Agricultural Sciences, China*



## 1 Introduction

It is essential to select the most favorable light environment for plant growth in vertical farms. Light, as a signal and energy source, is a vital factor affecting plant growth, development (photomorogenesis, photoperiodism, etc.), and nutritional quality (secondary metabolism), especially when artificial light is the sole light source for plant growth (Li et al., 2020; Metallo et al., 2018). Lighting is responsible for 60–80% of electric energy consumption in vertical farms (Graamans et al., 2018; Tong et al., 2014), so selecting the optimal light environment for plant growth and development, and reducing electric energy consumption, is essential to improve the economic benefits of vertical farms.

The light environment includes photosynthetic photon flux density (PPFD), light quality (wavelength or spectral distribution), light period and light direction. Compared with light quantity and light period, light quality has far more complex effects on plant morphology and physiology (Li et al., 2020; Xu et al., 2015). Therefore, numerous studies over the years have focused on the effects of light quality on plant growth and development (Li et al., 2021; Bian et al., 2018; Wang et al., 2016).

Solar light has a broad waveband with 97% within the range of 280–2800 nm. Solar light that reaches the ground is within the range of 300–2000 nm. Plants are capable of sensing and responding to a broad range of the light spectrum, from ultraviolet (280–400 nm) to far-red (FR) (720–780 nm) regions (Dou and Niu, 2020). For plant growth, the waveband of 300–800 nm is known

http://dx.doi.org/10.19103/AS.2023.0126.16

as biologically active radiation (BAR), and the waveband of 400–700 nm is called photosynthetically active radiation (PAR) (Tong and Fang, 2020). Compared with PAR, BAR has an extended wavelength that ranges from 300 nm to 400 nm for ultraviolet and 700 nm to 800 nm for FR. These two wavebands mainly act as signals for plant growth and affect plant morphogenesis and photosynthesis.

The optimal light environment varies among species and even cultivars, plant growth stages, and other environmental conditions such as temperature, nutrients, and $CO_2$ concentrations (Hesham et al., 2020; Kozai, 2013). Different light qualities have distinctly different biological effects on plants, including plant growth, morphology, photochemical compounds, photosynthesis, organ growth, development, and nutritional quality (Stuefer and Huber, 1998). In this chapter, the effects of monochromatic and polychromatic lights on plant growth, development, and qualities are described.

## 2 Monochromatic light

The development of light-emitting diode (LED) technology has increased interest in manipulating light quality to improve growth (biomass productivity and morphogenesis), and quality (bioactive secondary metabolites) of leafy greens in vertical farms (Li et al., 2021; Metallo et al., 2018; Wang et al., 2016). LED devices that are adjustable in terms of light quantity and quality are often used for research purposes in vertical farms to establish the biological effects of specific spectra on plants.

### *2.1 Red light*

The waveband of red light is 600-699 nm. The absorption peaks of chlorophyll for red light are between 640 nm and 660 nm. Red light is sensed in plants by phytochromes, which regulate responses related to seed germination, stem elongation, leaf expansion, photosynthetic apparatus and flowering induction. Red light is one of the most important types of artificial light for leafy plant growth and is often used in vertical farms.

Experiments have been conducted by using different monochromatic lights with a PPFD of 30 $\mu mol \cdot m^{-2} \cdot s^{-1}$ for 12 hours per day for 20 days. It was found that monochromatic red light could promote the leaf area and dry weight of tomato seedlings and increase the content of chlorophyll a and b (Pu et al., 2005). Du et al. (2002) suggested that red light (660 nm) could promote the growth of oilseed rape seedlings, which may be related to its ability to promote the transport of indoleacetic acid (IAA), thereby accelerating its synthesis. IAA has a positive effect on the secondary meristem and cell elongation of plants. Experiments also indicated that red light could reduce the accumulation of oxygen-containing free radicals. Moreover, the content of carbohydrate-derived tea polyphenols also increased, reaching a maximum value on day ten, and then began to decrease until the end of the trial on day 20 (Zhou et al., 2012). In terms of antioxidant capacity, Wen et al. (2011) and Du et al. (2002) proved that red light can enhance the activity of catalase and superoxide dismutase in tobacco and rape seedlings and increase their resistance to stress.

Although red light has many positive effects on plant growth, including an accumulation of sugars, phenols and chlorophyll, and increased antioxidant capacity, some studies have shown that monochromatic red light or too much red light can easily induce tumor-like protrusions on tomato or lettuce leaves. Moreover, monochromatic red light is not conducive to healthy plant growth (Tong and Fang, 2020).

## 2.2 Blue light

The waveband of blue light is 400-499 nm. The absorption peaks of blue light by chlorophyll are located between 430 nm and 450 nm. Blue light is sensed by cryptochromes and phototropins. Blue light also has regulatory effects on growth of the roots, stems and leaves, stomatal opening, seedling de-etiolation, phototropism, circadian rhythms, enzyme synthesis, biomass accumulation of plants, etc. Blue light plays an important role in plant photomorphogenesis, affecting phototropism, photomorphogenesis, stomatal opening and leaf photosynthesis.

Regulation of plant growth and morphogenesis by blue light may be related to changes in the concentration of plant hormones. The spectra of different wavelengths can affect the hormone balance in plants through their associated pigment effects, thereby triggering physiological and ecological changes in plants. Yu and Pan (1997) showed that, compared with white light, blue light inhibited the growth of rice seedlings and reduced the content of free IAA, gibberellin (GA), zeatin and dihydrozeatin, and increased abscisic acid and ethylene content.

Blue light has important effects on the growth and development of plant stems. Numerous studies have shown that blue light can inhibit stem elongation, but it varies by plant species. Mortensen and Strømme (1987) found that blue light significantly reduced the plant height of chrysanthemum and tomato compared with natural light. Hernández and Kubota (2016) showed that the stem elongation of cucumber seedlings increased significantly under blue light, while the stem elongation of lettuce was significantly reduced. Blue light can increase the activity of auxin (IAA) oxidase, reducing the level of auxin IAA, which in turn inhibits the elongation of plants. The effects of blue light on plant morphology thus vary by cultivar.

Studies have found that blue light has an inhibitory effect on the fresh weight and leaf area of pak choi (Tong and Fang, 2020) while the leaf area of *Arabidopsis thaliana* was reduced with an increasing proportion of blue light (Eskins, 1992). Blue light can stimulate the opening of stomata and promote the expansion of guard cells without using the energy produced by red light to drive chloroplasts (Assmann, 1988). Moreover, blue light is conducive to the growth and development of plant roots. A large number of hair roots and high biomass of plant seedlings were found when cultivated under blue light, and blue light can improve root vitality, total absorption area and active absorption area of plant seedling (Yang, 2019).

Blue light can regulate the nutritional quality of plants, such as promoting the accumulation of proteins, carotenoids, photosynthetic pigments, chlorophyll and vitamin C, accelerating the synthesis of anthocyanins, and inhibiting the synthesis of nitrates. Studies have confirmed that blue light ranging from 420 nm to 530 nm is the most effective monochromatic light to promote the synthesis of roselle anthocyanins. Anthocyanin, carotenoid and chlorophyll content in cabbage treated with blue light at 430 nm increased by 22.6%, 79.6% and 24.0% , respectively (Zheng, 2017). Adding 25–50% blue light can promote an increase in photosynthetic pigment, vitamin C and soluble sugar content in lettuce leaves, and can reduce nitrate content at the same time (Xu et al., 2015). The nitrate content of red-leaf cabbage also decreased significantly as the wavelength of blue light shortened.

## 2.3 Green light

The waveband of green light is 500-599 nm, which also belongs to PAR. Compared with red and blue light, chloroplasts have a lower absorption rate of green light, while leaves have higher reflection and projection rates of green light (Kozai et al., 2016). Green light is therefore generally considered to be a less effective light for plant photosynthesis. However, the average relative quantum efficiency value for broadband green light is 0.87, which is slightly lower than that of red light (0.91) and higher than that of blue light (0.73) (Sager et al., 1988). In addition, while blue and red light are strongly absorbed by upper-level leaves, green light can supply energy for the lower canopy and deep leaves of plants, especially those with overlapping leaves such as lettuce, which could potentially increase crop yield (Wang and Folta, 2013; Sun et al., 1998). Studies have shown that when a light source without green light is used for cultivation, the lower leaves are more likely to turn yellow at the end of the cultivation stage. When green light is added, the symptoms of yellowing lower leaves are alleviated since the degradation of chlorophyll in lower lettuce leaves is slowed by adding green light (Zhang et al., 2015).

For plant growth and development, green light acts as a signal, regulating plant nonphotosynthetic responses such as vegetative growth, anthocyanin accumulation and flowering initiation, likely via phytochromes and cryptochromes, although no specific green light photoreceptor has been detected in previous studies (Wang and Folta, 2013). Green light shifts cryptochromes from a semi-reduced active state caused by blue light to a fully reduced and inactive state. Green light can therefore deactivate blue-mediated responses, such as suppressing the induction of the stomata opening of plant leaves by blue light that is absorbed by cryptochromes (Bouly et al., 2007; Liu et al., 2010; Sellaro et al., 2010). Plant responses to green light are similar to FR light and share a general tendency to oppose blue or red light-induced responses (Talbott et al., 2006; Folta and Maruhnich, 2007). For instance, green light inhibits hypocotyl elongation, decreases anthocyanin concentrations and induces shade growth symptoms in Arabidopsis plants (Zhang and Folta, 2012; Anpo et al., 2018).

Meng et al. (2004) found that green light can promote flowering. *Arabidopsis thaliana* exposed to natural light during the day was exposed to green light for 7 hours at night. The first flowering time was significantly earlier than that of the control group, which might relate to cryptochrome.

## 2.4 Far-red light

FR light, also known as infrared light, has a wavelength ranging from 700 nm to 799 nm, which belongs to BAR rather than PAR. FR light is generally used as a supplemental light source for plant growth in vertical farms and can regulate plant flowering. For example, spinach flowering with a 12-hour light period can be inhibited by adding an additional 2 hours of FR light during both light and dark periods (Tong and Fang, 2020). FR light can thus inhibit/promote the flowering of long- and short-day plants.

The addition of FR light also increases the height of edible flower plants and promotes the accumulation of IAA and GA. FR light is reported to not only promote the accumulation of soluble metabolites, but also to improve the cold tolerance of plants.

Although it is not conducive to anthocyanin synthesis, it can effectively increase the internode height and bioaccumulation of plants and improve the taste and environmental resistance of vegetables.

## 2.5 Yellow-orange light

The wavelength range of yellow-orange light is from 575 nm to 625 nm, which belongs to PAR. Yellow-orange light helps increase clematis height, but causes plants to become thinner. Cucumber seedlings treated with yellow-orange light have short plants, yellow leaves and a smaller area. The chlorophyll content under yellow-orange light was the lowest, and its content decreased with the prolongation of irradiation time (Tong and Fang, 2020). The same result was obtained in tomato experiments. Plants in the yellow-orange light treatment group were short, the leaves were yellow, and fewer starch granules accumulated in the chloroplasts. However, yellow-orange light promoted the accumulation of plant flavonoids.

Evidence for yellow light (580–600 nm) suppression of lettuce growth has been reported by Dougher and Bugbee (2001a). Fan et al. (2013) reported that the dry mass, soluble sugar and protein concentrations decreased in nonheading Chinese cabbage under single yellow light. Yang et al. (2019) reported that yellow light had better effects on plant growth and bioactive flavonoids accumulation of *Epimedium pseudowushanense* than white light, red light or blue light. Furthermore, both RNAseq and qPCR analysis revealed that yellow light could highly upregulate the expression levels of flavonoid biosynthetic genes, in particular *CHS1*, *F3H1*, *PT_5* and *raGT_5*, which may enhance the accumulation of bioactive flavonoids in *Epimedium pseudowushanense*. Li et al. (2021) reported that supplemental orange radiation increased the fresh and dry weights of shoots and showed better photosynthetic capacity in lettuce, such as higher Pn, PSII and other photosynthetic parameters, while supplemental yellow radiation significantly decreased photosynthetic capacity but showed better nutritional quality performance, increased soluble sugar content and decreased nitrate accumulation. In practice, less attention has been paid to yellow-orange light, and further research is needed for optimal use of this wavelength for plant growth and development.

## 2.6 Ultraviolet light

The wavelength range of ultraviolet light is between 100 nm and 399 nm and is divided into three types:

- near-ultraviolet light (UV-A);
- medium-ultraviolet light (UV-B); and
- far-ultraviolet light (UV-C).

These are discussed next.

Near-ultraviolet light (UV-A) is a long-wave ultraviolet light with a wavelength range of 315–399 nm. Near-ultraviolet light regulates plant growth. The photoreceptors are the same as those in blue light , including cryptochromes and phototropins, but the two photoreceptors are separated *in vitro* and their absorption peaks are in the blue light region rather than the near-ultraviolet light region, so near-ultraviolet light has similar effects on plants as blue light but to different degrees.

Appropriate UV-A can promote the synthesis of plant secondary metabolites such as anthocyanins, flavonoids and ascorbic acid. Low doses of UV-A can increase biomass accumulation in plant roots, but many studies have used low doses of UV-A to avoid plant damage. UV-A can also change the structure of leaves, causing changes in the structure of cutin, leaf cells and palisade tissue, filtering part of the ultraviolet light, protecting leaves and thickening the epidermal cell layer of leaves. Studies have shown that UV-A can enhance the photoprotection of plant leaves and stimulate plants to enhance secondary metabolism, thereby producing more secondary metabolites, which can physically and biochemically defend plants against light damage (Yang, 2019). While secondary metabolites strengthen the blocking and filtering of high light intensity, they also eliminate toxic photoproducts such as peroxides and singlet oxygen, reducing the probability of photoinhibition. In general, UV-A can act as a signal to stimulate a series of physiological responses in plants to improve plant stress resistance, but in a less destructive way than UV-B, and does not damage plant tissue, organs or DNA.

Medium-ultraviolet (UV-B) wavelengths range from 280 nm to 315 nm. Studies have shown that, under UV-B, morphological changes in plants will occur, such as dwarfing and leaf area reduction, reduced hypocotyl elongation, small and thick leaves, short petioles, altered root to shoot ratio, while also causing increased accumulation of flavonoids, anthocyanins and peroxides (Bantis et al., 2018). These changes occur when plants sense external UV-B through their unique photoreceptor UV Resistance Locus 8 (UVR8). High-intensity or long-term UV-B irradiation will cause systemic damage to plants and reduce photosynthetic rate.

Some researchers have suggested that appropriate use of UV-B can improve the yield and quality of horticultural products. Schreiner et al. (2012) reported that UV-B causes cell wall rupture and accelerates fruit softening, while carotenoids and flavonoids can resist UV-B damage to plants. The UVR8 in *Arabidopsis thaliana* controls the transcription factor HY5 (a key factor in the UVR8 pathway), which is highly sensitive to UV-B, reducing damage by ultraviolet light to the photosynthetic system of plants and increasing disease resistance (Jansen and Bornman, 2012). Supplemental UV-B increases concentrations of anthocyanin, phenolics and flavonoids in green basil leaves by 9–18%, 28–126% and 80–169%, respectively, and the antioxidant capacity of basil leaves positively correlated to UV-B doses. However, supplemental UV-B generally decreased the crop yield in both green and purple basil plants (Dou et al., 2019).

Far-ultraviolet light (UV-C) has a wavelength range of 100–280 nm, and UV-C energy is stronger, which means UV-C is mostly used for sterilization of water or air. Rai et al. (2011) conducted a 14-day UV pretreatment on mugwort seedlings for 21 days after sowing and found that both UV-B and UV-C could increase anthocyanin content. UV-C had a better effect, and contents of artemisinin and flavonoids also increased. However, the study also showed that irradiation of 10–50 kJ $m^{-2}$ UV-C for just 2 hours can cause irreversible DNA damage in Arabidopsis. Determining how to use UV-B and UV-C correctly to increase disease resistance and bioflavonoids without causing serious damage to plant tissue is an important aspect of research in the field of photobiology.

UV chips can now be produced but they have a high unit price and heavy energy use. Therefore, UV light is still rarely used in plant cultivation. Correct use of low-dose continuous irradiation of UV-A and intermittent irradiation of UV-B is extremely important for high quality plant cultivation and further research is needed to find a balance between enhanced nutritional quality and yield reduction.

## 3 Polychromatic light

Specific spectrums stimulate different morphological and physiological responses. Research has indicated that exposure to monochromatic light may induce physiological disorders in some plant species. Some researchers have shown that full spectrum white LEDs are more suitable for leafy plant production in vertical farms. They not only ensure plants receive all the PAR required to optimize growth, quality and yield, but also provide a comfortable light environment for cultivators and allow for easy visual assessment of plants for any possible disorder. Common combinations of light spectra and their effects on plants are described.

### *3.1 Red and blue light*

Cucumber and tomato plants under monochromatic red light displayed abnormal leaf morphology, reduced photosynthetic capacity, decreased stomata density and impaired growth, defined as 'red light syndrome'. None of these effects occurred in leaves grown under white light or red light supplemented with blue light (Trouwborst et al., 2016; Savvides et al., 2012; Hogewoning et al., 2012). Hogewoning et al. (2010) reported that leaf photosynthetic machinery dysfunction appeared under red light alone, while just 7% blue light was sufficient to prevent any significant dysfunctional photosynthesis. This indicated that a certain amount of blue light is essential for maintaining normal plant growth. In addition, the use of blue light alone often produces compact plants with a small leaf area and low photosynthetic rate in many species, such as chrysanthemum plantlets (Kim et al., 2004) and *Withania somnifera* (L.) plantlets (Lee et al., 2007).

Nowadays, red light (660 nm) and blue light (450 nm) are widely considered to be the basic essential light combination for plant growth and development since their absorption by photosynthetic pigments are higher and more effective for plant production than other wavelengths (Pfündel and Baake, 1990; Olle and Virsile, 2013). A combination of red and blue light is often used in vertical farms and most leafy greens can grow healthily. Many previous studies focused on the effects of their different ratios on plant growth (Li et al., 2013; Hogewoning et al., 2010; Ohashi-Kaneko et al., 2006; Wang et al., 2016). Red and blue light act antagonistically in the regulation of leaf size and thickness. Blue light promotes leaf flattening by enhancing photons activity, while red light predominantly promotes branching by activating phyB (Anpo et al., 2018). Blue light involves inhibition of cell expansion or division, so a reduction in blue light could increase leaf area (Dougher and Bugbee, 2001b; Hernández and Kubota, 2014), which promotes light interception and dry weight accumulation. Careful attention must be paid to determine the balance between red and blue light so that plant architecture is regulated in accordance with the intended purpose.

Optimal R/B (red to blue light) ratio varies from plant to plant. For example, it was found to be 1:2 (Wang et al., 2016), 7:3 (Nhut et al., 2003), 1:3 (Li et al., 2013) and 9 (Hernández and Kubota, 2016), respectively, for fresh and dry weight accumulation in lettuce plants, strawberry plantlets, rapeseed (*Brassica napus* L.) plantlets *in vitro* and cucumber seedlings in vertical farms. Seedlings under the R/B ratio of 7:3 perform best in terms of leaf number, stem diameter, chlorophyll content and fresh weight, and have the best growth after transplanting. Compared with the R/B ratio of 8:2, the R/B ratio of 2:8 increased the anthocyanin concentration of lettuce but decreased the average fresh weight per plant, so the anthocyanin content of the whole plant is still higher under the R/B ratio of 8:2 (Tong and Fang, 2020). Based on the above analysis, there is no single optimal R/B ratio that fits all species. The optimal R/B ratio should be selected based on production purposes.

### 3.2 Blue and green light

Unlike red and blue light, green light is generally considered as a supplemental light for plant growth. The vital role of green light in affecting plant physiological activities has been established (Johkan et al., 2012; Wang and Folta, 2013; Materová et al., 2017). With red and blue light as the background light, adding a moderate amount of green light can promote the photosynthesis of the whole plant, slow the degradation of chlorophyll in the lower leaves of the plant and improve nutritional quality. Green light can also promote the elongation of seedlings and improve the seedling strength index of tomato seedlings. Many studies have indicated that green light, partially replacing red and/or blue light, has a significant regulatory effect on plant growth and qualities (Li et al., 2021; Kaiser et al., 2019). Kaiser et al. (2019) reported that the fresh and dry weights of tomato plants increased linearly with an increase in the percentage of green light.

The optimal G/B (green to blue light) ratio or the supplemental amount of green light is a key factor influencing plant growth. Son and Oh (2015) found that an 8% increase in green light resulted in greater leaf area and fresh weight in red leaf lettuce (cv. Sunmang), but not in green leaf lettuce. For green leaf lettuce, it seems that a higher percentage of green light is required to aid growth and development. Kim et al. (2004) found that a combination of red, green and blue light, with green light accounting for 24%, can promote the growth of lettuce and increase bioaccumulation compared with those of combination of red and blue light, but as the proportion of green light increases, its promotion effect decreases. Schenkels et al. (2020) showed that additional green light significantly increased the total fresh and dry weights of plant seedlings, whereas no significant increase was found in dry weight when replacing red and blue with green light. These inconsistent results are mainly caused by the different proportions of green light in the light source. The proportion of green light ranging from 15% to 30% is reported to be optimal for lettuce plant growth, resulting in higher yield and quality due to the increase in leaf expansion, stem length and total chlorophyll content in leaves and decreased nitrate content (Li et al., 2021).

Contrasting effects of green light and blue light on stomata characteristics, plant morphology, nutritional quality, etc. have been observed. Both blue and green light have the ability to regulate stomata. Blue light promotes stomata opening, while green light induces stomata closing. Green light can reverse stomatal openings stimulated by blue light, because the fully oxidized state of the flavin chromophore of cryptochrome in darkness is shifted to a semi-reduced form by blue light, and this semi-reduced form is converted to a reduced form particularly by absorbing green light (Bouly et al., 2007; Liu et al., 2010; Sellaro et al., 2010). However, stomatal opening could be restored if the green light was followed or replaced by a second blue light in a response analogous to the R/FR reversibility of phytochrome responses (Frechilla et al., 2000). Frechilla et al. (2000) found that green light at 540 nm promotes the closure of faba bean stomata. The higher the G/B value, the less stomata opened and the more growth was inhibited. However, it was reported by Son and Oh (2015) that stomatal density increased significantly when blue LEDs were replaced with green LEDs in green leaf lettuce. An increase in stomatal density from 47.2% to 70.6% and a decrease in the aperture area from 20.4% to 63.4% was observed when the relative amount of green light increased from 15% to 45% (Li et al., 2021).

Although the light absorption of photosynthetic pigments is relatively low within the green region compared with that within the blue region on the plant canopy (McCree, 1972; Smith et al., 2017), the relative quantum efficiency of green light once absorbed by plant leaves is higher than that of blue light, according to McCree (1972). Moreover, like the red/far-red (R/FR) ratio, the G/B ratio could also act as a shade signal within a plant canopy. The shade avoidance syndrome induced by green light was capable of triggering physiological and morphological changes of plants to best intercept available energy (Sellaro et al., 2010; Smith et al., 2017). The hypocotyl length of Arabidopsis seedlings increased linearly with increasing G/B ratios within the range of 0.5–1.0 while cultivated in a natural environment (Sellaro et al., 2010). The increase in stem length and leaf area induced by the green light for lettuce plant growth was also found by Li et al. (2021), Kang et al. (2016) and Johkan et al. (2012). These changes in plant morphology effectively enhance the photons captured by plant leaves, improving photosynthetic efficiency (Park and Runkle, 2017).

Chen et al. (2016) found that supplementing green light with white LEDs increased soluble sugar content and decreased nitrate content in green oak leaf lettuce. Li et al. (2020) also found that partially replacing red and blue light with green light decreased the nitrate content in lettuce leaves. Bian et al. (2018) agreed with the above result that green light could decrease the nitrate content in lettuce leaves by regulating the expression of various specific genes (e.g. NR and NiR).

### *3.3 Red and far-red light*

Red and FR light are important signals to plants and the R/FR ratio affects phytochrome-regulated responses such as seed germination, seedling de-etiolation, shade avoidance and reproduction responses. Phytochromes are reversibly photochromic biliproteins, which exist in a biologically inactive red light-absorbing inactive phytochrome ($P_r$) form and a biologically active FR light-absorbing activated phytochrome ($P_{fr}$) form. Figure 1 shows that the peak absorption wavelengths are different for $P_r$ and $P_{fr}$ (Sager at al., 1988). FR light is conducive to the conversion of $P_{fr}$ to $P_r$ (Fig. 2). The equilibrium between the two forms changes dynamically and strongly correlates with the R/FR ratio.

The optimum R/FR ratio can modify plant morphogenesis and promote plant growth. The lower R/FR ratio can promote stem elongation and induce shade avoidance effect. When the end of day FR light is used as a supplementary light for 14 days, the hypocotyl length of tomato seedlings is effectively promoted under an R/FR ratio of 0.05. When the R/FR ratio is decreased from 6.90 to 0.13, the internode length of tomatoes with different genotypes increases, which in turn increases the plant height. Park and Runkle (2017) found that FR light increased plant growth mainly through leaf expansion and an increase in the whole-plant net photosynthetic rate. Several studies have also found that FR decreased anthocyanins, carotenoids and chlorophyll concentrations of lettuce plants but increased soluble sugar and nitrate concentrations (Zou et al., 2019). A high R/FR (6.9) can promote the accumulation of anthocyanins. In other words, FR light is not conducive to the synthesis of anthocyanins.

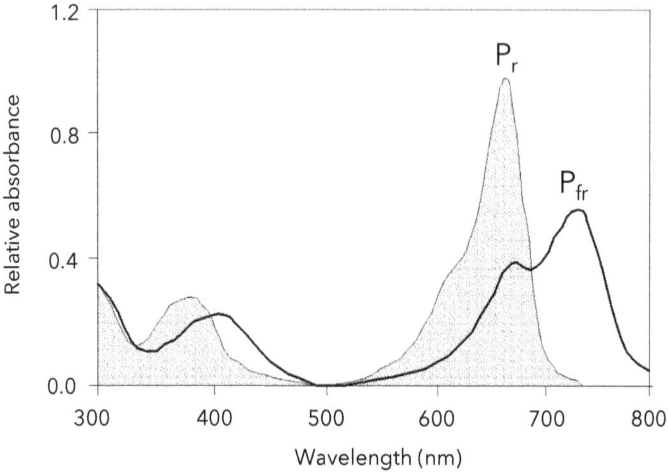


**Figure 1** Relative absorbances of phytochrome: far-red induced inactive phytochrome ($P_r$) and red induced activated phytochrome ($P_{fr}$).

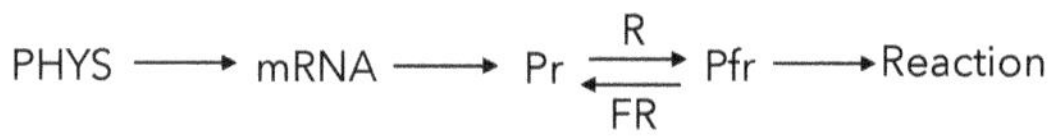


**Figure 2** The conversion of activated phytochrome ($P_{fr}$) to inactive phytochrome ($P_r$) (Wang, 2000).

Compared to red and blue light, FR light has limited contribution to photosynthesis due to the low leaf absorbance of its wavelength range. Studies have reported that the supplementation of FR light to shorter wavelength light (400–700 nm) can balance the excitation between photosystem (PS) I and PSII, which synergistically increased photochemistry and photosynthesis, further increased its bioaccumulation, and improved leaf color, thickness and taste (Park and Runkle, 2017; Zhen and van Iersel, 2017).

### 3.4 Red and white light

There is more red than white light in red and white light, and more green light than red and blue light. Increasing both red and green light is beneficial to the progress of photosynthesis, but the proportion of green light should not be too high. For common cool white fluorescent lamps or LED lamps, the proportion of green light is usually 45–55% of PPFD, which is obviously high. Chen et al. (2016) reported that 'Green Oak Leaf' lettuce grown under red and white LEDs appeared more compact and vigorous than under white LEDs. Mickens et al. (2018) also found that red and white light is more beneficial to increase the fresh and dry weights of plants than red and blue light. Yang et al. (2019) found that red and white light (red light accounting for 24.4%) treatment of purple leaf lettuce can increase the utilization efficiency of light and electricity by 22% and 7% respectively, compared with white light treatment.

## 4 Where to look for further information

For more detailed information, several books related to this section are suggested:

- https://link.springer.com/book/10.1007/978-981-10-1848-0 LEDLighting for Urban Agriculture | SpringerLink.
- https://www.mdpi.com/2223-7747/11/15/1911/htm Plants | Free Full-Text | The Effects of LED Light Spectra and Intensities on Plant Growth (mdpi.com).
- https://link.springer.com/book/10.1007/978-981-13-1065-2 Smart Plant Factory: The Next Generation Indoor Vertical Farms | SpringerLink.

The most relevant international conference is LightSys which is held every two years. the LightSym 2024 (X International Symposium on Light in Horticulture) will be held in Seoul, South Korea. http://lightsym2024.org/.

## 5 References

Ahmed, H. A., Yu-Xin, T. and Qi-Chang, Y. 2020. Optimal control of environmental conditions affecting lettuce plant growth in a controlled environment with artificial lighting: A review. *S. Afr. J. Bot.* 130:75–89.

Anpo, M., Fukuda, H. and Wada, T. 2018. *Plant Factory Using Artificial Light: Adapting to Environmental Disruption and Clues to Agricultural Innovation*. Elsevier Inc., Cambridge, MA.

Assmann, S. M. 1988. Enhancement of the stomatal response to blue light by red light, reduced intercellular concentrations of $CO_2$, and low vapor pressure differences. *Plant Physiol.* 87(1):226–231.

Bantis, F., Smirnakou, S., Ouzounis, T., Koukounaras, A., Ntagkas, N. and Radoglou, K. 2018. Current status and recent achievements in the field of horticulture with the use of light-emitting diodes (LEDs). *Sci. Hort.* 235:437–451.

Bian, Z., Cheng, R., Wang, Y., Yang, Q. and Lu, C. 2018. Effects of green light on nitrate reduction and edible quality of hydroponically grown lettuce (*Lactuca sativa L.*) under short-term continuous light from red and blue light-emitting diodes. *Environ. Exp. Bot.* 153:63–71.

Bouly, J. P., Schleicher, E., Dionisio-Sese, M., Vandenbussche, F., Van Der Straeten, D., Bakrim, N., Meier, S., Batschauer, A., Galland, P., Bittl, R. and Ahmad, M. 2007. Cryptochrome blue light photoreceptors are activated through interconversion of flavin redox states. *J. Biol. Chem.* 282(13):9383–9391.

Chen, X., Xue, X., Guo, W., Wang, L. and Qiao, X. 2016. Growth and nutritional properties of lettuce affected by mixed irradiation of white and supplemental light provided by light-emitting diode. *Sci. Hort.* 200:111–118.

Dou, H. and Niu, G. 2020. Chapter 9: Plant responses to light. In: Kozai, T., Niu, G. and Takagaki, M. (Eds.), *Plant Factory* (2nd ed.). Academic Press, pp. 153–166.

Dou, H., Niu, G. and Gu, M. 2019. Pre-Harvest UV-B Radiation and Photosynthetic Photon Flux Density Interactively Affect Plant Photosynthesis, Growth, and Secondary Metabolites Accumulation in Basil (Ocimum basilicum) Plants. *Agron.* 9(434):1–9.

Dougher, T. A. and Bugbee, B. 2001a. Differences in the response of wheat, soybean and lettuce to reduced blue radiation. *Photochem. Photobiol.* 73(2):199–207.

Dougher, T. A. and Bugbee, B. 2001b. Evidence for yellow light suppression of lettuce growth. *Photochem. Photobiol.* 73(2):208–212.

Du, J. F., Liao, X. R., Ye, B. Q. and Li, M. 2002. Effect of light quality on the growth and antioxidant enzyme activities of rape seedings. *Chin. Bull. Bot.* 19(6):743–745. Chinese With English Abstract.

Eskins, K. 1992. Light-quality effects on Arabidopsis development: Red, blue and far-red regulation of flowering and morphology. *Physiol. Plant.* 86(3):439–444.

Fan, X., Zhang, J., Xu, Z., Guo, S., Jiao, X., Liu, X. and Gao, Y. 2013. Effects of different light quality on growth, chlorophyll concentration and chlorophyll biosynthesis precursors of non-heading Chinese cabbage (*Brassica campestris* L.). *Acta Physiol. Plant.* 35(9):2721–2726.

Folta, K. M. and Maruhnich, S. A. 2007. Green light: A signal to slow down or stop. *J. Exp. Bot.* 58(12):3099–3111.

Frechilla, S., Talbott, L. D., Bogomolni, R. A. and Zeiger, E. 2000. Reversal of blue light-stimulated stomatal opening by green light. *Plant Cell Physiol.* 41(2):171–176.

Graamans, L., Baeza, E., Dobbelsteen, A. V. D., Tsafaras, I. and Stanghellini, C. 2018. Plant factories versus greenhouses: Comparison of resource use efficiency. *Agric. Syst.* 160:31–43.

Hernández, R. and Kubota, C. 2014. Growth and morphological response of cucumber seedlings to supplemental red and blue photon flux ratios under varied solar daily light integrals. *Sci. Hort.* 173:92–99.

Hernández, R. and Kubota, C. 2016. Physiological responses of cucumber seedlings under different blue and red photon flux ratios using LEDs. *Environ. Exp. Bot.* 121:66–74.

Hogewoning, S. W., Trouwborst, G., Meinen, E. and Van Ieperen, W. 2012. Finding the optimal growth-light spectrum for greenhouse crops. *Acta Hortic.* 956:357–363.

Jansen, M. A. and Bornman, J. F. 2012. UV-B radiation: From generic stressor to specific regulator. *Physiol. Plant.* 145(4):501–504.

Johkan, M., Shoji, K., Goto, F., Hahida, S. and Yoshihara, T. 2012. Effect of green light wavelength and intensity on photomorphogenesis and photosynthesis in Lactuca sativa. *Exp. Bot.* 75:128–133.

Kaiser, E., Weerherm, K., Schipper, R. and Dieleman, J. A. 2019. Partial replacement of red and blue by green light increases biomass and yield in tomato. *Sci. Hort.* 249:271–279.

Kang, W. H., Park, J. S., Park, K. S. and Son, J. E. 2016. Leaf photosynthetic rate, growth, and morphology of lettuce under different fractions of red, blue, and green light from light-emitting diodes (LEDs). *Hortic. Environ. Biotechnol.* 57(6):573–579.

Kim, H. H., Goins, G. D., Wheeler, R. M. and Sager, J. C. 2004. Green-light supplementation for enhanced lettuce growth under red-and blue-light-emitting diodes. *HortScience* 39(7):1617–1622.

Kim, S. J., Hahn, E. J., Heo, J. W. and Paek, K. Y. 2004. Effects of LEDs on net photosynthetic rate, growth and leaf stomata of chrysanthemum plantlets in vitro. *Sci. Hort.* 101(1-2):143-151.

Kozai, T. 2013. Resource use efficiency of closed plant production system with artificial light: Concept, estimation and application to plant factory. *Proc. Jpn Acad. Ser. B Phys. Biol. Sci.* 89(10):447-461.

Kozai, T., Fujiwara, K. and Runkle, E. S. 2016. *LED Lighting for Urban Agriculture*. Springer Nature, Germany.

Lee, S. H., Tewari, R. K., Hahn, E. J. and Paek, K. Y. 2007. Photon flux density and light quality induce changes in growth, stomatal development, photosynthesis and transpiration of *Withania somnifera* (L.) Dunal. plantlets. *Plant Cell Tiss. Organ Cult.* 90(2):141-151.

Li, H., Tang, C. and Xu, Z. 2013. The effects of different light qualities on rapeseed (*Brassica napus* L.) plantlet growth and morphogenesis in vitro. *Sci. Hort.* 150:117-124.

Li, L., Tong, Y., Lu, J., Li, Y. and Yang, Q. 2020. Lettuce growth, nutritional quality, and energy use efficiency as affected by red-blue combined with different monochromatic wavelengths. *Hortscience* 55(5):613-620.

Li, L., Tong, Y. X., Lu, J. L., Li, Y. M., Liu, X. and Cheng, R. F. 2021. Morphology, photosynthetic traits, and nutritional quality of lettuce plants as affected by green light substituting proportion of blue and red light. *Front. Plant Sci.* 12:627311.

Liu, B., Liu, H., Zhong, D. and Lin, C. 2010. Searching for a photocycle of the cryptochrome photoreceptors. *Curr. Opin. Plant Biol.* 13(5):578-586.

Materová, Z., Sobotka, R., Zdvihalová, B., Oravec, M., Nezval, J., Karlicky, V., Vrábl, D., Štroch, M. and Špunda, V. 2017. Monochromatic green light induces an aberrant accumulation of geranylgeranyled chlorophylls in plants. *Plant Physiol. Biochem.* 116:48-56.

McCree, K. J. 1972. The action spectrum, absorptance and quantum yield of photosynthesis in crop plants. *J. Agric. Meteorol.* 9:191-216.

Meng, X. C., Xing, T. and Wang, X. J. 2004. The role of light in the regulation of anthocyanin accumulation in Gerbera hybrida. *Plant Growth Regul.* 44(3):243-250.

Metallo, R. M., Kopsell, D. A., Sams, C. E. and Bumgarner, N. R. 2018. Influence of blue/red vs. white LED light treatments on biomass, shoot morphology, and quality parameters of hydroponically grown kale. *Sci. Hort.* 235:189-197.

Mickens, M. A., Skoog, E. J., Reese, L. E., Barnwell, P. L., Spencer, L. E., Massa, G. D. and Wheeler, R. M. 2018. A strategic approach for investigating light recipes for 'Outredgeous' red romaine lettuce using white and monochromatic LEDs. *Life Sci. Space Res. (Amst)* 19:53-62.

Mortensen, L. M. and Strømme, E. 1987. Effects of light quality on some greenhouse crops. *Sci. Hort.* 33(1-2):27-36.

Nhut, D. T., Takamura, T., Watanabe, H., Okamoto, K. and Tanaka, M. 2003. Responses of strawberry plantlets cultured in vitro under superbright red and blue light-emitting diodes (LEDs). *Plant Cell Tiss. Org. Cult.* 73(1):43-52.

Ohashi-Kaneko, K., Matsuda, R., Goto, E., Fujiwara, K. and Kurata, K. 2006. Growth of rice plants under red light with or without supplemental blue light. *Soil Sci. Plant Nutr.* 52(4):444-452.

Olle, M. and Virsile, A. 2013. The effects of light-emitting diode lighting on greenhouse plant growth and quality. *Agric. Food Sci.* 22(2):223-234.

Park, Y. and Runkle, E. S. 2017. Far-red radiation promotes growth of seedlings by increasing leaf expansion and whole-plant net assimilation. *Environ. Exp. Bot.* 136:41–49.

Pfündel, E. and Baake, E. 1990. A quantitative description of fluorescence excitation spectra in intact bean leaves greened under intermittent light. *Photosynth. Res.* 26(1):19–28.

Pu, G., Liu, S., Liu, L. and Ren, L. 2005. Effects of different light qualities on growth and physiological characteristics of tomato seedlings. *Acta Hortic. Sin.* 32(3):420–425. Chinese With English Abstract.

Rai, R., Meena, R. P., Smita, S. S., Shukla, A., Rai, S. K. and Pandey-Rai, S. 2011. UV-B and UV-C pre-treatments induce physiological changes and artemisinin biosynthesis in Artemisia annual.-An antimalarial plant. *J. Photochem. Photobiol. B* 105(3):216–225.

Sager, J., Smith, W., Edwards, J. and Cyr, K. 1988. Photosynthetic efficiency and phytochrome photoequilibria determination using spectral data. *Trans. ASAE* 31:1882–1889.

Savvides, A., Fanourakis, D. and Van Ieperen, W. 2012. Co-ordination of hydraulic and stomatal conductances across light qualities in cucumber leaves. *J. Exp. Bot.* 63(3):1135–1143.

Schenkels, L., Saeys, W., Lauwers, A. and Proft, M. P. D. 2020. Green light induces shade avoidance to alter plant morphology and increases biomass production in *Ocimum basilicum* L. *Sci. Hort.* 261:109002.

Schreiner, M., Mewis, I., Huyskens-Keil, S., Jansen, M. A. K., Zrenner, R., Winkler, J. B., O'brien, N. and Krumbein, A. 2012. UV-B-induced secondary plant metabolites-potential benefits for plant and human health. *Crit. Rev. Plant Sci.* 31(3):229–240.

Sellaro, R., Creyp, M., Trupkin, S. A., Karayekov, E., Buchovsky, A. S., Rossi, C. and Casal, J. J. 2010. Cryptochrome as a sensor of the blue/green ratio of natural radiation in arabidopsis. *Plant Physiol.* 154(1):401–409.

Smith, H. L., McAusland, L. and Murchie, E. H. 2017. Don't ignore the green light: Exploring diverse roles in plant processes. *J. Exp. Bot.* 68(9):2099–2110.

Son, K. H. and Oh, M. M. 2015. Growth, photosynthetic and antioxidant parameters of two lettuce cultivars as affected by red, green, and blue light-emitting diodes. *Hortic. Environ. Biotechnol.* 56(5):639–653.

Stuefer, J. F. and Huber, H. 1998. Differential effects of light quantity and spectral light quality on growth, morphology and development of two stoloniferous potentilla species. *Oecologia* 117(1–2):1–8.

Sun, J., Nishio, J. N. and Vogelmann, T. C. 1998. Green light drives $CO_2$ fixation deep within leaves. *Plant Cell Physiol.* 39(10):1020–1026.

Talbott, L. D., Hammad, J. W., Harn, L. C., Nguyen, V. H., Patel, J. and Zeiger, E. 2006. Reversal by green light of blue light-stimulated stomatal opening in intact, attached leaves of Arabidopsis operates only in the potassium-dependent, morning phase of movement. *Plant Cell Physiol.* 47(3):332–339.

Tong, Y. and Fang, W. 2020. *Digitization of the Plant Factory: Theory and Practice*. China Agricultural Science and Technology Press, Beijing. In Chinese.

Tong, Y., Yang, Q. and Shimamura, S. 2014. Analysis of electric-energy utilization efficiency in a plant factory with artificial light for lettuce production. *Acta Hortic.* 1037:277–284.

Trouwborst, G., Hogewoning, S. W., Van Kooten, O., Harbinson, J. and Van Ieperen, W. 2016. Plasticity of photosynthesis after the 'red light syndrome' in cucumber. *Environ. Exp. Bot.* 121:75–82.

Wang, J., Lu, W., Tong, Y. and Yang, Q. 2016. Leaf morphology, photosynthetic performance, chlorophyll fluorescence, stomatal development of lettuce (*Lactuca sativa* L.) exposed to different ratios of red light to blue light. *Front. Plant Sci.* 7:250.

Wang, Y. and Folta, K. M. 2013. Contributions of green light to plant growth and development. *Am. J. Bot.* 100(1):70–78.

Wang, Z. 2000. *Plant Physiology*. China Agricultural Science and Technology Press, Beijing. In Chinese.

Wen, J., Ke, X., Xu, C., Li, J. and Gong, M. 2011. Effects of light quality on antioxidant defense system during growth and development of tobacco leaves. *Acta Bot. Boreal.-Occident. Sin.* 31(9):1799–1804. Chinese With English Abstract.

Xu, D., Gao, W. and Ruan, J. 2015. Effects of light quality on plant growth and development. *Plant Physiol.* 51(8):1217–1234.

Xu, W., Liu, X., Jiao, X. and Xu, Z. 2015. Effect of blue light quantity on growth and quality of lettuce. *J. Nanjing Agric. Univ.* 38(6):890–895.

Yang, Q. 2019. *Plant Factory*. Tsinghua University Press, Beijing. In Chinese.

Yang, Q., Pan, J., Shen, G. and Guo, B. 2019. Yellow light promotes the growth and accumulation of bioactive flavonoids in *Epimedium pseudowushanense*. *J. Photochem. Photobiol. B* 197:111550.

Yu, R. and Pan, R. 1997. Effects of Blue light on the growth and levels of endogenous phytohormones in rice (*Oryza sativa* L.) seedlings. *Acta Phytophysiotogica Sin.* 23(2):175–180. Chinese With English Abstract.

Zhang, G., Shen, S., Takagaki, M., Kozai, T. and Yamori, W. 2015. Supplemental upward lighting from underneath to obtain higher marketable lettuce (*Lactuca sativa*) leaf fresh weight by retarding senescence of outer leaves. *Front. Plant Sci.* 6:1110.

Zhang, T. and Folta, K. M. 2012. Green light signaling and adaptive response. *Plant Signal. Behav.* 7(1):75–78.

Zhen, S. and Van Iersel, M. W. 2017. Far-red light is needed for efficient photochemistry and photosynthesis. *J. Plant Physiol.* 209:115–122.

Zheng, D. 2017. *Effects of supplemental blue and UV-A LED on the quality of pak choi and Chinese kale baby leaf*. Master Thesis. South China Agricultural University, Guangzhou, China. Chinese With English Abstract.

Zhou, L., Chen, Z., Wang, Y., Yang, L., Fang, W. and Li, X. 2012. Effect of light quality on tea polyphenol content and activities of anti-oxidative enzymes in Tea callus. *J. Tea Sci.* 32(3):210–216.

Zou, J., Zhang, Y., Zhang, Y., Bian, Z., Fanourakis, D., Yang, Q. and Li, T. 2019. Morphological and physiological properties of indoor cultivated lettuce in response to additional far-red light. *Sci. Hort.* 257:108725.

# Chapter 13

## Adjusting photosynthetic photon flux density (PPFD) to improve the quality of leafy vegetables in vertical farms

*Qianxixi Min, Leo F. M. Marcelis and Ernst J. Woltering, Wageningen University & Research, The Netherlands*



### 1 Introduction

To meet increasing consumer demand for healthy, tasty, plant-based food with low environmental impact, indoor vertical farming (IVF) has emerged as a fast-developing, feasible cultivation system that is attracting attention from researchers, investors, and consumers (Van Delden et al., 2021; SharathKumar et al., 2020). Using artificial light as the only light source is a key feature that differentiates IVF from other cultivation systems. The unique properties of light-emitting diodes (LEDs), including precise control of photosynthetic photon flux density (PPFD) and spectrum, low thermal emission, high photoelectric efficiency, and portable design, make them the primary light source in most indoor vertical farms (Kozai, 2013; Kusuma et al., 2020). These properties offer ample opportunities for designing specific lighting strategies that can

http://dx.doi.org/10.19103/AS.2023.0126.17

effectively influence phytochemical levels and thus regulate crop quality (Thilini Deepashika Perera et al., 2022). As many phytochemicals are health related and also function as antioxidants, they may affect nutritional quality as well as shelf life. Thus, light strategies are important for producing high-quality vegetables in indoor vertical farms.

Leafy vegetables (including leafy greens, herbs, and microgreens) are the prevailing crops grown in indoor vertical farms, as these fresh products are considered healthy diet sources and are in great market demand. However, they generally have a short shelf life due to their perishable characteristics (Mir et al., 2017). A definition of crop quality that incorporates nutritional quality (measured level of health-related phytochemicals) at harvest and during storage, consumer preferences (sensory parameters), and shelf life (the number of days that the product remains at a minimum quality level to meet consumer safety/quality requirements after harvest) were accepted by the horticulture industry. However, in the research, these aspects are often discussed separately. The level of phytochemicals in harvested leafy vegetables, including ascorbic acid (L-ascorbic acid, AsA, known as vitamin C), phenolic compounds, and carbohydrates, are associated with health-added values and may influence the product price. More importantly, their deterioration pattern following harvest is associated with sensory quality and determines shelf life.

After harvest, leafy vegetables can suffer from mechanical damage and fluctuating storage conditions, which accelerate senescence by promoting reactive oxygen species (ROS) (Jibran et al., 2013) and enzymatic oxidization (Degl'Innocenti et al., 2005; McEvily et al., 1992). The aforementioned phytochemicals can participate in ROS scavenging due to their antioxidant activities. Maintaining a sufficient amount of phytochemicals in harvested leafy vegetables is important to maintaining redox homeostasis, delaying leaf senescence, extending shelf life and thereby reducing waste in the food supply chain. For instance, the content of carbohydrates and ascorbic acid in harvested products is positively correlated with post-harvest sensory quality and shelf life, as observed in lettuce (Min et al., 2021), basil (Larsen et al., 2022), broccoli (Büchert et al., 2011), and microgreens (Mir et al., 2017).

Many studies have indicated the important role of light in plant growth and development (Massa et al., 2008), light-signaling pathways (Devlin et al., 2007), photo-sensory networks (Folta and Childers, 2008), crop production, and post-harvest microbiology management (D'Souza et al., 2015; Olle and Viršile, 2013). The biosynthesis and accumulation of health-related phytochemicals are also light-dependent processes that are involved in photosynthesis and photo-protection (Bian et al., 2015; Niyogi, 1999; Ntagkas et al., 2018; Ouzounis et al., 2015). The majority of current studies addressing light effects on crop quality examined the content of specific or groups of phytochemicals at harvest, while there is limited information regarding the impact of pre-harvest light on subsequent post-harvest quality. The bridge connecting pre-harvest lighting with post-harvest quality, sensory quality and shelf life, as well as linking phytochemical status at harvest with its decay during post-harvest, is required. Pre-harvest light strategies employed in IVF have promising potential to address the mentioned research questions and improve the integrated quality of vegetables and may provide valuable insights for breeding high-quality vegetables tailored for indoor vertical farm production.

In this chapter, we summarize recent research on the effects of pre-harvest lighting on the quality of leafy vegetables and discuss the following aspects:

1 A comprehensive definition of quality for leafy vegetables;
2 The effect of PPFD on quality-related phytochemicals, including ascorbic acid, phenolic compounds, and carbohydrates;
3 The connection between light-regulated phytochemicals and post-harvest quality and shelf life; and
4 The potential of end-of-production (EOP) lighting with high PPFD applied in indoor vertical farms.

Finally, we aim to put forward viewpoints regarding if and how pre-harvest lighting with high PPFD can improve both the nutritional quality and the shelf life of leafy vegetables.

## 2 Quality and shelf life

Fresh vegetables are an important part of a healthy diet. The World Health Organization (WHO and F. A. O., 2003) recommends at least 400 g daily intake of fresh fruit and vegetables per adult. Typically, leafy vegetables are sold fresh in whole crop heads, branches, leaves, and fresh-cut products that can be eaten raw or cooked. These products are expected to be rich in vitamins (e.g. vitamin C and vitamin E), soluble sugars, and attractive pigments (most of which are also antioxidants, e.g. anthocyanin and lycopene) but low in nitrate (Gangolli et al., 1994; Yahia et al., 2019). These vitamins and pigments are generally recognized as health-promoting compounds due to their antioxidant properties (De Rosa et al., 2017; Stavric, 1994). Although their health-promoting effects require further clinical and scientific evidence (Lee et al., 2003; Mellidou et al., 2019), their content is commonly used to define nutraceutical quality and evaluate the health-added value when product is traded. Additionally, when making a purchase, the consumer frequently evaluates the quality of leafy products by looking at sensory properties. Although numerous categories of leafy vegetables exhibit different quality characteristics, high-quality leafy vegetables frequently share a few common sensorial traits (Fig. 1). These include:

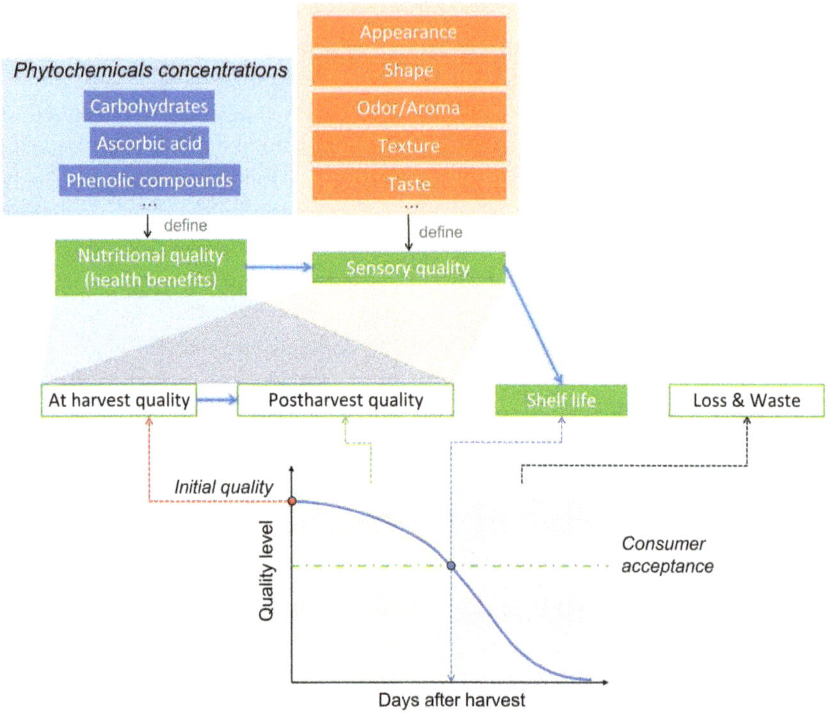


**Figure 1** A comprehensive definition of the quality of leafy vegetables, which includes nutritional quality, sensory quality, and shelf life (green rectangles with white letters). The parameters of nutritional quality and sensory quality are indicated by blue and orange backgrounds, respectively. The level of antioxidants is generally used to define nutraceutical quality, which is an aspect of nutritional quality that incorporates the carbohydrate content. A hypothetical curve shows the general quality decay following the harvest of leafy vegetables (solid blue curve). Both nutritional quality and sensory quality can be measured at harvest to indicate the initial quality level (indicated by the red dashed arrow). The number of days during which the quality drops to the consumer acceptance threshold indicates the shelf life (indicated by a blue dashed arrow). Once the post-harvest quality (green background in the hypothetical graph) drops below consumer acceptance (green dash-dotted line), products will be lost and wasted (grey background in the hypothetical graph). Shelf life is defined by sensory quality, which can be affected by nutritional quality. The level of post-harvest quality can be affected by the initial quality level (the relationship indicated by the blue solid arrow).

1 Good visual appeal, showing uniform, bright, and expected natural color without irregular light or dark patches and without discoloration (e.g. bright green leaf with white stems for fresh-cut iceberg lettuce, without brown spots, browning or pinking showing at the cutting surface);
2 A firm and undamaged shape with no shrunken or collapsed parts and an expected size or length (e.g. full and smooth surface with no wrinkles, abrasions and bruising at the leaf of butterhead lettuce);

3 A characteristic odor that is free of rancid, musty, fermented, and spoiled smells (e.g. typically rich fragrant aroma of sweet basil);
4 Distinctive texture without inconsistent soft or rough parts and watery or paste-like touch (e.g. succulent tissue and tender texture of lamb's lettuce); and
5 A good taste without mustiness or fermentation (e.g. sweet iceberg lettuce without a bitter or hay-like taste).

Some of these quality traits can be individually measured using instrumentation, but most are commonly evaluated by consumer and specialist panels to define sensory quality (Kader et al., 1973; Lin et al., 2013; Martínez-Sánchez et al., 2011).

Unfavorable climatic conditions during growth and storage and mechanical damage during processing reduce the product's nutritional quality at harvest and accelerate its deterioration after harvest. This may further induce visual symptoms that reduce the marketability of leafy products (Fig. 1) (Koukounaras, 2009). Most features of quality decay are imperceptible during growth and gradually become visible during storage, transport, and once on the shelf. Once they are expressed, most cannot be reversed and can only be prevented or delayed.

The shelf life of leafy vegetables may be influenced by their nutritional quality at harvest as well as by how much and how fast they deteriorate during storage and transportation (Fig. 1). Shelf life is commonly defined as the time it takes from harvest to reach the threshold of consumer acceptance (purchase willingness). Sensory quality scoring by specialist panels is often used to define consumer acceptance and monitor the shelf life and is based on a score scale specifically defined for a particular product. In order to standardize sensory evaluation, the selection of sensory quality features must reflect the typical quality concerns of the evaluated product. In addition, the scoring scale must be described in detail to reflect and capture crucial features during sensory quality decay. Both integrated sensory quality features and a single crucial sensory quality feature can be used to evaluate shelf life. The overall visual quality or overall acceptability that is associated with appearance (yellowing, senescence browning, and wound browning/pinking), texture (crispness), taste (bitterness), and odor can be used to evaluate the shelf life of lettuce (Bunning et al., 2010; Min et al., 2021); while in basil, dark spots as the key feature of chilling injury were emphasized in a sensorial evaluation (Larsen et al., 2022).

## 3 Antioxidants and quality

Ascorbic acid and phenolic compounds are the most abundant antioxidants in the plant cell (Yang et al., 2022). Dietary ascorbic acid has been proposed to be associated with a lower risk of various chronic diseases and cancers (Block, 1991; Palozza, 1998); hence, the ascorbic acid level is considered crucial for vegetables' commercial value. Phenolic compounds, including glutathione, tocopherol (known as vitamin E), phenolic acid, and flavonoids (e.g. anthocyanin), are a large group of antioxidants in leafy vegetables. The health benefits of phenolic compounds have been associated with free radical scavenging (de la Rosa et al., 2019; Panche et al., 2016), antidiabetic (Guo and Ling, 2015), reduced retinal damage (Wang et al., 2015), and anti-carcinogenic activity (Forbes-Hernandez et al., 2015; Nicolle et al., 2004). In addition, flavonoids, such as flavanols and anthocyanins, play a role in UV radiation absorption and as important pigments (red, blue, and purple) contributing to the colorful characteristics of vegetables (Alasalvar et al., 2001; Ignat et al., 2011; Liu et al., 2018; Naczk and Shahidi, 2006; Ryan et al., 1998). Furthermore, flavonoids and volatile phenols (e.g. eugenol and carvacrol) provide a rich aroma profile to herbs such as basil, parsley, coriander, and oregano (Kwee and Niemeyer, 2011; Santos et al., 2014; Vieira et al., 2016). Specific phenolics are also relevant to crop taste by providing the bitter or pungent taste in many herbs and spices (Tomás-Barberán and Espín, 2001). However, in lettuce, sesquiterpene lactones are mainly responsible for bitterness (Mai and Glomb, 2016) rather than phenolic compounds (Bunning et al., 2010).

Besides proposed human health-related benefits, antioxidants are also beneficial for plant health. When plants are subject to environmental stress, the metabolism of antioxidants is promoted to scavenge the ROS produced as a result of stress. Ascorbic acid and phenolic compounds are non-enzymatic antioxidants that work by donating a hydrogen atom to ROS, such as superoxide ($O_2^{\bullet-}$), hydroxyl ($OH^{\bullet}$), alkoxyl ($RO^{\bullet}$), or peroxyl ($ROO^{\bullet}$). They convert these radicals into stable molecules, which is often called ROS scavenging (Heim et al., 2002; Njus et al., 2020).

Ascorbic acid exhibits antioxidant ability through a consistent process of oxidation, transforming into monodehydroascorbate and dehydroascorbate (DHA), which are then recycled back to their original form (Ntagkas et al., 2018). Ascorbic acid is also a hydrogen peroxide ($H_2O_2$) scavenger that is employed in the ascorbate-glutathione cycle to reduce $H_2O_2$ to $H_2O$ in most cell compartments (Buettner, 1993; Noctor and Foyer, 1998; Yamasaki and Grace, 1998; Zhang, 2013). In addition, ascorbic acid indirectly preserves cell membranes by regenerating vitamin E and reduces the radical tocopheroxyl (Beyer, 1994).

The antioxidant properties of phenolic compounds largely depend on the configuration and number of hydroxyl groups (-OH), which can donate a hydrogen atom and electron (Balasundram et al., 2006; Burda and Oleszek, 2001; Hoch et al., 2001; Onslow, 2014). In an *in-vitro* study, Vinson et al. (1995)

Published by Burleigh Dodds Science Publishing Limited, 2023.

found that phenolic compounds (especially flavonoids) contributed more to antioxidant capacity than ascorbic acid or other antioxidants. A positive linear correlation has been found between antioxidant activity (trolox equivalent antioxidant capacity, ferric-reducing antioxidant power, and β-carotene bleaching %) and phenolic content in various fruits, leafy vegetables, and mint (Deighton et al., 2000; Kaur and Kapoor, 2002).

During growth, development, and storage, biotic and abiotic stress can promote the production of ROS, with the imbalanced redox state between ROS and scavenging capacity resulting in oxidative stress. Abiotic stress such as high PPFD, chilling temperature, salinity, and drought can lead to increased oxidative stress in leaves (Alscher et al., 1997), which is strongly linked to leaf senescence and induces DNA and cellular damage as well as cell death (Petrov et al., 2015; Wang et al., 2016). Navabpour et al. (2003) found that increased ROS stimulated the expression of genes that promote leaf senescence. As a result of being separated from the crop or root, leaf senescence is accelerated especially under adverse storage conditions and by mechanical damage (Ariffin et al., 2017; Koukounaras, 2009). Typical senescence symptoms such as yellowing (breakdown of chlorophyll) and wilting (membrane breakdown and ion leakage) severely impair the sensory quality of leafy vegetables.

Ascorbic acid is known to delay leaf senescence (Garg and Kapoor, 1972), increase the stability of cell membranes (Liang et al., 2018), and affect the phytohormones associated with senescence (Barth et al., 2006) in leaves. High levels of soluble sugars and phenolic compounds have also been reported to delay senescence in broccoli (Büchert et al., 2011; Hasperué et al., 2011). Other physiological processes, such as enzymatic browning, frequently occur at the cut surface of lettuce, seriously limiting its marketability. The brown color comes from *o*-quinones, which is the result of oxidized phenol catalyzed by polyphenol oxidase (López-Gálvez et al., 1996). Enzymatic browning hastens quality degradation of lettuce but can be inhibited by ascorbic acid (Degl'Innocenti et al., 2005; McEvily et al., 1992; Zhan et al., 2012).

## 4 Effects of photosynthetic photon flux density on ascorbic acid

Previous research has investigated the effects of PPFD on ascorbic acid biosynthesis and accumulation at the whole plant level, by examining light effects with a range of 5 μmol $m^{-2}s^{-1}$ to 1800 μmol $m^{-2}s^{-1}$. These studies have been conducted with various light sources (natural light, high-pressure sodium lamps, and LEDs) and with different crops. Increasing PPFD raises the ascorbic acid level in leaves; this has been observed in lettuce (Zhou et al., 2012, 2009), *Arabidopsis* (Golan et al., 2006), and tomato (Massot et al., 2012).

The mechanism of light regulation of the biosynthesis and recycling of ascorbic acid has been reviewed by Ntagkas et al. (2018), suggesting that light regulation through the biosynthesis pathway is closely related to both photosynthesis and respiration. According to Yabuta et al. (2007), *Arabidopsis'* ascorbic acid content decreases by 93% after 3 days of transitioning from light (100 μmol $m^{-2}s^{-1}$) to darkness. Samuolienė et al. (2012) studied the variability in ascorbic acid content over the seasons. From November to March, the ascorbic acid in lettuce increased by 53% with an increased average PPFD of natural light from 150 μmol $m^{-2}s^{-1}$ to 230 μmol $m^{-2}s^{-1}$. In some studies, however, ascorbic acid levels hit a maximum at a given PPFD level, after which further increasing PPFD had no effect or reduced ascorbic acid. When PPFD increased from 60 μmol $m^{-2}s^{-1}$ to 220 μmol $m^{-2}s^{-1}$, the content of ascorbic acid in lettuce changed in a parabolic pattern, peaking at 140 μmol $m^{-2}s^{-1}$ where the level is about two times higher than it is under 60 μmol $m^{-2}s^{-1}$ (Fu et al., 2017). Larsen et al. (2022) found that the total ascorbic acid (TAsA, which is DHA + AsA) content in basil doubled with increasing PPFD from 50 μmol $m^{-2}s^{-1}$ to 300 μmol $m^{-2}s^{-1}$ (5 days' EOP treatment) but showed no further increase when PPFD was further raised to 600 μmol $m^{-2}s^{-1}$.

In addition, a longer photoperiod and continuous illumination along with higher daily light integral (DLI) resulted in a higher level of ascorbic acid (Shen et al., 2014). In this study, extending the light period from 12, 16, to 20 h with 600 μmol $m^{-2}s^{-1}$ PPFD gradually increased ascorbic acid content by 30% in lettuce. The ascorbic acid content reached the maximum level at 20 h light period and 24 h continuous light. It is worth noting that prolonged high PPFD exposure declined the ratio of AsA to TAsA, although the TAsA significantly increased (same trends for glutathione and its oxidized form) (Zhou et al., 2009). This result suggests that the unbalanced redox homeostasis induced by excessive light might suppress ascorbic acid regeneration. The effect of PPFD on ascorbic acid regeneration is also influenced by other factors. Pérez-López et al. (2015) reported that in green lettuce, AsA increased by about 30% when the PPFD raised from 400 μmol $m^{-2}s^{-1}$ to 700 μmol $m^{-2}s^{-1}$ during 4 days' EOP light treatment. However, the TAsA level did not show an increase with higher PPFD unless the $CO_2$ concentration increased from 400 ppm to 700 ppm.

The upregulation of ascorbic acid by high PPFD is a fast process. When increasing PPFD from 210 μmol $m^{-2}s^{-1}$ to 470 μmol $m^{-2}s^{-1}$ in the EOP phase (6-day and 7-day treatment), the TAsA content in lettuce increased more than twice, and the TAsA increases began after the first-day treatment (Min et al., 2021). Similarly, the content of TAsA, AsA, and their biosynthetic genes were promoted after just 2-h illumination with increased PPFD from 200 μmol $m^{-2}s^{-1}$ to 500 μmol $m^{-2}s^{-1}$ (Zhou et al., 2021). Zhou et al. (2012) found that increasing PPFD in continuous illumination during EOP from 50 μmol $m^{-2}s^{-1}$ to 200 μmol $m^{-2}s^{-1}$ increased the ascorbic acid content in lettuce within 48 h. Other studies also reported that the light regulation of gene expression and enzyme activity in the ascorbic acid biosynthesis pathway is a quick response (Larsen et al., 2022; Pérez-López et al., 2015; Yabuta et al., 2007). As a result, applying a short-term light treatment (by the EOP) to evoke a burst of ascorbic acid has become a research trend for IVF scenarios.

## 5 Effects of photosynthetic photon flux density on phenolic compounds

High PPFD and continuous lighting increases the concentration of phenolics (de la Rosa et al., 2019) by inducing the biosynthesis pathway and ROS formation (Zoratti et al., 2014). The effect of PPFD often depends on light source spectrum, treatment duration, crops, and type of phenolic compounds (Liu et al., 2018; Maier and Hoecker, 2015; Samuolienė et al., 2012; Sousa Paiva et al., 2003).

Flavanol and anthocyanin levels increase under high PPFD, mainly through upregulating biosynthetic genes, accompanied by an increased maximum photosynthetic rate (Albert et al., 2009; Liu et al., 2018). Research has shown that the content of specific phenolic compounds leveled off or even decreased under excess PPFD. In red perilla, an increase in PPFD from 50 μmol $m^{-2}s^{-1}$ to 200 μmol $m^{-2}s^{-1}$ increased anthocyanin concentration, whereas 400 μmol $m^{-2}s^{-1}$ PPFD reduced its concentration (Yoshida et al., 2022). Similar results were found in basil, showing that the anthocyanins, flavonoids, and phenolic content increased with increasing PPFD but decreased when PPFD was over 290 μmol $m^{-2}s^{-1}$ (Dou et al., 2018). Continuous light (with no dark periods), alone or in combination with high PPFD, can also improve levels of phenolics in a short time and depending on the spectrum. According to Shiga et al. (2011), the total phenolic acid in basil increased by more than 50% following 14 days of white continuous lighting at 100 μmol $m^{-2}s^{-1}$, while blue light was found to be less effective than white light. In the same study, the level of rosmarinic acid, a phenolic acid, increased by approximately three times and reached saturation as the PPFD increased from 100 μmol $m^{-2}s^{-1}$ to 300 μmol $m^{-2}s^{-1}$ (resulting in DLI increase from 4.3 to 8.6 mol $m^{-2}$ $day^{-1}$).

The biosynthesis of anthocyanin responds quickly to the high PPFD. It has been reported that the anthocyanin pigmentation in petunia (leaves and stems) became visible within 12 h of exposure to high PPFD (750 μmol $m^{-2}s^{-1}$) (Albert et al., 2009). According to Bian et al. (2016), the total phenolic content quickly improved after 24 h of continuous lighting at 200 μmol $m^{-2}s^{-1}$, and content continued to rise over 60 h and thereafter decreased. In this research, the 1,1-diphenyl-2-picrylhydrazyl (DPPH) free radical scavenging activity was found to be in line with total phenolic content. This is consistent with the results of Pérez-López et al. (2015) who found that DPPH in red leaf lettuce increased in the same way as total phenolics and anthocyanin content when increasing

PPFD from 400 μmol $m^{-2}s^{-1}$ to 700 μmol $m^{-2}s^{-1}$ (4 days' EOP light treatment). As phenolic content increases, there is a corresponding rise in the levels of soluble sugars and starch. This suggests that the production of phenolic compounds in response to high PPFD or continuous lighting is linked to the dissipation of excess light energy, leading to an increase in the availability of sugar and an elevated risk of oxidative stress. In addition, according to Pérez-López et al. (2018), the content of total phenols, total flavonoids, and total free phenolic acid only increased with increased PPFD (from 400 μmol $m^{-2}s^{-1}$ to 700 μmol $m^{-2}s^{-1}$, 4 days' lighting by the EOP) under ambient $CO_2$ conditions (400 ppm). Under elevated $CO_2$ conditions (700 ppm), however, the increased PPFD reduced the level of the aforementioned phenolic compounds.

## 6 The sugar-antioxidant network

Carbohydrates, including soluble sugars (e.g. sucrose, glucose, and fructose), insoluble sugars (e.g. cellulose and starch), and sugar-related compounds (e.g. hexokinase and glucose-6-phosphate) play important roles in plant growth and development and serve as energy sources and storage and precursors for many phytochemicals metabolites (Yahia et al., 2019). D-glucose serves as a precursor for AsA biosynthesis through the Smirnoff-Wheeler pathway, where sugar-related compounds play important roles in the intermediate steps, including hexokinase and glucose-6-phosphate (Linster and Clarke, 2008; Wheeler et al., 1998). Additionally, glucose also serves as the initial precursor for carotenoid biosynthesis (Noctor et al., 1998; Pallet and Young, 1993) and act as a carbon provider for glutathione biosynthesis (Rolland et al., 2006). The nucleoside diphosphate sugar generated by hexokinase was reported as a substrate in glycosylated phenolic synthesis (Bolouri-Moghaddam et al., 2010; Hu et al., 2022).

In addition to being energy and carbon sources, carbohydrates have a dual function in ROS production and ROS scavenging (Bolouri-Moghaddam et al., 2010; Couée et al., 2006). Sugar starvation induces oxidative stress and upregulates general stress response genes through reduced electron transfer in mitochondria respiration, activated lipid mobilization, fatty acid transfer and peroxisomal β-oxidation (Contento et al., 2004; Couée et al., 2006). In this process, high sugar concentrations act as a direct ROS scavenger by repressing fatty acid mobilization and turning sugar into sugar radicals (Van Den Ende and Valluru, 2009). The sugar radicals can then be reverted back to sugar by anthocyanin (Solfanelli et al., 2006). The signaling effects of sugars are well documented as well, including glucose signaling involved in the feedback regulation of respiration and photosynthesis (Paul and Pellny, 2003), sucrose-signaling effects on anthocyanin biosynthesis (Solfanelli et al., 2006), and the sugar-signaling effects on interaction with hormones, such as abscisic acid, auxin, cytokine, and ethylene (Rolland et al., 2006). In addition, carbohydrates also contribute to the sweet flavor of leafy vegetables (Chadwick et al., 2016; Moskowitz, 1970) and delay crop texture deterioration (Yahia et al., 2019). Lin et al. (2013) found that harvested lettuce, which has a higher soluble sugar content, also achieved a higher sensory score in crispness and sweet taste. Additionally, Woltering and Seifu (2015) reported that fresh-cut lettuce with higher soluble sugar content (due to post-harvest lighting) had up to a 55% longer shelf life compared to the control.

## 7 Effects of photosynthetic photon flux density on the sugar-antioxidant network

Higher PPFD results in a higher photosynthesis rate until saturation levels are reached, although exposure to excess light can induce ROS production. Under high PPFD conditions, the production of carbohydrates is required to serve as both signal and substrate in antioxidant production, which are used to scavenge the produced ROS. In a greenhouse experiment, lettuce exposed to supplemental HPS (average PPFD 1100 μmol $m^{-2}s^{-1}$) showed significant increases in carbohydrate content (starch, hexose, and sucrose), which is accompanied by significantly higher ROS ($O2^{\bullet-}$ and $H_2O_2$) and antioxidant production (e.g. ascorbate acid, glutathione, phenolics, and α-tocopherol) (Zhou et al., 2009). Similar results were also found in indoor vertical farm conditions by Pérez-López et al. (2015), who found that the transfer of lettuce from 400 μmol $m^{-2}$ $s^{-1}$ to 700 μmol $m^{-2}$ $s^{-1}$ lighting for 4 days at EOP significantly increased both soluble sugars and starch levels. This increase was associated with increases in ascorbic acid, phenolics, glutathione, and carotenoids, although the antioxidant response varies by cultivar. Zhou et al. (2012) also reported that the soluble sugar content in lettuce significantly increased under 48 h of continuous light (at EOP). The relative increase of sugar steadily rises with PPFD from 50 μmol $m^{-2}s^{-1}$ to 200 μmol $m^{-2}s^{-1}$ and is accompanied by increased levels of ascorbic acid and reduced nitrate content.

## 8 Effects of photosynthetic photon flux density on sugar and nitrate

High nitrate levels are generally considered to be a quality-limiting factor for leafy vegetables, as they could easily convert to nitrite in the human body and induce health problems such as methemoglobinemia (Chan, 2011; Mensinga et al., 2003). Many studies have reported positive light effects in terms of reducing nitrate in leafy vegetables by inducing nitrate reductase. The synthesis of nitrate reductase can be activated by light, and the light-enhanced sugar concentration may stimulate nitrate reductase gene expression (Lillo, 1994; Lillo and Appenroth, 2001). Carbohydrates stimulated by high PPFD may also accelerate the transition from nitrate to amino acid (Price et al., 2004). With increased PPFD levels of up to 220 μmol $m^{-2}s^{-1}$, the nitrate content significantly decreased while the carbohydrate increased, especially when the nitrogen concentration in the root zone was high (Fu et al., 2017).

Short-term continuous lighting reduces nitrate significantly. An extended light period 12 h to 20 h with 600 μmol $m^{-2}s^{-1}$ PPFD significantly reduces nitrate and increases soluble sugars in lettuce, and this effect is even more pronounced following 24 h of continuous lighting (Shen et al., 2014). According to Bian et al. (2018), nitrate levels are reduced with an increase in soluble sugars under continuous lighting (24-h and 48-h EOP treatment, 200 μmol $m^{-2}s^{-1}$) compared to 12-h photoperiod light. More importantly, they found that nitrate reductase and related gene expressions are promoted quickly in the first 24 h but reduced in the next 24 h. In the same study, supplemental green light at 530 nm was found to be most effective in reducing nitrate under continuous light. Similar results are reported by Lin et al. (2013) with a long-term illumination at 210 μmol $m^{-2}s^{-1}$. In addition to the green spectrum, 3 days of EOP lighting with additional red spectrum at 638 nm were found to significantly reduce nitrate levels in lettuce, while the results are cultivar dependent. In the cultivar Muliblond2, the nitrate level increased six-fold after adding the red spectrum to EOP lighting at 300 μmol $m^{-2}s^{-1}$ (Samuolienė et al., 2011).

# 9 End-of-production light

Most research focused on the relationship between light and crop quality has been conducted with lighting applied throughout the whole growth period from transplanting. However, prolonged exposure to high PPFD can result in excessive production of ROS, posing a risk of exceeding the plant's antioxidative capacity. Unfavorable quality characteristics may appear under such conditions, for example, thicker leaves (Feng et al., 2019; Muller-Moule et al., 2004), tipburn (Sago, 2016) and yield loss (Gururani et al., 2015). EOP light treatment (Fig. 2) is applied for a few hours to a few days prior to harvest, offering a way to improve the quality of leafy vegetables without the negative effects of long-term illumination. Applying EOP lighting with high PPFD is based

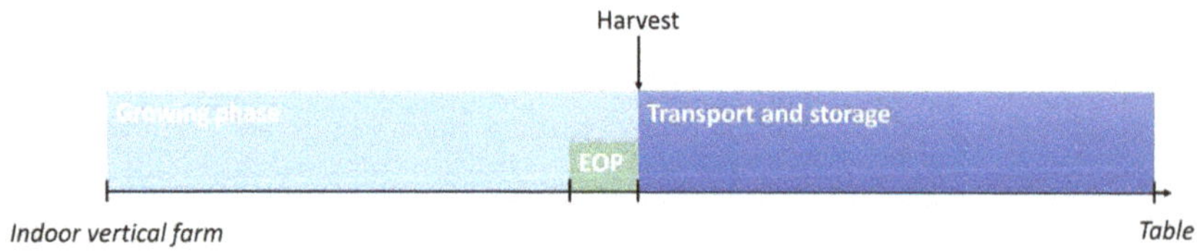


**Figure 2** The simplified schematic graph indicates the position of the end-Of-production (EOP) phase in the 'farm to table' process.

on the evidence that many phytochemicals can respond quickly to high PPFD through improved photosynthesis, stimulated biosynthesis, and increased need for ROS scavenging. In this way, exposing leafy vegetables shortly to high PPFD at EOP can quickly boost phytochemical levels just before harvest; will not interfere with standard growth, yield, and morphology; is cost and energy efficient compared to long-term lighting with high PPFD; and may readily fit in the existing lighting system of the indoor vertical farm (or as a modular room for EOP lighting).

During the post-harvest phase, sugars and antioxidants begin to degrade and are continually consumed in the respiration and senescence process. These processes are further enhanced by exogenous stresses such as temperature abuse, low humidity, prolonged darkness, and mechanical damage

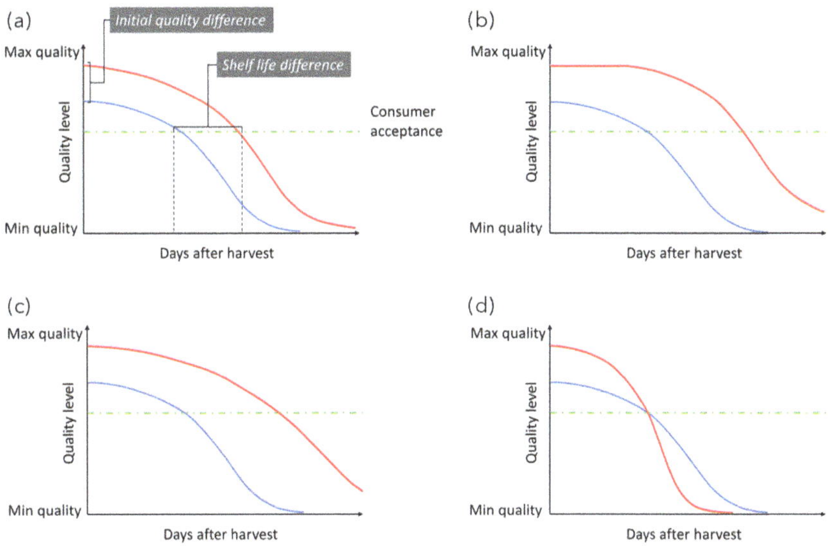


**Figure 3** Hypothetical general quality reduction patterns with different initial quality levels. The high and low initial quality scenarios are indicated by red and blue solid curves, respectively. The green dash-dotted line indicates the consumer acceptance threshold. The projection of consumer acceptance and quality curve on the *x*-axis represents the shelf life. (a) Different initial quality levels with a parallel deterioration pattern at a similar rate; the high initial level may lead to longer shelf life. (b) The high initial quality level stays steady for a period and drops in parallel in a similar rate with the low initial quality level; the high initial level leads to longer shelf life [e.g. the overall visual quality of fresh-cut red lettuce exposed to 35 days of post-harvest lighting, overall visual quality stayed steady in the first 12 days' storage and then quickly declined (Witkowska, 2013)]. (c) High initial quality level decline at a slower rate compared to the low initial quality level and leads to longer shelf life [e.g. high initial level of carbohydrates and TAsA in lettuce treated with high PPFD has slower decay during storage (Min et al., 2021)]. (d) High initial level has a faster decline rate compared to a low initial quality level and leads to a shorter shelf life [e.g. the relative fresh weight of carnation cut flowers drops dramatically at day 6 in vase life due to limited water uptake (Aalifar et al., 2020)].

Published by Burleigh Dodds Science Publishing Limited, 2023.

(Koukounaras, 2009; Santos et al., 2014). The implementation of EOP lighting with high PPFD may result in differences in the phytochemical content at the time of harvest. The high initial content of both antioxidants and carbohydrates can be beneficial to preserve the nutritional and sensory quality level during storage. Sufficient antioxidant capacity contributes to ROS scavenging under active respiration and senescence in storage, as well as preventing wound-induced enzymatic browning. A hypothetical general quality reduction pattern is shown in Fig. 3, indicating the potential impact of different initial quality levels, which may be influenced by EOP lighting, on the shelf life of leafy products.

## 10 Conclusion and future trends

An increase in PPFD promotes biosynthesis and accumulation of phytochemicals in leafy vegetables, including ascorbic acid and phenolic compounds (Fig. 4). The antioxidant properties of these phytochemicals have been suggested to improve nutraceutical quality (Fig. 4). More importantly, increased antioxidant capacity counteracts the unbalanced redox state caused by senescence and environmental stress during storage, thereby avoiding unfavorable visual traits and extending shelf life (Fig. 4). In addition, high PPFD during growth increases the sugar content in leafy vegetables through improved photosynthesis, which provides sufficient precursors for the biosynthesis of phytonutrients on the one hand, and its own antioxidant abilities participate in redox homeostasis on the other (Fig. 4). Additionally, carbohydrates contribute to the sweet taste of leafy vegetables, while phenolics may contribute to a bitter taste and provide specific aromatic qualities to herbs. High PPFD applied in the short term at EOP can quickly increase both antioxidant and carbohydrate concentration at harvest and, as a result, may provide sufficient antioxidants for post-harvest storage and benefits for nutritional quality, sensory quality, and shelf life of leafy vegetables.

High PPFD treatment can also be combined with other light parameters and environmental factors to influence the quality of leafy vegetables. The effect of light spectra on the quality of lettuce is discussed by Bian et al. (2015) and Thoma et al. (2020); the interaction of light spectra and PPFD on plant flowering and morphology is reported by Runkle et al. (2022); the interaction of PPFD and irrigation nitrogen concentration on lettuce quality is studied by Fu et al. (2017); and the interaction of PPFD with ambient $CO_2$ concentration is studied by Pérez-López et al. (2015). A different way of applying light, e.g. from underneath upwards, was also reported to delay senescence of lettuce (Zhang et al., 2015). Post-harvest lighting can be employed to continually increase the content or suppress the reduction of antioxidants and carbohydrates and control microbial growth after harvest, as the leaf serves as both the primary photosynthetic organ and the end product of leafy vegetables (Gogo et al.,

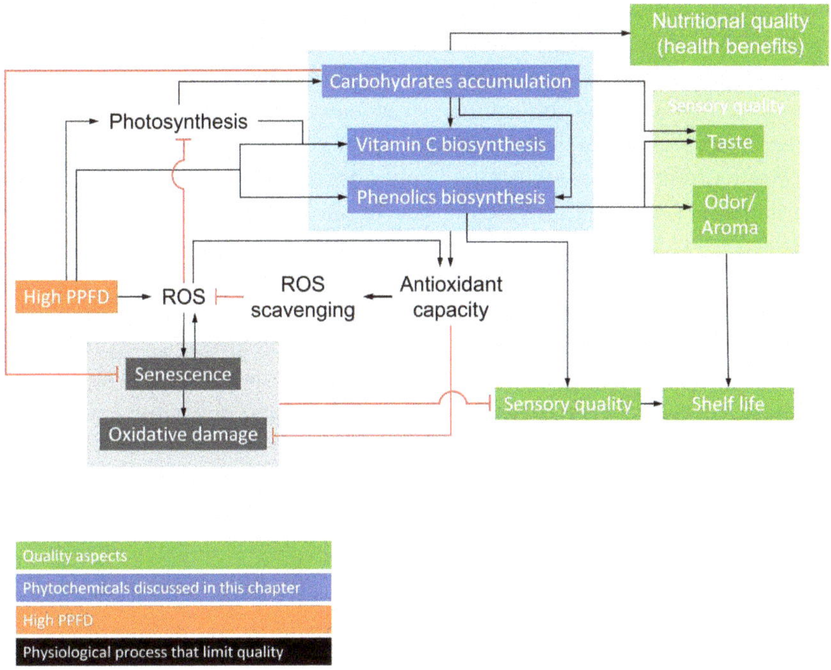


**Figure 4** The potential effect of high PPFD on leafy vegetable quality. The pointed arrow • in solid black indicates a positive influence and the blunt arrow T in solid red indicates a negative influence.

2017; Kasim and Kasim, 2017; Liao et al., 2016; Woltering and Witkowska, 2016; Zhan et al., 2012).

With the global population expected to reach 9.7 billion by 2050 (United Nations Department of Economic and Social Affairs, 2022), indoor vertical farms are believed to be a key part of future cultivation systems that can 'feed the world', due to their high efficiency in the use of land, water, $CO_2$, fertilizer, and labor (Despommier, 2013; Kozai, 2022; Bunge et al., 2022). More than one-third of fruits and vegetables are lost or wasted during transport and storage, particularly in developing countries (Food and Agriculture Organization of the United Nations (FAO), 2011). Thus, minimizing loss after harvest is another important way of ensuring global food supply. The initial investment and operating costs of indoor vertical farms are high, largely affected by the increasing price of fossil fuels. Minimizing the loss of vegetables that are already produced and extending shelf life by using light strategies during growth (with existing LED lighting systems) are significant advantages of indoor vertical farms. Another noteworthy advantage of indoor vertical farms is their potential for localized production, with some facilities strategically located in downtown areas. This shortened food supply chain demonstrated its advantages during the coronavirus disease 2019 pandemic by minimizing loss during storage and transport. Moreover, the implementation of modular indoor vertical farms that are designed to be close to warehouse and distribution centers may provide a good opportunity for the application of post-harvest lighting (e.g. in the sorting line and in storage space), extending the crop shelf life and reducing the wastes.

Most current studies focusing on light effects only investigate the content of a single or a group of phytochemicals at harvest and their corresponding physiology and molecular mechanisms. Further research is needed on how the light effect and light-induced phytochemicals affect the overall quality of vegetables, including nutritional quality, sensory quality, and shelf life. The comprehensive concept of integrated quality in this chapter (Fig. 1) provides a better reflection of consumer acceptance and preference towards vegetables. By establishing distinctive quality markers, shelf life and consumer preference can be better predicted with flavor models and shelf life estimate models (Morris et al., 2003; Tijskens et al., 1999). In addition, with the development of sensors and image analysis algorithms, crop phenomics and quality attributes can be measured non-destructively, for instance by employing hyperspectral imaging (Harfouche et al., 2022; Lara et al., 2013), electronic nose fingerprinting, and visible-near-infrared spectroscopy (Giovenzana et al., 2014). In this way, the quality of a large number of vegetable products can be either rapidly measured or measured non-destructively, and internal disorders can be recognized at an early stage.

## 11 Where to look for further information

International Society of Horticulture Science (ISHS)-organized conferences, held every 2–4 years, serve as global platforms for researchers and experts to present their scientific findings, research papers, and posters in controlled environmental agriculture and fresh crop quality. Some notable conferences in this field include:

- VertiFarm2024: The 3rd International Workshop on Vertical Farming (https://site.unibo.it/vertifarm2024/en).
- GreenSys2025 (International Symposium on Advanced Technologies and Management for Sustainable Greenhouse Systems) (https://www.ishs.org/symposium/724).
- International Conference Postharvest Unlimited (https://www.ishs.org/symposium/776).
- International Symposium on Urban Horticulture: From Vertical Farming to Planting Design (https://www-ishs-org.ezproxy.library.wur.nl/symposium/826).

Published by Burleigh Dodds Science Publishing Limited, 2023.

- IHC2026: International Horticultural Congress (https://www.ihc2026.org/).
- OHCEAC (Ohio Controlled Environment Agriculture Center) annual conference: Advancement of Sustainable Controlled Environment Crop Production Sciences and Technologies (https://ohceac.osu.edu/ohceac_annual_conference).

Several online courses and workshops offer free and accessible resources for gaining insights into novel research outcomes in vertical farming. These courses may also feature industry professionals sharing their experiences and discussing challenges related to this field.

- ISHS talks on vertical farming (https://vimeo.com/user135517524).
- OptimIA (Optimizing Indoor Agriculture) Indoor Ag Science Café (https://scri-optimia.org/index.php). The video recording of the previous workshop can be accessed on YouTube channel: Controlled Environment Plant Physiology and Technology (https://www.youtube.com/@controlledenvironmentplant4337/videos).

The WUR Academy organizes annual courses on vertical farms and crop postharvest technology, providing participants with in-depth knowledge through lectures and practical experiments.

- Course Vertical Farming (https://www.wur.nl/en/show/course-vertical-farming.htm).
- Course Postharvest Technology (https://www.wur.nl/en/show/course-postharvest-technology.htm).

Both Wageningen Food and Biobased Research Institute and UC Davis Postharvest Center offer comprehensive online databases for crop postharvest knowledge. These databases provide the latest research, practical solutions, and expert connections in the field. They serve as valuable resources for staying informed and finding solutions to postharvest challenges. Furthermore, these databases facilitate connections with experts, offering opportunities for collaboration and seeking expert guidance.

- The postharvest knowledge platform of WUR Food and biobased research center (https://www.freshknowledge.eu/en.htm), YouTube channel: A Postharvest platform of Wageningen University & Research (https://youtu.be/yBKjq3wYBe8).
- The postharvest center of UCDavis (https://postharvest.ucdavis.edu/).

## 12 References

Aalifar, M., Aliniaeifard, S., Arab, M., Zare Mehrjerdi, M., Dianati Daylami, S., Serek, M., Woltering, E. and Li, T. (2020). Blue light improves vase life of carnation cut flowers through its effect on the antioxidant defense system, *Frontiers in Plant Science* 11(May), 511, https://doi.org/10.3389/fpls.2020.00511.

Alasalvar, C., Grigor, J. M., Zhang, D., Quantick, P. C. and Shahidi, F. (2001). Comparison of volatiles, phenolics, sugars, antioxidant vitamins, and sensory quality of different colored carrot varieties, *Journal of Agricultural and Food Chemistry* 49(3), 1410-1416.

Albert, N. W., Lewis, D. H., Zhang, H., Irving, L. J., Jameson, P. E. and Davies, K. M. (2009). Light-induced vegetative anthocyanin pigmentation in Petunia, *Journal of Experimental Botany* 60(7), 2191-2202.

Alscher, R. G., Donahue, J. L. and Cramer, C. L. (1997). Reactive oxygen species and antioxidants: relationships in green cells, *Physiologia Plantarum* 100(2), 224-233.

Ariffin, S. H., Gkatzionis, K. and Bakalis, S. (2017). Leaf injury and its effect towards shelf-life and quality of ready-to-eat (RTE) spinach, Elsevier B.V., *Energy Procedia* 123, 105-112.

Balasundram, N., Sundram, K. and Samman, S. (2006). Phenolic compounds in plants and agri-industrial by-products: antioxidant activity, occurrence, and potential uses, *Food Chemistry* 99(1), 191-203.

Barth, C., De Tullio, M. and Conklin, P. L. (2006). The role of ascorbic acid in the control of flowering time and the onset of senescence, *Journal of Experimental Botany* 57(8), 1657-1665.

Beyer, R. E. (1994). The role of ascorbate in antioxidant protection of biomembranes: interaction with vitamin E and coenzyme Q, *Journal of Bioenergetics and Biomembranes* 26(4), 349-358.

Bian, Z., Cheng, R., Wang, Y., Yang, Q. and Lu, C. (2018). Effect of green light on nitrate reduction and edible quality of hydroponically grown lettuce (Lactuca sativa L.) under short-term continuous light from red and blue light-emitting diodes, Elsevier, *Environmental and Experimental Botany* 153(December), 63-71.

Bian, Z., Cheng, R., Yang, Q., Wang, J. and Lu, C. (2016). Continuous light from red, blue, and green light-emitting diodes reduces nitrate content and enhances phytochemical concentrations and antioxidant capacity in lettuce, *Journal of the American Society for Horticultural Science* 141(2), 186-195.

Bian, Z. H., Yang, Q. C. and Liu, W. K. (2015). Effects of light quality on the accumulation of phytochemicals in vegetables produced in controlled environments: a review, *Journal of the Science of Food and Agriculture* 95(5), 869-877.

Block, G. (1991). Vitamin C and cancer prevention: the epidemiologic evidence, *The American Journal of Clinical Nutrition* 53(1) (Suppl.), 270S-282S.

Bolouri-Moghaddam, M. R., Le Roy, K., Xiang, L., Rolland, F. and Van Den Ende, W. (2010). Sugar signalling and antioxidant network connections in plant cells, *FEBS Journal* 277(9), 2022-2037.

Büchert, A. M., Gómez Lobato, M. E., Villarreal, N. M., Civello, P. M. and Martínez, G. A. (2011). Effect of visible light treatments on postharvest senescence of broccoli (Brassica oleracea L.), *Journal of the Science of Food and Agriculture* 91(2), 355-361.

Buettner, G. R. (1993). The pecking order of free radicals and antioxidants: lipid peroxidation, α-tocopherol, and ascorbate, *Archives of Biochemistry and Biophysics* 300(2), 535-543.

Bunge, A. C., Wood, A., Halloran, A. and Gordon, L. J. (2022). A systematic scoping review of the sustainability of vertical farming, plant-based alternatives, food delivery services and blockchain in food systems, *Nature Food*. 3(11), 933–941. https://doi.org/10.1038/s43016-022-00622-8.

Bunning, M. L., Kendall, P. A., Stone, M. B., Stonaker, F. H. and Stushnoff, C. (2010). Effects of seasonal variation on sensory properties and total phenolic content of 5 lettuce cultivars, *Journal of Food Science* 75(3), S156–S161.

Burda, S. and Oleszek, W. (2001). Antioxidant and antiradical activities of flavonoids, *Journal of Agricultural and Food Chemistry* 49(6), 2774–2779.

Chadwick, M., Gawthrop, F., Michelmore, R. W., Wagstaff, C. and Methven, L. (2016). Perception of bitterness, sweetness and liking of different genotypes of lettuce, Elsevier Ltd, *Food Chemistry* 197(A), 66–74.

Chan, T. Y. K. (2011). Vegetable-borne nitrate and nitrite and the risk of methaemoglobinaemia, *Toxicology Letters, Netherlands* 200(1–2), 107–108.

Contento, A. L., Kim, S. J. and Bassham, D. C. (2004). Transcriptome profiling of the response of Arabidopsis suspension culture cells to Suc starvation, *Plant Physiology* 135(4), 2330–2347.

Couée, I., Sulmon, C., Gouesbet, G. and El Amrani, A. (2006). Involvement of soluble sugars in reactive oxygen species balance and responses to oxidative stress in plants, *Journal of Experimental Botany* 57(3), 449–459.

D'Souza, C., Yuk, H.-G., Khoo, G. H. and Zhou, W. (2015). Application of light-emitting diodes in food production, postharvest preservation, and microbiological food safety, *Comprehensive Reviews in Food Science and Food Safety* 14(6), 719–740.

de la Rosa, L. A., Moreno-Escamilla, J. O., Rodrigo-García, J. and Alvarez-Parrilla, E. (2019). Phenolic compounds, *Postharvest Physiology and Biochemistry of Fruits and Vegetables*, Elsevier, pp. 253–271.

De Rosa, L. A., Alvarez-parrilla, E. and Gonz, G. A. (2017). *Fruit and Vegetable Phytochemicals, Fruit and Vegetable Phytochemicals*, https://doi.org/10.1002/9781119158042.

Degl'Innocenti, E., Guidi, L., Pardossi, A. and Tognoni, F. (2005). Biochemical study of leaf browning in minimally processed leaves of lettuce (Lactuca sativa L. var. acephala), *Journal of Agricultural and Food Chemistry* 53(26), 9980-9984.

Deighton, N., Brennan, R., Finn, C. and Davies, H. V. (2000). Antioxidant properties of domesticated and wild Rubus species, *Journal of the Science of Food and Agriculture* 80(9), 1307–1313.

Despommier, D., (2013). Farming up the city: the rise of urban vertical farms, *Trends in Biotechnology* 31(7), 388–389. https://doi.org/10.1016/j.tibtech.2013.03.008.

Devlin, P. F., Christie, J. M. and Terry, M. J. (2007). Many hands make light work, *Journal of Experimental Botany* 58(12), 3071–3077.

Dou, H., Niu, G., Gu, M. and Masabni, J. G. (2018). Responses of sweet basil to different daily light integrals in photosynthesis, morphology, yield, and nutritional quality, *HortScience* 53(4), 496–503.

Feng, L., Raza, M. A., Li, Z., Chen, Y., Khalid, M. H., Bin, D., J., Liu, W., Wu, X., Song, C., Yu, L., Zhang, Z., Yuan, S., Yang, W. and Yang, F. (2019). The influence of light intensity and leaf movement on photosynthesis characteristics and carbon balance of Soybean, *Frontiers in Plant Science* 9(January), 1–16.

Folta, K. M. and Childers, K. S. (2008). Light as a growth regulator: controlling plant biology with narrow-bandwidth solid-state lighting systems, *HortScience* 43(7), 1957–1964.

Food and Agriculture Organization of the United Nations (FAO) (2011). *Global Food Losses and Food Waste: Extent, Causes and Prevention, Save Food!*

Forbes-Hernandez, T. Y., Gasparrini, M., Afrin, S., Bompadre, S., Mezzetti, B., Quiles, J. L., Giampieri, F. and Battino, M., (2015). The healthy effects of strawberry polyphenols: which strategy behind antioxidant capacity?, *Critical Reviews in Food Science and Nutrition* 56, S46–S59. https://doi.org/10.1080/10408398.2015.1051919.

Fu, Y., Li, H. Y., Yu, J., Liu, H., Cao, Z. Y., Manukovsky, N. S. and Liu, H. (2017). Interaction effects of light intensity and nitrogen concentration on growth, photosynthetic characteristics and quality of lettuce (Lactuca sativa L. Var. youmaicai), Elsevier B.V., *Scientia Horticulturae* 214, 51–57.

Gangolli, S. D., van den Brandt, P. A., Feron, V. J., Janzowsky, C., Koeman, J. H., Speijers, G. J. A., Walker, R., et al. (1994). Nitrate, nitrite and N-nitroso compounds, *International Journal of Cancer* 86(292), 1–38.

Garg, O. P. and Kapoor, V. (1972). Retardation of leaf senescence by ascorbic acid, *Journal of Experimental Botany* 23(3), 699–703.

Giovenzana, V., Beghi, R., Buratti, S., Civelli, R. and Guidetti, R. (2014). *"Monitoring of fresh-cut Valerianella locusta Laterr. shelf life by electronic nose and VIS-NIR spectroscopy"*, *Talanta*, Elsevier (vol. 120), pp. 368–375.

Gogo, E. O., Opiyo, A. M., Hassenberg, K., Ulrichs, C. and Huyskens-Keil, S. (2017). Postharvest UV-C treatment for extending shelf life and improving nutritional quality of African indigenous leafy vegetables, *Postharvest Biology and Technology* 129(April), 107–117.

Golan, T., Müller-Moulé, P. and Niyogi, K. K. (2006). Photoprotection mutants of Arabidopsis thaliana acclimate to high light by increasing photosynthesis and specific antioxidants, *Plant, Cell and Environment* 29(5), 879–887.

Guo, H. and Ling, W. (2015). The update of anthocyanins on obesity and type 2 diabetes: experimental evidence and clinical perspectives, *Reviews in Endocrine and Metabolic Disorders* 16(1), 1–13.

Gururani, M. A., Venkatesh, J. and Tran, L. S. P. (2015). Regulation of photosynthesis during abiotic stress-induced photoinhibition, Elsevier Ltd, *Molecular Plant* 8(9), 1304–1320.

Harfouche, A. L., Nakhle, F., Harfouche, A. H., Sardella, O. G., Dart, E. and Jacobson, D. (2022). A primer on artificial intelligence in plant digital phenomics : embarking on the data to insights journey, *Trends in Plant Science*, Elsevier Ltd 27(1), 1–3.

Hasperué, J. H., Chaves, A. R. and Martínez, G. A. (2011). End of day harvest delays postharvest senescence of broccoli florets, *Postharvest Biology and Technology* 59(1), 64–70.

Heim, K. E., Tagliaferro, A. R. and Bobilya, D. J. (2002). Flavonoid antioxidants: chemistry, metabolism and structure-activity relationships, *Journal of Nutritional Biochemistry* 13(10), 572–584.

Hoch, W. A., Zeldin, E. L. and McCown, B. H. (2001). Physiological significance of anthocyanins during autumnal leaf senescence, *Tree Physiology* 21(1), 1–8.

Hu, W., Sarengaowa, W., Guan, Y. and Feng, K. (2022). Biosynthesis of phenolic compounds and antioxidant activity in fresh-cut fruits and vegetables, *Frontiers in Microbiology* 13(May), 906069.

Ignat, I., Volf, I. and Popa, V. I. (2011). A critical review of methods for characterisation of polyphenolic compounds in fruits and vegetables, *Food Chemistry* 126(4), 1821–1835.

Jibran, R., A Hunter, D. and P Dijkwel, P. (2013). Hormonal regulation of leaf senescence through integration of developmental and stress signals, *Plant Molecular Biology* 82(6), 547–561.

Kader, A. A., Lipton, W. J. and Morris, L. L. (1973). Systems for scoring quality of harvested lettuce, *HortScience* 8(5), 408–409.

Kasim, M. U. and Kasim, R. (2017). While continuous white LED lighting increases chlorophyll content (SPAD), green LED light reduces the infection rate of lettuce during storage and shelf-life conditions, *Journal of Food Processing and Preservation* 41(6), 1–7.

Kaur, C. and Kapoor, H. C. (2002). Anti-oxidant activity and total phenolic content of some Asian vegetables, *International Journal of Food Science and Technology* 37(2), 153–161.

Koukounaras, A. (2009). Senescence and quality of green leafy vegetables, *Stewart Postharvest Review* 5(2), 1–5.

Kozai, T. (2013). Resource use efficiency of closed plant production system with artificial light: concept, estimation and application to plant factory, *Proceedings of the Japan Academy. Series B, Physical and Biological Sciences* 89(10), 447–461.

Kozai, T. (2022). Chapter 4: Contribution of PFALs to the sustainable development goals and beyond. In: Kozai, T., Niu, G. and Masabni, J. (Eds). *Plant Factory Basics, Applications and Advances*, Academic Press, pp. 57–79.

Kusuma, P., Pattison, P. M. and Bugbee, B. (2020). From physics to fixtures to food: current and potential LED efficacy, *Horticulture Research* 7(1), 56, https://doi.org/10.1038/s41438-020-0283-7.

Kwee, E. M. and Niemeyer, E. D. (2011). Variations in phenolic composition and antioxidant properties among 15 basil (Ocimum basilicum L.) cultivars, Elsevier Ltd, *Food Chemistry* 128(4), 1044–1050.

Lara, M. A., Lleó, L., Diezma-Iglesias, B., Roger, J. M. and Ruiz-Altisent, M. (2013). Monitoring spinach shelf-life with hyperspectral image through packaging films, *Journal of Food Engineering* 119(2), 353–361.

Larsen, D. H., Li, H., van de Peppel, A. C., Nicole, C. C. S., Marcelis, L. F. M. and Woltering, E. J. (2022). High light intensity at end-of-production improves the nutritional value of basil but does not affect postharvest chilling tolerance, *Food Chemistry* 369(April), 130913.

Lee, K. W., Lee, H. J., Surh, Y. J. and Lee, C. Y. (2003). Vitamin C and cancer chemoprevention: reappraisal, *American Journal of Clinical Nutrition* 78(6), 1074–1078.

Liang, D., Shen, Y., Ni, Z., Wang, Q., Lei, Z., Xu, N., Deng, Q., Lin, L., Wang, J., Lv, X. and Xia, H. (2018). Exogenous melatonin application delays senescence of kiwifruit leaves by regulating the antioxidant capacity and biosynthesis of flavonoids, *Frontiers in Plant Science* 9(April), 426.

Liao, C., Liu, X., Gao, A., Zhao, A., Hu, J. and Li, B. (2016). *"Maintaining postharvest qualities of three leaf vegetables to enhance their shelf lives by multiple ultraviolet-C treatment", Lwt*, Elsevier Ltd (vol. 73), pp. 1–5.

Lillo, C. (1994). Light regulation of nitrate reductase in green leaves of higher-plants, *Physiologia Plantarum* 90(3), 616–620.

Lillo, C. and Appenroth, K.-J. (2001). Light regulation of nitrate reductase in higher plants: which photoreceptors are involved?, *Plant Biology* 3(5), 455–465.

Lin, K., Huang, M., Huang, W., Hsu, M., Yang, Z. and Yang, C. (2013). The effects of red, blue, and white light-emitting diodes on the growth, development, and edible quality of hydroponically grown lettuce (Lactuca sativa L. var. capitata), *Scientia Horticulturae* 150, 86–91. https://doi.org/10.1016/j.scienta.2012.10.002.

Linster, C. L. and Clarke, S. G. (2008). l-ascorbate biosynthesis in higher plants: the role of VTC2, *Trends in Plant Science* 13(11), 567–573.

Liu, Y., Tikunov, Y., Schouten, R. E., Marcelis, L. F. M., Visser, R. G. F. and Bovy, A. (2018). Anthocyanin biosynthesis and degradation mechanisms in Solanaceous vegetables: a review, *Frontiers in Chemistry* 6, 52, https://doi.org/10.3389/fchem.2018.00052.

López-Gálvez, G., Saltveit, M. and Cantwell, M. (1996). Wound-induced phenylalanine ammonia lyase activity: factors affecting its induction and correlation with the quality of minimally processed lettuces, *Postharvest Biology and Technology* 9(2), 223–233.

Mai, F. and Glomb, M. A. (2016). Structural and sensory characterization of novel sesquiterpene lactones from iceberg lettuce, *Journal of Agricultural and Food Chemistry* 64(1), 295–301.

Maier, A. and Hoecker, U. (2015). COP1/SPA ubiquitin ligase complexes repress anthocyanin accumulation under low light and high light conditions, *Plant Signaling and Behavior* 10(1), e970440-1–e970440-3.

Martínez-Sánchez, A., Tudela, J. A., Luna, C., Allende, A. and Gil, M. I. (2011). Low oxygen levels and light exposure affect quality of fresh-cut Romaine lettuce, *Postharvest Biology and Technology* 59(1), 34–42.

Massa, G. D., Kim, H. H., Wheeler, R. M. and Mitchell, C. A. (2008). Plant productivity in response to LED lighting, *HortScience* 43(7), 1951–1956.

Massot, C., Stevens, R., Génard, M., Longuenesse, J. J. and Gautier, H. (2012). Light affects ascorbate content and ascorbate-related gene expression in tomato leaves more than in fruits, *Planta* 235(1), 153–163.

McEvily, A. J., Iyengar, R. and Otwell, W. S. (1992). Inhibition of enzymatic browning in foods and beverages, *Critical Reviews in Food Science and Nutrition* 32(February) 2013, 253–273.

Mellidou, I., Georgiadou, E. C., Kaloudas, D., Panagiotis, K., Fotopoulos, V. and Kanellis, A. K. (2019). Vitamins, *Postharvest Physiology and Biochemistry of Fruits and Vegetables*, Elsevier, pp. 395–383.

Mensinga, T. T., Speijers, G. J. A. and Meulenbelt, J. (2003). Health implications of exposure to environmental nitrogenous compounds, Springer, *Toxicological Reviews* 22(1), 41–51.

Min, Q., Marcelis, L. F. M., Nicole, C. C. S. and Woltering, E. J. (2021). High light intensity applied shortly before harvest improves lettuce nutritional quality and extends the shelf life, *Frontiers in Plant Science* 12(January), 615355.

Mir, S. A., Shah, M. A. and Mir, M. M. (2017). Microgreens: production, shelf life, and bioactive components, *Critical Reviews in Food Science and Nutrition* 57(12), 2730–2736.

Morris, S. C., Jobling, J. J., Tanner, D. J. and Forbes-Smith, M. R. (2003). Prediction of storage or shelf life for cool stored fresh produce transported by reefers, *Acta Horticulturae* 604(604), 305–311.

Moskowitz, H. R. (1970). Ratio scales of sugar sweetness, *Perception and Psychophysics* 7(5), 315–320.

Muller-Moule, P., Golan, T. and Niyogi, K. K. (2004). Ascorbate-deficient mutants of Arabidopsis Grow in high light despite chronic photooxidative stress, *Plant Physiology* 134(3), 1163–1172.

Naczk, M. and Shahidi, F. (2006). Phenolics in cereals, fruits and vegetables: occurrence, extraction and analysis, *Journal of Pharmaceutical and Biomedical Analysis* 41(5), 1523–1542.

Navabpour, S., Morris, K., Allen, R., Harrison, E., A-H-Mackerness, S. and Buchanan-Wollaston, V. (2003). Expression of senescence-enhanced genes in response to oxidative stress, *Journal of Experimental Botany, England* 54(391), 2285–2292.

Nicolle, C., Cardinault, N., Gueux, E., Jaffrelo, L., Rock, E., Mazur, A., Amouroux, P. and Rémésy, C. (2004). Health effect of vegetable-based diet: lettuce consumption improves cholesterol metabolism and antioxidant status in the rat, *Clinical Nutrition* 23(4), 605–614.

Niyogi, K. K. (1999). Photoprotection revisited: genetic and molecular approaches, *Annual Review of Plant Physiology and Plant Molecular Biology* 50, 333–359.

Njus, D., Kelley, P. M., Tu, Y. J. and Schlegel, H. B. (2020). Ascorbic acid: the chemistry underlying its antioxidant properties, Elsevier B.V., *Free Radical Biology and Medicine* 159(July), 37–43.

Noctor, G., Arisi, A.-C. M., Jouanin, L., Kunert, K. J., Rennenberg, H. and Foyer, C. H. (1998). Glutathione: biosynthesis, metabolism and relationship to stress tolerance explored in transformed plants, *Journal of Experimental Botany* 49(321), 623–647.

Noctor, G. and Foyer, C. H. (1998). Ascorbate and glutathione: keeping active oxygen under control, *Annual Review of Plant Physiology and Plant Molecular Biology* 49, 249–279.

Ntagkas, N., Woltering, E. J. and Marcelis, L. F. M. (2018). Light regulates ascorbate in plants: an integrated view on physiology and biochemistry, *Environmental and Experimental Botany* 147(July), 271–280.

Olle, M. and Viršile, A. (2013). The effects of light-emitting diode lighting on greenhouse plant growth and quality, *Agricultural and Food Science* 22(2), 223–234.

Onslow, M. W. (2014). *The Anthocyanin Pigments of Plants*, Cambridge University Press.

Ouzounis, T., Razi Parjikolaei, B., Frette, X., Rosenqvist, E. and Ottosen, C. O. (2015). Predawn and high intensity application of supplemental blue light decreases the quantum yield of PSII and enhances the amount of phenolic acids, flavonoids, and pigments in Lactuca sativa, *Frontiers in Plant Science* 6(February), 19.

Pallet, K. E. and Young, A. J. (1993). Carotenoids. In: Alscher, R. G. and Hess, J. L. (Eds), *Antioxidants in Higher Plants*, pp. 59–89.

Palozza, P. (1998). Prooxidant actions of carotenoids in biologic systems, *Nutrition Reviews* 56(9), 257–265.

Panche, A. N., Diwan, A. D. and Chandra, S. R. (2016). Flavonoids: an overview, *Journal of Nutritional Science* 5, e47, https://doi.org/10.1017/jns.2016.41.

Paul, M. J. and Pellny, T. K. (2003). Carbon metabolite feedback regulation of leaf photosynthesis and development, *Journal of Experimental Botany* 54(382), 539–547.

Pérez-López, U., Miranda-Apodaca, J., Muñoz-Rueda, A. and Mena-Petite, A. (2015). Interacting effects of high light and elevated CO2 on the nutraceutical quality of two differently pigmented Lactuca sativa cultivars (Blonde of Paris Batavia and Oak Leaf), *Scientia Horticulturae* 191, 38–48.

Pérez-López, U., Sgherri, C., Miranda-Apodaca, J., Micaelli, F., Lacuesta, M., Mena-Petite, A., Quartacci, M. F. and Muñoz-Rueda, A. (2018). Concentration of phenolic compounds is increased in lettuce grown under high light intensity and elevated CO2, Elsevier, *Plant Physiology and Biochemistry* 123(October), 233–241.

Petrov, V., Hille, J., Mueller-Roeber, B. and Gechev, T. S. (2015). ROS-mediated abiotic stress-induced programmed cell death in plants, *Frontiers in Plant Science*, FEB 6, 1–16.

Price, J., Laxmi, A., St Martin, S. K. and Jang, J. C. (2004). Global transcription profiling reveals multiple sugar signal transduction mechanisms in Arabidopsis, *Plant Cell* 16(8), 2128–2150.

Rolland, F., Baena-Gonzalez, E. and Sheen, J. (2006). SUGAR sensing and signaling in plants: conserved and novel mechanisms, *Annual Review of Plant Biology* 57(1), 675–709.

Runkle, E. S., Park, Y. and Meng, Q. (2022). High photosynthetic photon flux density can attenuate effects of light quality, *Acta Horticulturae* 1337(1337), 333–340.

Ryan, K. G., Markham, K. R., Bloor, S. J., Bradley, J. M., Mitchell, K. A. and Jordan, B. R. (1998). UVB radiation induced increase in Quercetin: Kaempferol ratio in wild-type and transgenic lines of Petunia, *Photochemistry and Photobiology* 68(3), 323–330.

Sago, Y. (2016). Effects of light intensity and growth rate on tipburn development and leaf calcium concentration in butterhead lettuce, *HortScience* 51(9), 1087–1091.

Samuolienė, G., Brazaityte, A., Sirtautas, R., Novičkovas, A. and Duchovskis, P. (2011). Supplementary red-LED lighting affects phytochemicals and nitrate of baby leaf lettuce, *Journal of Food, Agriculture and Environment* 9(3–4), 271–274.

Samuolienė, G., Brazaitytė, A., Sirtautas, R., Sakalauskienė, S., Jankauskienė, J., Duchovskis, P. and Novičkovas, A. (2012). The impact of supplementary short-term red led lighting on the antioxidant properties of microgreens, *Acta Horticulturae* 956(956), 649–656.

Samuolienė, G., Sirtautas, R., Brazaitytė, A., Viršilė, A. and Duchovskis, P. (2012). Supplementary red-LED lighting and the changes in phytochemical content of two baby leaf lettuce varieties during three seasons, *Journal of Food, Agriculture and Environment* 10, 701–706.

Santos, J., Herrero, M., Mendiola, J. A., Oliva-Teles, M. T., Ibáñez, E., Delerue-Matos, C. and Oliveira, M. B. P. P. (2014). Fresh-cut aromatic herbs: nutritional quality stability during shelf-life, *LWT: Food Science and Technology* 59(1), 101–107.

SharathKumar, M., Heuvelink, E. and Marcelis, L. F. M. (2020). Vertical farming: moving from genetic to environmental modification, *Trends in Plant Science* 25(8), 724–727.

Shen, Y. Z., Guo, S. S., Ai, W. D. and Tang, Y. K. (2014). Effects of illuminants and illumination time on lettuce growth, yield and nutritional quality in a controlled environment, *Life Sciences in Space Research* 2, 38–42.

Shiga, T., Shoji, K., Goto, E., Hashida, S., Goto, F. and Yoshihara, T. (2011). Effect of light quality on the polyphenol content and antioxidant activity of sweet basil (Ocimum basilicum L.), *Acta Horticulturae* 259(907), 95–99.

Solfanelli, C., Poggi, A., Loreti, E., Alpi, A. and Perata, P. (2006). Sucrose-specific induction of the anthocyanin biosynthetic pathway in Arabidopsis, *Plant Physiology* 140(2), 637–646.

Sousa Paiva, É. A. S., Dos Santos Isaias, R. M., Aguiar Vale, F. H. A. and De Senna Queiroz, C. G. (2003). The influence of light intensity on anatomical structure and pigment

contents of Tradescantia pallida (Rose) Hunt. cv. purpurea boom (Commelinaceae) leaves, *Brazilian Archives of Biology and Technology* 46(4), 617-624.
Stavric, B. (1994). Role of chemopreventers in human diet, *Clinical Biochemistry* 27(5), 319-332.
Thilini Deepashika Perera, W. P., Navaratne, S. and Wickramasinghe, I. (2022). Impact of spectral composition of light from light-emitting diodes (LEDs) on postharvest quality of vegetables: a review, Elsevier B.V., *Postharvest Biology and Technology* 191(December), 111955.
Thoma, F., Somborn-Schulz, A., Schlehuber, D., Keuter, V. and Deerberg, G. (2020). Effects of light on secondary metabolites in selected leafy greens: a review, *Frontiers in Plant Science* 11(April), 497.
Tijskens, L. M. M., Van Schaik, A. C. R., Hertog, M. L. A. T. M. and De Jager, A. (1999). Modelling the firmness of 'elstar' apples during storage and transport, *Acta Horticulturae* (485), 363-372.
Tomás-Barberán, F. A. and Espín, J. C. (2001). Phenolic compounds and related enzymes as determinants of quality in fruits and vegetables, *Journal of the Science of Food and Agriculture* 81(9), 853-876.
United Nations Department of Economic and Social Affairs and P. D. (2022). *World Population Prospects 2022: Summary of Results* (vol. 2022).
Van Delden, S. H., SharathKumar, M., Butturini, M., Graamans, L. J. A., Heuvelink, E., Kacira, M., Kaiser, E., Klamer, R. S., Klerkx, L., Kootstra, G., Loeber, A., Schouten, R. E., Stanghellini, C., van Ieperen, W., Verdonk, J. C., Vialet-Chabrand, S., Woltering, E. J., van de Zedde, R., Zhang, Y. and Marcelis, L. F. M. (2021). Current status and future challenges in implementing and upscaling vertical farming systems, *Nature Food*, Springer 2(12), 944-956.
Van Den Ende, W. and Valluru, R. (2009). Sucrose, sucrosyl oligosaccharides, and oxidative stress: scavenging and salvaging?, *Journal of Experimental Botany* 60(1), 9-18.
Vieira, A. R., Abar, L., Vingeliene, S., Chan, D. S. M., Aune, D., Navarro-Rosenblatt, D., Stevens, C., Greenwood, D. and Norat, T. (2016). Fruits, vegetables and lung cancer risk: a systematic review and meta-analysis, *Annals of Oncology* 27(1), 81-96.
Vinson, J. A., Dabbagh, Y. A., Serry, M. M. and Jang, J. (1995). Plant flavonoids, especially tea flavonols, are powerful antioxidants using an in vitro oxidation model for heart disease, *Journal of Agricultural and Food Chemistry* 43(11), 2800-2802.
Wang, F., Liu, J., Zhou, L., Pan, G., Li, Z., Zaidi, S. H. and Cheng, F. (2016). Senescence-specific change in ROS scavenging enzyme activities and regulation of various SOD isozymes to ROS levels in psf mutant rice leaves, *Plant Physiology and Biochemistry* 109, 248-261.
Wang, Y., Zhang, D., Liu, Y., Wang, D., Liu, J. and Ji, B. (2015). The protective effects of berry-derived anthocyanins against visible light-induced damage in human retinal pigment epithelial cells, *Journal of the Science of Food and Agriculture* 95(5), 936-944.
Wheeler, G. L., Jones, M. A. and Smirnoff, N. (1998). The biosynthetic pathway of Vitamin C in higher plants, *Nature* 393(6683), 365-369.
WHO and F. A. O. (2003). *Diet, Nutrition, and the Prevention of Chronic Diseases (Report of a Joint WHO and FAO Expert Consulation), WHO Technical Report Series* (vol. 916).
Witkowska, I. M. (2013). *Factors Affecting the Postharvest Performance of Fresh-Cut Lettuce*. Doctor of Philosophy, Wageningen University, [S.l. https://edepot.wur.nl/259987.

Woltering, E. J. and Seifu, Y. W. (2015). Low intensity monochromatic red, blue or green light increases the carbohydrate levels and substantially extends the shelf life of fresh-cut lettuce, *Acta Horticulturae* 1079(1079), 257–264.

Woltering, E. J. and Witkowska, I. M. (2016). Effects of pre- and postharvest lighting on quality and shelf life of fresh-cut lettuce, *Acta Horticulturae* 1134(1134), 357–366.

Yabuta, Y., Mieda, T., Rapolu, M., Nakamura, A., Motoki, T., Maruta, T., Yoshimura, K., Ishikawa, T. and Shigeoka, S. (2007). Light regulation of ascorbate biosynthesis is dependent on the photosynthetic electron transport chain but independent of sugars in Arabidopsis, *Journal of Experimental Botany* 58(10), 2661–2671.

Yahia, E. M., Carrillo-López, A. and Bello-Perez, L. A. (2019). Carbohydrates, *Postharvest Physiology and Biochemistry of Fruits and Vegetables*, Elsevier, pp. 175–205.

Yahia, E. M., García-Solís, P. and Celis, M. E. M. (2019). Contribution of fruits and vegetables to human nutrition and health, *Postharvest Physiology and Biochemistry of Fruits and Vegetables*, Elsevier, pp. 19–45.

Yamasaki, H. and Grace, S. C. (1998). EPR detection of phytophenoxyl radicals stabilized by zinc ions: evidence for the redox coupling of plant phenolics with ascorbate in the H2O2-peroxidase system, *FEBS Letters, England* 422(3), 377–380.

Yang, X., Gil, M. I., Yang, Q. and Tomás-Barberán, F. A. (2022). Bioactive compounds in lettuce: highlighting the benefits to human health and impacts of preharvest and postharvest practices, *Comprehensive Reviews in Food Science and Food Safety* 21(1), 4–45.

Yoshida, H., Ishigami, Y., Hikosaka, S. and Goto, E. (2022). Changes in the levels of bioactive compounds in red perilla at various leaf positions and ages under different photosynthetic photon flux densities, *Acta Horticulturae* 1337(1337), 179–186.

Zhan, L., Hu, J., Li, Y. and Pang, L. (2012). Combination of light exposure and low temperature in preserving quality and extending shelf-life of fresh-cut broccoli (Brassica oleracea L.), *Postharvest Biology and Technology* 72, 76–81.

Zhan, L., Li, Y., Hu, J., Pang, L. and Fan, H. (2012). Browning inhibition and quality preservation of fresh-cut romaine lettuce exposed to high intensity light, *Innovative Food Science and Emerging Technologies* 14, 70–76.

Zhang, G., Shen, S., Takagaki, M., Kozai, T. and Yamori, W. (2015). Supplemental upward lighting from underneath to obtain higher marketable lettuce (Lactuca sativa) leaf fresh weight by retarding senescence of outer leaves, *Frontiers in Plant Science* 6(December), 1110.

Zhang, Y. (2013). *Ascorbic Acid in Plants. Biosynthesis, Regulation and Enhancement*, Springer, New York, https://doi.org/10.1007/978-1-4614-4127-4.

Zhou, C., Shao, M., Liu, W., Li, B., Wang, Q., Liu, J., Wen, Y. and Yang, Q. (2021). Regulation of ascorbate accumulation and metabolism in lettuce by end-of-production high light irradiation provided by red and blue LEDs, Elsevier B.V., *Environmental and Experimental Botany* 189(June), https://doi.org/10.1016/j.envexpbot.2021.104567.

Zhou, W. L., Liu, W. K. and Yang, Q. C. (2012). Quality changes in hydroponic lettuce grown under pre-harvest short-duration continuous lightof different intensities, *Journal of Horticultural Science and Biotechnology* 87(5), 429–434.

Zhou, Y. H., Zhang, Y. Y., Zhao, X., Yu, H. J., Shi, K. and Yu, J. Q. (2009). Impact of light variation on development of photoprotection, antioxidants, and nutritional value in Lactuca sativa L, *Journal of Agricultural and Food Chemistry* 57(12), 5494–5500.

Zoratti, L., Karppinen, K., Luengo Escobar, A., Häggman, H. and Jaakola, L. (2014). Light-controlled flavonoid biosynthesis in fruits, *Frontiers in Plant Science* 5(October), 534.

# Part 5

## Spectral manipulations for controlling the growth and quality of ornamentals, fruit vegetables and herbs

# Chapter 14

# Effect of far-red light on improving yields of tomatoes produced in vertical farms

*Yongran Ji and Michele Butturini, Wageningen University & Research, The Netherlands*



## 1 Introduction

Tomato (*Solanum lycopersicum* L.) is one of the major crops cultivated globally and one of the most important vegetable crops in the world. Global tomato production has increased substantially from ~35 million tonnes in the 1970s to ~186 million tonnes in 2020 (FAOSTAT, 2023). With a significant production area being temperate zones, tomatoes are also produced in tropical and subtropical climate zones around the globe. China (65 million tonnes), India (21 million tonnes), Turkey (13 million tonnes), the United States (12 million tonnes), and Egypt (7 million tonnes) are the primary producers of tomatoes ranked by total production (Table 1). Tomato varieties differ in their shapes and sizes. They can be categorised as follows: classic round, plum and baby plum, beefsteak, cherry and cocktail, vine or truss, and regional varieties and landraces. Tomatoes are produced for fresh consumption or food processing and can be grown in the open field or a protected environment such as greenhouses.

http://dx.doi.org/10.19103/AS.2023.0126.19

**Table 1** Major tomato-producing countries (areas) ranked by total production in 2020

| Area | Total Production (million tonnes) | Total harvested area (thousand ha) |
|---|---|---|
| China (Mainland) | 64.8 | 1107. 5 |
| India | 20.6 | 812.0 |
| Turkey | 13.2 | 181.9 |
| United States of America | 12.2 | 110.4 |
| Egypt | 6.7 | 170.9 |
| Italy | 6.2 | 99.8 |
| Iran (Islamic Republic of) | 5.8 | 129.1 |
| Spain | 4.3 | 55.5 |
| Mexico | 4.1 | 84.9 |
| Brazil | 3.8 | 52.0 |

*Source*: FAOSTAT.

Although tomatoes are produced mainly in open fields, the production of tomatoes in a protected environment has expanded rapidly in recent decades. Compared to conditions in an open field, growth conditions in protected horticulture ensure more stable crop production and enable year-round production. Optimising growth conditions and breeding for cultivars suitable for production in a protected environment, the average yield of tomatoes has continuously increased over the years. In the Netherlands, for example, where tomatoes are mainly produced in greenhouses, the average yield is 60–70 kg $m^{-2}$ and is substantially higher than the global average (Heuvelink, 2018).

## 1.1 Yield determination in tomato

Tomato yield, which is evaluated by the fresh mass of the fruits, can be divided into multiple components (Fig. 1). Fruit fresh mass is determined by the fruit water content and fruit dry mass. The dry mass of the fruit is the result of the accumulation of sugars transported into the fruit from the leaves. Fruit dry mass is divided into total plant dry mass and dry mass partitioning to fruits. Total plant dry mass is determined by leaf net photosynthesis and leaf area index, where leaf area index is further defined by leaf number per plant and leaf area per leaf. Dry mass partitioning in fruits is determined by relative fruit sink strength and total fruit number, further defined by truss appearance rate and fruit number per truss. Each component positively correlates with the plant's fresh fruit mass (yield). For example, yield increase due to higher fruit water content results in fruits with more water, while yield increases due to higher plant dry mass lead to fruits with more fruit tissues.

Yield can also be increased by a larger number of fruits per plant due to an increased truss appearance rate or increased fruit number per truss. However, characteristics can cancel each other out. For example, a fresh fruit mass may

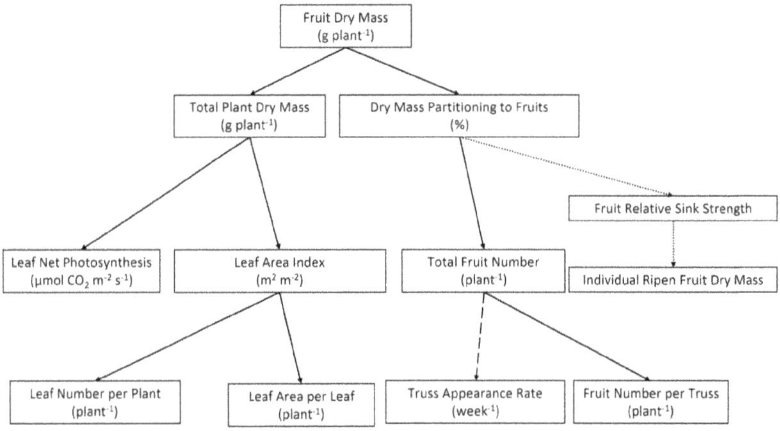


**Figure 1** Yield component analysis used in this experiment (adapted from Ji et al. (2019)). Dashed lines indicate that the component was not directly measured but was indirectly estimated.

remain unaffected if the increase in the fruit dry mass is similar to the decrease in fruit water content.

With a detailed understanding of components affecting yield, it is possible to quantify changes in each component between treatments, such as environmental factors and genetic differences, and evaluate their influences on the final yield. For example, Higashide and Heuvelink (2009) analysed morphological and physiological changes in breeding of nine varieties to explain that the yield improvement of these cultivars through 50 years of breeding, finding improvements were mainly due to higher leaf photosynthesis and lower light extinction coefficient.

## 1.2 Supplemental lighting in tomato greenhouse production

Improved control of climate parameters, such as temperature, humidity, and $CO_2$, has substantially improved recent yields of tomatoes in greenhouses. In addition to this improvement, supplementary lighting is a common practice for tomato production in greenhouse areas with limited solar radiation during winter. Traditionally, gas discharge-type lamps, such as high-pressure sodium (HPS) lamps, were used to provide supplementary lighting. There has been a recent increase in the popularity of using light-emitting diodes (LEDs) as an alternative light source. LED lighting features several advantages, such as a higher efficacy in converting electricity to photon energy, extended longevity, and higher flexibility in the placement on top or within the plant canopy. Katzin et al. (2021) showed that replacing HPS lamps (efficacy = 1.8 μmol $J^{-1}$) with LEDs (efficacy = 3 μmol $J^{-1}$) can reduce annual total energy consumption by up to 27% even when heating is required.

LED lighting fixtures offer the additional benefit of controlling the light spectrum, enabling greenhouse growers to use fully customised light recipes for supplementary lighting. Furthermore, LEDs can provide photons within 400–700 nm (traditionally known as photosynthetically active radiation, PAR) and can also supply photos within the far-red range (FR, 700–800 nm). Unlike photons from the PAR range, FR photons have little activity in driving photosynthesis independently. FR acts synergistically with photons in the PAR range to improve the overall quantum yield (Zhen et al., 2021). Adding FR alters the ratio between red (R) and FR photons.

A cassette of photoreceptors called phytochromes senses this change. Phytochromes exist as two isoforms:

- The Pr form (red-absorbing, biologically inactive); and
- The Pfr form (far-red-absorbing, biologically active).

The light spectrum thereby determines the equilibrium between Pr and Pfr. The Pfr isoform then translocates into the nucleus and mediates a series of photomorphogenic responses collectively known as the shade avoidance syndrome (SAS). Typical SAS responses such as stem elongation, leaf hyponasty, reduced branching, and accelerated flowering have been intensively studied in model species such as *Arabidopsis thaliana* (Casal, 2012). FR may also be supplied as additional lighting in the greenhouse to promote crop production. Several authors reported improved crop growth in leafy crops like lettuce when grown with additional FR (Li and Kubota, 2009; Zou et al., 2019; Zhen and Bugbee, 2020), and such increase was further studied in vertical farming conditions at different planting densities (Jin et al., 2020).

Yield improvement was also reported in tomato crops grown with additional FR (Hao et al., 2016; Kalaitzoglou et al., 2019; Zou et al., 2019). Hao et al. (2016) observed increased yield when adding 8, 16, and 24 μmol $m^{-2}$ $s^{-1}$ of FR to a background HPS lighting of 165 μmol $m^{-2}$ $s^{-1}$. Kalaitzoglou et al. (2019) experimented with LED lighting and reported increased tomato yield when FR was added during and at the end of the photoperiod. Zou et al. (2019) reported a similar positive effect of FR provided as overhead lighting while the background LED lighting was provided as intra-canopy lighting.

In this chapter, we summarise our recent research published in Ji et al. (2019, 2020, 2021) on the physiological and molecular mechanisms in which FR increased tomato yield. Subsequently, we discuss the possible application of this knowledge to vertical farming and some key aspects for the future improvement of tomatoes grown in vertical farms.

## 2 Far-red light affects dry mass production and shoot: root ratio in young tomato plants

We conducted an experiment, as has been published by Ji et al. (2021), to explore the genotypic variation in tomatoes' growth responses to FR. Thirty-three different tomato genotypes with significant differences in morphology, growth patterns, and growth rates were grown at 0, 25, or 100 μmol $m^{-2}$ $s^{-1}$ of FR added to a red + white LED light background at μmol $m^{-2}$ $s^{-1}$. Ten days after sowing, uniform seedlings were transplanted to the experimental climate chamber and were grown for 21 days, after which the plants were destructively harvested, and traits including the dry mass of leaves, stems, roots, as well as plant height, were measured. The total dry mass of the plant responded strongly and positively within one-third of the tested genotypes, while the other genotypes showed moderate or no responses. For almost all genotypes, a significant increase in stem dry mass and plant height was observed FR increased.

Interestingly, more photosynthetic assimilates were partitioned to sustain the growth of stems and/or leaves, as demonstrated by the increased shoot: root ratio. This increase by FR in shoot: root ratio is in line with previous studies in tomatoes (Cao et al., 2018; Kasperbauer, 1987; Lee et al., 2016). Despite all the genotypes showing a higher shoot: root ratio under FR, the underlying processes leading to this increase differed across different genotypes. For some genotypes, this increase correlated with an increase in the dry mass of leaves, stems, and roots with a stronger increase in the shoot dry mass. For some others, however, leaves, and root dry mass both decreased, with a stronger relative decrease of root dry mass. Adding FR suppresses the activity of phytochromes in mediating growth and development. This suggests that phytochromes and hormonal regulation may play a crucial role in mediating the shoot: root ratio responses.

## 3 Far-red light increases tomato yield by increasing dry mass partitioning to fruits

We conducted an experiment, as has been published by Ji et al. (2019), where tomato plants (*S. lycopersicum* L. cv Moneymaker) were transplanted into the experimental greenhouse three weeks after sowing. With a background of a red + blue (RB) LED lighting at 150 μmol $m^{-2}$ $s^{-1}$, two FR treatments ~30 and 50 μmol $m^{-2}$ $s^{-1}$ were included, providing light conditions with phytochrome photostationary state (PSS) of 0.84 and 0.80, respectively, compared to the PSS of 0.88 in the RB treatment. We measured the growth and development of tomato plants grown under different light conditions according to a modified yield component analysis scheme (Fig. 1). Comparisons were made between

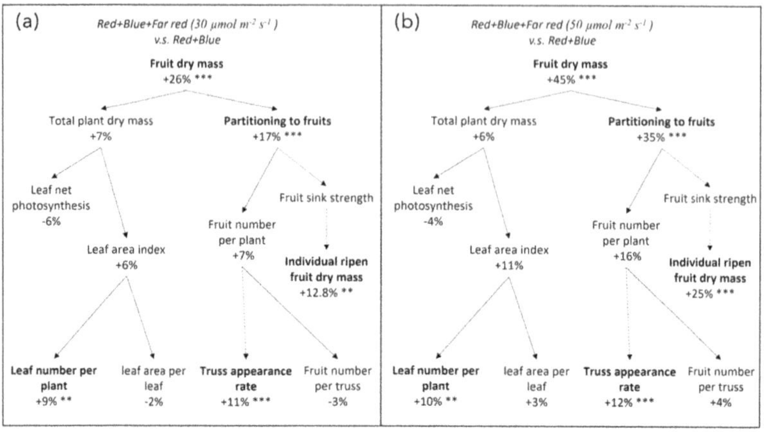


**Figure 2** Effects of (a) 30 µmol m$^{-2}$ s$^{-1}$ and (b) 50 µmol m$^{-2}$ s$^{-1}$ of additional far red in a red and blue background light on the growth components. Dashed lines indicate that the parameter was based on estimation. Asterisks denote significant effects of additional FR as tested by linear regression (*$P<0.05$, **$P<0.01$, ***$P<0.001$). Fruit sink strength was not measured. (Adapted from Ji et al. (2019)).

each of the FR and RB reference treatments for each component and tested for their statistical differences.

The yield component analysis showed that, when compared to plants grown without FR, fruit dry mass was significantly increased by 26% at 30 µmol m$^{-2}$ s$^{-1}$ of FR and was further increased by 45% at 50 µmol m$^{-2}$ s$^{-1}$ of FR (Fig. 2). In both cases, total plant dry mass was 6–7% higher, but their differences were not statistically significant. FR significantly increased dry mass partitioning to fruits by 17% and 35% in the two FR treatments. It is clear from the yield component analysis that the increase in fruit dry mass was mainly the result of an increased fraction of dry mass partitioned to the fruits, accompanied by a slight increase in total plant dry mass. An improved light distribution may explain the increase in total plant dry mass in the canopy as FR increased the internode length and led to a more open canopy architecture which, in turn, allowed a deeper radiation penetration into the canopy (Zou et al., 2019). Sarlikioti et al. (2011) used model simulation and showed that an increase in internode length indeed increased the canopy photosynthesis rate by up to 6–8% due to increased canopy light absorption in the upper part of the tomato plant canopy.

Interestingly, FR also increases in the individual fruit dry mass of ripened fruits, which may indicate that the sink strength of individual fruits was increased. Sink strength is the intrinsic capacity of an organ to attract photosynthetic assimilates from the assimilate source (Marcelis, 1996). In tomatoes, the sink strength of the fruit is quantified as the growth rate of tomato fruits under non-limiting assimilate supply, which is also often termed the potential growth rate of the fruit. At a given time during growth, the fraction of dry mass partitioned to the fruits is calculated as total fruit sink strength (sum of individual fruit sink strength existing on a plant at a given time divided by the total plant sink strength, i.e. the sum of fruit sink strength and vegetative sink strength).

Similar to the previous experiment, we set up a new experiment (as published by Ji et al. (2020)) to determine the effect of FR on fruit sink strength and to evaluate whether this effect can explain the previously observed effect in dry mass partitioning. To determine the potential growth rate of tomato fruits, we measured the height and diameter of the fruits on the second, third, and fourth truss of plants pruned to one or two fruits per truss. This measurement was taken non-destructively twice a week. Here, we assumed that the fruits with only one fruit per truss grew without any source limitation. We calculated the size of tomato fruits based on the routine non-destructive measurement of their height and diameter every 2–3 days. Fruit fresh mass was converted to fruit dry mass according to Li et al. (2015) and Maaike Wubs et al. (2012) and then fitted with a non-linear mixed model as a function of fruit age by a Gompertz function. The derivative of this function gives the fruit growth rate as a function of fruit age (days after anthesis, DAA).

We compared the fruit growth on plants with one and two per truss, and there was no significant difference even when the fruit load was doubled hence confirming that fruit growth at one fruit per truss was indeed a potential growth. Additional FR significantly promoted potential fruit growth and led to a higher growth rate throughout the fruit growth period. It reached the maximum growth rate at an earlier fruit age (Fig. 3). A higher potential fruit growth rate under additional FR means increased individual fruit sink strength. This result strongly indicated that the increased fruit sink strength may explain the increased dry mass partitioning under additional FR. To quantify this effect, we needed a growth model to simulate the growth of tomato plants in the above-mentioned experimental conditions.

We then used a growth model to simulate a whole tomato plant's daily dry mass production using data of daily average $CO_2$ partial pressure during the photoperiod, total daily PAR integral, and the leaf area index (Haverkort et al., 2015). The calculated daily dry mass production was then simulated to be partitioned into the fruits and accumulated throughout the experimental season. The simulated fruit dry mass fraction increased from 33% to 40% for the simulation scenarios with five fruits per truss and increased from 40% to 47% for no pruning. These simulated results agreed very well with the measured dry mass partitioning. Considering that only the individual fruit sink strength was altered during the simulation, it can be concluded that the higher fruit sink strength was the major factor explaining the higher dry mass partitioning to fruits.

To summarise, in the two experiments presented here, we combined experimental research and modelling to demonstrate that additional FR increased dry mass partitioning to fruit by increasing individual fruit sink

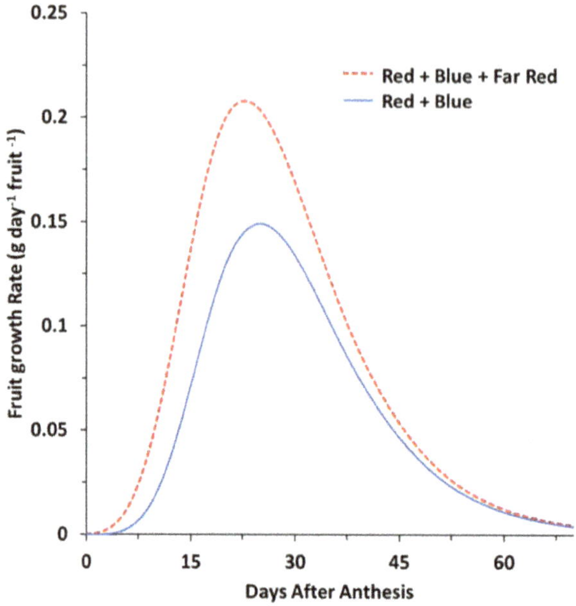


**Figure 3** Potential fruit growth rate of tomato (*Solanum lycopersicum*) plants grown with or without additional far red (Ji et al., 2020).

strength. This increase, combined with a slight increase in total plant dry mass, possibly due to a better light distribution within the canopy, led to a higher yield of tomatoes.

## 4 Physiological and molecular regulation of dry mass partitioning as affected by far-red

Sugar is the building block for plant growth and development. Sugar metabolism in the sink organs is closely linked with the sink strength of these organs (Osorio et al., 2014). We monitored the changes in four major sugars in tomato fruits: starch, fructose, glucose, and sucrose. Tomato fruits typically accumulate starch until around 15–20 days after anthesis before switching to hexose accumulation (Ruan et al., 1997). We observed that starch concentration increased in the fruits until 20 days after the anthesis and decreased to almost zero at ripening.

Additional FR substantially increased the fruits' starch, fructose, and glucose concentrations. Sucrose concentration was not significantly affected by FR. However, considering that a higher concentration of starch, fructose, and glucose was observed, it was reasoned that FR elevated sugar metabolism. Sucrose is the form of sugar being translocated between organs, but FR did not significantly affect its concentration. Considering the elevated sugar metabolism and the higher individual fruit dry mass under FR, it is reasonable to argue that an FR-elevated import rate of sucrose is required to maintain sucrose concentration in the fruit while ensuring a higher fruit growth rate.

In order to further investigate the molecular regulation of sugar metabolism as affected by FR, we studied the key genes involved in sugar transportation and metabolism at different stages of fruit growth. Matching the dynamics of starch concentration during fruit growth, we observed an increased expression of starch synthesis genes during the early days of fruit development, while starch catabolism genes were upregulated after 30 days after anthesis. The upregulation genes encoding sucrose synthase explain the higher fructose and glucose concentration. Invertase genes from *LIN* gene family were also upregulated to further enhance fruit sugar metabolism. Interestingly, the expression of *INVINH1*, which encodes the inhibitor of the invertase Lin5, was also upregulated. The expression of invertase inhibitor caps cell-wall invertase activities to regulate fruit development (Jin et al., 2009). We reason that FR-induced higher sucrose transportation into the fruit to sustain the enhanced sugar metabolism, which supported the upregulation of sugar transporter genes, especially at the flowering stage.

Several key genes involved in sugar transportation and metabolism were previously reported to be regulated by FR or FR-regulatory networks. For example, knocking down phytochromes in the fruit significantly upregulated the expression of cell-wall invertase genes *LIN5* and *LIN6* (Ernesto Bianchetti et al., 2018). Also, Chen et al. (2016) reported that expressions of genes encoding sucrose transporters SWEET11 and SWEET12 were regulated by HY5, a transcription factor directly regulated by FR (van Gelderen et al., 2018).

To summarise, additional FR increased the expression of key genes responsible for fruit sugar transportation and metabolism. Such an increase contributed to a substantial increase in fruit sink strength which quantitatively explained the increase of dry mass partitioning to fruit under additional FR.

## 5 Current status of tomato production in vertical farming

As one of the most important vegetable crops grown globally, tomatoes are primarily produced in open fields or greenhouses. Greenhouse production typically uses the indeterminant type of tomato growing on the high-wire system to ensure a production period of 9–11 months. Recent years have seen increased interest in using completely enclosed systems (no solar radiation) with total environmental control and artificial lighting to grow crops traditionally grown in open fields or greenhouses. Such facilities, often termed vertical farms or plant factories, are theoretically capable of growing any conventionally farmed crops, and their applications are dictated mainly by the profitability of the products. Currently, the most popular crops for vertical farming are mostly leafy vegetables such as lettuce, basil, and microgreens (van Delden et al., 2021). These crops typically take up little vertical space and can be grown in stacked layers and hence can vastly increase the crop yield per ground area.

Tomatoes are not commonly grown on vertical farms. Having said that, several trials and attempts by researchers and commercial companies have been made to grow tomatoes using vertical farming. For example, the horticultural lighting company Signify (the Netherlands) initiated a long-term study to grow high-wire tomatoes in their GrowWise research centre in 2019 (Mooren, 2021). Certhon (the Netherlands), another Dutch tech company active in the greenhouse and vertical farming business, conducted a 5-year experiment with over 25 cultivation rounds to demonstrate the possibilities of growing cherry tomatoes in vertical farms with a high grade of automation (Breure, 2022). Vertical farming company Infarm (Germany) recently teamed up with Wageningen University and Research (the Netherlands) to sponsor three research projects focusing on the improvement of the production of tomatoes in vertical farms (WUR announces collaboration on crop research with Infarm–WUR, 2022). Similarly, the Foundation for Food & Agricultural Research (United States)'s Precision Indoor Plants consortium awarded a US$2 million grant to the collaboration between the University of Florida (United States), the University of California Riverside (United States), Wageningen University, and Research and Aerofarms (United States) to enhance traits in tomato to make it suitable for indoor production.

Despite active trials and research projects, few companies worldwide have attempted commercial indoor tomato production (PIP Consortium Promotes Indoor Tomato Farming- Foundation for Food & Agriculture Research, 2022). In 2020, Madar Farms (United Arab Emirates) announced the development of a large-scale vertical farm in the desert city of Abu Dhabi (World's first commercial-scale indoor tomato farm to open in Abu Dhabi - Caterer Middle East, 2020). PlantLab (the Netherlands), a Dutch scale-up company, also announced commercial production of tomatoes (Dutch scale-up PlantLab launches first production site in USA, 2020). 80 Acres Farms (United States) has established indoor tomato production facilities in Ohio (Coolidge, 2020). To our best knowledge, a commercially viable production and business model for vertically farmed tomatoes is yet to be demonstrated.

## 6 Ideotype of vertically farmed tomato

A niche market for vertically farmed tomatoes exists where some customers will prefer premium products at a higher price. The future of tomato production in vertical farms will require a dedicated selection of varieties suitable for the distinct growth condition provided by vertical farms, which differs substantially from the growth conditions in a greenhouse or an open field. However, almost all tomato cultivars currently available on the market are selected and bred in conventional growth conditions, which means that a good performance in these conditions does not guarantee the same performance in vertical farms. The unique nature of vertical farms, such as LED lighting and increased automation, requires new or different traits from existing breeding strategies. This section will discuss some critical traits of an ideotype of a vertically farmed tomato variety.

Several aspects of crop architecture can be optimised for vertical farming. The leaf architecture and branching of the stems should be optimised to allow light penetration throughout the entire vertical profile of the canopy to ensure a uniform light distribution to maximise canopy photosynthesis among all leaf ranks. Similar to varieties grown in greenhouses, breeding for higher harvest index and light interception are also important. Space is expensive and limited in vertical farms; hence good candidate crops for vertical farming maximises space use efficiency. This means that compactness is a key trait for the breeding of dwarf tomatoes to be grown in a stacked layer vertical farming system.

It also remains relevant for high-wire tomatoes to have shorter internodes to allow more trusses to grow within the vertical space limit. This trait may also help reduce labour costs for crop maintenance. It is worth noticing that a shorter internode length may restrict canopy photosynthesis and hence limit dry mass production and limit yield (Sarlikioti et al., 2011). Also, applying FR may lead to a longer internode and a faster increase in plant height and leaf number (Ji et al., 2019). This may lead to more frequent crop maintenance in activities such as crop rotation and pruning and, as a consequence, increase operational costs in high-wire tomatoes.

Other improvements can be achieved in crop physiology, especially in tomatoes' response to light because a significant proportion of the already high energy demand of vertical farming is attributed to lighting. Hence, vertically farmed tomatoes should be selected for their high light use efficiency and capacity to maintain efficiency during a long photoperiod and high photon flux density. Velez-Ramirez et al. (2014) e.g. demonstrated that a single locus conferred tolerance against continuous lighting in tomatoes and showed the potential to overcome the negative responses under continuous lighting. This should be accompanied by adapting varieties to grow in artificial lighting provided by LEDs.

Since it is not affected by the outdoor climate, a vertical farm allows for non-stop operation, and a short growth cycle will allow more production rounds. For this reason, the quick establishment of leaf area during the early stages of vegetative growth, rapid transition from vegetative to the generative stage, and earlier flowering and fruiting are desired (van Delden et al., 2021). Cultivars bred for conventional production often need to consider resistance against biotic and abiotic stress. Often, an improved resistance to pests and pathogens, or an improved tolerance to extreme climate conditions such as drought, heat, or cold, will lead to a trade-off in yield (Lazzarin et al., 2021; Malhotra et al., 2022). For vertical farms, however, the high degree of climate control and improved hygiene standard makes pests and pathogens or suboptimal climates rare. For this reason, breeding for vertically farmed tomatoes can favour selecting for better yield without the limitation of having to be robust for resistance-related traits.

Growers of conventionally produced tomatoes rely heavily on a consistent supply of high quantities at a competitive yield-to-cost ratio to remain profitable. While a high yield remains relevant for vertically farmed tomatoes, it is more important to focus on increasing the consistency and efficiency of supply. The target market for vertically farmed tomatoes may also differ from those produced conventionally. Consumers may pay a premium price for fruit with higher sugar content, vitamin content, and more intense flavour. In addition, vertically farmed tomatoes may also target specific niche markets and customers that require unique colours, shapes, or flavours.

Compared to tomato production in the open field and greenhouses, vertical farming enables a higher degree of automation using sensors to monitor plant growth and robotic machines for crop handling and harvest. Vertically farmed tomato plants should have an architecture that facilitates and benefits from automation. Such traits may include a longer truss and pedicel for easier fruit detection by sensors and cameras. Better fruit shape and texture will allow easy handling by harvesting robots while reduced axillary shoot growth will avoid pruning. Ideally, fruits should be sufficiently distant from the leaves while the flowering and ripening process should be synchronised in the vertical farms to allow maximum efficiency of automated detection and harvesting of fruits.

Knowledge of flowering and fruit development has been well studied in tomatoes (Seymour et al., 2013) and can allow targeted genotyping for ripening-related traits. Advances in understanding the control of branching and fruit truss architecture have provided exciting possibilities to manipulate crop architecture for vertical farming (Kwon et al., 2020).

## 7 Conclusion and future trends

In this chapter, we have summarised the mechanisms behind increased tomato yield resulting from the application of additional FR in vertical farms, as demonstrated in studies by Ji et al. (2019, 2020). Adding FR during growth substantially increased individual fruit sink strength, allowing fruits to attract more sugar from the pool of photosynthetic assimilates. This increase leads to a higher dry mass partitioning to fruits and consequently increases tomato yield. This study was conducted in the greenhouse and was demonstrated by a climate chamber experiment with artificial LED lighting only (Ji et al., 2020), showing that the yield improvement of FR is also applicable to vertically farmed tomatoes. The high investment and operation costs largely limit crop choice in vertical farming. Nevertheless, it is often argued that fruiting crops such as tomatoes are a logical future step to attempt to scale up vertical farming.

Although the genetics and physiology of tomatoes have been intensively studied for decades, the successful production of vertically farmed tomatoes still requires the dedicated breeding of cultivars suitable for indoor production. Such targeted breeding needs to be conducted in vertical farms and should not be limited to currently existing breeding populations. Revisiting old and wild genotypes and heirloom genotypes for traits lost in the selection for conventional production, as assisted by modern breeding concepts such as *de novo* domestication, will allow for accurate and rapid manipulation of target traits. Breeding may be accelerated even further with speed breeding and genome editing. Further understanding of plant physiology and photobiology and their interaction with vertical farming production environments will support breeding and help establish efficient growing strategies for tomatoes in vertical farming. With optimised engineering and advanced artificial intelligence, coupled with plant sensors and automation, the production of tomatoes and other fruit crops in vertical farms will pose exciting commercial and scientific potential as a significant part of the modern food supply chain.

## 8 Where to look for further information

Plant's responses to FR has been extensively studied in the past years and several comprehensive reviews have been published on this topic. For example, the review of "Shade Avoidance" (see Casal, 2012 in reference list) provided a complete overview of the mode-of-action of the shade avoidance responses. Further, Demotes-Mainard et al. (2016) summarized the FR responses of plants and their application in horticulture (https://doi.org/10.1016/j.envexpbot.2015.05.010). For a good understanding of tomato and its production, the book "Tomatoes" is recommended (see Heuvelink, 2018 in reference list).

Several universities and research centers are active in the research of FR and its application in indoor crop production. For example, Wageningen University & Research in the Netherlands, China Academy of Agricultural Science in China, Utah State University, Ohio State University, Michigan State University in the United States. Several active research projects are focused on the indoor production of crops such as tomato. For example, the Sky-High project founded by the Dutch Research Council (NWO) and the indoor tomato project founded by the Foundation for Food & Agriculture Research (FFAR).

Also, several important international conferences are of great interest in the field. For example, the International Workshop on Vertical Farming (VertiFarm, held annually), the International Symposium on New Technologies for Greenhouse (GreenSys, held bi-annually), the International Symposium on

Light in Horticulture (LightSym, held every four years) and the International Horticultural Cognress (IHC, held every four years).

# 9 References

Breure, A. (2022). Indoor growing system for cherry tomatoes where no labor is needed. Available at: https://www.hortidaily.com/article/9442012/indoor-growing-system-for-cherry-tomatoes-where-no-labor-is-needed/ (accessed 21 February 2023).

Cao, K., Yu, J., Xu, D., Ai, K., Bao, E. and Zou, Z. (2018). Exposure to lower red to far-red light ratios improve tomato tolerance to salt stress. *BMC Plant Biology* 18(1): 92. doi: 10.1186/s12870-018-1310-9.

Casal, J. J. (2012). Shade avoidance. *The Arabidopsis Book*. American Society of Plant Biologists: 10: e0157. doi: 10.1199/tab.0157.

Chen, X., Yao, Q., Gao, X., Jiang, C., Harberd, N. P. and Fu, X. (2016). Shoot-to-root mobile transcription factor HY5 coordinates plant carbon and nitrogen acquisition. *Current Biology* 26(5): 640–646. doi: 10.1016/j.cub.2015.12.066.

Coolidge, A. (2020). A tomato and produce farm 80 acres flourishes - indoors amid coronavirus. Available at: https://eu.cincinnati.com/story/money/2020/05/03/tomato-and-produce-farm-80-acres-flourishes-indoors-amid-coronavirus/3039525001/ (accessed 22 February 2023).

Dutch scale-up PlantLab launches first production site in USA (2020). Available at: https://www.verticalfarmdaily.com/article/9274840/dutch-scale-up-plantlab-launches-first-production-site-in-usa/ (accessed 22 February 2023).

Ernesto Bianchetti, R., Silvestre Lira, B., Santos Monteiro, S., Demarco, D., Purgatto, E., Rothan, C., Rossi, M. and Freschi, L. (2018). Fruit-localized phytochromes regulate plastid biogenesis, starch synthesis, and carotenoid metabolism in tomato. *Journal of Experimental Botany* 69(15): 3573–3586. doi: 10.1093/jxb/ery145.

FAOSTAT (2023). Available at: http://www.fao.org/faostat/en/#data/QC (accessed 15 February 2023). Food and Agriculture Organization of the United Nations.

Hao, X., Little, C., Zheng, J. M. and Cao, R. (2016). Far-red LEDs improve fruit production in greenhouse tomato grown under high-pressure sodium lighting. *Acta Horticulturae* 1134(1134): 95–102. doi: 10.17660/ActaHortic.2016.1134.13.

Haverkort, A. J., Franke, A. C., Steyn, J. M., Pronk, A. A., Caldiz, D. O. and Kooman, P. L. (2015). A robust potato model: LINTUL-POTATO-DSS. *Potato Research* 58(4): 313-327. doi: 10.1007/s11540-015-9303-7.

Heuvelink, E. (2018). *Tomatoes*. Boston, MA : CABI Publishing.

Higashide, T. and Heuvelink, E. (2009). Physiological and morphological changes over the past 50 years in yield components in tomato. *Journal of the American Society for Horticultural Science* 134(4): 460–465. doi: 10.21273/JASHS.134.4.460.

Ji, Y., Ouzounis, T., Courbier, S., Kaiser, E., Nguyen, P. T., Schouten, H. J., Visser, R. G. F., Pierik, R., Marcelis, L. F. M. and Heuvelink, E. (2019). Far-red radiation increases dry mass partitioning to fruits but reduces Botrytis cinerea resistance in tomato. *Environmental and Experimental Botany* 168: 103889. doi: 10.1016/j.envexpbot.2019.103889.

Ji, Y., Nuñez Ocaña, D., Choe, D., Larsen, D. H., Marcelis, L. F. M. and Heuvelink, E. (2020). Far-red radiation stimulates dry mass partitioning to fruits by increasing fruit sink strength in tomato. *New Phytologist* 228(6): 1914–1925. doi: 10.1111/nph.16805.

Ji, Y., Ouzounis, T., Schouten, H. J., Visser, R. G. F., Marcelis, L. F. M. and Heuvelink, E. (2021). Dissecting the genotypic variation of growth responses to Far-red radiation in tomato. *Frontiers in Plant Science* 11: 614714. https://doi.org/10.3389/fpls.2020.614714.

Jin, W., Urbina, J. L., Heuvelink, E. and Marcelis, L. F. M. (2020). Adding Far-red to red-blue light-emitting diode light promotes yield of lettuce at different planting densities. *Frontiers in Plant Science* 11: 609977 doi: 10.3389/fpls.2020.609977.

Jin, Y., Ni, D. A. and Ruan, Y. L. (2009). Posttranslational elevation of cell wall invertase activity by silencing its inhibitor in tomato delays leaf senescence and increases seed weight and fruit hexose Level. *Plant Cell* 21(7): 2072–2089. doi: 10.1105/tpc.108.063719.

Kalaitzoglou, P., van Ieperen, W., Harbinson, J., van der Meer, M., Martinakos, S., Weerheim, K., Nicole, C. C. S. and Marcelis, L. F. M. (2019). Effects of continuous or end-of-day far-red light on tomato plant growth, morphology, light absorption, and fruit production. *Frontiers in Plant Science* 10: 322. doi: 10.3389/fpls.2019.00322.

Kasperbauer, M. J. (1987). Far-red light reflection from green leaves and effects on phytochrome-mediated assimilate partitioning under field conditions. *Plant Physiology* 85(2): 350–354. doi: 10.1104/pp.85.2.350.

Katzin, D., Marcelis, L. F. M. and van Mourik, S. (2021). Energy savings in greenhouses by transition from high-pressure sodium to LED lighting. *Applied Energy* 281: 116019. doi: 10.1016/j.apenergy.2020.116019.

Kwon, C. T., Heo, J., Lemmon, Z. H., Capua, Y., Hutton, S. F., Van Eck, J., Park, S. J. and Lippman, Z. B. (2020). Rapid customization of Solanaceae fruit crops for urban agriculture. *Nature Biotechnology* 38(2): 182–188. doi: 10.1038/s41587-019-0361-2.

Lazzarin, M., Meisenburg, M., Meijer, D., van Ieperen, W., Marcelis, L. F. M., Kappers, I. F., van der Krol, A. R., van Loon, J. J. A. and Dicke, M. (2021). LEDs make it resilient: effects on plant growth and defense. *Trends in Plant Science* 26(5): 496–508. doi: 10.1016/J.TPLANTS.2020.11.013.

Lee, M. J., Son, K. H. and Oh, M. M. (2016). Increase in biomass and bioactive compounds in lettuce under various ratios of red to far-red LED light supplemented with blue LED light. *Horticulture, Environment, and Biotechnology* 57(2): 139–147. doi: 10.1007/s13580-016-0133-6.

Li, Q. and Kubota, C. (2009). Effects of supplemental light quality on growth and phytochemicals of baby leaf lettuce. *Environmental and Experimental Botany* 67(1): 59–64. doi: 10.1016/j.envexpbot.2009.06.011.

Li, T., Heuvelink, E. and Marcelis, L. F. M. (2015). Quantifying the source-sink balance and carbohydrate content in three tomato cultivars. *Frontiers in Plant Science* 6(June): 416. doi: 10.3389/fpls.2015.00416.

Maaike Wubs, A. M., Ma, Y. T., Heuvelink, E., Hemerik, L. and Marcelis, L. F. M. (2012). Model selection for nondestructive quantification of fruit growth in pepper. *Journal of the American Society for Horticultural Science* 137(2): 71–79. doi: 10.21273/JASHS.137.2.71.

Malhotra, B., Kumar, P. and Bisht, N. C. (2022). Defense versus growth trade-offs: insights from glucosinolates and their catabolites. *Plant, Cell and Environment* 1–21. doi: 10.1111/PCE.14462.

Marcelis, L. F. M. (1996). Sink strength as a determinant of dry matter partitioning in the whole plant. *Journal of Experimental Botany* 47 (Spec No): 1281–1291. doi: 10.1093/jxb/47.Special_Issue.1281(SPEC. ISS.).

Mooren, J. (2021). 5 learnings for growing top-quality tomatoes in a vertical farm. Available at: https://www.lighting.philips.com/main/products/horticulture/hortiblog/vertical-farming/growing-top-quality-tomatoes-in-a-vertical-farm (accessed 21 February 2023).

Osorio, S., Ruan, Y. L. and Fernie, A. R. (2014). An update on source-to-sink carbon partitioning in tomato. *Frontiers in Plant Science* 5: 1–11. doi: 10.3389/fpls.2014.00516.

PIP consortium promotes indoor tomato farming - foundation for food & agriculture research (2022). Available at: https://foundationfar.org/news/pip-consortium-promotes-indoor-tomato-farming/ (accessed 22 February 2023).

Ruan, Y.-L., Patrick, J. W. and Brady, C. (1997). Protoplast hexose carrier activity is a determinate of genotypic difference in hexose storage in tomato fruit. *Plant, Cell and Environment* 20(3): 341–349. doi: 10.1046/j.1365-3040.1997.d01-73.x.

Sarlikioti, V., De Visser, P. H. B., Buck-Sorlin, G. H. and Marcelis, L. F. (2011). How plant architecture affects light absorption and photosynthesis in tomato: towards an ideotype for plant architecture using a functionalstructural plant model. *Annals of Botany* 108(6): 1065–1073. doi: 10.1093/aob/mcr221.

Seymour, G. B., Chapman, N. H., Chew, B. L. and Rose, J. K. (2013). Regulation of ripening and opportunities for control in tomato and other fruits. *Plant Biotechnology Journal* 11(3): 269–278. doi: 10.1111/J.1467-7652.2012.00738.X.

van Delden, S. H., SharathKumar, M., Butturini, M., Graamans, L. J. A., Heuvelink, E., Kacira, M., Kaiser, E., Klamer, R. S., Klerkx, L., Kootstra, G., Loeber, A., Schouten, R. E., Stanghellini, C., van Ieperen, W., Verdonk, J. C., Vialet-Chabrand, S., Woltering, E. J., van de Zedde, R., Zhang, Y. and Marcelis, L. F. M. (2021) Current status and future challenges in implementing and upscaling vertical farming systems. *Nature Food* 2(12): 944–956. doi: 10.1038/s43016-021-00402-w.

van Gelderen, K., Kang, C., Paalman, R., Keuskamp, D., Hayes, S. and Pierik, R. (2018). Far-red light detection in the shoot regulates lateral root development through the HY5 transcription factor. *Plant Cell* 30(1): 101–116. doi: 10.1105/tpc.17.00771.

Velez-Ramirez, A. I., van Ieperen, W., Vreugdenhil, D., van Poppel, P. M., Heuvelink, E. and Millenaar, F. F. (2014). A single locus confers tolerance to continuous light and allows substantial yield increase in tomato. *Nature Communications* 5(1): 4549. doi: 10.1038/ncomms5549.

World's first commercial-scale indoor tomato farm to open in Abu Dhabi - Caterer Middle East (2020). Available at: https://www.caterermiddleeast.com/ingredients/82896-worlds-first-commercial-scale-indoor-tomato-farm-to-open-in-abu-dhabi (accessed 22 February 2023).

WUR announces collaboration on crop research with Infarm - WUR (2022). Available at: https://www.wur.nl/en/research-results/research-institutes/plant-research/show-wpr/wur-announces-collaboration-on-crop-research-with-infarm.htm (accessed 21 February 2023).

Yating, Z., Yuqi, Z., Yang, Q. and Li, T. (2019). Overhead supplemental far-red light stimulates tomato growth under intra-canopy lighting with LEDs. *Journal of Integrative Agriculture* 18(1): 62–69. doi: 10.1016/S2095-3119(18)62130-6.

Zhen, S. and Bugbee, B. (2020). Far-red photons have equivalent efficiency to traditional photosynthetic photons: implications for redefining photosynthetically active radiation. *Plant, Cell and Environment* 43(5): 1259–1272. doi: 10.1111/pce.13730.

Zhen, S., van Iersel, M. and Bugbee, B. (2021). Why Far-red photons should be included in the definition of photosynthetic photons and the measurement of horticultural fixture efficacy. *Frontiers in Plant Science* 0: 1158. doi: 10.3389/FPLS.2021.693445.

Zou, J., Zhang, Y., Zhang, Y., Bian, Z., Fanourakis, D., Yang, Q. and Li, T. (2019). Morphological and physiological properties of indoor cultivated lettuce in response to additional far-red light. *Scientia Horticulturae* 257: 108725. doi: 10.1016/j.scienta.2019.108725.

# Chapter 15

## Growth and nutritional contents of medicinal plants and herbs as affected by light and root zone environments in plant factories with artificial lighting

*Na Lu, Wenshuo Xu and Duyen T. P. Nguyen, Center for Environment, Health and Field Sciences, Chiba University, Japan*



## 1 Introduction

A number of global trends and challenges, including a growing population, accelerated urbanization, rapid climate change, and a dramatic increase in the number and intensity of natural disasters, are influencing food supply structure, production systems, and security. Plant factory with artificial lighting (PFAL) systems, which enable high productivity and resource-use efficiency, are being developed to cope with the global food supply challenge. Research into food production in PFALs is attracting more attention worldwide.

Demand for healthy products derived from plants has increased globally in recent years. Many herbs that contain a large number of bioactive compounds have been reported to have positive effects on human health in protecting against diseases (Javed et al., 2012; Igarashi and Miyazaki, 2013; Ong et al., 2013). The best single way to prevent or fight disease is to boost the human immune system. To function well, the immune system requires balance and harmony, including a healthy, balanced diet based on natural ingredients. During the COVID-19 pandemic, people became more aware of the importance of healthy food and fresh ingredients to human health (Arshad et al., 2020; Wimalawansa, 2020). There is an urgent need for research to

http://dx.doi.org/10.19103/AS.2023.0126.22

develop technologies that enable rapid and efficient plant production for the production of nutraceutical vegetables, herbs and medicinal plants in PFALs.

Plant growth and secondary metabolite accumulation are affected by various environmental factors, including light, air temperature, relative humidity, carbon dioxide concentration, and the root zone environment. In a PFAL, the cost of the electrical energy required for artificial lighting can account for more than 20% of total operation costs (Kozai and Niu, 2016). Optimizing the light environment for the crop is therefore the first priority to maximize the economic output of plant production. The lighting environment, including photosynthetic photon flux density (PPFD), light spectrum, photoperiod, and the daily light integral (DLI – the total amount of light received by plants within 24 h), influences plant growth, development, and secondary metabolism (Dou et al., 2018; Lu et al., 2018; Xu et al., 2021).

In addition, among various environmental conditions, the root zone environment is another important environmental factor that affects plant growth and is relatively easy to be regulated at low cost. For example, adjusting the air temperature or light intensity, regulating fertilizer supply with different electrical conductivity (EC), and regulating the root zone temperature (RZT) are relatively simple and cheap tasks. It has been reported that the EC level of the nutrient solution (Bertram, 1991; Dewir et al., 2005; Lu et al., 2017; Ren et al., 2022a), nutrient and water availability (Loach, 1988; Puri and Thompson, 2002; López-Bucio et al., 2003; Ren et al., 2022b), and RZT (Shibuya et al., 2014; Nguyen et al., 2019; Nguyen et al., 2020b) greatly influence the growth and development of plants. RZT influences physiological processes in roots and plants, such as the uptake of water and mineral nutrients, metabolic absorption, photosynthesis, stem and leaf growth, secondary metabolite accumulation, and the activity of various enzymes (Cooper, 1973; Nxawe et al., 2010; Yan et al., 2012; Sakamoto and Suzuki, 2015; He et al., 2016; Sun et al., 2016; Al-Rawahy et al., 2018).

This chapter introduces recent research on the effects of the light and root zone environment on the growth and nutritional components of various medicinal plants and herbs grown in PFALs. Some technologies developed in the research are applicable to commercial plant production. Different regulation strategies can be selected or combined to produce plants for different applications, such as for fresh consumption, extractions, or medicinal purposes.

# 2 Perilla

## 2.1 Introduction

Perilla (*Perilla frutescens*, Labiatae) is one of the most popular medicinal plants and is used as a culinary ingredient and a medicine for healthcare and treatment of disease (Brenner, 1993; Yu, 1997). Secondary metabolites in perilla plants, such as perillaldehyde (PA) and rosmarinic acid (RA), reportedly have the potential to prevent disease, particularly due to their anti-allergic, anti-inflammatory, and antidepressant effects (Makino et al., 2003; Sanbongi et al., 2004; Zhang et al., 2008; Ito et al., 2011; Igarashi and Miyazaki, 2013). As a medicinal herb, perilla extract is used in clinical applications and is listed in pharmacopoeia in Japan, the Republic of Korea, and the People's Republic of China.

PA is a major compound in the essential oil extracted from perilla plants (Ito et al., 1999). As the unique flavour of perilla originates from PA and other monoterpenes, the concentration of PA in perilla leaves is important as a food-flavouring culinary herb. In addition, the concentration of PA is pharmaceutically important for antibacterial activity, and thus designated in the Japanese Pharmacopeia.

RA is commonly found in Labiatae family plants and possesses various types of phenolic bioactivity, including antiviral, antibacterial, anti-inflammatory, and antioxidant activities. Perilla extracts enriched with RA have beneficial anti-allergic effects (Takano et al., 2004) so the concentrations of PA and RA in perilla are important for their culinary and clinical application.

Apigenin (AG) and luteolin (LT) are the main flavonoids present in perilla and are also known to be antioxidant, anti-inflammatory, and anticarcinogenic agents. As such, they have attracted researchers' attention due to their potential as chemopreventive agents (Patel et al., 2007; Gupta et al., 2010; Liu-Smith and Meyskens, 2016; George et al., 2017). AG and LT have been candidate compounds for many pharmaceuticals and nutraceuticals from plant-derived dietary agents for the development of new therapies (Patel et al., 2007; Gupta et al., 2010; Liu-Smith and Meyskens, 2016; George et al., 2017).

Perilla plants are mainly obtained from outdoor cultivation sites or wild areas. However, the quality of plants is difficult to control, and the concentrations of the bioactive compounds may vary according to cultivation conditions, such as location, season (Natsume et al., 2006; Iwai et al., 2010), cultivar (e.g. green or red perilla) (Martinetti et al., 2012), and plant part (Chauhan et al., 2013). Production in a PFAL can ensure the supply of perilla plants with uniform quality.

## 2.2 Effects of photosynthetic photon flux density and electrical conductivity on growth and secondary metabolite accumulation of perilla

Lu et al. (2017, 2018) investigated the growth, photosynthesis, and accumulation of secondary metabolites in red and green perilla plants cultivated under nine treatments comprised of three levels of PPFD (100, 200, and 300 μmol $m^{-2}$ $s^{-1}$) combined with three EC levels of nutrient solution (1.0 dS $m^{-1}$, 2.0 dS $m^{-1}$, and 3.0 dS $m^{-1}$).

In terms of growth, there were interactive effects between PPFD and EC for fresh and dry weights of green perilla, but not for red perilla. Fresh and dry weights of shoots were affected more by EC than by PPFD in green perilla, whereas they were affected more by PPFD than by EC in red perilla (Fig. 1). Leaf net photosynthetic rates increased as PPFD increased in both perilla varieties, regardless of EC, and in general the plant's biomass was enhanced by high PPFD (Lu et al., 2017).

Looking at the accumulation of secondary metabolites, PA concentration (mg $g^{-1}$) in red perilla was unaffected by the treatments, but PA accumulation per plants (mg per plant) was enhanced as the weight increased. PA concentrations in green perilla showed significant differences between combinations of the highest PPFD + the highest EC and the lowest PPFD + the lowest EC. RA concentration (mg $g^{-1}$) increased in a combination of low EC and high PPFD conditions (Fig. 2).

The AG and LT accumulation in green and red perilla were also compared (Fig. 3). The results show that the effect of EC levels on the accumulation of AG and LT was quite different and more significant than the effects of PPFD on AG and LT concentrations. The concentration of AG was highest for an EC of 3.0 dS $m^{-1}$, whereas the LT was highest for an EC of 1.0 dS $m^{-1}$. In addition, the AG concentration was lowest under an EC of 1.0 dS $m^{-1}$, whereas the LT was lowest under an EC of 3.0 dS $m^{-1}$. These contrasting responses between AG and LT

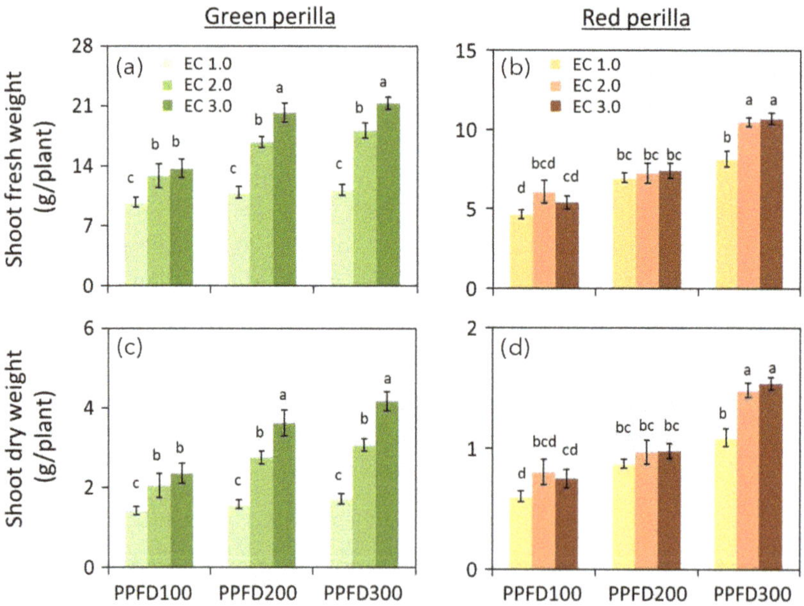


**Figure 1** Shoot fresh and dry weights per plant in green and red perilla plants after 5 weeks of cultivation under different PPFDs and EC treatments.

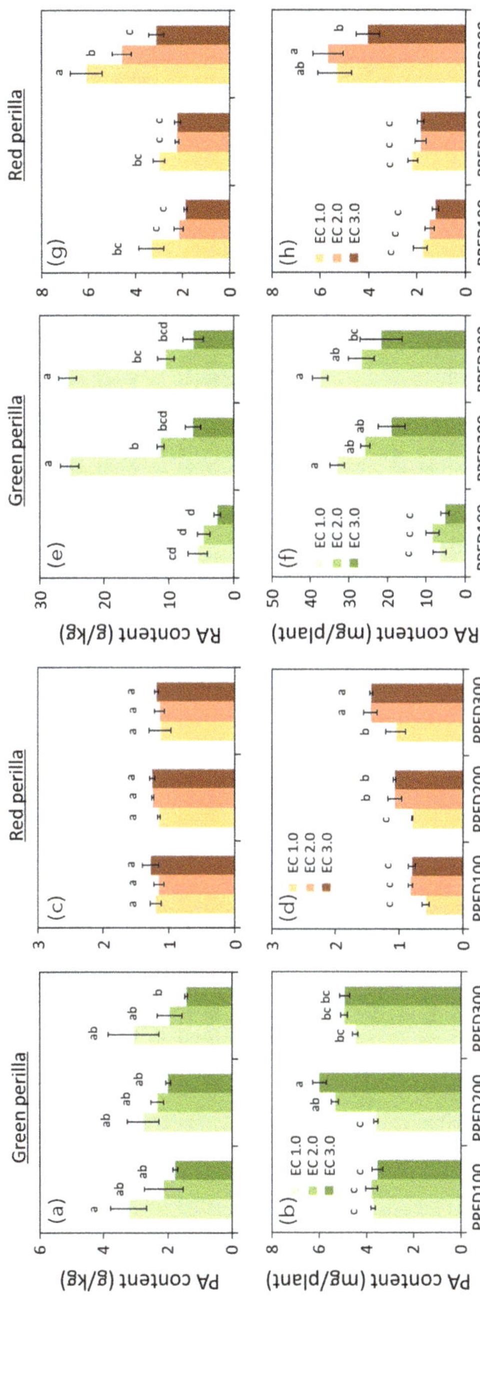


**Figure 2** Contents of PA and RA per leaf dry weight and contents of PA and RA per plant in green and red perilla plants after five weeks of cultivation under PPFD and EC treatments.

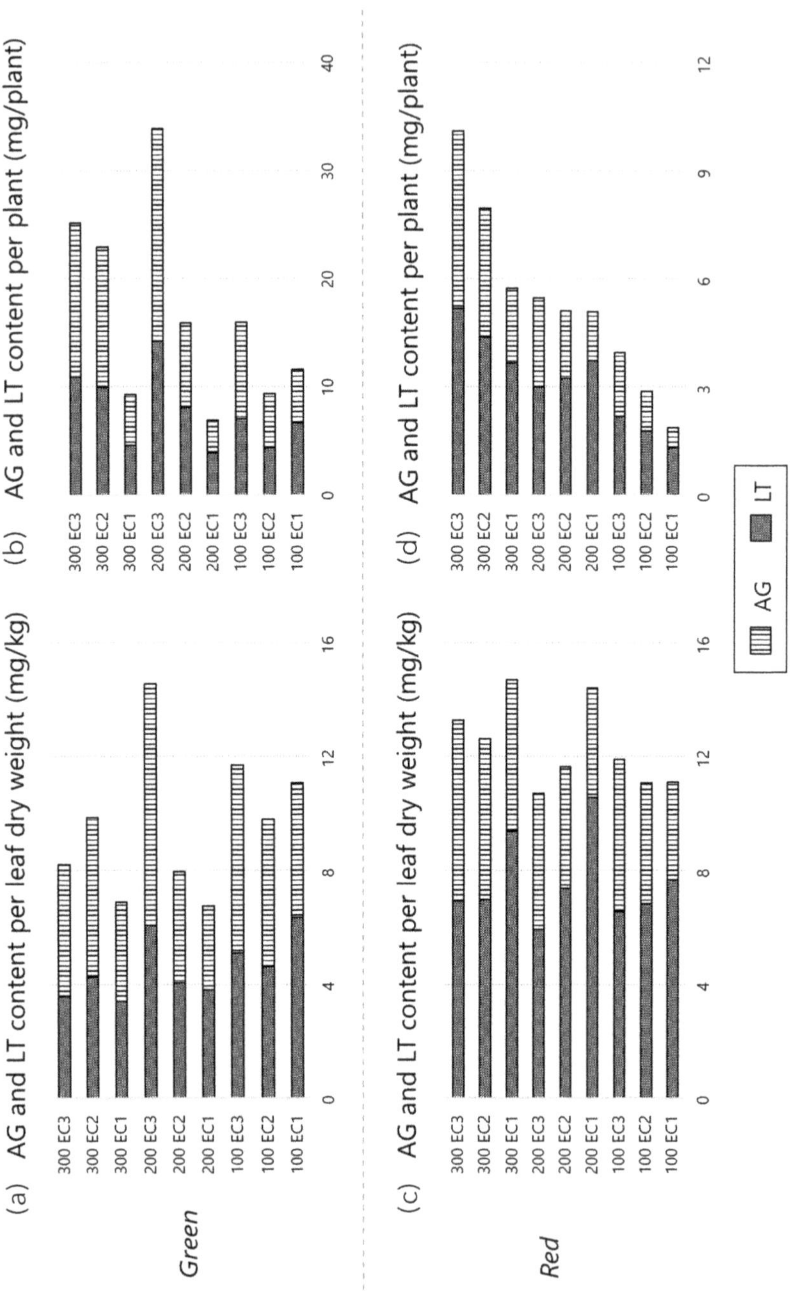


**Figure 3** Total contents of AG and LT per leaf dry weight (a, c) and total contents of AG and LT per plant (b, d) in green and red perilla plants after 5 weeks of cultivation under different PPFD and EC treatments.

could reflect their competitive relationship in the sharing of naringenin, which relates to the connection between the pathways to AG and anthocyanins.

Differences between the two varieties were also observed. In red perilla, LT accumulation was always higher than that of AG, whereas there was no such trend in green perilla. In green perilla, the highest concentration of AG was obtained under treatment of a PPFD at 200 μmol $m^{-2}$ $s^{-1}$ with an EC at 3.0 dS $m^{-1}$, whereas in red perilla, the highest concentration of AG was achieved from the highest levels of PPFD and EC (300 μmol $m^{-2}$ $s^{-1}$ and 3.0 dS $m^{-1}$). The yield of AG and LT per plant affected the biomass yield of the plants. Overall, green perillas had a higher total yield of PA, RA, AG, and LT than red perillas. These results suggest that a lower EC could enhance RA and LT concentrations in perilla plants, although the biomass of the plants is reduced. Applying a PPFD of 200 μmol $m^{-2}$ $s^{-1}$ with an EC of 3.0 dS $m^{-1}$ for green perilla, and a PPFD of 300 μmol $m^{-2}$ $s^{-1}$ with an EC of 2.0 or 3.0 dS $m^{-1}$ for red perilla are necessary to obtain a relatively sufficient quantity and quality of perilla plants in a PFAL.

## 3 Coriander

### *3.1 Introduction*

Coriander (*Coriandrum sativum* L.) is an annual herbaceous plant and its fresh leaves are widely used in food flavouring. All parts of this herb are used as traditional remedies for the treatment of different disorders in the folk medicine systems of different civilizations. Among the herbs that have the potential for efficient cultivation in plant factories, coriander has gained attention due to its various uses and low waste.

Coriander leaves are rich in vitamins and minerals, such as vitamin C, vitamin $B_2$, calcium, potassium, and zinc (Bhat et al., 2014). Coriander plants are also well known for their antioxidant properties, derived from volatile components, flavonoids, linolenic acid, β-carotene, and phenolic compounds (Kohara et al., 2006; Divya et al., 2014; Hwang et al., 2014; El-Zaeddi et al., 2017). The concentrations of these secondary metabolic compounds in herbs are important indexes for their quality assessment.

In coriander plants, chlorogenic acid (CA) and rutin (quercetin-3-rutinoside; QR) are the predominant phenolic compounds in leaves and stems (Barros et al., 2012; Divya et al., 2014). CA is well known for its strong anti-diabetic and anti-lipidemic effects (Ong et al., 2013) while QR demonstrates a number of pharmacological activities, including antioxidant, cytoprotective, vasoprotective, anticarcinogenic, neuroprotective, and cardioprotective effects (Javed et al., 2012). Another important compound is *trans*-2-decenal (DC), which is one of the major volatile compounds identified in coriander leaves and stems, and is an essential oil that produces the oily, sweet, or grassy odour of the plant (Kohara et al., 2006; Shahwar et al., 2012; Pacheco et al., 2016).

Understanding the effects of the cultivation environment on the content of CA, QR, DC, and total phenolic compounds in coriander plants is important for its production and culinary and clinical applications.

### 3.2 Effects of photosynthetic photon flux density and root zone temperature on growth and secondary metabolite accumulation of coriander

Different light intensities (PPFD of 100 μmol $m^{-2}$ $s^{-1}$, 200 μmol $m^{-2}$ $s^{-1}$, and 300 μmol $m^{-2}$ $s^{-1}$) and RZTs (20, 25, and 30°C) were applied to produce coriander plants in a plant factory. The optimal combinations were determined to gain the highest biomass and/or high concentration of secondary metabolites (Nguyen et al., 2019). To simplify the treatments and reduce operation costs, the effects of short-term RZT treatments (15–35°C for 3 or 6 days before harvest) on plant growth and nutritional components were also estimated (Nguyen et al., 2020b).

In the first study, the biomass of coriander plant was enhanced by increasing PPFD from 100 μmol $m^{-2}$ $s^{-1}$ to 300 μmol $m^{-2}$ $s^{-1}$ under all RZT conditions. Water content of both stems and leaves in coriander were decreased with increases in PPFD. Increasing PPFD might foster photosynthesis and produce carbohydrates, thus enhancing dry mass accumulation, which could cause a decrease in the water content in leaf and stem. An RZT of 25°C had the most positive influence on the biomass of coriander plants, as can be seen in Fig. 4. The poor development of the root system under 30°C RZT resulted in a small uptake of water and minerals through the roots and led to a relatively low water content in the plant. RZT also affected biomass distribution in the shoot and root of the coriander plant. The results also revealed that the growth of coriander plants was affected by interaction between RZT and PPFD and that the combination of mid-RZT (25°C) and high levels of PPFD (300 μmol $m^{-2}$ $s^{-1}$) led to the greatest plant growth (Fig. 4 and Table 1).

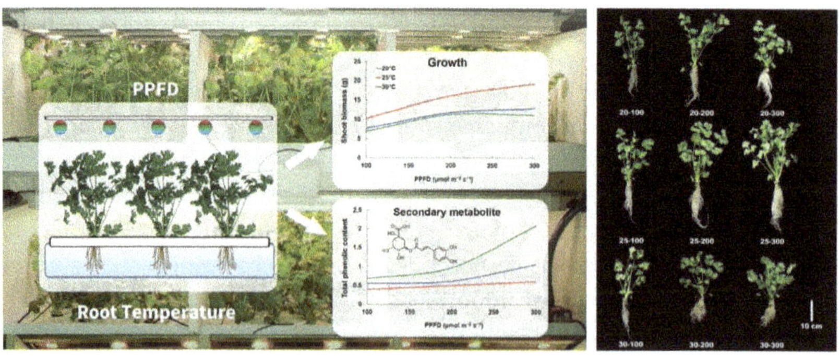


**Figure 4** Effects of PPFD and RZT on growth and secondary metabolites accumulation of coriander in a PFAL.

Phenolic compounds are the main class of natural antioxidants and predominantly secondary metabolites in plants. In this study, total phenolic content (TPC) in coriander had a positive correlation with DPPH free radical scavenging activity ($P \leq 0.01$), exhibiting coefficient $R^2 = 0.902$. The positive, liner relationship between TPC and antioxidant capacity indicates that phenolic compounds are responsible for the antioxidant activity in coriander plants, as reported in some other medicinal herbs (Cai et al., 2004).

The highest amount of TPC and DC was achieved when the highest RZT (30°C) and the highest PPFD (300 μmol $m^{-2}$ $s^{-1}$) were applied (Fig. 5). The environmental stimuli inducing oxidative stress to coriander plants may promote the production and/or accumulation of phenolic compounds in plant tissues. Furthermore, a drought-stress-like reduction was observed in leaf and stem water content under high PPFD and/or high RZT. RZT at 30°C, combined with PPFD of 300 μmol $m^{-2}$ $s^{-1}$, showed the highest DC concentration with significant differences to that of any other treatment. This suggests that coriander plants responded to the environment with a combination of the highest RZT and the highest PPFD in this experiment and had increased accumulation of DC in cells. It appears that there are limits to these levels, at less than 30°C of RZT and less than 300 μmol $m^{-2}$ $s^{-1}$ in PPFD, for coriander to not react to these as harmful conditions and to maintain the general accumulation of DC.

The contents of CA and QR in coriander were influenced by the effects of RZT, PPFD, and the interaction between RZT and PPFD (Fig. 6). In leaves, the accumulation of CA was higher than that of QR under RZT of 30°C or

**Table 1** Biomass of coriander plants after 19 days of cultivation under different PPFD and RZT

| Treatment code | Leaf FW (g) | Stem FW (g) | Leaf DW (g) | Stem DW (g) | Root DW (g) |
|---|---|---|---|---|---|
| 20-100 | 3.75 e | 3.99 d | 0.40 d | 0.24 f | 0.14 e |
| 20-200 | 5.41 cd | 5.97 c | 0.65 c | 0.39 de | 0.25 d |
| 20-300 | 6.13 bc | 6.68 c | 0.83 b | 0.50 bc | 0.48 a |
| 25-100 | 4.77 d | 5.56 c | 0.49 d | 0.30 ef | 0.13 e |
| 25-200 | 6.80 b | 9.31 b | 0.76 bc | 0.54 bc | 0.29 cd |
| 25-300 | 8.55 a | 10.7 a | 1.06 a | 0.70 a | 0.40 ab |
| 30-100 | 3.34 e | 3.59 d | 0.43 d | 0.26 f | 0.13 e |
| 30-200 | 5.44 cd | 5.82 c | 0.74 bc | 0.48 cd | 0.25 d |
| 30-300 | 5.15 d | 5.97 c | 0.84 b | 0.58 b | 0.34 bc |
| ANOVA | | | | | |
| RZT (A) | *** | *** | *** | *** | ** |
| PPFD (B) | *** | *** | *** | *** | *** |
| A × B | *** | *** | ** | ** | ** |

Different lowercase letters represent significant differences among treatments at $p < 0.05$ (Turkey's test). Asterisks show the significance of factors/interaction for: ***$P < 0.001$; **$P < 0.01$; * $P < 0.05$ (Two-way ANOVA).

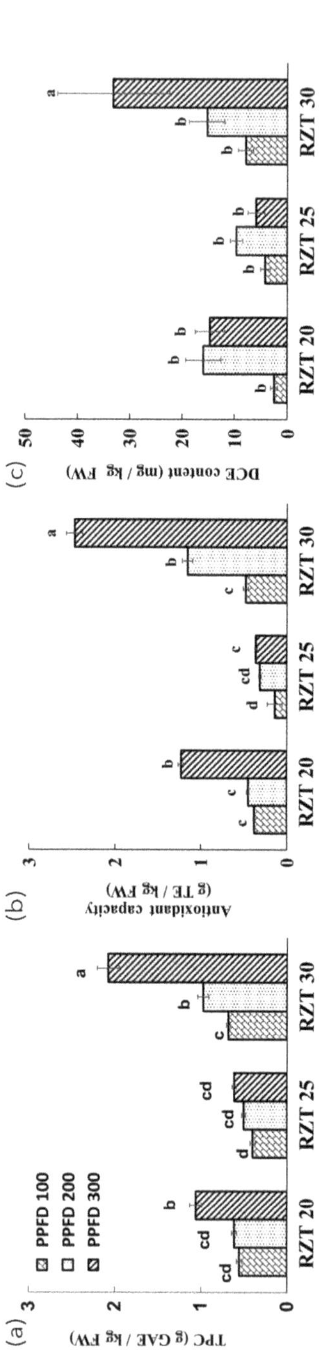


**Figure 5** TPC (a), antioxidant capacity (b), and *trans*-2-DC content (c) per shoot fresh weight in coriander plants after 19 days of cultivation under different PPFDs and RZTs.

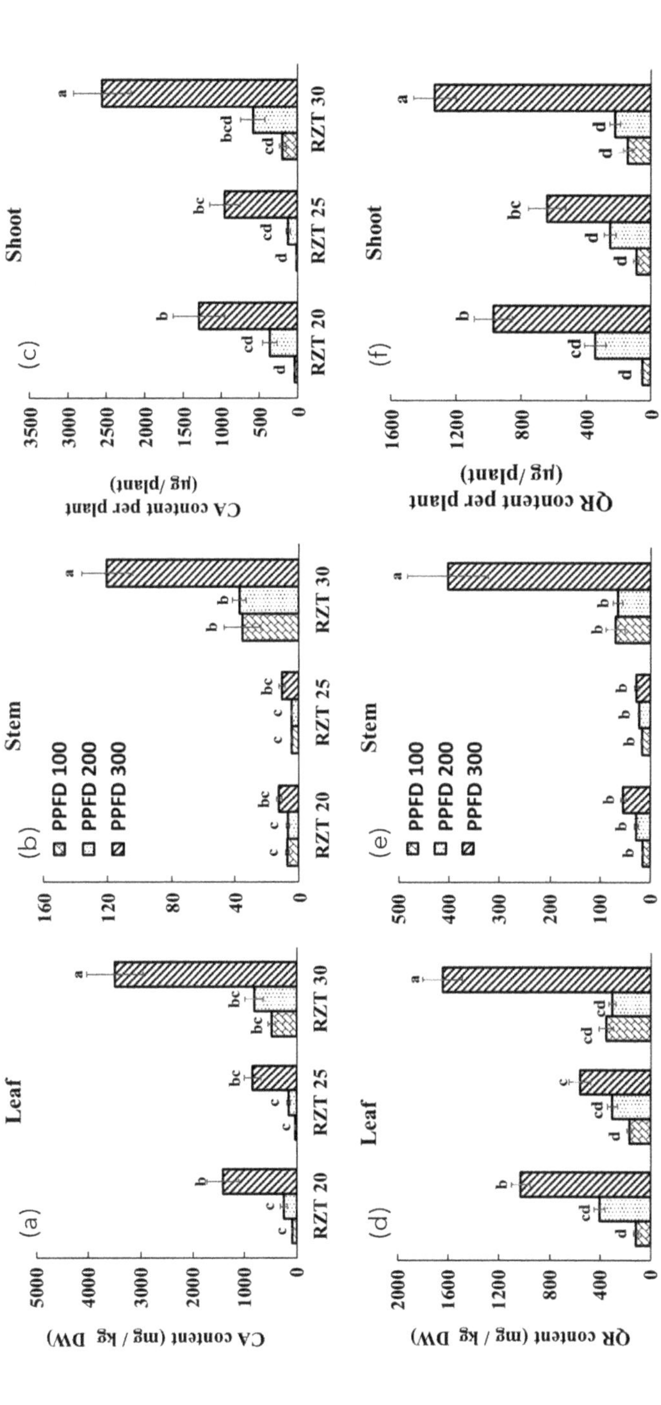


**Figure 6** CA concentration per leaf dry weight (a), CA concentration per stem dry weight (b), and CA content per plant (c) in coriander plants after 19 days of cultivation under different PPFDs and RZTs. (d), (e), and (f) are the results of rutin (QR) concentration in leaf, stem, and QR content per plant, respectively.

PPFD of 300 μmol $m^{-2}$ $s^{-1}$, and the ratio of CA and QR content differed among treatments. In addition, the content of CA in leaves showed 7- to 23-fold increases over the levels of PPFD at the same RZT, while the content of QR in leaves showed 4- to 8-fold increases. The content of CA showed large increases over the levels of the RZT at the same PPFD, while the content of QR showed relatively small increases. Therefore, the CA content in leaves differed more widely with treatments than did QR content in leaves. Statistical analysis shows that the interaction between the two factors, PPFD and RZT, made a significant difference in the content of both CA and QR in leaves, and the effect of the interaction of these two factors was the main factor in this trend. These outcomes indicate that CA metabolism in coriander leaves is more sensitive to PPFD and RZT than QR metabolism. Thus, the biosynthesis of CA in coriander leaves could be easily activated by applying such environmental stimuli.

In stems, the opposite metabolic response to the environmental changes was observed with accumulation of CA and QR. The accumulation of QR in stems was always higher than that of CA, regardless of environmental conditions. The content of QR in stems showed 2- to 6-fold increases over the levels of the PPFD at the same RZT, while the content of CA showed 2- to 3-fold increases. The content of QR in stems showed 3- to 14-fold increases over the levels of RZT at the same PPFD, while the content of CA showed 8- to 12-fold increases. The QR content in stems varied more widely than the CA content in stems. The interaction between the two factors was estimated to contribute to this trend. It seems likely that stems need to store and consume much more QR than CA. Both CA and QR content in coriander leaves were higher than those in stems. This might indicate that the enzymes involved in the biosynthetic pathways of QR and CA would be induced and expressed more sensitively with the effects of PPFD and RZT in leaf cells than in stem cells.

In the second study, the effects of short-term RZT on the growth and quality of coriander plants were investigated. The results show that plant fresh weight decreased by RZT of 30°C, 20°C, and 15°C treatments for a short stress period (3 or 6 days), compared with the control (25°C) (Fig. 7a and b). The highest CA content per plant was obtained in coriander plants grown under an RZT of 35°C for 6 days and reached 5.41 mg per plant in dry weight, higher than any of those in RZTs of 20°C, 25°C, and 30°C (Fig. 7c). The QR content in stems was raised with stressful treatments but the content per plant did not differ significantly among treatments. These results confirmed the findings in the first study that CA biosynthesis in coriander is more sensitive to environmental changes than QR synthesis. Thus, stimuli from RZT for a short duration before harvesting is an effective method to induce the biosynthesis of CA in coriander plants.

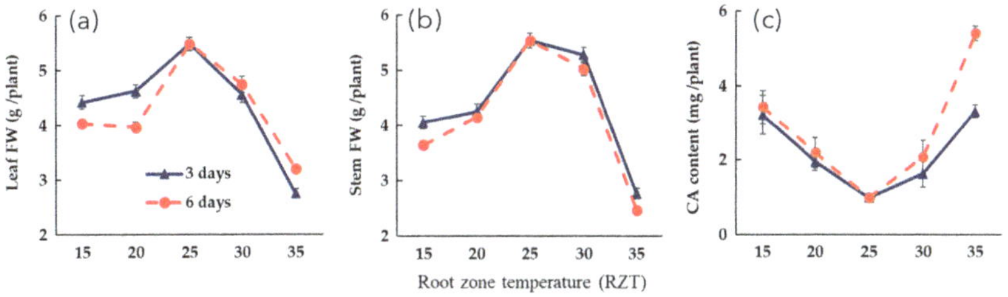


**Figure 7** Leaf fresh weight (a), stem fresh weight (b), and CA content (c) per plant in coriander under different short-term RZT treatments.

### 3.3 Effects of light spectrum on growth, mineral content and secondary metabolite accumulation of coriander

The light spectrum is an important factor that influences plant growth, development, and phytochemical accumulation (Bian et al., 2015). Red light is the most efficient light driving photosynthesis, but other wavelengths are necessary to maximize growth and maintain normal plant physiology and morphology. For example, blue light regulates stomatal opening and transpiration and is also required to prevent 'red light syndrome' through suboptimal morphology (Davis and Burns, 2016). Green light can penetrate deep into the leaf due to its high transmittance and reflectance. Far-red light regulates a variety of photomorphogenic responses, including stem extension, enlarged leaf area, and increased whole-plant net assimilation (Park and Runkle, 2017). Light spectrum also affects the contents of flavonoids, soluble sugars, free amino acids, vitamin C, total polyphenol, and chlorophyll content, minerals, and cardenolides (Verma et al., 2018; Wang et al., 2018; Zhang et al., 2018; Kitayama et al., 2019; Nguyen et al., 2020a).

Five different light qualities - red (R), blue (B), green (G), red:blue (RB, ratio of 87:13), and red:blue:far-red (RBFr, ratio of 82:12:6) - were applied to identify the optimal light quality for enhancing the growth and accumulation of minerals and bioactive metabolites in coriander. The results showed that biomass, chlorophyll index, and ascorbic acid (ASA) content of coriander plants were higher in RB or RBFr light treatments than in other light treatments. For example, the plant fresh weight and leaf area were promoted by RB and RBFr light in comparison with monochromatic light (Fig. 8).

The antioxidant capacity and TPC (per unit fresh weight) were the highest under blue light. The ASA content per plant and TPC per plant were the highest under RBFr light due to the high growth rate and biomass production of the plants (Fig. 9). In this study, RB and RBFr were found to be effective for producing coriander with high phosphorus and low potassium levels (Table 2).

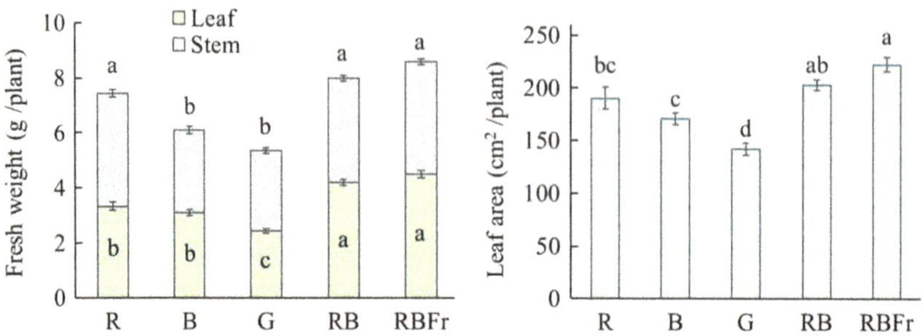


**Figure 8** Fresh weight and leaf area of coriander plant under different light spectra treatments for 18 days.

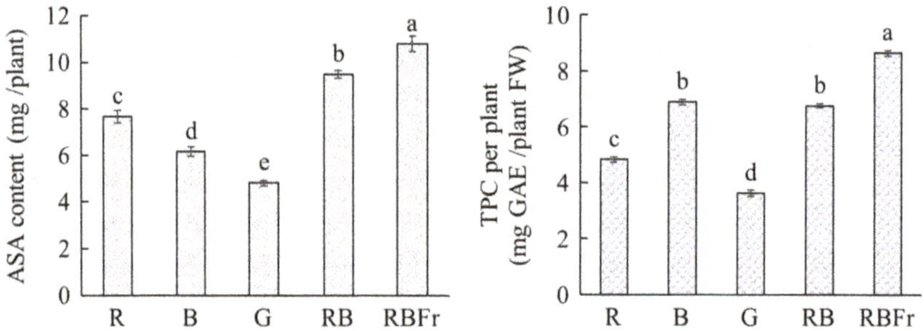


**Figure 9** ASA content and TPC per plant in coriander under different light spectra treatments for 18 days.

**Table 2** Element contents of coriander under different light spectra treatments for 18 days

| Treatment code | Element content (g/kg DW) | | | | | | | |
|---|---|---|---|---|---|---|---|---|
| | C | N | P | K | Ca | Mg | Fe | Zn |
| R | 375.2 b | 57.1 b | 12.9 b | 94.3 a | 10.4 a | 3.2 | 0.18 ab | 0.04 |
| B | 381.7 b | 60.1 a | 13.7 ab | 84.2 b | 10.8 a | 3.2 | 0.20 a | 0.06 |
| G | 382.2 b | 56.2 b | 14.5 ab | 81.3 b | 8.7 b | 2.7 | 0.16 ab | 0.05 |
| RB | 397.1 a | 58.0 ab | 16.7 a | 69.7 c | 10.4 a | 2.9 | 0.13 ab | 0.05 |
| RBFr | 394.7 a | 58.1 ab | 15.8 ab | 72.1 c | 9.8 ab | 2.9 | 0.12 b | 0.04 |

Different lowercase letters represent significant differences among treatments at $p < 0.05$ (Turkey's test).

# 4 Nasturtium

## 4.1 Introduction

Nasturtium (*Tropaeolum majus* L.) belongs to the Tropaeolaceae family and is cultivated as a landscaping, culinary, and medicinal plant worldwide. It is an indeterminate plant, which can grow and keep flowering for a long time under suitable environmental conditions. Its flowers are trumpet-shaped and yellow, orange, or red in colour, and its leaves are green and round-shaped. Given its light spicy flavour, the flowers and leaves of nasturtium are used for salads and sandwiches, and its seeds are used as seasonings (Fig. 10).

The nasturtium plant contains high levels of bioactive compounds and therefore has various therapeutic effects, such as anti-cancer, anti-tumour, anti-cardiovascular disease, anti-inflammation, anti-virus, anti-oxidation, anti-mutation, anti-allergy, and anti-radiation properties (Brondani et al., 2016; Jakubczyk et al., 2018).

Polyphenols and glucotropaeolin (GT) are the main chemical constituents isolated from nasturtium. The types of polyphenols identified from nasturtium are mainly phenolic acids and flavonoids. GT belongs to the group of aromatic glucosinolates (GS) and is the only GS found in nasturtium. The content of GT in nasturtium plants allows for the direct preparation of a mono-drug without any pre-extraction procedures, thereby reducing production costs. Nasturtium has a higher GS content than other common cruciferous vegetables. These properties have increased the commercial value of nasturtium flowers and leaves and enhanced the need to increase nasturtium production to meet rapidly increasing demand.

Most of the current research focuses on how to preserve and efficiently extract the chemical compositions in nasturtium seeds, leaves, and flowers (Chisholm and Hopkins, 1953; Garzón and Wrolstad, 2009; Schreiner et al., 2009), rather than how to produce large quantities of nasturtium plants

**Figure 10** Morphological characters of nasturtium plants grown in PFAL, nasturtium leaves and flowers, and salads with nasturtium leaves and flowers.

efficiently. Therefore, the technological development for mass production in plant factories has become particularly important.

## 4.2 Effects of electrical conductivity on rooting of nasturtium cuttings from different nodes

One of the most important aspects when designing the technology for mass production of nasturtium is to establish a rapid and efficient method of plant propagation. Broadly speaking, seed propagation and vegetative propagation are two methods of plant propagation. Many researchers use the seed propagation method to obtain nasturtium plants as samples for experiments. However, the germination of nasturtium seeds is difficult, time consuming, and uneven. Some researchers use in vitro techniques to propagate nasturtium plants, but the cost of in vitro propagation is high, and this technique is more sophisticated than seed propagation. Since nasturtium is an infinitely growing type of vine plant, the cutting propagation method could be better suited to the production of nasturtium seedlings than seed propagation. Compared with seed propagation, the time needed for the seedling stage of nasturtium may be reduced.

Cutting propagation is certainly less costly and easier to operate than in vitro propagation. The rooting process of cutting propagation is influenced by internal and external factors. Internally, the content of auxin and its distribution varies with parental parts of a plant. Externally, root zone environments, such as RZT, nutrient and water availability, and EC of the nutrient solution, also have a significant influence on the growth and development of plant roots.

An experiment was conducted to investigate the effects of node position and EC of nutrient solution on the root formation of the nasturtium cuttings. Cuttings from five node positions (apical bud, second node, third node, fourth node, and fifth node) were subjected to water and five EC (1.0, 2.0, 3.0, 4.0, and 5.0 dS $m^{-1}$) treatments with a hydroponic cultivation system in a PFAL (Xu et al., 2021a). The results show that root cuttings began to appear 7 days after planting, and the rooting rate of the cuttings under all treatments reached 100% within 2 weeks (Fig. 11). Cuttings taken from the apical bud tended to develop more roots more rapidly, and cuttings taken from lower node positions tended to have longer roots and heavier root FW and DW, particularly for lower EC levels (Fig. 12).

The rooting characteristics of cuttings grown under water and different EC levels showed an increasing trend from water to EC of 1.0 dS $m^{-1}$ and then a decreasing trend from EC of 1.0–5.0 dS $m^{-1}$ when the node positions were controlled. The increase in concentration of nutrient solution (when EC > 2.0 dS $m^{-1}$ in this case) showed an inhibitory effect on the rooting of nasturtium cuttings. This trend is believed to result from the ability of plants to absorb water from the nutrient solution, which decreases as the EC level increases. This is because high EC levels can cause a certain degree of water

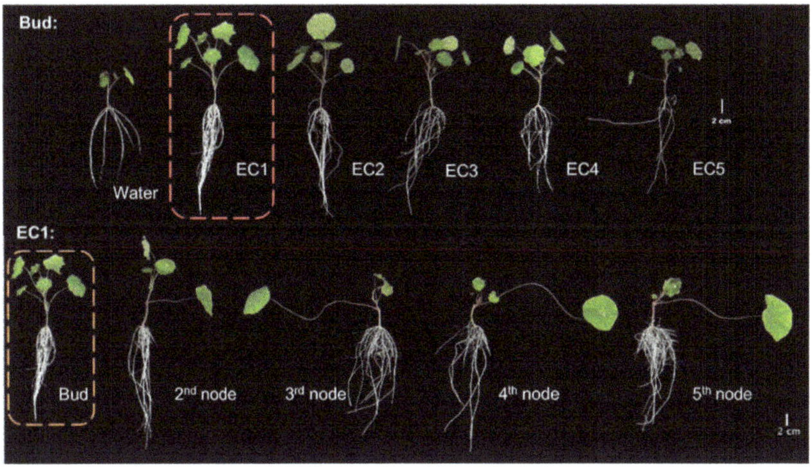


**Figure 11** Effects of different EC levels on rooting of nasturtium cuttings taken from different node positions of mother plants (photo was taken after 14 days of treatment).

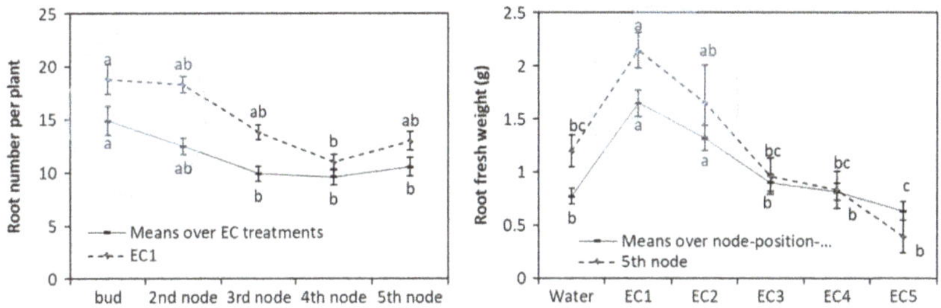


**Figure 12** Root number and root fresh weight per plant on day 14 after planting.

stress (physiological drought) on plants and water stress is not conducive to the rooting of cuttings. The rooting of cuttings grown under water and EC1 occurred earlier than that of cuttings grown under higher EC levels. Under the same EC levels, cuttings taken from the apical bud reached the rooting rate of 100% earlier and the apical bud developed the highest root numbers. This was attributed to the higher auxin concentration in the apical bud compared with other node positions (Fig. 13). No significant difference was found in the root number and the rooting rate among cuttings taken from the second to the fifth nodes, likely because the variation in the auxin allocation was not significant across the node positions from the second to the fifth nodes (Fig. 13).

This research has shown that the cutting propagation of nasturtium is rapid and easy. It is an effective and better alternative plant propagation method compared with seed propagation. Cutting propagation would be less expensive than in vitro and seed propagations and easier to operate. Given that the later growth of the seedlings propagated from cuttings is also

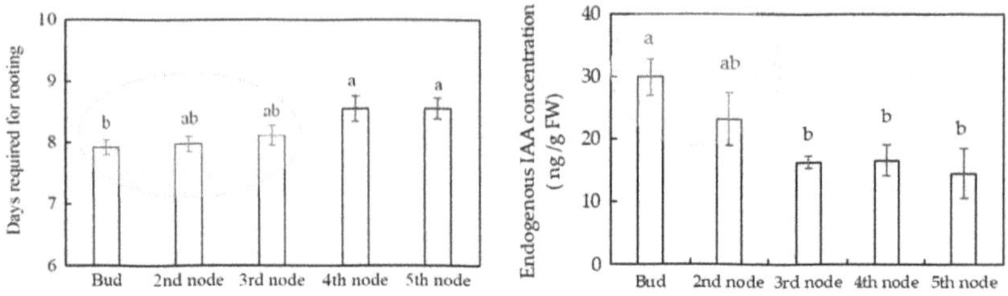


**Figure 13** Mean number of days required for rooting (the means over the EC treatments) and endogenous indole-3-acetic acid concentrations of cuttings taken from different node positions.

important for nasturtium production in PFALs, the shoot growth performance of the seedlings produced from different node positions needs to be investigated in future research. In addition, the optimal EC level could be in the region of 1.0 dS $m^{-o}$, so the performance of cuttings within the narrow EC range of 0–2.0 dS $m^{-}$ needs to be explored.

## 4.3 Effects of continuous lighting and daily light integral on nasturtium plants

As mentioned previously, lighting control is one of the most important environmental control methods in a PFAL to regulate the growth and nutritional components of plants, such as controlling light periods, light intensities, and light spectra.

DLI is the total amount of light received by plants within 24 h, which is an important index combining both light intensity and photoperiod, calculated by multiplying PPFD by lighting period. DLI is generally proportional to plant biomass within a certain range. The same DLI conditions can be generated either by a long light period with a low light intensity level or by a short light period with a high light intensity level. For the accumulation of plant biomass, extending the photoperiod with a low light intensity level is more effective than increasing light intensity with a short photoperiod under the same DLI (Sysoeva et al., 2010). Continuous lighting (CL) maximizes the light period while minimizing light intensity, at the same DLI. CL mode reduces the quantity of photons applied per unit of time, the electric power peaks required by the system, and the size of both the illumination and cooling subsystems. Therefore, the application of CL in a PFAL is considered an effective method to save the initial investment cost of lighting systems and operating costs for air conditioning and management (Kitaya et al., 1998; Ohyama et al., 2005a; Ohyama et al., 2005b). Both the negative effects of CL (such as leaf chlorosis, accelerated senescence, and development inhibition) on CL-sensitive plants, as well as the positive effects (such as breeding acceleration, increased

**Table 3** Two lighting treatments for the same level of DLI created by different combinations of light intensity and light period in Experiment 1 and three lighting treatments for different levels of DLI created by different light intensities under the condition of continuous lighting (CL) in Experiment 2

| Experiment | Treatment symbols | Light intensity (μmol $m^{-2}$ $s^{-1}$) | Light period (h $d^{-1}$) | DLI (mol $m^{-2}$ $d^{-1}$) |
|---|---|---|---|---|
| 1 | T16-300 | 300 | 16 | 17.3 |
| | T24-200 | 200 | 24 | |
| 2 | CL-200 | 200 | 24 | 17.3 |
| | CL-300 | 300 | 24 | 25.9 |
| | CL-400 | 400 | 24 | 34.6 |

productivity, and quality improvements) on CL-tolerant plants have been widely recognized (Sysoeva et al., 2010; Velez-Ramirez et al., 2011).

As a medicinal plant, nasturtium has a higher antioxidant concentration than ordinary leafy vegetables such as lettuce (Avio et al., 2017) and may have a high tolerance to abiotic stress. Therefore, by applying CL, more plant biomass and accumulation of bioactive compounds of nasturtium plants could be achieved compared with shorter lighting durations under the same DLI. A study investigated the effects of CL and different DLIs on the growth of nasturtium, its nutrient content, and the light use efficiency (LUE) in growing the plant, to find the most effective light conditions for nasturtium production in PFALs (Xu et al., 2021b). Two experiments were conducted on the seedlings of nasturtium plants. In the first, a 16 h light period under a PPFD of 300 μmol $m^{-2}$ $s^{-1}$ and CL (24 h light period) under the PPFD of 200 μmol $m^{-2}$ $s^{-1}$ (hereafter: T16-300 and T24-200, respectively), at the same DLI level of 17.3 mol $m^{-2}$ $d^{-1}$, were applied for 3 weeks. In the second, the seedlings were grown under CL with three different DLI treatments: 17.3, 25.9, and 34.6 mol $m^{-2}$ $d^{-1}$ for 3 weeks. These three DLI treatments were created at three different PPFD levels: 200, 300, and 400 μmol $m^{-2}$ $s^{-1}$ (denoted hereafter as CL-200, CL-300, and CL-400, respectively) (Table 3).

The growth rate of the above-ground biomass increased almost linearly during the cultivation period (Fig. 14). More importantly, the leaf and shoot dry weights were both higher for T24-200 than for T16-300, with all the mean differences that were statistically significant. For these two growth parameters, the rate of difference between T16-300 and T24-200 decreased substantially from 30% to 40% at 7 days after treatment (DAT) to 18–20% at 21 DAT. Therefore, under the same DLI (17.28 mol $m^{-2}$ $d^{-1}$), the dry mass of nasturtium plant was enhanced by extending the light period from 16 h to 24 h, rather than increasing light intensity from 200 μmol $m^{-2}$ $s^{-1}$ to 300 μmol $m^{-2}$ $s^{-1}$. The leaf number and leaf area were also higher for T24-200 than for T16-300 during the earlier growth stages of the first and the second week, whereas there were no significant differences between the treatments at the end of the third week (Fig. 14).

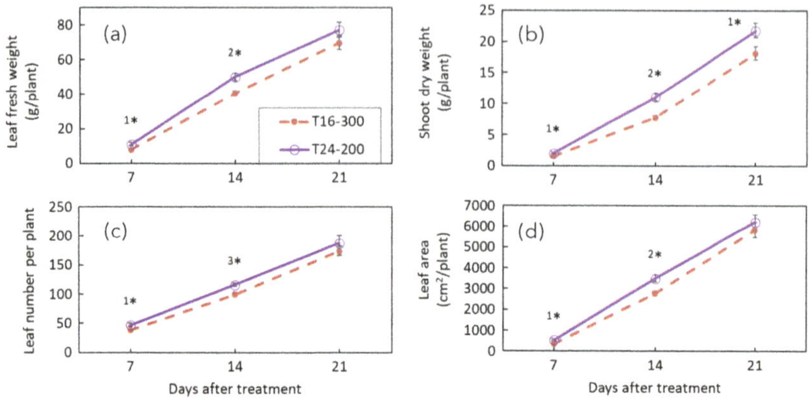


**Figure 14** Leaf fresh weight (a), shoot dry weight (b), leaf numbers (c), and leaf area (d) per nasturtium plant grown under T16-300 and T24-200 at 7, 14, and 21 days after treatment. Asterisks indicate significant differences between treatments ($^{1*}$, $P < 0.05$; $^{2*}$, $P < 0.01$; $^{3*}$, $P < 0.001$), determined by the *t*-test.

**Table 4** $F_v/F_m$ and secondary metabolites of nasturtium leaves under T16-300 and T24-200 treatments

| Parameter | Unit | DAT | Treatment | | Significance |
|---|---|---|---|---|---|
| | | | T16-300 | T24-200 | |
| *Fv/Fmax* | (ratio) | 21 | 0.84 | 0.82 | |
| Antioxidant capacity | (g TE/kg leaf FW) | 14 | 1.7 | 2.3 | * |
| | | 21 | 1.4 | 2.0 | * |
| Total phenolic content | (g GAE/kg leaf FW) | 14 | 2.0 | 2.3 | * |
| | | 21 | 2.1 | 2.4 | |

*Represents a significant difference between the two treatments (p < 0.05, t-test).

This study found that nasturtium grown under CL (T24-200) attained a heavier plant biomass and faster RGR compared to T16-300 without causing any physiological leaf injury ($F_v/F_{max} > 0.8$; Table 4). For each light-period treatment, the antioxidant capacity of nasturtium decreased from the second to the third growth stage, and for each growth stage, it was higher under CL than under T16 (Table 4). The TPC showed a similar trend and, especially at the second week after treatment, TPC was higher under CL than under T16 (Table 4).

The LUE of nasturtium cultivation was higher under T24 than under T16 at 7, 14, and 21 DAT (Fig. 15).

In the second experiment, plants grew better under higher DLI than under lower ones. After 3 weeks of treatment, the DLI of 34.6 mol $m^{-2}$ $d^{-1}$ enhanced the leaf and shoot biomasses and leaf area, except for the leaf number, compared to those under lower DLIs (Fig. 16). This study also found that, at the end of the third growth stage, there was a highly linear relationship between DLI and

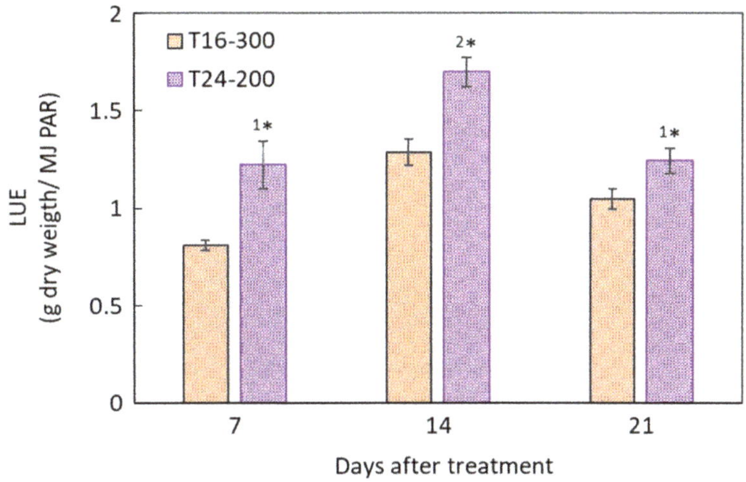


**Figure 15** LUEs of nasturtium grown under T16-300 and T24-200, measured at 7, 14, and 21 days after treatment. Asterisks indicate significant difference between treatments (1*, $P < 0.05$; 2*, $P < 0.01$), determined by the *t*-test.

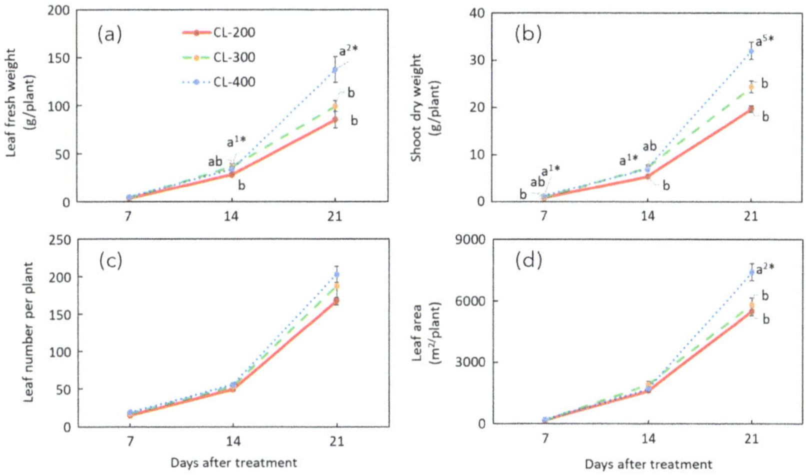


**Figure 16** Leaf fresh weight (a), shoot dry weight (b), leaf numbers (c), and leaf area (d) of nasturtium grown under CL-200, CL-300, and CL-400 at 7, 14, and 21 days after treatment. Different alphabet letters indicate significant differences between the treatments, the signs, 1*, 2*, 3*, 4*, or 5* indicating that the probability of the ANOVA (the multiple-mean comparison by Tukey) was $P < 0.05$, $P < 0.01$, $P < 0.001$, $P < 1.0E\text{-}04$, or 1.0E-05, respectively.

the biomass of nasturtium plants. For both shoot and root, the DW increased linearly as the DLI increased. However, in contrast to the biomass, the LUE of nasturtium cultivation was higher under CL-200 than under CL-300 and CL-400 at 21 DAT (Fig. 17).

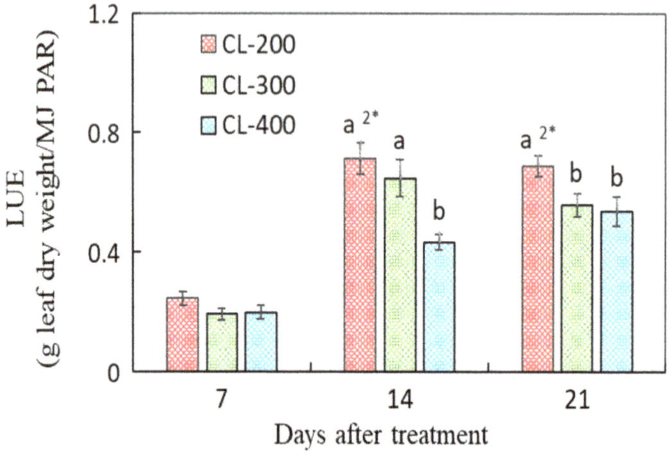


**Figure 17** LUEs of nasturtium grown under CL-200, CL-300, and CL-400 at 7, 14, and 21 days after treatment.

In the first experiment, the values of $F_v/F_{max}$ of nasturtium grown under T24-200 as well as under T16-300 were higher than 0.8 at 21 DAT, which indicated that nasturtium plants were not subjected to light stress under CL conditions. A reversible photoinhibition caused by light stress was observed in the second experiment (Fig. 18). At 6 DAT, the values of $F_v/F_{max}$ of nasturtium plants under all treatments were lower than 0.8, indicating that plants were suffering from light stress. At this early growth stage, $F_v/F_{max}$ and PhiPS2 decreases as DLI increases, suggesting a greater degree of photoinhibition for higher DLIs. The decline in $F_v/F_{max}$ is indicative of photodamaged PSII and due to a decrease in the proportion of open PSII reaction centres (qP). In addition, the decrease in PhiPS2 could be attributed to both the decrease in the proportion of open PSII reaction centres (lower qP) and the increase in light energy dissipated in the form of heat (larger qN). At 20 DAT, an elevation of $F_v/F_{max}$ to over 0.8 for all DLI treatments and diminished the among-treatment differences for $F_v/F_{max}$ were observed, which implies that the photoinhibition was recovered at 20 DAT. The elevation of $F_v/F_{max}$ could be explained by the increase in qP, which allowed more absorbed energy to be used for photochemistry, rather than dissipated in the form of heat (qN). As a result, the quantum yield of the PSII electron transport (PhiPS2) improved. The diminished difference in $F_v/F_{max}$ among treatments could be explained by the larger differences in qP and qN. The levels of qP under CL-300 and CL-400 were lower than those under CL-200, whereas it was the other way around for the qN levels (Fig. 18). These results indicate that under high DLIs, the photosynthetic apparatus was protected from light stress by the greater ability of nasturtium plants to dissipate excess excitation energy as heat. In addition, in the second experiment, the levels of

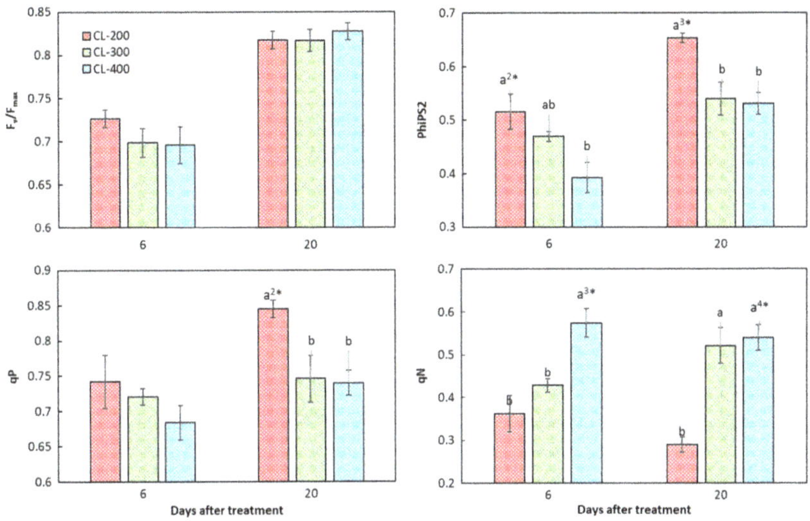


**Figure 18** The maximum quantum yield of the PSII primary photochemistry ($F_v/F_{max}$) (a), quantum yield of the PSII electron transport (PhiPS2) (b), the coefficient of photochemical quenching (qP) (c), and non-photochemical quenching (qN) (d) of nasturtium grown under CL-200, CL-300, and CL-400 at 6 and 20 days after treatment.

antioxidant capacity and TPC were higher at the earlier growth stage (week 1) than at the later growth stages (weeks 2 and 3), whereas there were no significant differences among the different DLI treatments at the same growth stage (see detailed data in Xu et al., 2021b). The former observation could be understood as the young seedlings at the first growth stage need time to acclimate to CL. The latter observation suggests that CL under these levels of DLI does not create extra harmful stress on the growth of nasturtium plants.

Through these two experiments, it is clear that CL, compared with shorter light periods, could improve the quality of nasturtium while not causing photosynthetic damage under a certain range of DLI. CL is a feasible and beneficial technology for high-quality mass production of nasturtium in PFALs, and an increase in DLI under CL increases nasturtium biomass linearly up to the light intensity of 400 μmol $m^{-2}$ $s^{-1}$. The effects of DLI higher than this level on nasturtium plants requires further study.

## 5 Conclusion and future trends

In summary, the regulation of light and root zone environments can effectively control growth and development of medicinal plants grown in PFALs. PPFD, light spectrum, EC level, and RZT can be adjusted to increase biomass and accumulation of secondary metabolites. These factors can be adjusted to change the absorption of minerals and promote rooting. Short-term pre-harvest treatment of the root zone environment can improve plant quality at low cost. It was also found that different plant varieties respond differently to levels of regulation. Therefore, figuring out the mechanism of different environmental factors on plant growth and development will help to achieve plant production at low cost and high productivity in PFALs. Moreover, plants are living, total-life systems that require totality or whole-system thinking when grown in PFALs. All environmental factors, species of plant, and their interdependence should be considered, to understand how they are related and how they influence one another within a whole system. The research introduced in this chapter only revealed a very few of these cause-and-effect relationships.

Global climate change has become more intense and frequent. Celestial movements and solar activity are likely to increase the frequency of volcanic and seismic eruptions, causing unstable crop production in various countries and regions. In addition, these changes may result in a large-scale reduction in food production for a long period of time. Volcanic ash from eruptions of large volcanos (e.g. Tonga in 2022) can also block sunlight and cause insufficient light and a drop in temperature, which will directly influence plant growth and food production. Tsunamis and increasing rainfall will make food production unstable in some regions. The impact of these natural changes on agricultural production should not be taken lightly or ignored. More research on stable and controllable production of vegetables and medicinal plants is becoming more and more important. Current PFAL technologies can ensure stable and predictable plant production but there remains a need to improve efficiency and reduce costs. Technologies for the commercial production of common medicinal plants and herbs, as well as for the reproduction of rare and high-value medicinal plants (especially those threatened by climate change and facing extinction) are urgently needed. Therefore, in the future, plant factory technology should be used to determine the key environmental factors required for the growth of scarce species for species conservation.

## 6 References

Al-Rawahy, M. S., Al-Rawahy, S. A., Al-Mulla, Y. A. and Nadaf, S. K. 2018. Effect of cooling root-zone temperature on growth, yield and nutrient uptake in cucumber grown in hydroponic system during summer season in cooled greenhouse. *J. Agric. Sci.* 11(1), 47.

Arshad, M. S., Khan, U., Sadiq, A., Khalid, W., Hussain, M., Yasmeen, A., Asghar, Z. and Rehana, H. 2020. Coronavirus disease (COVID-19) and immunity booster green foods: A mini review. *Food Sci. Nutr.* 8(8), 3971–3976.

Avio, L., Sbrana, C., Giovannetti, M. and Frassinetti, S. 2017. Arbuscular mycorrhizal fungi affect total phenolics content and antioxidant activity in leaves of oak leaf lettuce varieties. *Sci. Hortic. (Amsterdam)* 224, 265–271.

Barros, L., Dueñas, M., Dias, M. I., Sousa, M. J., Santos-Buelga, C. and Ferreira, I. C. F. R. 2012. Phenolic profiles of in vivo and in vitro grown Coriandrum sativum L. *Food Chem.* 132(2), 841-848.

Bertram, L. 1991. Vegetative propagation of Hibiscus rosa-sinensis L. in relation to nutrient concentration of the propagation medium. *Sci. Hortic. (Amsterdam)* 48(1-2), 131-139.

Bhat, S., Kaushal, P., Kaur, M. and Sharma, H. K. 2014. Coriander (Coriandrum sativum L.): processing, nutritional and functional aspects. *Afr. J. Plant Sci.* 8(1), 25-33.

Bian, Z. H., Yang, Q. C. and Liu, W. K. 2015. Effects of light quality on the accumulation of phytochemicals in vegetables produced in controlled environments: a review. *J. Sci. Food Agric.* 95(5), 869-877.

Brenner, D. M. 1993. Perilla: botany, uses and genetic resources. In: *New Crops*, eds. J. Janick and J. E. Simon; New York, NY: Wiley, 322-328.

Brondani, C., Cuelho, F., Helena, C., Marangoni, D., Guex, G., Bonilha, D. F. and Manfron, P. 2016. Traditional usages, botany, phytochemistry, biological activity and toxicology of Tropaeolum majus L. - A review. Boletín Latinoamericano y del Caribe de Plantas Medicinales y Aromáticas 15(4), 264-273. Universidad de Santiago de Chile Santiago, Chile.

Cai, Y., Luo, Q., Sun, M. and Corke, H. 2004. Antioxidant activity and phenolic compounds of 112 traditional Chinese medicinal plants associated with anticancer. *Life Sci.* 74(17), 2157-2184.

Chauhan, N. K., Singh, S., Zafar Haider, S., Lohani, H. and Kushwaha, B. L. 2013. Compositional variability in volatiles from different plant organs of Perilla frutescens L. cultivated in Uttarakhand (India). *J. Pharm. Res.* 6(3), 361-363.

Chisholm, M. J. and Hopkins, C. Y. 1953. Fatty acids of filbert oil and nasturtium seed oil. *Can. J. Chem.* 31(11), 1131-1137.

Cooper, A. J. 1973. Root temperature and plant growth: a review. In: *Research Review* 4, Commonwealth Bureau of Horticulture and Plantation Crops; Commonwealth Agriculture Bureaux: East Malling, Kent, UK.

Davis, P. A. and Burns, C. 2016. Photobiology in protected horticulture. *Food Energy Secur.* 5(4), 223-238.

Dewir, Y. H., Chakrabarty, D., Ali, M. B., Hahn, E. J. and Paek, K. Y. 2005. Effects of hydroponic solution EC, substrates, PPF and nutrient scheduling on growth and photosynthetic competence during acclimatization of micropropagated Spathiphyllum plantlets. *Plant Growth Regul.* 46(3), 241-251.

Divya, P., Puthusseri, B. and Neelwarne, B. 2014. The effect of plant regulators on the concentration of carotenoids and phenolic compounds in foliage of coriander. *LWT Food Sci. Technol.* 56(1), 101-110.

Dou, H., Niu, G., Gu, M. and Masabni, J. G. 2018. Responses of sweet basil to different daily light integrals in photosynthesis, morphology, yield, and nutritional quality. *Hort.Sci.* 53(4), 496-503.

El-Zaeddi, H., Calín-Sánchez, Á., Nowicka, P., Martínez-Tomé, J., Noguera-Artiaga, L., Burló, F., Wojdyło, A. and Carbonell-Barrachina, Á. A. 2017. Preharvest treatments with malic, oxalic, and acetylsalicylic acids affect the phenolic composition and antioxidant capacity of coriander, dill and parsley. *Food Chem.* 226, 179-186.

Garzón, G. A. and Wrolstad, R. E. 2009. Major anthocyanins and antioxidant activity of Nasturtium flowers (*Tropaeolum majus*). *Food Chem.* 114(1), 44-49.

George, V. C., Dellaire, G. and Rupasinghe, H. P. V. 2017. Plant flavonoids in cancer chemoprevention: role in genome stability. *J. Nutr. Biochem.* 45, 1–14.

Gupta, S. C., Kim, J. H., Prasad, S. and Aggarwal, B. B. 2010. Regulation of survival, proliferation, invasion, angiogenesis, and metastasis of tumor cells through modulation of inflammatory pathways by nutraceuticals. *Cancer Metastasis Rev.* 29(3), 405–434.

Hwang, E., Lee, D. G., Park, S. H., Oh, M. S. and Kim, S. Y. 2014. Coriander leaf extract exerts antioxidant activity and protects against UVB-induced photoaging of skin by regulation of procollagen type I and MMP-1 expression 1. *J. Med. Food* 17(9), 985–995.

Igarashi, M. and Miyazaki, Y. 2013. A review on bioactivities of perilla: progress in research on the functions of perilla as medicine and food. *Evid. Based Complement. Altern. Med.* 2013, 1–7.

Ito, M., Toyoda, M. and Honda, G. 1999. Chemical composition of the essential oil of Perilla frutescens. *Nat. Med.* 53, 32–36.

Ito, N., Nagai, T., Oikawa, T., Yamada, H. and Hanawa, T. 2011. Antidepressant-like effect of l-perillaldehyde in stress-induced depression-like model mice through regulation of the olfactory nervous system. *Evid. Based Complement. Alternat. Med.* 2011, 512697.

Iwai, M., Ohta, M., Tsuchiya, H. and Suzuki, T. 2010. Enhanced accumulation of caffeic acid, rosmarinic acid and luteolin-glucoside in red perilla cultivated under red diode laser and blue LED illumination followed by UV-A irradiation. *J. Funct. Foods* 2(1), 66–70.

Jakubczyk, K., Janda, K., Watychowicz, K., Łukasiak, J. and Wolska, J. 2018. Garden nasturtium (*Tropaeolum majus* L.) - a source of mineral elements and bioactive compounds. *Rocz. Panstw. Zakl. Hig.* 69(2), 119–126.

Javed, H., Khan, M. M., Ahmad, A., Vaibhav, K., Ahmad, M. E., Khan, A., Ashafaq, M., Islam, F., Siddiqui, M. S., Safhi, M. M. and Islam, F. 2012. Rutin prevents cognitive impairments by ameliorating oxidative stress and neuroinflammation in rat model of sporadic dementia of alzheimer type. *Neuroscience* 210, 340–352.

Kitaya, Y., Niu, G., Kozai, T. and Ohashi, M. 1998. Photosynthetic photon flux, photoperiod, and $CO_2$ concentration affect growth and morphology of lettuce plug transplants. *Hortscience* 33(6), 988–991.

Kitayama, M., Nguyen, D. T. P., Lu, N. and Takagaki, M. 2019. Effect of light quality on physiological disorder, growth, and secondary metabolite content of water spinach (*Ipomoea aquatica* forsk) cultivated in a closed-type plant production system. *Hortic. Sci. Technol.* 37, 206–218.

Kohara, K., Sakamoto, Y., Hasegawa, H., Kozuka, H., Sakamoto, K. and Hayata, Y. 2006. Fluctuations in volatile compounds in leaves, stems, and fruits of growing coriander (*Coriandrum sativum* L.) plants. *J. Jpn. Soc. Hortic. Sci.* 75(3), 267–269.

Kozai, T. and Niu, G. 2016. Role of the plant factory with artificial lighting (PFAL) in urban areas. In: Kozai, T., Niu, G. and Takagaki, M. (Eds). *Plant Factory–an Indoor Vertical Farming System for Efficient Quality Food Production*; San Diego, CA: Academic Press, 7–33.

Liu-Smith, F. and Meyskens, F. L. 2016. Molecular mechanisms of flavonoids in melanin synthesis and the potential for the prevention and treatment of melanoma. *Mol. Nutr. Food Res.* 60(6), 1264–1274.

Loach, K. 1988. Water relations and adventitious rooting. In: *Adventitious Root Ormation in Cuttings*, eds. T. D. Davis, B. E. Haissing, and N. Sankhla. Portland, OR: Dioscorides Press, 102–116.

López-Bucio, J., Cruz-Ramírez, A. and Herrera-Estrella, L. 2003. The role of nutrient availability in regulating root architecture. *Curr. Opin. Plant Biol.* 6(3), 280–287.

Lu, N., Bernardo, E. L., Tippayadarapanich, C., Takagaki, M., Kagawa, N. and Yamori, W. 2017. Growth and accumulation of secondary metabolites in perilla as affected by photosynthetic photon flux density and electrical conductivity of the nutrient solution. *Front. Plant Sci.* 8, 1–12.

Lu, N., Takagaki, M., Yamori, W. and Kagawa, N. 2018. Flavonoid productivity optimized for green and red forms of Perilla frutescens via environmental control technologies in plant factory. *J. Food Qual.* 2018, 1–9.

Makino, T., Ono, T., Matsuyama, K., Nogaki, F., Miyawaki, S., Honda, G. and Muso, E. 2003. Suppressive effects of Perilla frutescents on IgA nephropathy in HIGA mice. *Nephrol. Dial. Transplant.* 18(3), 484–490.

Martinetti, L., Ferrante, A., Bassoli, A., Borgonovo, G., Tosca, A. and Spoleto, P. 2012. Characterization of some qualitative traits in different perilla cultivars. *Acta Hortic.* (939), 301–308.

Natsume, M., Muto, Y., Fukuda, K., Tokunaga, T. and Osakabe, N. 2006. Determination of rosmarinic acid and luteolin in Perilla frutescens Britton (*Labiatae*) 901, 897–901.

Nguyen, D. T. P., Kitayama, M., Lu, N. and Takagaki, M. 2020a. Improving secondary metabolite accumulation, mineral content, and growth of coriander (*Coriandrum sativum* L.) by regulating light quality in a plant factory. *J. Hortic. Sci. Biotechnol.* 95(3), 356–363.

Nguyen, D. T. P., Lu, N., Kagawa, N., Kitayama, M. and Takagaki, M. 2020b. Short-term root-zone temperature treatment enhanced the accumulation of secondary metabolites of hydroponic coriander (*Coriandrum sativum* L.) grown in a plant factory. *Agronomy* 10(3), 413.

Nguyen, D. T. P., Lu, N., Kagawa, N. and Takagaki, M. 2019. Optimization of photosynthetic photon flux density and root-zone temperature for enhancing secondary metabolite accumulation and production of coriander in plant factory. *Agronomy* 9(5), 224.

Nxawe, S., Ndakidemi, P. A. and Laubscher, C. P. 2010. Possible effects of regulating hydroponic water temperature on plant growth, accumulation of nutrients and other metabolites. *Afr. J. Biotechnol.* 9, 9128–9134.

Ohyama, K., Manabe, K., Omura, Y., Kozai, T. and Kubota, C. 2005a. Potential use of a 24-hour photoperiod (continuous light ) with alternating air temperature for production of tomato plug transplants in a closed system. *HortSci.* 40(2), 374–377.

Ohyama, K., Omura, Y. and Kozai, T. 2005b. Effects of air temperature regimes on physiological disorders and floral development of tomato seedlings grown under continuous light. *HortSci.* 40(5), 1304–1306.

Ong, K. W., Hsu, A., Kwong, B. K. and Tan, H. 2013. Anti-diabetic and anti-lipidemic effects of chlorogenic acid are mediated by AMPK activation. *Biochem. Pharmacol.* 85(9), 1341–1351.

Pacheco, L., N. A., Cano-Sosa, J., Poblano, C., F., Rodríguez-Buenfil, I. M. and Ramos-Díaz, A. 2016. Different responses of the quality parameters of Coriandrum sativum to organic substrate mixtures and fertilization. *Agronomy* 6, 1–10.

Park, Y. and Runkle, E. S. 2017. Far-red radiation promotes growth of seedlings by increasing leaf expansion and whole-plant net assimilation. *Environ. Exp. Bot.* 136, 41–49.

Patel, D., Shukla, S. and Gupta, S. 2007. Apigenin and cancer chemoprevention: progress, potential and promise (review) (Review). *Int. J. Oncol.* 30(1), 233-245.

Puri, S. and Thompson, F. B. 2002. *Relationship of Water to Adventitious Rooting in Stem Cuttings of Populus species Agroforestry Systems* 58, 1-9.

Ren, X., Lu, N., Xu, W., Zhuang, Y. and Takagaki, M. 2022a. Optimization of the yield, total phenolic content, and antioxidant capacity of basil by controlling the electrical conductivity of the nutrient solution. *Horticulturae* 8(3), 216.

Ren, X., Lu, N., Xu, W., Zhuang, Y., Tsukagoshi, S. and Takagaki, M. 2022b. Growth and nutrient utilization in basil plant as affected by applied nutrient quantity in nutrient solution and light spectrum. *Biology* 11(7), 991.

Sakamoto, M. and Suzuki, T. 2015. Effect of root-zone temperature on growth and quality of hydroponically grown red leaf lettuce (Lactuca sativa L. cv. R. Wave). *Am. J. Plant Sci.* 06(14), 2350-2360.

Sanbongi, C., Takano, H., Osakabe, N., Sasa, N., Natsume, M., Yanagisawa, R., Inoue, K. I., Sadakane, K., Ichinose, T. and Yoshikawa, T. 2004. Rosmarinic acid in perilla extract inhibits allergic inflammation induced by mite allergen, in a mouse model. *Clin. Exp. Allergy* 34(6), 971-977.

Schreiner, M., Krumbein, A., Mewis, I., Ulrichs, C. and Huyskens-Keil, S. 2009. Short-term and moderate UV-B radiation effects on secondary plant metabolism in different organs of nasturtium (Tropaeolum majus L.). *Innov. Food Sci. Emerg. Technol.* 10(1), 93-96.

Shahwar, M. K., El-ghorab, A. H., Anjum, M., Butt, M. S., Hussain, S., and Nadeem, M. 2012. Characterization of coriander (Coriandrum sativum L.) seeds and leaves: volatile and non volatile extracts. *Int. J. Food Prop.* 15, 736-747.

Shibuya, T., Taniguchi, T., Tsukuda, S., Shiozaki, S. and Itagaki, K. 2014. Adventitious root formation of Japanese cedar (Cryptomeria japonica D. Don) cuttings is stimulated by soaking basal portion of cuttings in warmed water while cooling their apical portion. *New Forests* 45(4), 589-602.

Sun, J., Lu, N., Xu, H., Maruo, T. and Guo, S. 2016. Root zone cooling and exogenous spermidine root-pretreatment promoting Lactuca sativa L. growth and photosynthesis in the high-temperature season. *Front. Plant Sci.* 7, 368.

Sysoeva, M., Markovskaya, E. and Shibaeva, T. 2010. Plants under continuous light: a review. *Plant Stress* 4, 5-17.

Takano, H., Osakabe, N., Sanbongi, C., Yanagisawa, R., Inoue, K. I., Yasuda, A., Natsume, M., Baba, S., Ichiishi, E. I. and Yoshikawa, T. 2004. Extract of Perilla frutescens enriched for rosmarinic acid, a polyphenolic phytochemical, inhibits seasonal allergic rhinoconjunctivitis in humans. *Exp. Biol. Med. (Maywood)* 229(3), 247-254.

Velez-Ramirez, A. I., Van Ieperen, W., Vreugdenhil, D. and Millenaar, F. F. 2011. Plants under continuous light. *Trends Plant Sci.* 16(6), 310-318.

Verma, S. K., Gantait, S., Jeong, B. R. and Hwang, S. J. 2018. Enhanced growth and cardenolides production in Digitalis purpurea under the influence of different LED exposures in the plant factory. *Sci. Rep.* 8(1), 18009.

Wang, W., Su, M., Li, H., Zeng, B., Chang, Q. and Lai, Z. 2018. Effects of supplemental lighting with different light qualities on growth and secondary metabolite content of Anoectochilus roxburghii. *PeerJ* 6, e5274.

Wimalawansa, S. J. 2020. Fighting against COVID-19: boosting the immunity with micronutrients, stress reduction, physical activity, and vitamin D. *Nutr. Food Sci. J. (Sci. Lit.)* 3, 126.

Xu, W., Lu, N., Kikuchi, M. and Takagaki, M. 2021a. Effects of node position and electric conductivity of nutrient solution on adventitious rooting of nasturtium (*Tropaeolum majus* L.) cuttings. *Agronomy* 11(2), 363.

Xu, W., Lu, N., Kikuchi, M. and Takagaki, M. 2021b. Continuous lighting and high daily light integral enhance yield and quality of mass-produced nasturtium (*Tropaeolum majus* L.) in plant factories. *Plants (Basel)* 10(6), 1203.

Yu, H.-C. 1997. Introduction. In *Perilla the Genus Perilla*, eds. Yu, H.-C., Kosuna, K., and Haga, M.. Amsterdam: Harwood Academic Publishers, 1–8.

Zhang, T., Shi, Y., Piao, F. and Sun, Z. 2018. Effects of different LED sources on the growth and nitrogen metabolism of lettuce. *Plant Cell Tiss. Organ Cult.* 134(2), 231–240.

Zhang, Y., Seeram, N. P., Lee, R., Feng, L. and Heber, D. 2008. Isolation and identification of strawberry phenolics with antioxidant and human cancer cell antiproliferative properties. *J. Agric. Food Chem.* 56(3), 670–675.

# Part 6

## Business case studies

# Chapter 16

# Lessons learned from operational and shuttered vertical plant farms

*Francis Baumont De Oliveira, University of Liverpool, UK; and Ronald Dyer, University of Sheffield, UK*



## 1 Introduction

Within this chapter, a plant factory or vertical plant farm is defined as an operation that cultivates plants using vertical space in a controlled environment. Plants can be stacked on top of each other using multi-tier racking or vertical towers and can be grown with natural or artificial lighting. When a vertical plant farm specifically grows plants without sunlight, it is categorised as an indoor vertical farm (IVF) or plant factory with artificial lighting (PFAL). These terms will be used interchangeably. Typically, a vertical plant farm uses soilless hydroponic techniques. This study predominantly focuses on IVFs with a couple of exceptions and all of the farms utilised hydroponic techniques. The benefits of utilising these techniques are explored elsewhere in this book.

Recent advances in light-emitting diode, automation and greenhouse technologies have catalysed the VPF industry (van Delden et al., 2021). Deep expertise is required to reverse-engineer nature's processes to achieve optimal results in a controlled environment. The practice is still in its innovative stages, and many VPF companies have been developing farm technology whilst operating their farms. The risk and investment required mean there has been some secrecy around business models and lessons learned (Kummer, 2018).

http://dx.doi.org/10.19103/AS.2023.0126.05

Some insiders complain that VPF is mostly 'smoke and mirrors' (Brodwin, 2019a,b), implying the route to business viability may not be as clear cut as first appears.

VPF has reached an inflexion point where industry operations begin to scale and strive for positive unit economics. However, banks and traditional investors highlight their cautious optimism and hesitancy to invest in this high-risk sector (Baumont de Oliveira, 2021). In addition, a lack of benchmarking data, best practices and proven business models are barriers to access to funding (Baumont de Oliveira, 2021). These factors reflect a prominent industry survey that found the top reported challenges in the sector were labour, raising funding, and scaling (Agritecture, 2019).

Since the accelerated development of commercial vertical plant farms in 2010, many farms have shut down. Although there is no official record of the number of shuttered farms, the failure rate is thought to be high. In addition, high operational costs and complexity will probably result in more costly failures to come. Figure 1 shows several examples of projects that have not survived due to reasons explored later in this discourse.

There is no data on what drives operations to close but only anecdotal reports, which will be examined more closely in Section 3. As a new sector, there are unknown risk factors that the literature is yet to explore, and therefore entrepreneurs are left to speculate what the risks may be. This issue is amplified by a lack of experience, with 49% of entrants having no agricultural experience (Agritecture, 2020). Additionally, there is no standard method for building vertical farms, and there is a wide diversity of designs and practices. On the other end of the spectrum, there are profitable farms but the core factors driving their success have yet to be explored. Figure 2 shows two operational and profitable VPF operations.

Identifying lessons learned as a post-project review system can help improve performance, reduce effort, and prevent the repetition of similar

**Figure 1** Shuttered vertical plant farm projects: (a) Local Garden (Nasseri, 2022), (b) FarmedHere (Hardej, 2022), (c) Wigan UTC (Macaulay, 2022). Used with written permission.

**Figure 2** Profitable indoor vertical plant farm projects: (a) 808 Factory (808 Factory, 2022), (b) greenLand (greenLand, 2022). Used with permission.

mistakes (Soltész and Berényi, 2021). This review system has been applied in many industries, including agriculture (Butler et al., 2021). It is relevant to vertical farming because of three aspects:

- the unprecedented regulatory territory of urban food-water-energy projects;
- the high failure rate; and
- the lack of proven business models that are capable of scalability.

We aim to derive lessons from VPF start-ups for best practices. We will do this through reflexive thematic analysis applied to a series of interviews and secondary data items (Braun and Clarke, 2012; Maguire and Delahunt, 2014). The objectives are as follows:

1. review related works and anecdotal learning experiences of building and operating vertical farms;
2. identify thought-leaders viewpoints on aspects that have aided or hindered vertical farming's emergence;
3. use a taxonomical hierarchy to categorise viewpoints into core themes;
4. identify specific lessons learned by themes;
5. identify failure modes of vertical farms; and
6. develop guidelines to support vertical farms.

There has been no formal literature examining lessons learned from vertical farming projects, although there are discussions for greenhouses with some cross-applicability (De Zwart, 2012; Thomas and Stephen, 1984). Discussions of lessons from vertical farming have been exchanged in blog posts, online videos, and industry conferences (Agritecture, 2021; Colangelo, 2021; Liotta et al., 2017; Michael, 2021), but there has been no academic analysis. In this research, we integrate findings from both the formal and informal literature to inform an initial thematic taxonomy (see Fig. 3).

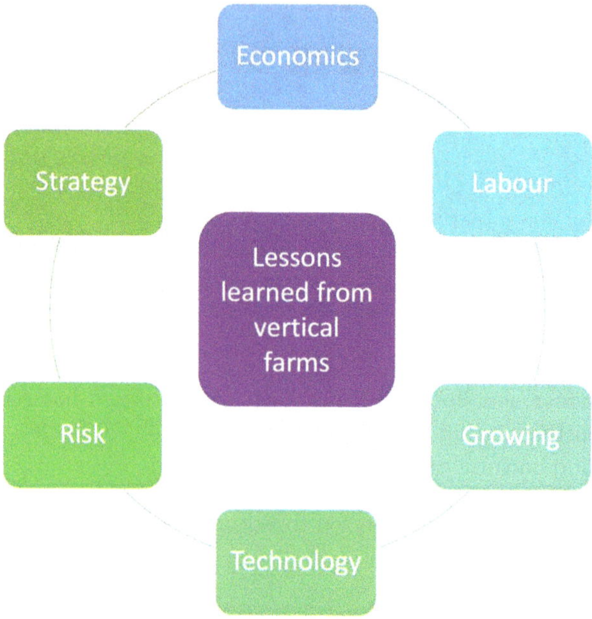


**Figure 3** Initial thematic taxonomy based on related work.

## 2 Methodology

This study uses a mixed-method reflexive thematic analysis (TA) drawing from Braun and Clarke (2012). Reflexive TA is an open-ended approach that facilitates identifying and analysing patterns (themes) within data (Braun and Clarke, 2006). We adopted an inductive approach as we code mainly from the data and draw upon some existing concepts from the primary researcher's knowledge. The focus of this study is experiential and exploratory, giving voice to the experiences of the practitioners (interviewees) of vertical farming businesses.

Interviews occurred either at the participant's place of work or through online video calls. Interviews were semi-structured. We employed a purposeful sampling strategy (Palinkas et al., 2015) by selecting a combination of farm operators, business owners (of both closed and operating farms), vendors, and consultants that have helped set up multiple vertical farms. Overall, we recruited 20 participants across 18 companies for this study. In addition, ten secondary sources were used within the analysis to account for other shuttered farm operators' and practitioners' experiences (Agritecture, 2021; Brodwin, 2019a; Colangelo, 2021; Crumpacker, 2018; DHNews, 2017; Levenston, 2014; Liotta et al., 2017; McGoran, 2020; Michael, 2017; Sykes, 2017). Figure 4 and Table 1 show the sample population characteristics including relevant secondary sources.

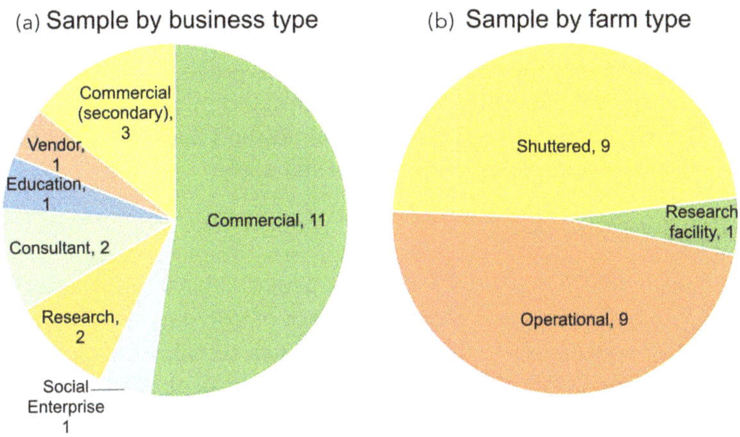


**Figure 4** Sample characteristics for 21 businesses and 19 farms.

**Table 1** Population information for farms

| Characteristic | Data |
|---|---|
| Notable crop types: | Leafy vegetables: lettuce, chard, kale, spinach, rocket, salad mix, cabbages, mizuna, and mustard greens. Herbs: basil, coriander, thyme, mint, dill, and parsley. Microgreens: pea sprouts, cabbages, sunflower sprouts, coriander, basil, broccoli, brassicas, amaranth, wheatgrass, cress, kale, and radish. Edible flowers: nasturtiums, pansies, and violas |
| Production quantity | Between 260 kg/year and 584 000 kg/year, mean: 173 000 kg/year, standard deviation: 229 000 kg/year |
| Facility farm size | Between 20 $m^2$ and 10 000 $m^2$, mean: 1680 $m^2$, standard deviation: 3140 $m^2$ |
| Farm launch year | Between 2008 and 2020, mode: 2018 |
| Company lifespan | Between 1 and 10 years. Mean: 4.71 years |
| Configurations | Greenhouse vertical farm (10.5%), indoor vertical farm (63.2%), container (15.8%), façade vertical farm (10.5%), and multi-floor farm (15.7%) |
| Irrigation systems (for definitions see (Baumont de Oliveira, 2022; Lu and Shimamura, 2018)). | Vertical growing towers (5-foot and 8-foot, ZipGrow™ towers (ZipGrow, 2022))<br>Nutrient film technique (NFT) racks (bespoke and 5-layer racks from V-Farm™ (V-Farm, 2022)<br>Deep flow technique (DFT) racks (bespoke)<br>Ebb and flow systems (bespoke)<br>Aeroponic spray system (bespoke)<br>Modified hybrid NFT/DFT system |
| Locations | United Kingdom, France, United States of America, Sweden, Denmark, Canada, Australia, Japan |
| Profitability of commercial farms | 10.5% profitable, 89.5% unprofitable |

# 3 Results

This section presents the analysed results of reflexive TA from coding 20 interview transcripts alongside 10 news articles, blog posts, and conference panels. The six subsections address the themes and sub-themes generated from the data with analysis grounded in the literature.

## 3.1 Theme 1: Economics

The economics of VPF is a central element explored within the interviews. There is substantial uncertainty about the economics of VPF due to a lack of peer-reviewed and real-life production or financial data available on a commercial scale (Baumont de Oliveira et al., 2022; Valle de Souza et al., 2021). This section highlights participants' path towards economic viability, capital funding, and cost components (see sub-themes shown in Table 2 with the percentage of occurrence within the dataset).

### 3.1.1 Getting to economic viability

Economic viability is critical for vertical farms to expand globally and improve consumer access through affordability. Two of the 19 farms considered in this sample were profitable. This reflects the difficulty in developing financially sustainable farms. Although IVFs can be economically viable (Agritecture, 2020; Kozai and Niu, 2019; Valle de Souza et al., 2021), there are no business models that have demonstrated scalability at the time of writing this chapter. Rising electricity prices and inflation across the globe after the coronavirus pandemic and Russia's invasion of Ukraine will have no doubt influenced the delayed scaling of companies in the VPF sector.

Many farms struggled with financial forecasting as there was too much uncertainty in the early stages to accurately predict their return on investment (see Section 3.4.3 for further detail). Several farm operators underlined that alternative revenue buffered their farm operations so that they were not solely reliant on sales to be viable. Successful farm owners were able to benefit from consulting to leverage their expertise, which paid for research and development costs. It was also clear from responses that the economic viability of VPF is closely tied to technology selection. There were mixed responses about whether technology was currently economically viable (see Section 3.4.3). However, two profitable PFALs revealed that they reached their break-even point within 2–3 years of operation. They attributed this to high productivity, strict management costs (their core management team was small and competent), and a strong customer focus (solving a problem for their customer and marketing).

From the perspective of economies of scale, operating smaller farms and larger farms have different trade-offs depending on their customers. Operators

**Table 2** Economics sub-themes and codes

| Sub-theme | Codes | Proportion of occurrence (%) |
|---|---|---|
| Challenges with capital | Finding the right funding mechanism | 9.7 |
| | High capital requirement | 7.4 |
| | Misalignment of investors' expectations | 6.5 |
| | Access to funding | 4.7 |
| | Insufficient capital runway | 4.7 |
| | Insufficient returns | 2.9 |
| | Burning capital on experimentation | 2.1 |
| | Debt | 0.3 |
| Getting to economic viability | Profitability | 9.1 |
| | Economies of scale | 8 |
| | Freebies, incentives, and benefits | 4.4 |
| | Economics coupled with technology | 3.8 |
| | Alternative revenue | 3.5 |
| | Unit economics before commitment | 3.2 |
| | Realistic projections | 1.8 |
| Core cost components | Technology | 8.3 |
| | Labour | 5.6 |
| | Lighting | 5.6 |
| | Rent | 5.3 |
| | Depreciation | 1.8 |
| | Electricity | 1.2 |

and consultants reported that a viable farm targeting wholesale customers should have a minimum footprint of 1000m$^2$ and up to 6000m$^2$ and focus on a small number of crops. One profitable and long-standing PFAL operator revealed that multiple modularised growing rooms of 500 m$^2$ provided the best labour efficiency and ease of maintenance, which can then be scaled by adding adjacent growing rooms. A growing module has the primary benefit of having systems serviced without comprising the entire farm to pests or hygiene issues. Beyond 500 m$^2$, operators experienced diminishing benefits of labour efficiency due to excess movement. This modularised approach supports the discussion of cultivation module systems in Chapter 3 as well as the latest industry trends (Japan Plant Factory Association, 2022a). On the other hand, smaller farms can produce a wider variety of crops and target retail customers. It was affirmed that a size between 50 m$^2$ and 200 m$^2$ was suitable for smaller farms. Locations for respective farm types are discussed in Section 3.5.3.

32% of farm operators discussed their experience benefiting from subsidised rent, capital costs, or electricity and how it was vital to sustaining their business. In these cases, farms report being able to negotiate free rent and subsidised utilities by providing added value to a location. Moreover, farms can benefit from strategic placement next to a power plant to uptake excess energy. Some regional governments may support this with subsidies (this was the case for one Japanese farm). Location partnerships were the primary reason several small farms reported that they could function economically long term; however, one operator achieved profitability without any subsidies or incentives.

### *3.1.2 Challenges with financing*

Challenges with capital were the most discussed aspect of economics, which is expected considering the high costs required to build and operate vertical farms. In recent industry surveys, controlled environment agriculture (CEA) owners' most commonly reported issue was funding and 71% of founders were either unsuccessful in raising capital or could not rely on traditional sources like commercial lenders for financing (Agritecture, 2020; Agritecture, 2019). Within the interviews, farm owners discussed access to funding, accumulating debt, and burning capital on experimentation.

The growing system and lighting technology required to build and operate a vertical farm are notoriously expensive, and this was the most common complaint across all the interviews. Operators reported that when they paid less upfront (CAPEX), they ended up having higher costs to run the business (OPEX) (Sykes, 2017). This high capital requirement for economic viability is a considerable challenge, as practitioners' complained that access to funding is their primary barrier to growth. This is due to a lack of proven business models and benchmarking data as well as numerous failures which have made investors and bankers hesitant to invest. However, many farm operators alleviated their capital requirements through government grants and subsidies. One profitable farm was funded by its parent company. A common trend is that acquiring funding with low or zero interest rates contributes heavily towards achieving profitability.

32% of VPF operators expressed that conducting research and development (R&D) takes substantial time and capital. Therefore, operators need to have a sufficient capital runway for commissioning the farm and creating standard operating procedures (SOPs) that they did not plan. In hindsight, many wished that they had raised more capital and communicated this to investors. Balancing R&D and commercial farming is discussed further in Section 3.3.3. If a business plans to operate a commercial farm, it may be worth outsourcing the building of the farm to an experienced developer to avoid costly efforts to integrate all the different technologies involved.

> I think the top reason why vertical farms crash and burn is they're undercapitalised. When you're building a farm, as you're operating it, you don't know how much capital you need. - Robert Colangelo, CEO of Greensense Farms (Colangelo, 2021).

The capital market is highly stratified. Equity and private investments have funded many vertical farms whilst seeking high returns. However, vertical farms cannot achieve high returns until the cost of technology reduces, productivity improves, and farms can scale with the same fixed costs (Baumont de Oliveira et al., 2022; Valle de Souza et al., 2021). Problems finding the right capital source and misaligned investor expectations came up frequently as a common failure mode, as investors can push the wrong decisions. Farm operators require patient forms of capital that understand the long development cycle, similar to agricultural seed development and biotechnology. Careful planning, a good understanding of risk management, and effective communications with funding sources are mandatory and can help reduce this issue.

> He's out of patience on this, and he's all of a sudden frustrated that he's tied himself up into a capital-intensive business that grows $5 lettuces with $50000 machines. He wants to see returns are in the order of 10 to 15 to 20 times what he's put in. - CEO of a commercial vertical plant farm.

### *3.1.3 Cost components*

The interviews discussed core cost components for CAPEX and OPEX to identify considerations and typical values. These considerations have been collated and synthesised in Table 3.

## *3.2 Theme 2: Labour*

> I assure you that you are underestimating labour, and it will be the biggest cost and the most likely failure of your farm. - Matt Liotta, Founder of PodPonics (Liotta et al., 2017).

Labour was the top reported problem for CEA growers in 2019 in a prominent industry survey (Agritecture, 2019) and is typically the highest operating cost (Agrilyst, 2018; Kozai and Niu, 2019; Uraisami, 2018). These interviews revealed some of the challenges, tactics, and considerations of vertical farms through the lens of labour. The ideas were disaggregated into two sub-themes: team and process flow (coding is shown in Table 4 in proportion to occurrence within the dataset).

**Table 3** Cost component considerations

| Type | Cost component | Consideration |
|---|---|---|
| CAPEX | Rent | The cost of rent can be very high if placed in an urban context without any cost deduction resulting in an unviable business case. Profitable farms found that being placed on the outskirts of cities or in the countryside was economically preferable due to lower land costs. A long-term agreement should be sought to ensure security and avoid rent increases. |
| | Technology | High technology costs are a significant barrier for entrepreneurs, and they feel the costs do not justify purchasing unprofitable hardware. On the other hand, a higher CAPEX can lower OPEX using the best available equipment. The cost of lighting was the highest technology cost and was reported to be a constraint for some farms scaling. |
| | Depreciation | Business plans often overlook depreciation. It is suggested to factor in at least a 30% CAPEX refit after 10 years (Sykes, 2017). Depreciation typically accounts for 12% to 30% of OPEX (Kozai and Niu, 2019; Liaros et al., 2016; Uraisami, 2018). Depreciation for a farm is reported by Kozai and Niu (2019): 15 years for the facility, 10 years for equipment, and 5 years for lighting. |
| OPEX | Electricity | Electricity accounts for 21% to 52.5% of OPEX (Avgoustaki and Xydis, 2020; Hayashi and Kozai, 2018). The electrical efficiency of lighting plays a pivotal role in the profitability of a VPF. Methods to reduce energy costs by 20–30% and a theoretical 50–80% exist (Kozai and Niu, 2019). In addition, it is possible to negotiate rates from energy providers and utilise load shifting to optimise prices (Avgoustaki and Xydis, 2021). |
| | Labour | Labour is likely to be the highest operational cost and is typically underestimated. Costs range from 26% to 56% of OPEX (Agrilyst, 2018; Uraisami, 2018). Labour costs can be reduced by 50% within several years of farm operation through semi-automation/full-automation, production processes, equipment layout, and human resource development (Hayashi and Kozai, 2018) which we explore further in Section 3.2.3. |

### 3.2.1 Team

> With vertical farms, you are balancing horticulture, engineering, and economic, and it's like a three-legged stool; if one of those legs is weak, the whole thing goes down.
> - Vertical farming and greenhouse consultant.

Vertical farms are operationally complex, requiring interdisciplinary expertise. Even with a viable business plan, a qualified and experienced team is needed in the skills illustrated in Fig. 5. These skills can also be outsourced, but they

**Table 4** Labour sub-themes and codes

| Sub-theme | Codes | Proportion of occurrence (%) |
|---|---|---|
| Process flow | Iterating process workflows | 11.4 |
| | Automation is a necessary element | 10.3 |
| | Farm design and ergonomics | 10.3 |
| | Labour is a limiting factor | 8.9 |
| | Estimating labour | 6.2 |
| | Identifying process bottlenecks | 3.4 |
| | Standard operating procedures | 3 |
| | Scaling labour | 2.4 |
| | Manufacturing expertise | 1.6 |
| | Crop selection | 0.8 |
| Team | The right team skills and knowledge | 8.7 |
| | Lack of available expertise | 6.8 |
| | You pay peanuts, you get monkeys | 6.5 |
| | Growing experience | 4.3 |
| | Time management | 4.4 |
| | Ensuring quality of life | 3.3 |
| | Work schedule | 3.3 |
| | Connection between management and growing teams | 3 |
| | Surge labour | 1.6 |

must be considered. There are currently no standards and a lack of education in this new industry, making hiring and developing best management practices challenging for companies. Some operators report finding employees from similar industries (managers from mechanised food facilities, greenhouse growers and people, with cannabis farming experience). Smaller farms struggle with the broad skillset required due to increased labour costs.

For vertical farming to scale, there is a need for an educated workforce for senior positions. Quality farm labour requires farm education. Leveraging local universities or colleges to build training programmes is one successful approach that has been applied (Liotta et al., 2017).

Maintaining a good connection between the managerial team and the growing team is imperative. Due to R&D, there can be a lot of uncertainty, and experienced leadership can help mitigate associated difficulties. One of the challenges across many farms was a disconnect between management and the shop-floor employees. This can result in low productivity, poor product quality, and a lack of data collection. Developing SOPs was an integral way to

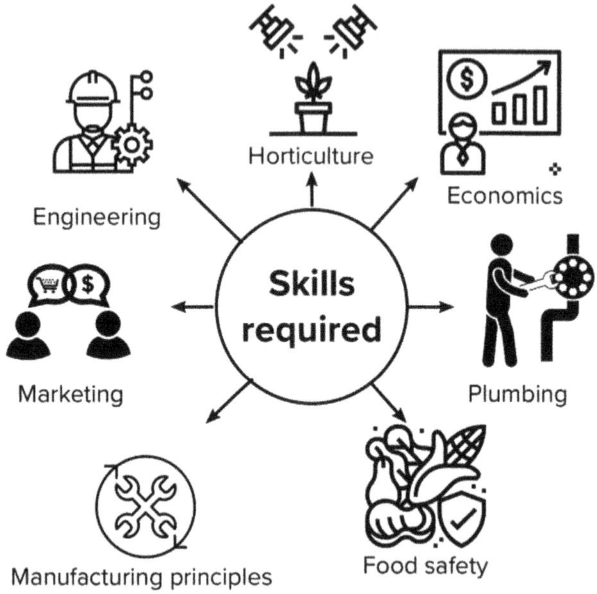


**Figure 5** Skills required for a successful VPF operation.

reduce miscommunication. The goal is to design a farm that is not dependent on the competency, memorisation, or capability of any one individual. An example of good management was discussing with shop-floor employees how to remove their 'pain points' and bottlenecks. They listen and act on their concerns whilst providing autonomy to apply their skills. Managers that did this reported higher than average labour satisfaction and engagement. All of the farms interviewed followed a weekly schedule for planning and delegating tasks.

An industry survey reports that 49% of CEA founders had no prior experience or knowledge in farming before starting their businesses (Agritecture, 2020). Several of the farm operators in this study reported no experience either and struggled with learning the basics of growing. Growing experience significantly affected the consistency of product quality, yields, and waste rates.

> 'A key lesson that I learned early on is that you pay peanuts, and you get monkeys, right? If you pay a little bit more than minimum wage, you're retaining [your employees], cost of training will go down, your cost of recruiting will go down.' - CEO of a commercial vertical plant farm.

Vertical farms can provide manual work that is relatively simple that can be accomplished by elderly and disabled workers. Due to its repetitive nature, some farms decide to bring in volunteers, interns, or low-cost workers, which benefit the cost perspective of a VPF, however present challenges in terms of consistency, product quality, and risks. Total costs may be higher from the cost of training and recruiting new employees frequently. Farm managers reported finding it challenging to correct nuanced mistakes (i.e. careless seeding) that volunteers made because they felt they were doing them a favour, resulting in crop loss. Low-paid staff may miss work more frequently due to sickness or court dates, take longer to execute tasks, and potentially ruin crop cycles due to accidents (see Section 3.6.1). Providing financial incentives for farmhands to become team leaders improves performance.

Most vertical farms have a small core team working on a longer-term strategy, with a team of casual workers for farmhands. Operators used casual workers at a higher intensity during seeding, transplanting, and harvesting, especially when they had to fulfil customer orders in the mornings. This can provide job opportunities for students, stay-at-home parents, or other demographics. Profitable and large-scale farm operators interviewed had highly competent and small management teams (having people with overlapping skills shown in Fig. 5) supported by a large part-time workforce.

### 3.2.2 Process flow

Practitioners have labelled VPF as a combination of manufacturing processes and food production due to consistent control and production. Therefore, practitioners can treat VPF as a process suited to applying manufacturing principles (Baumont de Oliveira et al., 2020). The process flow is defined as sequential tasks that guide people to get work done. Its main objective is to streamline and standardise business processes.

A lens from which to view process flow is through accessibility, maintainability, and repairability. System designers and operators often neglect these aspects and end up with higher operating costs over the long term when equipment fails or parts are challenging to clean (explored in Section 3.4.3). Multi-layered systems with high grow beds that need to be accessed can become problematic as growers navigate the grow space. Operators who used equipment to access the high levels found that scissor lifts were cumbersome, expensive, dangerous and blocked access resulting in higher waiting times. Two profitable farms had two floors to access up to 12 layers of cultivation shelves and found this adequate.

From a farm layout perspective, operators should consider how people will move around the facility efficiently without collisions. There has to be sufficient space for operations and access between racks. It is recommended that 50% of floor space be allocated to the growing area whilst leaving the remaining area for operations, walkways, seedling production, processing, and equipment (Kozai and Niu, 2019). This may increase or decrease depending on certain factors (i.e. cold storage or mobile racking). As the facility scales, labour will scale differently from other factors. Two profitable PFAL operators explained that smaller growing rooms are better for labour productivity, minimise movement, and are easier to manage. These operators aimed to design their farms in line with their workers' potential to achieve a certain work speed per unit of crop (i.e. 60 seconds per lettuce head) (Japan Plant Factory Association, 2022a). Equation 1 shows how to calculate work speed. 500 $m^2$ partitioned cultivation rooms are reported to be a suitable size by operators; however, this requires further research. Kozai (2019) suggests guidelines on farm layout

$$\text{Work speed} = \text{No. of workers present} \times \frac{\text{Portal to portal hours}}{\text{No. of plants harvested}} \qquad (1)$$

Automation is not necessary for all vertical farms, according to Kozai (2018a), even though many operators thought it was. It can bring challenges such as high costs and highly-skilled workers to manage the technology, which is why it is important to assess cost-benefit before making decisions on whether to invest in automated technologies. For smaller farms, it is imperative to implement robust systems and processes that can reduce labour costs and not rely solely upon automation. All the profitable farm operators interviewed were only partially automated and relied upon manual labour for seeding and harvesting. The activities that should be automated according to the size of the facility are shown in Fig. 6.

In 38% of the data sources, practitioners described underestimating labour, and 53% discussed how labour was a limiting factor in their production. Labour was universally the highest cost across all the participants that reported their operational cost allocations. Addressing bottlenecks on a farm can improve efficiency and production output. For example, one farm operator recounted overcoming process bottlenecks by switching from growing in pots to trays, cutting with hedge trimmers instead of scissors, and cumulatively shifting their gross margins from -30% to +30% (see Fig. 7a). Introducing machinery for cleaning, packaging, or seeding can yield similar step changes that improve productivity and labour satisfaction. We asked interviewees to draw their perspectives on the change of performance in their farm since becoming operational to understand the learning curve and the impacts of bottlenecks. Figure 7 shows two farms with bottlenecks in their production.

Operators found it helpful to set a target time per crop (e.g. 60 seconds of work per head of lettuce). Additionally, using surveillance cameras can help managers to understand where mistakes are being made to adapt SOPs and improve the flow of operations. Successful farms implemented visual signals and management alongside environment 'nudging' to create or sustain desired employee behaviour.

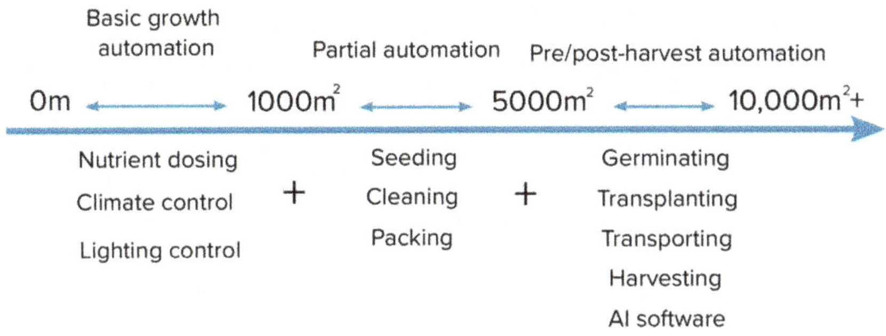


**Figure 6** Automation types according to size (see Kozai, 2019, 2018).

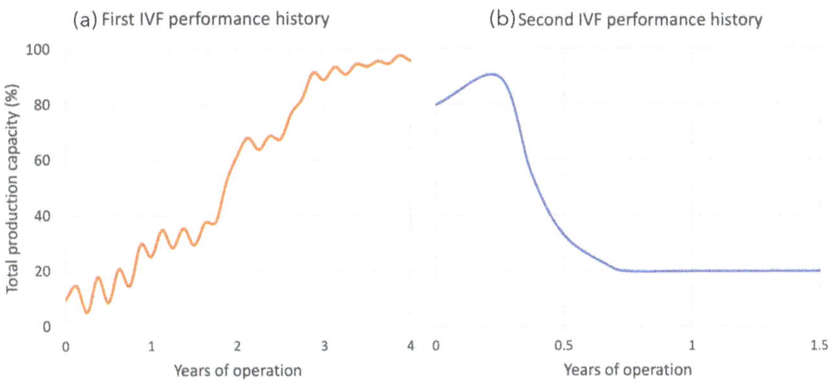


**Figure 7** The production performance of two vertical plant farms: (a) a multi-container $1000m^2$ IVF that overcame three bottlenecks in production and (b) a $450m^2$ basement IVF that had a process bottleneck that inhibited production.

## 3.3 Theme 3: Growing

The interviews revealed considerations for achieving the best possible product quality whilst ensuring food safety and balancing R&D. The codes were grouped into three sub-themes: growing factors, food safety, and effort dedicated to R&D (shown in Table 5 with coding represented in proportion to occurrence within the dataset).

### 3.3.1 Growing factors

The interconnected factors that contribute towards growing produce were explored and are visualised in Fig. 8. The growing factors influence biomass yield, produce quality, crop cycle duration, waste rates, and the crop limiting factor. Table 6 links the growing factors to relevant literature that can aid growers to implement practical changes. The anecdotal reports

**Table 5** Growing sub-themes and codes

| Sub-theme | Codes | Proportion of occurrence (%) |
|---|---|---|
| Growing factors | Produce quality | 8 |
| | Climate | 6.7 |
| | Waste | 7.1 |
| | Growing media | 4.9 |
| | Nutrients | 4.5 |
| | Crop density | 3.1 |
| | Crop limiting factor | 3.1 |
| | Crop cycle duration | 3.1 |
| | Propagation | 2.2 |
| | Carbon dioxide | 1.8 |
| | Human benefit | 1.3 |
| | Microbial communities | 0.4 |
| Food safety | Biosecurity practices | 12.5 |
| | Food hygiene | 8 |
| | Sterilisation protocols | 4.9 |
| | Isolate the growing area | 3.6 |
| Efforts dedicated to R&D | Crop growth recipes | 7.1 |
| | Crop varietal optimisation | 6.3 |
| | Space and time allocated to research | 6.3 |
| | Growing whilst building | 4.9 |

from interviews are discussed to glean any insights and expose voids in the literature.

### 3.3.1.1 Crop limiting factor

Understanding crop requirements to identify and improve the limiting factor is key to successful VPF system design. Operators can focus on low-hanging fruit by handling the macro-requirements of crops such as temperature, vapour pressure deficit, $CO_2$, and daily light integral before finetuning micro-elements like light spectra, nutrient composition, microbial communities, and oxygenating water. For example, several operators reported that they did not see noticeable improvements from a change in light spectra because their climate control was not optimised. A challenge for VPF is understanding Liebig's Law of the Minimum, which states that growth is dictated not by the total resources available but by the scarcest resource (the limiting factor) (Sinclair, 1994). Strategies to identify the scarcest resource are needed. For profitable farms, they stated that the limiting factor in their production was presented as tip burn in leafy green production.

**Figure 8** Growing factors are shown as outputs (orange) and inputs (yellow).

### 3.3.1.2 Crop density, climate control and pests

> All the heat, humidity, and pests that result from these processes are amplified in a denser growing environment. Vertical farming is not about how much production you can possibly cram into a space. It's about growing better food closer to market and maximising your production as a function of the resources you invest, such as capital, light, water, energy, and labour. - Chris Michael, co-founder of Bright Agrotech (Michael, 2017).

There was a contrast among some farm owners' thoughts around crop density. Some interviewees thought that the highest crop density possible was desirable, whilst others recognised the risk posed by increasing density to adequately control micro-climates to the detriment of plant quality. Heat and humidity easily increase with higher densities, increasing the risk of pests and reducing yields and quality. Managing the growing environment becomes increasingly complex, with higher crop densities resulting in physiological disorders.

An air temperature between 18°C and 25°C is suitable for most crops in IVFs (Lu and Shimamura, 2018). The vapour pressure deficit should be set between 0.8 kPa and 0.95 kPa, and optimally 0.85 kPa (Niu et al., 2019). Air

**Table 6** Growing factors

| Growing factor | Relevant literature |
|---|---|
| Nutrient solution pH | (Gillespie et al., 2020; Roosta, 2011; Tyson et al., 2008) |
| Air temperature | (Carotti et al., 2020; Choi et al., 2000; Lee et al., 2013) |
| Water temperature | (Hendrickson et al., 2022; Nxawe et al., 2009; Thakulla et al., 2021) |
| Vapour pressure deficit (humidity) | (Gates et al., 1998; Graamans et al., 2017; Grossiord et al., 2020; Katsoulas and Kittas, 2008; Rawson et al., 1977) |
| Carbon dioxide | (Esmaili et al., 2020; Kozai, 2022a; Singh et al., 2020) |
| Airflow | (Kozai, 2022b; Nishikawa et al., 2013; Peiro et al., 2020) |
| Light spectra | (Paradiso and Proietti, 2022; Park and Runkle, 2018; Rahman et al., 2021) |
| Daily light integral | (Esmaili et al., 2020; Gao et al., 2020; Xu et al., 2021) |
| Light uniformity | (Balázs et al., 2022; Hwa-Soo et al., 2014; Van Der Meer et al., 2021) |
| Microbial communities | (Compant et al., 2005; Sheridan et al., 2017) |
| Nutrient intensity and flow | (Baiyin et al., 2021; Lee et al., 2017) |
| Nutrient composition | (Khodijah et al., 2021; Nazarideljou et al., 2019) |
| Water oxygenation | (Cherif et al., 1997; Goto et al., 1996) |
| Propagation | (Japan Plant Factory Association, 2022a; Nunomura et al., 2019) |
| Growing media | (Kowalczyk et al., 2016; Wortman et al., 2016) |
| Growing density | (Lam et al., 2019; Lim et al., 2020; Maboko and Du Plooy, 2013) |
| Micro-climates | (Katsoulas and Kittas, 2008; LI et al., 2021) |
| Growing experience | (Baumont de Oliveira et al., 2022; Kozai, 2018b) |
| Pests | (Donatelli et al., 2017) |

current speed should be set between 0.5 $ms^{-1}$ and 1.0 $ms^{-1}$ to disrupt the boundary layer flow in the plant canopy and promote gas exchange (Lu and Shimamura, 2018).

### *3.3.1.3 $CO_2$ enrichment*

Many smaller operators rely on atmospheric $CO_2$ (400–420 ppm) because they compared their operation to open-field farming that does not require $CO_2$ enrichment. However, for airtight facilities with higher crop densities, adding $CO_2$ will benefit production substantially (Kozai and Niu, 2019). In Japan, most operations use $CO_2$ enrichment as they realise this is critical for profitability. For lettuce, 1200 ppm achieves the maximum net photosynthetic rate (Esmaili et al., 2020); however, this can be limited due to standards of ventilation requirements

(Kozai, 2022a). Japanese standards which are based on body odour perception are counter-productive to PFAL operation, as there is no evidence suggesting $CO_2$ at such concentrations is detrimental to comfort or health (Kozai, 2022a). Therefore, there are opportunities by adapting such standards to improve the productivity of IVF.

### 3.3.1.4 Crop cycle duration

> Most of these [vertical farming] operations produce baby leaf [salads]; they produce very small plants at a very high density, that are cut and removed at a very young stage. That's why most of these companies are not producing full-sized plants, because managing the environment is difficult, and they start to have nutritional and physiological disorders. – Vertical plant farming and greenhouse consultant.

Farm managers noted that a short crop cycle duration benefited production. Smaller crops (like baby leaf salads or microgreens) were quicker to produce, preferred by the consumer, and less prone to physiological disorders and diseases. This was counter-intuitive to some growers who initially assumed a bigger crop would be better. Larger crops would require more labour and present challenges with managing micro-climates. Microgreen farmers felt that their cultivation was more manageable and perhaps a suitable starting point to understand the basics of VPF and hydroponic cultivation.

### 3.3.1.5 Nutrient solution, sterilisation and microbe management

Many practitioners used a general nutrient solution for their crops. Lu and Shimamura (2018) provide guidelines on proven nutrient solution formulae for leafy vegetables. Finetuning composition to match the absorption characteristics of the crops can yield improvements in quality. In the future, the concentration of individual ions should be measured and managed separately, and Kozai et al. (2018) provide guidelines to achieve this. Maintaining a pH value of 5.0–7.0 and a root zone temperature of 18–25°C is suitable for most crops in IVF (Lu and Shimamura, 2018).

When asked about sterilisation procedures, most farms used ultraviolet light filters, Ozone, or hydrogen peroxide to reduce contamination risk within the growing system. However, sterilisation can also kill beneficial microbes, reducing the overall quality of the produce. Some operators also report unexplained imbalances in the nutrient solution that affected plant growth, which they suspected was from not deep cleaning the systems often enough. They explained that this has received little attention in the academic literature which could be fertile territory for more research.

### 3.3.1.6 Germination and propagation

VPF operators reported that seeding and propagation were the most critical steps in the growing cycle to achieve the best possible harvest. However, several farms found that propagation was their limiting factor. For example, large yield fluctuations depended on who was seeding the trays/pots at a given time. Standardised propagation or automated seeding resulted in more consistent and better yields. Guidelines on propagation management have been taught as part of a training course (Japan Plant Factory Association, 2022a).

### 3.3.1.7 Waste

Waste, defined as shrinkage, unsalable produce, unsold produce, and inedible biomass, varied considerably. Waste and shrinkage can minimise profitability for a farm and occur at any point in the growing chain. Growers reported waste rates ranging from 2% to 22% in the growing process. Two profitable farms reported 10–15% wastage rates depending on cultivars and quality standards.

### 3.3.2 Food safety

Food safety is an essential part of any agricultural operation. However, there was much ambiguity about best practices for VPF. There were two camps of interviewees. On one side, there is the cautious approach with almost sterile environments with employees wearing protective clothing. These farms have robust sterilisation protocols and follow strict compliance schemes.

On the other hand, some operations treat their farm less seriously, viewing an indoor farm as a traditional farm that does not need to be sterile and clean but maintains light food hygiene standards. Figure 9 shows the reported practices. This section breaks down views on sterilisation protocols and food hygiene.

Most farms visited and interviewed mainly used diluted hydrogen peroxide for sterilisation protocols for cleaning equipment, disinfecting seeds, and recirculating the growing system. As the hydrogen peroxide breaks down into the water quickly, this renders the chemical harmless and compatible with organic production – this chemical reduces the risk of fungal outbreaks. The application of hydrogen peroxide in hydroponics is relatively unstudied. However, one recent study found that dosing is necessary for crop quality and can potentially improve the viability of organic hydroponic fertilisers (Lau and Mattson, 2021). Other sterilisation methods include ultraviolet lamps, Ozone, heat sterilisation, silver/titanium oxide utilisation, sand filtration, and oxygen bubbling (Son et al., 2019). There is no ideal solution, and each method has its pros and cons (Kubota, 2019; Lu and Shimamura, 2018). One consultant emphasised that sterilisation kills beneficial microbes and harmful microbes, hindering crop quality. This requires further investigation.

Hygiene education should be given to operators to inform practices that mitigate contamination risk. There are no standards for VPF. However, the practices in Fig. 9 help prevent the worst-case scenario of any pathogen or pest outbreaks. Many farms did not use any pesticides at all in their production. Additionally, one of the value propositions of VPF produce is cleanliness which does not require washing and has a longer shelf-life. Strict hygiene practices achieve these benefits. The highest contamination risk comes from the nutrient solution and water; therefore, it is best to avoid direct contact between the water and the crop leaves. Kubota (2019) gives examples of conducting microbiological testing for water quality and crop leaves.

Hiring a food safety consultant before designing a facility is advised to meet compliance demands by wholesalers and mitigate any contamination risk or potential liability issues. However, each farm had its own practices because there is a lack of literature in this area.

### 3.3.3 The efforts dedicated to research and development

> One of the biggest challenges… was really that while we were producing and scaling, we had to conduct R&D. Not only did we have to have time for R&D, but also money and space. So I think it was a killer of a formula. - Pawel Hardej, CTO of FarmedHere.

Unanimously, farms developed technology (software and hardware) whilst trialling crop cultivars. Consequently, farms are raising capital, hiring skilled employees, identifying customers, selecting crops, packaging, finetuning processes, and getting the product to their customers at a high level of quality. Juggling all these tasks burns cash faster, as farms require space, effort, and

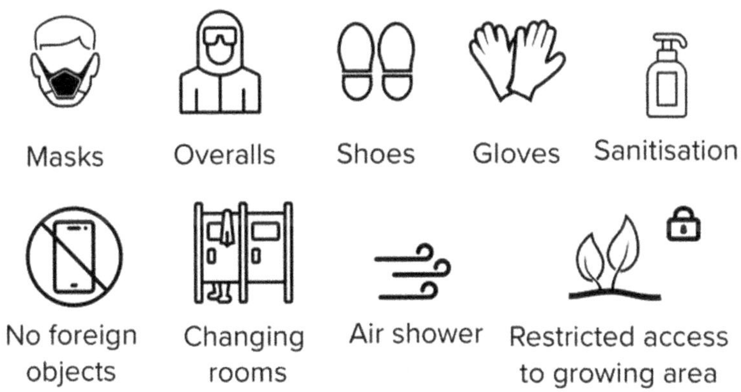


**Figure 9** Reported food safety practices.

time dedicated to research and development to improve all these aspects. In some cases, operators allocated 50% of farm space to R&D. This presents disorganisation, as operators dilute their focus to build and operate their farms simultaneously. Eventually, some begin to shift their business focus towards technology development as the technology is 'evolving too quickly to jump in as an operator'. Profitable PFAL operators interviewed focused more on well-researched crops and less on experimentation, dedicating only 5% of their facilities and efforts to R&D and new crop types.

Discovering and selecting the appropriate crop cultivars is part of VPF R&D. Crops with low labour demands or easily automatable present better unit economics. Moreover, crop breeding is presenting cultivars suitable for VPF production. Profitable farms were working with seed companies to develop new cultivars. New lettuce varieties are currently being scaled which are capable of larger harvest weights with tip-burn resistance and therefore reduced labour requirements (Leaflab, 2022). Before designing their facility, many farms were unsure what they wanted to grow, experimenting with a large diversity of crops and then whittling down to the most profitable crop varietals that best meet the local market conditions.

### 3.4 Theme 4: Technology

Technology selection involves many costly decisions that determine the economics and scalability of the operation. This section examines the sub-themes generated from the interviews: pushing a square peg into a round hole, being intentional with data, and system considerations (coding is shown in Table 7 in proportion to occurrence within the dataset).

#### 3.4.1 Pushing a square peg into a round hole

A typically reported mistake is purchasing technology before identifying the problem that the farm hopes to solve. A market analysis should determine what crops to grow that may be challenging to grow traditionally and are subject to risks in the distribution process, system, or time (temperature, seasonality). The product can then be defined alongside a target customer and price point. Then farmers can reverse-engineer the most appropriate technology and farm design to achieve competitive unit economics and product quality for their customers. The solution may not be a vertical farm but, in fact, a standard greenhouse or hoop house. Selecting the most appropriate technology is essential.

#### 3.4.2 Data

> When I looked at vertical farming, there is just very little data out there that's published and reliable. - Robert Colangelo, CEO of Greensense Farms (Colangelo, 2021).

**Table 7** Technology sub-themes and codes

| Sub-theme | Codes | Proportion of occurrence (%) |
|---|---|---|
| System considerations | System connectivity | 29.8 |
| | Proving technologies | 21.1 |
| | Specific system challenges | 21.1 |
| | Keep it simple | 3.6 |
| | Energy | 3.2 |
| | Accessibility, maintainability, and repairability | 2.3 |
| | System flexibility | 2.1 |
| Being intentional with data collection | Paucity of data collected or available | 4.4 |
| | Optimising inputs-outputs | 3 |
| | Putting data to use | 3.8 |
| | Data quality | 2.3 |
| Pushing a square peg into a round hole | Reverse-engineering from the customer and crop | 3.8 |
| | Choosing the right technology | 1.5 |

An IVF's outputs can be highly precise due to controlled and measured inputs and environmental control. Crop growth and phenotyping can be optimised with millions of parameter combinations. Data can augment production, improve farm yields, and reduce labour time through better decision-making when applied appropriately. However, some farms try to collect everything possible without a coherent data strategy, spending capital and effort without leveraging the data. Converting data into meaningful and actionable insight is paramount. Optimising the production as a function of the resources invested is where IVF has the potential to excel due to the unprecedented feedback loops (seed to harvest) achievable within a year. These feedback loops make artificial intelligence (AI) compelling for IVF, and almost all interviewees wanted to incorporate AI, but the costs were prohibitive.

> It's very important to build up like the foundation for use of machine learning [AI], because, at some point, you understand that you cannot calculate it yourself anymore. - Manager of a commercial vertical plant farm.

Many farm operators interviewed did not keep track of standard metrics because they were overwhelmed with operations and delivering sales. Lack of standardisation and lack of available data present a significant obstacle to comparing and establishing a performance baseline. Many participants had to build their proprietary software platform to store and manage data to iterate crop growth recipes and link harvests to sales. They noted an absence of available data, finding data from the literature as indicative.

At a minimum, farms should measure the metrics in Fig. 10 for each crop type. This will allow them to build an operational history. In addition, by measuring these metrics, farms will be more likely to manage these variables and show investors their performance history if seeking funding.

### 3.4.3 System considerations

Several farm operators felt the category was still immature and that many technologies were unproven. As a result, farm operators shifted systems repeatedly, trialling different growing systems or lights. Lighting is a prohibitive cost for many, and some farms found through trials that when comparing cheaper lighting solutions to reputable brands, their yield and quality were the same with a higher ROI (see crop limiting factor in Section 3.3.1).

> The incentive for the technology vendors to appear more efficient and productive than their competitors fosters a culture of greed and dishonesty. - Vertical farming and greenhouse consultant.

Unanimously, all the interviewees asked about yield estimations, and productivity claims found them to be overestimated and inaccurate. With so many interconnected factors at play, this is not surprising. In some cases, operators reported achieving only 10–50% yields stated by their system vendor because of labour workflow and environmental control challenges. Therefore, it is worth adding a contingency margin to projections given by vendors.

> Don't buy it if you can't clean it. - Matt Liotta, Founder of Podponics (Liotta et al., 2017)

For container farms, operators reported an antagonistic relationship between light, layout, heat, and workflow issues and that HVAC was the limiting factor in production. For vertical towers, five operators were either using or had tried

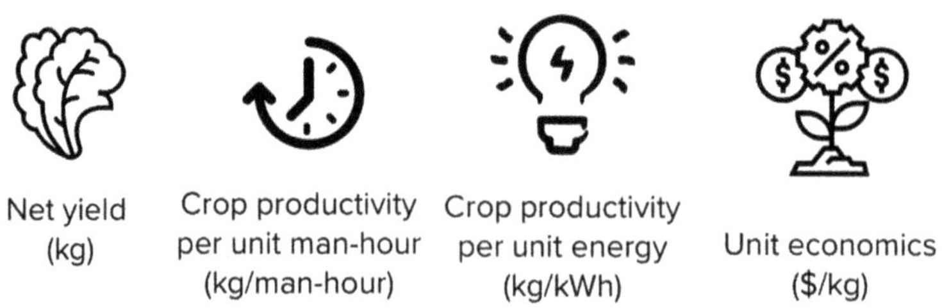


**Figure 10** Suggested key performance metrics.

vertical tower systems and experienced leaks, dripping of nutrient solution onto lower plants (a health risk in Section 3.3.2), physiological plant disorders (plant orientation effect on plant quality is unknown in the literature), and ergonomics issues (neck and back pains). These are some examples of how system performance is exaggerated.

> LED lights are going to fizzle out, mechanical systems are going to fail, and pumps will need to be rebuilt. It would not be crazy to suggest that up to 70% of a vertical farm system will undergo some level of replacement or repair within a 10-year window. - Chad Sykes, Founder of Indoor Harvest (Sykes, 2017)

Many interviewees expressed the need to keep technology simple and avoid reinventing the wheel. Overengineered systems can have many failure points, resulting in high repair costs. High humidity can cause mechanical system failures due to rust. Ease of access, maintenance, and repair are fundamental aspects of viable VPF technologies. Many farm owners found that their systems were more problematic to upkeep than anticipated and most challenges began to arise after several years of operation. A preventative maintenance plan can mitigate the risk of equipment failure and its impact on finance or production. Profitable farm operators that were interviewed planned for this in advance, and used simple yet proven technologies whilst introducing automation only when they were able to pay back the costs through labour reduction savings.

Smaller farm operators reported that system flexibility was an important factor in decision-making for technology. Agility can be significant for smaller farms that target a retail market. Specifically, mobile racking systems allow for higher space utilisation, and deep flow technique (DFT) systems can allow multiple crop types.

When asked about the most challenging aspect of developing a farm, system integration (synchronising systems to interact and adapt to data and track compliance) was at the top for most interviewees. Additionally, many VPF operators overlooked HVAC, with engineering companies using inappropriate assumptions from other industries to size the requirements (Agritecture, 2021). An adequately engineered HVAC system can remove humidity and heat whilst enabling airflow management across the plant canopy and becomes increasingly important as crop density increases. When profitable farm operators were asked what they would do differently for their next farm, efficient and agricultural-grade building materials were emphasised as an opportunity to reduce building costs by 60%. This could encourage further custom-built PFALs as opposed to using old structures that usually come with complex challenges such as accessibility and food safety.

### 3.5 Theme 5: Strategy

Strategic decisions around location, market, and business goals influence the success of a VPF. In this section, we examine the sub-themes generated from the interviews: market, location, focus business goals, build collaboration, create a backup plan, and start with a pilot (coding is shown in Table 8 in proportion to occurrence within the dataset).

#### 3.5.1 Market

The target market and local context are pivotal in a viable strategy. VPF differentiates itself due to benefits discussed elsewhere in this book, and operators reported that customers were most interested in the pesticide-free quality and are increasingly willing to pay a premium for local produce (Adams and Salois, 2010; Faires and Charlebois, 2020).

Participants found that the most challenging part of the production was not growing but selling produce and getting it to the customer. The industry is segmented between large-scale vertical farms producing large quantities of leafy greens for wholesale markets, medium-scale farms focusing on herbs and niches (like edible flowers, microgreens, and animal fodder), and many small farms serving retail markets a range of crops. Each customer type had its respective trade-offs. Operators said that although retail customers paid more, they had comparable work requirements and costs as wholesale customers to sell dramatically less produce. A more sophisticated and large farm is needed for wholesale customers, but distributors can take a high percentage of sales

**Table 8** Strategy sub-themes and codes

| Sub-theme | Codes | Proportion of occurrence (%) |
|---|---|---|
| Market | Growing is easy, selling is difficult | 32.7 |
| | Competition | 7.5 |
| Choose a location | Choosing a flexible site | 13.9 |
| | Geography | 7.5 |
| | Placement in relation to the customer | 3.8 |
| Focus business goals | Goal clarity | 8.9 |
| | Solving a problem | 4.5 |
| | Seek continuous improvement | 4 |
| | Pivot to a technology provider | 1.6 |
| | Model scenarios | 1.6 |
| Start with a pilot | | 8.9 |
| Build collaboration | | 2.6 |
| Create a backup plan | | 2.4 |

(scale and customer acquisition save costs). Profitable farms stated that access to a big market was a key factor in their success, and this is likely why Japan has some of the most successful IVFs. Japan has a health- and hygiene-conscious population that is used to paying a premium for high-quality fruit and vegetables (up to 320% price increase for PFAL lettuce compared to conventional). Some farms are distributed in their customers' locations where crops are transplanted from a warehouse, achieving the benefits of wholesale production with the power of retail sales with lower distribution costs – Section 3.5.3 details suitable locations for these customer models.

Forming a pricing strategy based on the unique value of VPF is crucial to ensuring sustainable unit economics. Food produced indoors has fewer unknown variables, unlike the outdoors, and therefore the price should reflect the added security. In several cases, where vertical farms found product-market fit in their local context, their sales outgrew their production. Even when charging a 20% premium over organic produce, their produce sold out in supermarkets. Learning that organic farming uses pesticides helped drive customer sales. Most farms noted that a growing contract was the best way to mitigate marketing risks but that they were hard to negotiate as most customers were reluctant. They also enabled operators to insure their crops, reduce their risk profile, and reduce waste.

VPF can produce unique combinations of crops with synchronised harvest times. One operator claims that the only way to compete is by offering customers a product they cannot find elsewhere. To achieve this, it is necessary to integrate continuous feedback from the target market, marketing team, and the growing team. In a couple of cases, competitor traditional and vertical farms would engage in predatory pricing strategies (undercutting their product at loss-making levels). Market dynamics were highly correlated to the seasonality of conventional produce.

### 3.5.2 Build collaborations

Farms are increasingly finding that collaboration amongst technology providers, farms, landlords, academics, energy providers, and governments will be essential to advance the sector. This is echoed in the literature (Burwood-Taylor, 2018; Kozai, 2015; Kuack, 2019; Kummer, 2018; Shao et al., 2016; Valle de Souza et al., 2021) and is also becoming apparent through an increase in the membership of vertical farming industry groups such as FarmTech Society (2022) or Japan Plant Factory Association (2022b). Additionally, vertical farms can reduce their risk profile by drawing from expertise without the associated human resource costs by building business collaborations. For example, farms can work with lighting providers and system providers to trial new systems. They can synergise with their location to provide ecosystem services for their landlord in exchange for subsidised rent. Working with educational institutions can access university researchers to conduct R&D.

### *3.5.3 Choosing the right location*

Site selection is key to the philosophy behind VPF, especially in urban environments and is one of the most important decisions that will determine the success of a project. Fundamentally, the goal of VPF is to disrupt an agricultural system more efficiently and sustainably. Three core factors determine whether a location is suitable for a vertical farm: (i) placement in relation to the customers, (ii) the building and its characteristics, and (iii) the geography and environmental conditions.

Each farm is wholly unique, and there is no one-size-fits-all approach. However, based on the learnings reported, some common trends amongst desirable sites are represented in Fig. 11.

Planning permission can be challenging due to a lack of zoning codes. Potential operators should be careful to identify power availability prior to construction. Accessibility will reduce logistical difficulties and improve labour efficiency. Looking for synergy with the building and other businesses is a way to build collaboration, improve resilience, and reduce overheads. VPF should

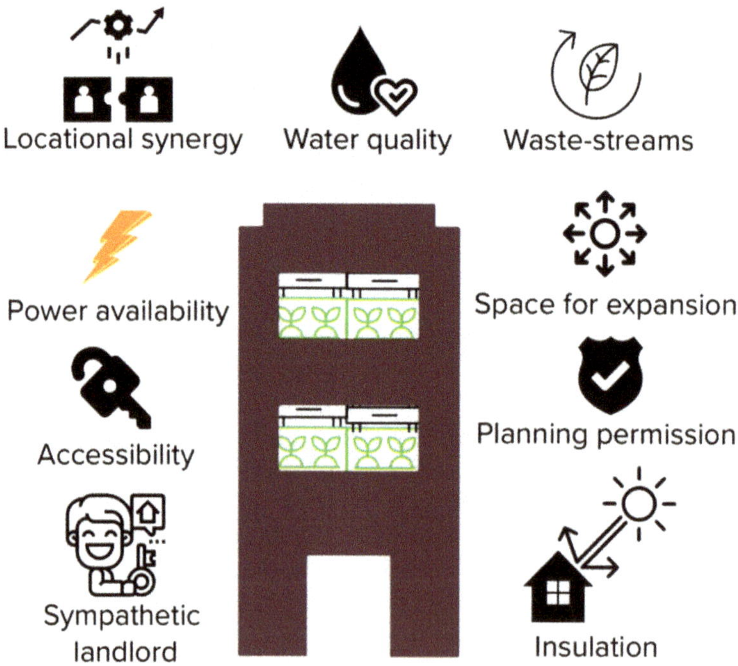


**Figure 11** Building considerations.

synergise with the building location as much as possible, taking advantage of opportunities to recycle heat waste, $CO_2$, or waste. Other opportunities include providing education to local schools or universities, being placed near power plants to utilise electricity during off-peak hours, sub-letting expansion space to other companies, and being next to a clean water supply. Companies that place their farms in locations with strategic alignment with government initiatives (near a new powerplant facility or with sustainability priorities) will also improve their likelihood of receiving financial support.

Participants expressed that not every region is suitable for a vertical farm, whether indoors or using natural lighting. Successful farms pick a location where operators can solve a problem for their local market. For a vertical farm that does not have proper climate control or is exposed to natural light in a greenhouse set-up, it is essential to check environmental numbers with geographers regarding temperature and daily light integral. Lastly, developers that would like to incorporate renewable energy should consider the regional prospects. One farm reported that they could recapture 86% of their energy through waste heat and convert this to geothermal power. Other farms reported capturing solar energy to power their entire farm. Renewable energy-powered farms are becoming increasingly important to address the environmental impact of artificial lighting which is discussed further in Chapter 2.

VPF can vertically integrate the supply chain and take over distribution responsibilities; however, this complicates logistics for a farm. Therefore, some operators choose to outsource distribution. There are three distribution trends amongst vertical farms illustrated in Fig. 12.

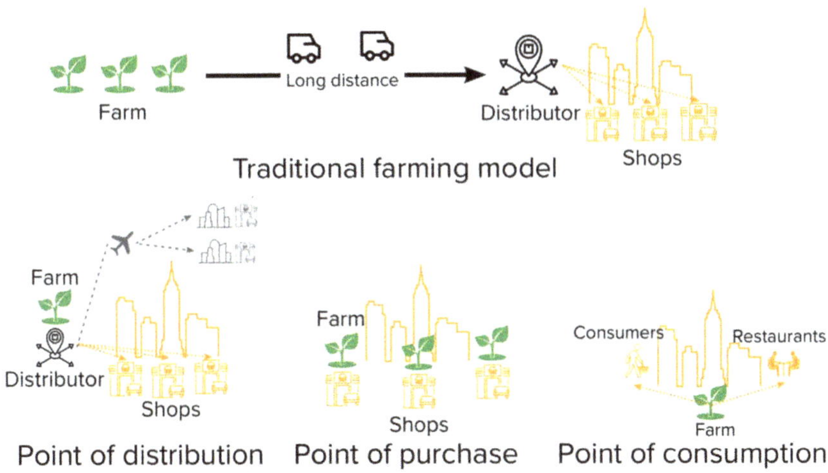


**Figure 12** Vertical farming distribution models.

Placing a farm at the point of consumption is suitable for small farms targeting retail customers that benefit from a partnership with their landlord or co-sharing a space to reduce the high cost of the rent. VPF can also be co-located with traditional farms and provide fodder for livestock. Placing a farm at the distribution point, typically in a peri-urban environment, is suitable for larger farms that target wholesale markets. Placing a farm at the point of purchase eliminates distribution costs and middle management, improving unit economics.

### 3.5.4 Focused business goals

VPF is an exciting endeavour with opportunities to tackle many issues, but practitioners must ground their vision in reality and pick a problem they want to solve. According to several farm owners, those that try to grow food and develop technology simultaneously struggle to do well and eventually have to pick one (Liotta et al., 2017). VPF companies should know what their goal is and apply the 80/20 rule to identify the company's best assets and create maximum value. Farms should choose their priorities amongst the following values: aesthetic, social, education, health, economic, environment, research, and technology development.

On the other hand, only a few farms focused solely on growing. Some claimed they would fail if they had focused entirely on growing, most likely because they were small vertical farms that did not benefit from economies of scale. Smaller farms likely require alternative revenue streams to compensate, as described in Section 3.1.1. Therefore, it is advisable to model and test different business models and growing scenarios to compare the most appropriate strategy.

### 3.5.5 Start with a pilot

A pilot farm is a small-scale version of a planned facility enabling teams to apply their knowledge, make small mistakes, and find a product-market fit. A pilot reduces financial risk by keeping a project low-cost and validating the business plan. It also allows operators to select specific crops and test the appropriate system whilst gaining customer feedback without investing large sums of capital. The goal is to highlight any problems in system operation, plant growth, and marketing.

## 3.6 Theme 6: Risks

VPF is a financially high-risk endeavour with a high chance of failure due to vast uncertainty regarding labour, growth, and economics. Despite this, there is little understanding of the risks associated with vertical farming specifically and what causes them to close down. IVF protects agriculture from outdoor risks that are typically hard to control. For example, a greenhouse is subject to weather and climate risks and a traditional farm is even more so. IVF has managed to de-risk many of these factors from agriculture. However, it has traded these for new risks that have yet to be fully understood. We asked participants about the challenges and risks they faced when operating their farm. If the farm had closed down, the objective was to explore the root cause. The theme is split into a risk taxonomy and a discussion of the failure modes.

### 3.6.1 Risk taxonomy

From all the data items, reported risks were categorised into financial (Table 9), environmental (Table 10), production (Table 11), labour (Table 12), technological (Table 13), political (Table 14), and market risks (Table 15). A risk taxonomy can enable management teams to control risks more efficiently, plan risk-reduction actions, and avoid knee-jerk reactions. However, this risk taxonomy is exploratory and does not categorise risks in terms of likelihood or impact. Each farm considered was a unique case, and they did not have robust risk reporting protocols to provide detailed information. Moreover, as a young industry there will likely be risks that operators have not experienced yet. The 'percentage of farms' is relative to the sample size that reported the risk. If the percentage is 0%, that indicates the risk is known, but no farm operator within the sample had encountered it.

**Table 9** Financial risks descriptions

| Type | Source of risk | Description | % of farms |
|---|---|---|---|
| Internal | Debt risk | Unable to repay loans resulting in penalty fees and interest. | 3.4 |
| | Misalignment of investor's expectations | Investor's expectations do not align with the goals of the business resulting in diluted strategic focus. | 19.6 |
| | Funding risk | Unsuccessful in acquiring funding and inability to sustain capital runway potentially resulting in bankruptcy. | 16.8 |
| External | Electricity price | Increased energy costs resulting in lower profit margins. | 6.7 |
| | Rent price | Increased rent costs resulting in lower profit margins. | 3.4 |
| | Liquidity risk | Difficulty in converting assets into cash. | 0 |
| | Nutrient price | Increased consumable costs resulting in lower profit margins. | 6.7 |

**Table 10** Environmental risk descriptions

| Type | Source of risk | Description | % of farms |
|---|---|---|---|
| Internal | Daily light integral | Insufficient light intensity and duration to support the crop type grown. | 3.4 |
| | Temperature fluctuations | The farm is susceptible to heat waves or freezing temperatures from inadequate climate control. | 13.3 |
| | Airflow/humidity | Temperature hot spots and lack of sufficient airflow and humidity control prohibit uniform plant quality. | 3.4 |
| | Irrigation risk | Problems with irrigation systems providing water and nutrients to plants (clogged pipes or filters). | 33.6 |
| | Organic acids | Plant hormones affecting crop quality and growth requiring frequent disposal of nutrient solution. | 0 |
| | Fire | A fire may result from a faulty electrical hook-up, a lighting malfunction, or other reasons. | 0 |
| External | Pandemic/Epidemic | The effects of a pandemic could require higher food hygiene practices, farm closure due to employee sickness, and loss of customers. | 13.1 |
| | Earthquake | An earthquake which could cause structural damage, electrical problems, or cracked pipes. | 6.7 |
| | Flooding | Flooding of the facility from leakages or external areas resulting in a lack of workforce or reduced productivity. | 0 |
| | Typhoon | Typhoon in the region affecting electricity availability, building, or workforce. | 0 |

None of the risks described are specific to VPF. However, the weighting of risks compared to traditional agriculture lies towards labour, finance, production, and technology instead of the environment. IVFs lean more heavily towards financial risk due to the required capital and operating costs. There is also a trade-off because an IVF can operate independently of the natural environment and loses the buffer that nature provides, requiring constant control. Any problem in an indoor environment is likely to be magnified as it affects the entire operation. Technology plays a more prominent role in growing crops, introducing more points of failure throughout the entire operation. 33.6% of

**Table 11** Production risk descriptions

| Type | Source of risk | Description | % of farms |
|---|---|---|---|
| Internal | Contamination | Improper cleaning protocols or human contact can result in unwanted bacteria, fungi, or pests. | 13.3 |
| | Process bottleneck | Manual work or limited machinery capacity may result in scaling back production capacity. | 6.6 |
| | Shrink or waste | Waste from mishandling, improper storage, or delays in the supply chain can result in reduced yield and sales. | 10.3 |
| | Plant diseases | Plant diseases could result in crop failure, reduced yield, and reduced sales. | 16.8 |
| | Human pathogen | Adverse health impacts on customers that could cause serious liability consequences. | 0 |
| | Pest outbreak | Pests may infest the facility resulting in crop failure, reduced yield, and a deep clean of equipment. | 23.3 |
| | Physiological disorders | Combination of growing factors may result in defects such as tip burn, necrosis, chlorosis, and shortened stem rotting. This results in reduced product quality and sales. | 30.0 |
| | Crop uniformity | The crops produced are not consistent due to challenges with airflow, temperature, lighting, nutrients, etc. | 6.6 |
| External | Electrical outages | Light, irrigation, climate control, and automated systems are dependent on power availability. May result in crop loss. | 6.6 |
| | Delivery issue | Product is damaged in transit (shrinkage, left at room temperature, or damage) or distribution is delayed. | 3.4 |

operators reported equipment failure, which could hamper crop quality and have high repair costs. Labour issues, predominantly human error, were the most commonly reported risk (53.3%) and could result in 100% crop loss due to an accident like incorrect dosing of nutrients or pH. Similarly, pest outbreaks and physiological disorders have led to crop failure. Electricity price is the largest risk at this point in time for many operators; however, the interviews were conducted over several years before the coronavirus pandemic until after Russia's invasion of Ukraine which has indirectly increased global energy costs. Therefore, the risk of electricity price increase did not affect most farms at the time of interviews.

**Table 12** Labour risk descriptions

| Type | Source of risk | Description | % of farms |
|---|---|---|---|
| Internal | Allergic reactions | Reactions due to crops, insects, or chemicals could cause severe and dangerous reactions in some people. | 6.6 |
| | Chemical exposure | Exposure to Ozone, hydrogen peroxide, or other chemicals that could harm employees. | 3.4 |
| | Human error | Improper seeding, overdosing nutrients or pH, careless harvesting, etc. could impact crop quality and yield. | 53.3 |
| | Staffing shortage | Reduced productivity and capacity to meet customer orders. | 13.1 |
| | Industrial accident | An injury when operating machinery or a collision of objects. | 6.6 |
| | Overexertion and repetitive strain | Manual work leading to employee injury that can result in the loss of employees and a liability issue. | 6.6 |
| | Slips, trips, and falls | Falling over can harm to employees and pose a liability issue. | 6.6 |
| | Loss of expertise | Loss of a skilled employee, such as the head grower, could impact production and expected improvements. | 0 |
| | Lack of management qualities | Management may not be sufficiently experienced. Lack of connection between management and growing team can also result in reduced productivity. | 0 |
| Exter-nal | Shortage of skilled workers | Not enough skilled workers in the local area to work on the farm for specific roles. | 13.1 |

Profitable farms attributed some of their success to a good understanding of risk management and conducting a risk register before building their facility. Long-standing operators said that after running IVF for several years, many of the problems arise. Specifically, after 5–10 years when LED lighting requires replacing and many farms are not designed for ease of maintenance.

The risks are interconnected and can affect one another. For example, a pest outbreak may result from improper growing conditions due to a HVAC failure. Technology advancements could result in a competitor risk with higher unit economics and better plant quality. A geopolitical crisis may increase the electricity price. Fig. 13 shows the interconnectivity of the main risks detailed within this study and how they connect. Further investigation is required to determine the risks of the greatest concern.

**Table 13** Technology risk descriptions

| Type | Source of risk | Description | % of farms |
|---|---|---|---|
| Internal | Equipment failure | Problems with HVAC, lighting, refrigeration, growing systems, automation, or software issues. | 33.6 |
| | Overengineering | Using technology that is above the required specification or adds unnecessary complexity. | 6.6 |
| External | Lack of continuity from supplier or vendor | Initial provider is not available for guidance or part replacement causing challenges for maintenance and troubleshooting. | 6.6 |
| | False information provided by suppliers/vendors | Reduction in production capacity and increase in labour costs compared to estimates. | 13.1 |
| | Technological advancements | Advancements lead to obsolete technology which is no longer competitive. | 6.6 |

**Table 14** Political risk descriptions

| Type | Source of risk | Description | % of farms |
|---|---|---|---|
| Internal | Internal company disagreement | Lack of strategic alignment leading to loss of employees, productivity, or product quality. | 6.6 |
| | Landlord dispute | Change in terms of agreement leading to the requirement to relocate. | 10.3 |
| External | Governmental policy | Change of policy could benefit or hurt production. | 3.4 |
| | Government ban | Government bans a crop type due to a food safety scare. | 3.4 |
| | Geopolitical issues | Challenges associated with geopolitical issues. | 6.7 |

### 3.6.2 Failure modes

VPF is operationally complex, and there are many potential modes of failure. The failure rate is high, yet there is no data for the number of vertical farms globally, let alone the failure rate. Even as VPF technology and production advance, the sector suffers from a knowledge gap which forces operators to seek information from outdoor growing practices or greenhouses. This study draws upon the experiences of nine shuttered farms (47% of the farms examined) and three consultants to understand failure modes. Figure 14 diagrammatically represents the reported causes of failure alongside indirect causes.

**Table 15** Market risk descriptions

| Type | Source of risk | Description | % of farms |
|---|---|---|---|
| External | Price risk | Misestimating the market price of the product to be sold. | 6.7 |
| | Commercial risk | Excess supply without enough customers. | 3.4 |
| | Competitor risk | Comparable products at a lower price or higher quality result in a lower market share. | 10.3 |
| | Supply chain risk | Unable to source certain supplies or receiving damaged consumables. | 13.1 |
| | Customer withdrawal | A customer retracts their purchase order. | 3.4 |
| | Customer complaint | Customer complaints tarnish the reputation of the business. | 3.4 |

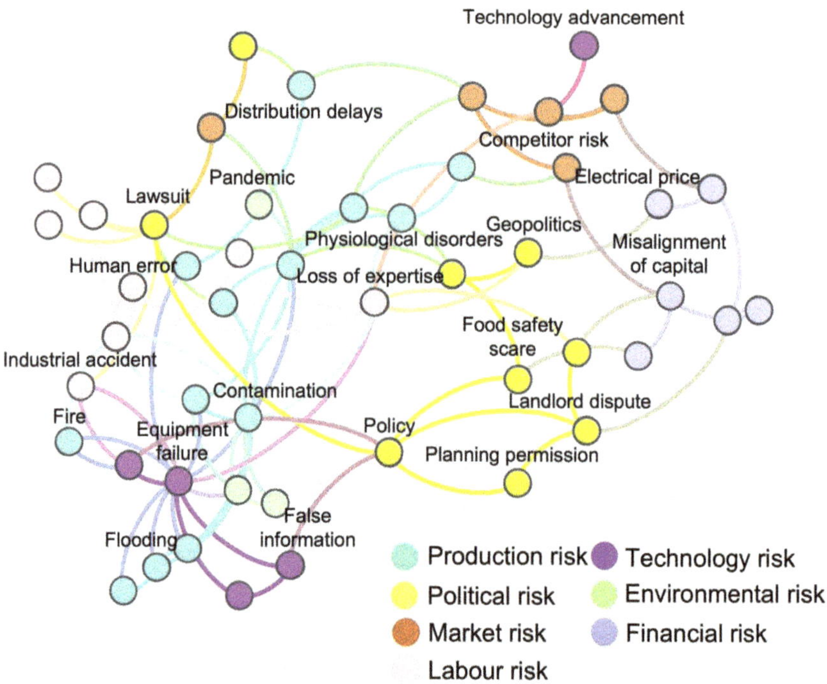


**Figure 13** Risk interconnections map.

The most common modes of failure were related to funding (70% of closed farms). Either farms were undercapitalised and at risk of not raising further funding or misaligned between investors' expectations and project goals. Due to unfavourable unit economics during the development cycle,

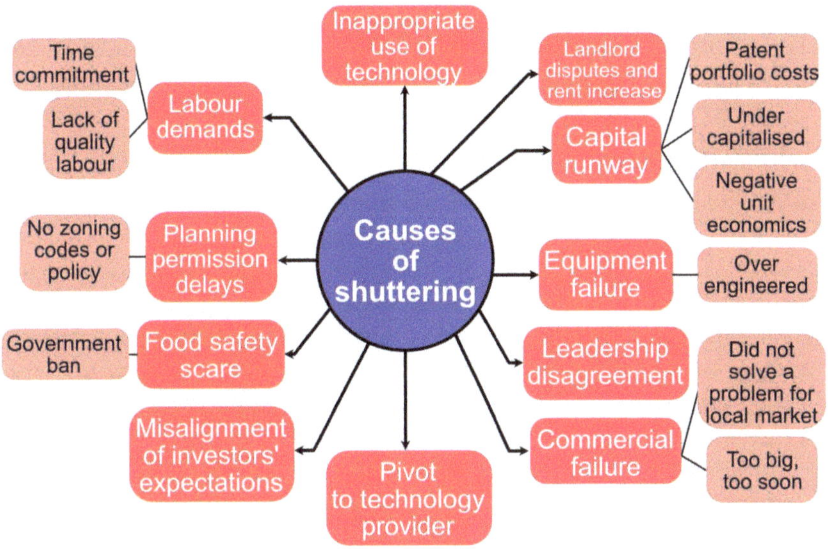


**Figure 14** Modes of failure reported by farm owners.

farm owners could not bootstrap a business and required sufficient capital runway. As a result, they are dependent on capital investment (as described in Section 3.1.2).

The second highest reported failure mode was due to labour, or lack thereof. Farm owners struggled to find skilled labour to sustain their operations and could not provide the time commitment required themselves. This resulted in the shuttering of two farms and prohibited the scaling of others.

Another element in failure was building an ambitious project without first conducting a pilot to find a product-market fit. Big projects that have not validated the business model and technology at scale seem to run into costly obstacles that are difficult to overcome once decisions are locked in place. Some systems were overengineered, which can have higher failure points and, therefore, more repair costs that bankrupted a farm. According to one consultant, inappropriate technology use was the most frequent cause of failure that they had witnessed. In other cases, they experienced commercial failure with insufficient demand to justify their business case.

Unforeseen planning permission delays due to a lack of zoning codes can also hinder the development of farms. In one particular case, environmentalists protested against constructing a multi-tier farm resulting in a 4-year delay in construction. In addition, the lack of government awareness has been detrimental in other ways, banning certain crop sales due to a food safety scare that meant one farm had to close down operations entirely even when VPF was unaffected by the particular risk. Other challenges that were reported but are not unique to VPF were leadership disputes, rent increases, commercial failure, and founder burnout due to operational complexity and opportunity cost.

## 4 Conclusion

The VPF industry is uniquely placed to disrupt the distribution methods of the perishable produce industry and can provide unique benefits (high-quality, consistent, and local produce with no pesticides). Given its potential as a secure and sustainable source of fresh food, vertical farms are rising globally. However, there are many challenges that VPF has to overcome to become mainstream, and there are learnings from many dissolved projects. This study analysed the combined experiences of 20 VPF practitioners and 10 secondary sources through mixed-method TA to investigate lessons learned.

Growing factors and how they interact are nuanced, and there was a lack of understanding amongst operators about addressing the limiting factor. This is compounded by the lack of knowledge about crop growth recipes, with many farms having to develop their own through trial and error. Literature and guidance on how to approach addressing the limiting factor are needed. VPF is capable of higher food safety measures from being in a controlled environment, unaffected by external food safety issues from soil-based agriculture. Many operators do not meet this potential, primarily due to a lack of understanding, guidance, and literature. This is a crucial element to address so that VPF is taken seriously.

The experiences highlighted the still-experimental nature of the practice, where almost all practitioners had to engage in R&D alongside commercial growing resulting in a high cash burn rate. Profitable farms were careful to limit their R&D efforts in terms of space and finance. Some decided to become technology developers solely because the technology was advancing too quickly to remain competitive as commercial growers. The precise control of IVF and rapid crop cycles enables the transfer of scientific findings between academia and farms, which removes an obstacle in greenhouses and open-field farms. However, data aggregation requires collaboration amongst farms to develop standardised metrics and definitions. Definitions have been presented that now need to be integrated into business operations (Kozai, 2022c; Kozai et al., 2022). More metrics and indices are sorely needed and will accelerate R&D, avoid the need for duplicative efforts, and eventually enable growers to focus primarily on growing solely.

Energy efficiency and automation are two vital areas for improvement. Moreover, VPF technology may not always be an appropriate choice, and operators should carefully analyse the selected product's CAPEX and OPEX requirements relative to other growing methods. Finally, they should prioritise listening to their target customer, and solving a problem for them by using technology. If they choose to create an entirely new product not available elsewhere, then they will need to invest capital and time to build the market which opens a business to substantial commercial risk.

The most significant barrier for VPF was funding, and to substantially contribute to agri-food supply chains, the set-up costs must come down. Vertical farms are capital-intensive, and because of this, investors expect significant profit margins and cash flow. Therefore, it is essential to communicate the reality of VPF to acquire suitable types of investment. Producing an operational history showing continuous improvement and reducing volatility alongside unit economics and a large total addressable market are steps that can help when seeking funding.

Labour was another primary challenge for farms. It was the highest cost for all the farms involved and was the most frequently reported limiting factor in production. VPF was most susceptible to labour and technology risks, whereby one mistake from an employee could result in crop failure for the entire farm. It is important to design a farm and processes that are not dependent on an individual's competence to avoid such accidents. Compared to traditional agriculture, where the environment provides a buffer but is also the primary source of risk, VPF appears to magnify labour and technology risks substantially. A competent and experienced team in food production is helpful to mitigate this. Improving process flow through tried and tested methods of manufacturing principles could be vital to improving profitability and reducing risk. Further work is required to apply these principles in an indoor farm environment.

The study is limited by a relatively small sample size across leafy greens, herbs, microgreens, and edible flowers. Further investigation is required amongst larger-scale producers and other produce such as plantceuticals, mushrooms, berries, transplants, and fodder (as described elsewhere in the book).

This study is the first to examine and report risk factors and failure modes for VPF and was conducted in conjunction with a financial risk assessment study (Baumont de Oliveira et al., 2022). However, further work is required to engage VPF operators to prioritise and rank risks to evaluate likelihood, impact, and mitigation measures. In addition, longitudinal production data will enable more precise financial and risk estimation.

## 5 Guidelines

### *5.1 Economics*

- Seek subsidised or free rent and utilities from a sympathetic landlord or governmental subsidies where possible.

- Create alternative revenue streams to buffer the farm from dependence on sales.
- Find a patient funding provider with aligned expectations to the project's goals.
- Account for depreciation and repair costs in projections.
- Raise enough capital to allow for commissioning and experimentation phases.
- Estimate financial projections with validated data and add contingency margins on yield, waste, and costs.
- Purchase the maximum possible light efficiency at the best value when purchasing lights.
- Consider a 50–200 $m^2$ facility for the retail market or a 1000-6000 $m^2$ facility with modularised cultivation rooms for the wholesale market.
- Labour is the highest cost, and estimations should account for challenges, workflow bottlenecks, and slower working speeds (see Section 3.2).
- Calculate unit economics and tie them to customer sales to drive decision-making.

## 5.2 Labour

- Hire a team with food production experience and predominant skills in horticulture, engineering, economics, and marketing.
- Hire or outsource knowledge in food safety, manufacturing, and plumbing.
- Pay over minimum wage with incentives to improve labour satisfaction and engagement.
- Hire flexible labour for peak work periods.
- Apply manufacturing principles to improve the process flow of the farm (see Baumont de Oliveira et al., 2020).
- Cultivate connection between management and growing teams.
- Consider accessibility, maintainability, and repairability of the farm and systems.
- Purchase automation only when necessary (see Fig. 6).

## 5.3 Growing

- Create a strategy for identifying the limiting factor that considers all interconnected growing factors in Fig. 8 and Table 6.
- Ensure proper food safety practices to minimise contamination risk.
- Avoid building a farm whilst operating it unless wanting to pivot into technology development.

## 5.4 Technology

- Conduct trials and ensure appropriate technology selection that combines competitive plant quality and economics.
- Select the technology by reverse-engineering from the crop and the customer.
- Measure net yield (kg/month), unit economics (£/kg), and crop productivity per unit of energy (kg/kWh) and labour (kg/man-hour).
- Have a deliberate data collection and action strategy to build the foundation for machine learning for managing growing factors.
- Create an operational history of yield per crop type since starting the farm, ideally showing increasing performance with reducing volatility to help seek funding.
- Enhance agility with flexible systems like DFT and mobile racks, especially for smaller farms.
- Keep systems simple and avoid overengineering the facility, focusing on the customer and the crop type.
- Consider accessibility, maintainability, and repairability of the farm and systems as critical criteria.
- Bring in a HVAC expert with experience in indoor horticulture to properly size requirements.
- Integrate systems early with an experienced farm designer or connecting systems will require substantial cost and time.

## 5.5 Strategy

- Start with a pilot to validate the technology and business model.
- Select and prioritise the values and goals of the business with the 80/20 rule.
- Create a backup plan.
- Build collaborations with academia, system suppliers, location partners, and other farms.
- Choose a site with characteristics shown in Fig. 11.
- Place the farm in the right area relative to the customer (see Fig. 12).
- Form a unique pricing strategy based on the value of VPF produce.
- Secure a growing contract to reduce waste and insure the crop.

## 5.6 Risk

- Create a risk management plan before building a PFAL to avoid knee-jerk reactions and costly mistakes based on Section 3.6.
- Consider the common failure modes and strategies to mitigate these challenges.

## 6 Ethics statement

The study received full ethical approval from the Liverpool University ethics committee. All participants provided informed verbal or written consent to participate and were advised that they were free to withdraw if they wished to do so. All interviews were recorded with the participants' permission and responses were anonymised.

## 7 Acknowledgements

Thank you to all the generous participants of this study who shared their time and experience to bring to light the lessons within this paper to benefit the entire industry. The special opportunity to interview Japanese plant factory operators was only possible because of support from Eri Hayashi, Na Lu, and the Japan Society for Promotion of Science Summer 2022 Programme for international research fellows. Thank you to Jens Thomas for his unwavering mentorship and helpful editing. This research was supported by the Low Carbon Eco-Innovatory (funded by European Regional Development Fund, Grant number: 22R16P00045).

## 8 References

808 Factory. 2022. [IMG] 808 Factory webpage. Available at: https://www.808factory.jp/ (accessed 8.30.22).

Adams, D. C. and Salois, M. J. 2010. Local versus organic: A turn in consumer preferences and willingness-to-pay. *Renewable Agriculture and Food Systems* 25(4), 331–341. https://doi.org/10.1017/S1742170510000219.

Agrilyst. 2018. State of indoor farming. Available at: https://artemisag.com/guides_reports/ (accessed 3.27.22) (vol. 2017).

Agritecture. 2021. 8 Mistakes vertical farms make. Agritecture blog. Available at: https://www.agritecture.com/blog/2021/7/15/8-mistakes-vertical-farms-make (accessed 7.20.21).

Agritecture. 2019. Global CEA census 2019. Available at: https://www.agritecture.com/census (accessed 9.14.22).

Agritecture. 2020. Global CEA census report. Available at: https://www.agritecture.com/census (accessed 9.14.22).

Avgoustaki, D. D. and Xydis, G. 2020. Indoor vertical farming in the urban nexus context: Business growth and resource savings. *Sustainability* 12(5), 1–18. https://doi.org/10.3390/su12051965.

Avgoustaki, D. D. and Xydis, G. 2021. Energy cost reduction by shifting electricity demand in indoor vertical farms with artificial lighting. *Biosystems Engineering* 211, 219–229. https://doi.org/10.1016/j.biosystemseng.2021.09.006.

Baiyin, B., Tagawa, K., Yamada, M., Wang, X., Yamada, S., Shao, Y., An, P., Yamamoto, S. and Ibaraki, Y. 2021. Effect of nutrient solution flow rate on hydroponic plant growth and root morphology. *Plants* 10(9), 1–14. https://doi.org/10.3390/plants10091840.

Balázs, L., Dombi, Z., Csambalik, L. and Sipos, L. 2022. Characterizing the spatial uniformity of light intensity and spectrum for indoor crop production. *Horticulturae* 8(7), 644. https://doi.org/10.3390/horticulturae8070644.

Baumont de Oliveira, F. 2021. *Personal Correspondence via Video Call with Private Investors and Banks (Confidential).*

Baumont de Oliveira, F. 2022. A typology review of vertical farming: Classifications, configurations, business models and economic analyses. Available at: https://www.researchgate.net/profile/Francis-Baumont-De-Oliveira/research (accessed 9.14.2022).

Baumont de Oliveira, F., Forbes, H., Schaefer, D. and Milisavljevic, J. 2020. Lean principles in vertical farming: A case study. *CIRP Procedia* 93, 712–717.

Baumont de Oliveira, F. J., Ferson, S., Dyer, R. A. D., Thomas, J. M. H., Myers, P. D. and Gray, N. G. 2022. How high is high enough? Assessing financial risk for vertical farms using Imprecise Probability. *Sustainability* 14(9), 5676. https://doi.org/10.3390/su14095676.

Braun, V. and Clarke, V. 2006. Using thematic analysis in psychology. *Qualitative Research in Psychology* 3(2), 77–101. https://doi.org/10.1191/1478088706qp063oa.

Braun, V. and Clarke, V. 2012. Thematic analysis. APA handbook of research methods in psychology. *Research Designs: Quantitative, Qualitative, Neuropsychological, and Biological* 2, 57–71. https://doi.org/10.1037/13620-004.

Brodwin, E. 2019a. A futuristic farming startup raised $260 million from Jeff Bezos and SoftBank on the promise of upending agriculture. Insiders are raising questions. *Business Insider*. Available at: https://www.businessinsider.com/bezos-softbank-farming-startup-plenty-insiders-2019-10?r=US&IR=T (accessed 10.2.21).

Brodwin, E. 2019b. The Epstein-funded MIT lab has an ambitious project that purports to revolutionize agriculture. Insiders say it's mostly smoke and mirrors. *Business Insider*. Available at: https://www.businessinsider.com/mit-media-lab-personal-food-computers-dont-work-fake-staff-say-2019-9?r=US&IR=T (accessed 10.2.21).

Burwood-Taylor, L. 2018. Japan's indoor AG sector is becoming more collaborative. *AgFunder*. Available at: https://agfundernews.com/japans-indoor-ag-sector-is-becoming-more-collaborative.html (accessed 6.25.18).

Butler, J. R. A., Davila, F., Alders, R., Bourke, R. M., Crimp, S., McCarthy, J., McWilliam, A., Palo, A. S. M., Robins, L., Webb, M. J., van Wensveen, M., Sanderson, T. and Walker, D. 2021. A rapid assessment framework for food system shocks: Lessons learned from COVID-19 in the Indo-Pacific region. *Environmental Science and Policy* 117, 34–45. https://doi.org/10.1016/j.envsci.2020.12.011.

Carotti, L., Graamans, L., Puksic, F., Butturini, M., Meinen, E., Heuvelink, E. and Stanghellini, C. 2020. Plant factories are heating up: Hunting for the best combination of light intensity, air temperature and root-zone temperature in lettuce production. *Frontiers in Plant Science* 11, 592171. https://doi.org/10.3389/fpls.2020.592171.

Cherif, M., Tirilly, Y. and Belanger, R. R. 1997. Effect of oxygen concentration on plant growth, lipid peroxidation, and receptivity of tomato roots to Pythium under hydroponic conditions. *European Journal of Plant Pathology* 103(3), 255–264.

Choi, K. Y., Paek, K. Y. and Lee, Y. B. 2000. Effect of air temperature on tipburn incidence of butterhead and leaf lettuce in a plant factory. *Transplant Production in the 21st Century 166–171*. https://doi.org/10.1007/978-94-015-9371-7_27.

Colangelo, R. 2021. Indoor ag Sci Café #34 - why vertical farms succeed?. Indoor ag sci cafe YouTube channel. *YouTube*. Available at: https://www.youtube.com/watch?v=G6XfQ1OGM5U&t=8s&ab_channel=ControlledEnvironmentPlantPhysiologyandTechnology (accessed 11.30.21).

Compant, S., Duffy, B., Nowak, J., Clément, C. and Barka, E. A. 2005. Use of plant growth-promoting bacteria for biocontrol of plant diseases: Principles, mechanisms of action, and future prospects. *Applied and Environmental Microbiology* 71(9), 4951-4959. https://doi.org/10.1128/AEM.71.9.4951-4959.2005.

Crumpacker, M. 2018. A look at the reasons why vertical farms fail. Medium. Available at: https://medium.com/@MarkCrumpacker/a-look-at-the-reasons-why-vertical-farms-fail-b6a57caae731 (accessed 2.14.22).

De Zwart, H. F. 2012. Lessons learned from experiments with semi-closed greenhouses. *Acta Horticulturae* 952, 583–588. https://doi.org/10.17660/ActaHortic.2012.952.74.

DH. 2017. News. Downtown Vancouver's vertical urban farm closes, files for bankruptcy. *Daily Hive*. Available at: https://dailyhive.com/vancouver/downtown-vancouvers-rooftop-garden-closes-files-bankruptcy (accessed 4.9.22).

Donatelli, M., Magarey, R. D., Bregaglio, S., Willocquet, L., Whish, J. P. M. and Savary, S. 2017. Modelling the impacts of pests and diseases on agricultural systems. *Agricultural Systems* 155, 213–224. https://doi.org/10.1016/j.agsy.2017.01.019.

Esmaili, M., Aliniaeifard, S., Mashal, M., Ghorbanzadeh, P., Seif, M., Gavilan, M. U., Carrillo, F. F., Lastochkina, O. and Li, T. 2020. $CO_2$ enrichment and increasing light intensity till a threshold level, enhance growth and water use efficiency of lettuce plants in controlled environment. *Notulae Botanicae Horti Agrobotanici Cluj-Napoca* 48(4), 2244–2262. https://doi.org/10.15835/nbha48411835.

Faires, S. and Charlebois, S. 2020. Canada food price report. *Engineering* 2019, 1–5.

FarmTech Society. 2022. FarmTech webpage. Available at: https://farmtechsociety.org/ (accessed 9.1.22).

Gao, W., He, D., Ji, F., Zhang, S. and Zheng, J. 2020. Effects of daily light integral and LED spectrum on growth and nutritional quality of hydroponic spinach. *Agronomy* 10(8), 1082. https://doi.org/10.3390/agronomy10081082.

Gates, R. S., Zolnier, S. and Buxton, J. 1998. Vapor pressure deficit control strategies for plant production. *IFAC Proceedings Volumes* 31(12), 271–276. https://doi.org/10.1016/S1474-6670(17)36076-7.

Gillespie, D. P., Kubota, C. and Miller, S. A. 2020. Effects of Low pH of hydroponic nutrient solution on plant growth, nutrient uptake, and root rot disease incidence of Basil (Ocimum basilicum L.). *Hortscience* 55(8), 1251–1258. https://doi.org/10.21273/HORTSCI14986-20.

Goto, E., Both, A. J., Albright, L. D., Langhans, R. W. and Leed, A. R. 1996. Effect of dissolved oxygen concentration on lettuce growth in floating hydroponics. *Acta Horticulturae* 440, 205-210. https://doi.org/10.17660/ActaHortic.1996.440.36.

Graamans, L., van den Dobbelsteen, A., Meinen, E. and Stanghellini, C. 2017. Plant factories; crop transpiration and energy balance. *Agricultural Systems* 153, 138–147. https://doi.org/10.1016/j.agsy.2017.01.003.

greenLand. 2022. [IMG] greenLand webpage. Available at: https://greenland-farm.com/ (accessed 8.30.22).

Grossiord, C., Buckley, T. N., Cernusak, L. A., Novick, K. A., Poulter, B., Siegwolf, R. T. W., Sperry, J. S. and McDowell, N. G. 2020. Plant responses to rising vapor pressure deficit. *New Phytologist* 226(6), 1550–1566. https://doi.org/10.1111/nph.16485.

Hardej, P. 2022. [IMG]. FarmedHere vertical farm. Available at: https://images.fascompany.net/image/upload/w_596,c_limit,q_auto:best,f_auto/fc/3053217-inline-i-2-farmedhere-wants-to-bring-a-vertical-farm-to-your-city-copy.jpg/ (accessed 9.14.22).

Hayashi, E. and Kozai, T. 2018. *Smart Plant Factory* (1st edn.). Springer, Singapore. https://doi.org/10.1007/978-981-13-1065-2.

Hendrickson, T., Dunn, B. L. and Hall, A. 2022. Effects of elevated water temperature on growth of basil using nutrient film technique. HortScience 57, 925–932.

Hwa-Soo, L., Sook-Youn, K. and Jae-Hyun, L. 2014. Improvement of light uniformity by lighting arrangement for standardized crop production. *Journal of Central South University* 21(11), 4311–4319. https://doi.org/10.1007/s11771-014-2430-5.

Japan Plant Factory Association. 2022a. JPFA online introductory course (English). Available at: https://select-type.com/s/JPFA-Training (accessed 8.1.22).

Japan Plant Factory Association. 2022b. Japan plant factory association Webpage. Available at: https://npoplantfactory.org/en/ (accessed 9.1.22).

Katsoulas, N. and Kittas, C. 2008. Impact of greenhouse microclimate on plant growth and development with special reference to the Solanaceae. *The European Journal of Plant Science and Biotechnology-Global Science Books*, 31–44.

Khodijah, N. S., Santi, R., Kusmiadi, R. and Asriani, E. 2021. The growth rate of hydroponic lettuce at various nutrient compositions from liquid synthetic, solid synthetic, and liquid organic fertilizers. *Anjoro: International Journal of Agriculture and Business* 2(2), 41–49. https://doi.org/10.31605/anjoro.v2i2.993.

Kowalczyk, K., Mirgos, M., Baczek, K., Niedzińska, M. and Gajewski, M. 2016. Effect of different growing media in hydroponic culture on the yield and biological quality of lettuce (Lactuca sativa var. capitata). *Acta Horticulturae* 1142, 10–110. https://doi.org/10.17660/ActaHortic.2016.1142.17.

Kozai, T. 2015. *PFAL Business and R & D in the World: Current Status and Perspectives: Current Status and Perspectives, Plant Factory: An Indoor Vertical Farming System for Efficient Quality Food Production*. Elsevier Inc., Amsterdam, Netherlands. https://doi.org/10.1016/B978-0-12-801775-3.00003-2.

Kozai, T. 2018a. Current status of plant factories with artificial lighting (PFALs) and smart PFALs. In: Kozai, T. (Ed.), *Smart Plant Factory: The Next Generation Indoor Vertical Farms*. Springer, Singapore.

Kozai, T. 2018b. Designing a cultivation system module (CSM) considering the cost performance: A step toward smart PFALs. In: Kozai, T. (Ed.), *Smart Plant Factory: The Next Generation Indoor Vertical Farms*. Springer, Singapore, pp. 57–80. https://doi.org/10.1007/978-981-13-1065-2.

Kozai, T. 2019. Plant production process, floor plan, and layout of PFAL. In: *Plant Factory: An Indoor Vertical Farming System for Efficient Quality Food Production* (2nd edn.). Elsevier Inc., Amsterdam, Netherlands, pp. 261–271. https://doi.org/10.1016/B978-0-12-816691-8.00019-4.

Kozai, T. 2022a. Balances and use efficiencies of $CO_2$, water, and energy. In: Kozai, T., Niu, G. and Masabni, J. (Eds.), *Plant Factory Basics, Applications and Advances*. Academic Press, Cambridge, Massachusetts. https://doi.org/10.1016/b978-0-323-85152-7.00003-3.

Kozai, T. 2022b. Requirements and features of the cultivation system modules in advanced plant factories with artificial lighting. In: Hayashi, E. and Kozai, T. (Eds.), *Advances in Plant Factories: New Technologies in Indoor Vertical Farming*. Burleigh Dodds Science Publishing, Cambridge, UK.

Kozai, T. 2022c. *Terms Related to PFALs*. In: Kozai, T., Niu, G. and Masabni, J. (Eds.), *Plant Factory Basics, Applications and Advances*. Academic Press, Cambridge, Massachusetts. pp. 11-23. https://doi.org/10.1016/b978-0-323-85152-7.00007-0.

Kozai, T. and Niu, G. 2019. Role of the plant factory with artificial lighting (PFAL) in urban areas. In: *Plant Factory: An Indoor Vertical Farming System for Efficient Quality Food Production* (2nd edn.). Elsevier Inc., Amsterdam, Netherlands, pp. 7–34. https://doi.org/10.1016/B978-0-12-816691-8.00002-9.

Kozai, T., Tsukagoshi, S. and Sakaguchi, S. 2018. Toward nutrient solution composition control in hydroponic system. In: Kozai, T. (Ed.), *Smart Plant Factory: The Next Generation Indoor Vertical Farms*. Springer, Singapore, pp. 395–403. https://doi.org/10.1007/978-981-13-1065-2_24.

Kozai, T., Uraisami, K., Kai, K. and Hayashi, E. 2022. Productivity: Definition and application. In: Kozai, T., Niu, G. and Masabni, J. (Eds.), *Plant Factory Basics, Applications and Advances*. Academic Press, Cambridge, Massachusetts, pp. 197–216. https://doi.org/10.1016/B978-0-323-85152-7.00009-4.

Kuack, D. 2019. Could open source agriculture revolutionize the industry? *Urban AG News*. Available at: https://urbanagnews.com/blog/exclusives/food-for-thought-could-open-source-agriculture-revolutionize-the-industry/ (accessed 9.14.2022).

Kubota, C. 2019. Biological factor management. In: *Plant Factory: An Indoor Vertical Farming System for Efficient Quality Food Production* (2nd edn.), pp. 347–356. https://doi.org/10.1016/B978-0-12-816691-8.00024-8.

Kummer, C. 2018. Trouble in the high-rise hothouse. *NEO*. Life Technologies (Corporation). Available at: https://medium.com/neodotlife/vertical-farming-paul-gauthier-76e81-ace79d0 (accessed 10.3.21).

Lam, V. P., Choi, J., Kim, S., Park, J. and Hernandez, R. 2019. Optimizing plant spacing and harvest time for yield and glucosinolate accumulation in watercress (Nasturtium officinale l.) grown in a hydroponic system. *Horticultural Science and Technology* 37, 733–743. https://doi.org/10.7235/HORT.20190073.

Lau, V. and Mattson, N. 2021. Effects of hydrogen peroxide on organically fertilized hydroponic lettuce (*Lactuca sativa* l.). *Horticulturae* 7(5), 106. https://doi.org/10.3390/horticulturae7050106.

Leaflab. 2022. Leaflab webpage. Available at: https://leaflab.tech/ (accessed 8.30.22).

Lee, J. Y., Rahman, A., Azam, H., Kim, H. S. and Kwon, M. J. 2017. Characterizing nutrient uptake kinetics for efficient crop production during Solanum Lycopersicum var. Cerasiforme Alef. Growth in a closed indoor hydroponic system. *PLoS ONE* 12(5), e0177041. https://doi.org/10.1371/journal.pone.0177041.

Lee, S. G., Choi, C. S., Lee, J. G., Jang, Y. A., Lee, H. J., Lee, H. J., Chae, W. B. and Um, Y. C. 2013. Influence of air temperature on yield and phytochemical content of red chicory and garland chrysanthemum grown in plant factory. *Horticulture, Environment, and Biotechnology* 54(5), 399–404. https://doi.org/10.1007/s13580-013-0095-x.

Levenston, M. 2014. Vancouver's rooftop farm "local garden" bankrupt. *City Farmer News*. Available at: https://cityfarmer.info/vancouvers-rooftop-farm-local-garden-bankrupt/ (accessed 4.9.22).

Li, K., Fang, H., Zou, Z., Cheng, R. and Feng, R. 2021. Optimization of rhizosphere cooling airflow for microclimate regulation and its effects on lettuce growth in plant factory. *Journal of Integrative Agriculture* 20(10), 2680–2695. https://doi.org/10.1016/S2095-3119(20)63382-2.

Liaros, S., Botsis, K. and Xydis, G. 2016. Technoeconomic evaluation of urban plant factories: The case of basil (*Ocimum basilicum*). *Science of the Total Environment* 554–555, 218–227. https://doi.org/10.1016/j.scitotenv.2016.02.174.

Lim, M. Y., Jeong, H. J., Choi, S. H., Choi, G. L. and Kim, S. H. 2020. Effect of planting density by cultivars on the growth and yield of melons (*Cucumis melo* L.) in hydroponics using coir substrates. *Horticultural Science and Technology* 38, 850–859. https://doi.org/10.7235/HORT.20200077.

Liotta, M., Hardej, P. and Nasseri, M. 2017. An examination of shuttered vertical farm facilities. *YouTube*. Aglanta Conference Panel. Available at: https://www.youtube.com/watch?v=1mB_8TE-t2E (accessed 10.3.21).

Lu, N. and Shimamura, S. 2018. Protocols, issues and potential improvements of current cultivation systems. In: Kozai, T. (Ed.), *Smart Plant Factory: The Next Generation Indoor Vertical Farms*. Springer, Singapore, pp. 31–49. https://doi.org/10.1007/978-981-13-1065-2_3.

Maboko, M. M. and Du Plooy, C. P. 2013. High-plant density planting of basil (*Ocimum basilicum*) during summer/fall growth season improves yield in a closed hydroponic system. *Acta Agriculturae Scandinavica, Section B – Soil and Plant Science* 63(8), 748–752. https://doi.org/10.1080/09064710.2013.861921.

Macaulay, L. 2022. [IMG]. Wigan UTC vertical farm. *Personal Correspondance*.

Maguire, M. and Delahunt, B. 2014. Doing a thematic analysis: A practical, step-by-step guide for learning and teaching scholars. *IEEE Transactions on Industry Applications* 50, 3135–3140. https://doi.org/10.1109/TIA.2014.2306979.

McGoran, C. 2020. Things to think about before purchasing A freight farm. Medium. Available at: https://medium.com/@connormcgoran/things-to-think-about-before-purchasing-a-freight-farm-963b6d53cfa3 (accessed 2.14.22).

Michael, C. 2017. The good, the bad, and the ugly of container farms. Available at: https://medium.com/bright-agrotech/the-good-the-bad-and-the-ugly-of-container-farms-d32f4f13f31b (accessed 2.14.22). Farms [WWW Document]. Medium.

Michael, C. 2021. The good, the bad, and the ugly of container farms. *Medium*. Available at: https://medium.com/bright-agrotech/the-good-the-bad-and-the-ugly-of-container-farms-d32f4f13f31b (accessed 4.30.21).

Nasseri, M. 2022. [IMG]. Local garden vertical farm. *Personal Correspondance*.

Nazarideljou, M. J., Haghshenas, M., Jaberian Hamedan, H. and Ferrante, A. 2019. Growth, yield and antioxidant capacity of strawberry under various K+:Ca++ ratios in hydroponic culture. *Acta Agriculturae Scandinavica, Section B – Soil and Plant Science* 69(2), 105–113. https://doi.org/10.1080/09064710.2018.1506500.

Nishikawa, T., Fukuda, H. and Murase, H. 2013. Effects of airflow for lettuce growth in the plant factory with an electric turntable. *IFAC Proceedings Volumes* 46(4), 270–273. https://doi.org/10.3182/20130327-3-JP-3017.00062.

Niu, G., Kozai, T. and Sabeh, N. 2019. *Physical Environmental Factors and Their Properties, Plant Factory: An Indoor Vertical Farming System for Efficient Quality Food Production* (2nd edn.). Elsevier Inc., Amsterdam, Netherlands, https://doi.org/10.1016/B978-0-12-816691-8.00011-X.

Nunomura, O., Kozai, T., Shinozaki, K. and Oshio, T. 2019. Seeding, seedling production and transplanting. In: *Plant Factory: An Indoor Vertical Farming System for Efficient Quality Food Production* (2nd edn.). Elsevier Inc., Amsterdam, Netherlands, pp. 285–297. https://doi.org/10.1016/B978-0-12-816691-8.00021-2.

Nxawe, S., Laubscher, C. P. and Ndakidemi, P. A. 2009. Effect of regulated irrigation water temperature on hydroponics production of Spinach (*Spinacia oleracea* L). *African Journal of Agricultural Research* 4, 1442–1446.

Palinkas, L. A., Horwitz, S. M., Green, C. A., Wisdom, J. P., Duan, N. and Hoagwood, K. 2015. Purposeful sampling for qualitative data collection and analysis in mixed method implementation research. *Administration and Policy in Mental Health* 42(5), 533–544. https://doi.org/10.1007/s10488-013-0528-y.

Paradiso, R. and Proietti, S. 2022. Light-quality manipulation to control plant growth and photomorphogenesis in greenhouse horticulture: The state of the art and the opportunities of modern LED systems. *Journal of Plant Growth Regulation* 41(2), 742–780. https://doi.org/10.1007/s00344-021-10337-y.

Park, Y. and Runkle, E. S. 2018. Spectral effects of light-emitting diodes on plant growth, visual color quality, and photosynthetic photon efficacy: White versus blue plus red radiation. *PLoS ONE* 13(8), e0202386. https://doi.org/10.1371/journal.pone.0202386.

Peiro, E., Pannico, A., Colleoni, S. G., Bucchieri, L., Rouphael, Y., De Pascale, S., Paradiso, R. and Gòdia, F. 2020. Air distribution in a fully-closed higher plant growth chamber impacts crop performance of hydroponically-grown lettuce. *Frontiers in Plant Science* 11, 537. https://doi.org/10.3389/fpls.2020.00537.

Rahman, M. M., Field, D. L., Ahmed, S. M., Hasan, M. T., Basher, M. K. and Alameh, K. 2021. LED illumination for high-quality high-yield crop growth in protected cropping environments. *Plants* 10(11), 2470. https://doi.org/10.3390/plants10112470.

Rawson, H. M., Begg, J. E. and Woodward, R. G. 1977. The effect of atmospheric humidity on photosynthesis, transpiration and water use efficiency of leaves of several plant species. *Planta* 134(1), 5–10. https://doi.org/10.1007/BF00390086.

Roosta, H. R. 2011. Interaction between water alkalinity and nutrient solution ph on the vegetative growth, chlorophyll fluorescence and leaf magnesium, iron, manganese, and zinc concentrations in lettuce. *Journal of Plant Nutrition* 34(5), 717–731. https://doi.org/10.1080/01904167.2011.540687.

Shao, Y., Heath, T. and Zhu, Y. 2016. Developing an economic estimation system for vertical farms. *International Journal of Agricultural and Environmental Information Systems* 7(2), 26–51. https://doi.org/10.4018/IJAEIS.2016040102.

Sheridan, C., Depuydt, P., De Ro, M., Petit, C., Van Gysegem, E., Delaere, P., Dixon, M., Stasiak, M., Aciksöz, S. B., Frossard, E., Paradiso, R., De Pascale, S., Ventorino, V., De Meyer, T., Sas, B. and Geelen, D. 2017. Microbial community dynamics and response to plant growth-promoting microorganisms in the rhizosphere of four common food crops cultivated in hydroponics. *Microbial Ecology* 73(2), 378–393. https://doi.org/10.1007/s00248-016-0855-0.

Sinclair, T. R. 1994. Limits to crop yield? In: *NAS Colloquium. Plants and Population: Is There Time?* Irvine, CA, pp. 509–532.

Singh, H., Poudel, M. R., Dunn, B. L., Fontanier, C. and Kakani, G. 2020. Effect of greenhouse $CO_2$ supplementation on yield and mineral element concentrations of leafy greens grown using nutrient film technique. *Agronomy* 10(3), 323. https://doi.org/10.3390/agronomy10030323.

Soltész, L. and Berényi, L. 2021. *Utilization of Lessons Learned in Product Development, Lecture Notes in Mechanical Engineering*. Springer, Singapore. https://doi.org/10.1007/978-981-15-9529-5_25.

Son, J. E., Kim, H. J. and Ahn, T. I. 2019. Hydroponic systems. In: *Plant Factory: An Indoor Vertical Farming System for Efficient Quality Food Production* (2nd edn.). Elsevier Inc., pp. 273–283. https://doi.org/10.1016/B978-0-12-816691-8.00020-0.

Sykes, C. 2017. *Is the Vertical Farming Industry Learning From Its Mistakes*. Available at: https://www.linkedin.com/pulse/vertical-farming-industry-learning-from-its-mistakes-chad-sykes/ (accessed 11.22.19).

Thakulla, D., Dunn, B., Hu, B., Goad, C. and Maness, N. 2021. Nutrient solution temperature affects growth and brix parameters of seventeen lettuce cultivars grown in an NFT hydroponic system. *Horticulturae* 7(9), 321. https://doi.org/10.3390/horticulturae7090321.

Thomas S. and Stephen, G. 1984. Solar greenhouses and sunspaces: Lessons learned. https://doi.org/DOE/CE/15095-8.

Tyson, R. V., Simonne, E. H., Treadwell, D. D., Davis, M. and White, J. M. 2008. Effect of water pH on yield and nutritional status of greenhouse cucumber grown in recirculating hydroponics. *Journal of Plant Nutrition* 31(11), 2018–2030. https://doi.org/10.1080/01904160802405412.

Uraisami, K. 2018. Business planning on efficiency, productivity, and profitability. In: Kozai, T. (Ed.), *Smart Plant Factory*. Springer, Singapore. https://doi.org/978-981-13-1065-2.

V-Farm. 2022. V-Farm. Available at: https://v-farm.co.uk/ (accessed 8.23.22).

Valle de Souza, S., Peterson, H. C. and Seong, J. 2021. Emerging economics and profitability of PFALs. In: Kozai, T., Niu, G. and Masabni, J. (Eds.), *Plant Factory Basics, Applications and Advances*. Academic Press, Cambridge, Massachusetts. pp. 251–270. https://doi.org/10.1016/b978-0-323-85152-7.00025-2.

van Delden, S. H., SharathKumar, M., Butturini, M., Graamans, L. J. A., Heuvelink, E., Kacira, M., Kaiser, E., Klamer, R. S., Klerkx, L., Kootstra, G., Loeber, A., Schouten, R. E., Stanghellini, C., van Ieperen, W., Verdonk, J. C., Vialet-Chabrand, S., Woltering, E. J., van de Zedde, R., Zhang, Y. and Marcelis, L. F. M. 2021. Current status and future challenges in implementing and upscaling vertical farming systems. *Nature Food* 2(12), 944–956. https://doi.org/10.1038/s43016-021-00402-w.

Van Der Meer, M., De Visser, P. H. B., Heuvelink, E. and Marcelis, L. F. M. 2021. Row orientation affects the uniformity of light absorption, but hardly affects crop photosynthesis in hedgerow tomato crops. *In Silico Plants* 3(2), diab025. https://doi.org/10.1093/insilicoplants/diab025.

Wortman, S. E., Douglass, M. S. and Kindhart, J. D. 2016. Cultivar, growing media, and nutrient source influence strawberry yield in a vertical, hydroponic, high tunnel system. *hortTechnology* 26(4), 466–473. https://doi.org/10.21273/HORTTECH.26.4.466.

Xu, W., Lu, N., Kikuchi, M. and Takagaki, M. 2021. Continuous lighting and high daily light integral. *Plants* 10, 1–16.

ZipGrow. 2022. ZipGrow Website. Available at: https://zipgrow.com/zipfarm/ (accessed 9.2.22).

# Chapter 17

# Design and management of globally-networked plant factories: commercial application and future opportunities

*Viviana Correa Galvis, Pádraic J. Flood and Pavlos Kalaitzoglou, Infarm - Indoor Urban Farming B.V., The Netherlands*



## 1 Introduction

In our current time of climatic uncertainty, the development of plant factories with artificial lighting (PFAL) is key to climate-resilient food production. PFALs around the world can be found in a wide range of production models such as indoor-controlled growing warehouses, shipping containers, in-store located PFALs, and plug-and-play in-home/office use PFALs (Butturini and Marcelis, 2019). Growing systems used in PFALs can also vary from vertically/horizontally grown, hydro/aeroponic, and with/without substrate, to name a few. In spite of this heterogeneity, there is one characteristic that unifies most PFAL systems, they are centralized systems. A centralized PFAL is located in one specific place where production is monitored, and even though one company can have various PFALs in different locations they are maintained and managed locally as independent growing units. This chapter discusses the implementation of

http://dx.doi.org/10.19103/AS.2023.0126.24

a decentralized PFAL system that can be connected globally. To do so, we present the specific case of Infarm, a company founded in 2013 that became the first world's fastest-growing global, cloud-connected PFAL network. In partnership with more than 30 of the world's top retail chains in 10 countries in North America, Asia, and Europe, Infarm's product is currently available in more than 1850 stores worldwide.

Based on Infarm's case, a description of specific requirements to implement decentralized globally networked PFALs is presented, from the business model to modular software and hardware designs. Furthermore, we discuss opportunities and challenges for data-driven optimization of plant growth and the introduction of new crops in modular decentralized PFAL systems.

## 2 A globally networked business model

Infarm's business model integrates the modularity of the production systems into a highly flexible and dynamic product offering. The business approach is called Farming as a Service (FaaS). In brief, the customer rents a given area of growth space from which a wide range of crops and products can be produced. The customer can choose what is grown and can change both the area rented and the products grown at any point during the year. This flexibility allows the customer to take better advantage of market opportunities, i.e. holidays and product promotions. In addition to flexibility in production capacity and crop composition, Infarm works with customers to tailor the taste and morphology of the product to their particular needs.

This flexibility requires zero upfront investment from the customer; they decide how much they want and when and only pay for the final product. Thus the customer gets the benefits of a year-round dynamic supply of fresh produce without any risk from their side. Infarm can accommodate this due to optimized growth recipes and a plant/environment monitoring system that allows coordination of sowing and harvesting to meet customer demand.

In addition to an on-demand supply of high-quality fresh produce Infarm also offers branding and marketing support to customers ensuring that the advantages of vertically farmed produce are visible to the final consumer. To achieve this a range of different sales agreements is provided, catering for a wide range of businesses from restaurants and supermarkets to online retailers.

## 3 Infarm's product catalog

One of the advantages of plant factories located in or near urban centers is the possibility to adjust production to demand, thus negating the need for long supply distances and times. Consequently, the quality of the product is less affected by storage and transportation conditions (van Delden et al., 2021). Based on a supply-by-demand scheme, Infarm has established a catalog of

**Figure 1** Representative products from the Infarm catalog (June 2022). (a) Unless specified all Infarm products are sold as a whole plant including the roots in substrate. Products are sold in different packaging types and sizes such as (b) paper cone, (c) tray seal, and (d) pots. Other products include (e) cut product and (f) microgreens.

products that can be sold either directly after harvest, when the PFAL is located inside the supermarket (Fig. 1a and b) or transported from production centers located near stores (Fig. 1c-f). The current catalog includes various crops commonly produced in PFAL and mushrooms (Appendix 1).

## 4 System configuration and description

A key concept to establishing globally networked PFALs is modularity. Infarm has kept system and unit modularity as a key consideration in the development of all hardware and software. This scalable modularity is, to our knowledge, unique in the PFAL industry. Modularity allows for a wide range of production sizes and both speed and flexibility in deployment to new and existing markets. In larger growing centers, the independence of the different growing modules allows for compartmentalization of the growing environment, enabling the creation of different growing environments, hence it is possible to produce a broad variety of crops that require different growth conditions in the same location.

### 4.1 Hardware

The ability to connect PFALs from different locations is highly dependent on hardware design. Independently of the size, having a modular unit that can be installed everywhere provides the possibility to create a linked PFAL network. There are two main categories of hardware design depending on the type of facilities in which the PFALs can be located. At Infarm, these are defined as InStore and Infarm Growing Centers (IGC). InStore comprises a four-level semi-closed climate-controlled PFAL installed inside the customer facilities. From this InStore, the product is harvested and displayed next to the PFAL (Fig. 2a). IGCs (Fig. 2b) are production centers near or inside urban areas that distribute

**Figure 2** Infarm growing systems. (a) InStore are four-level semi-closed climate control growing systems located inside supermarkets. (b) Infarm Growing Centers constitute the industrial-scale PFAL system, whose main growing systems are (c) Growth Chambers, and (d) ACRE growing towers. Growth chambers are used as (e) plant nurseries or (f) for exotic mushroom production.

products by demand and contain two main types of hardwares: growth chambers (Fig. 2c, e, f) and Automated Controlled Remote Environment (ACRE)-growing towers (Fig. 2d). All InStore and ACRE hardware systems incorporate the concept of modularity in their design; however, as the ACRE-growing towers constitute the industrial-scale PFAL system, modularity in this system has provided the added value that makes a difference in the PFAL industry.

The ACRE towers have been designed based on the station concept (Fig. 3). Plants are transplanted into a bench that is not fixed inside the tower. The bench

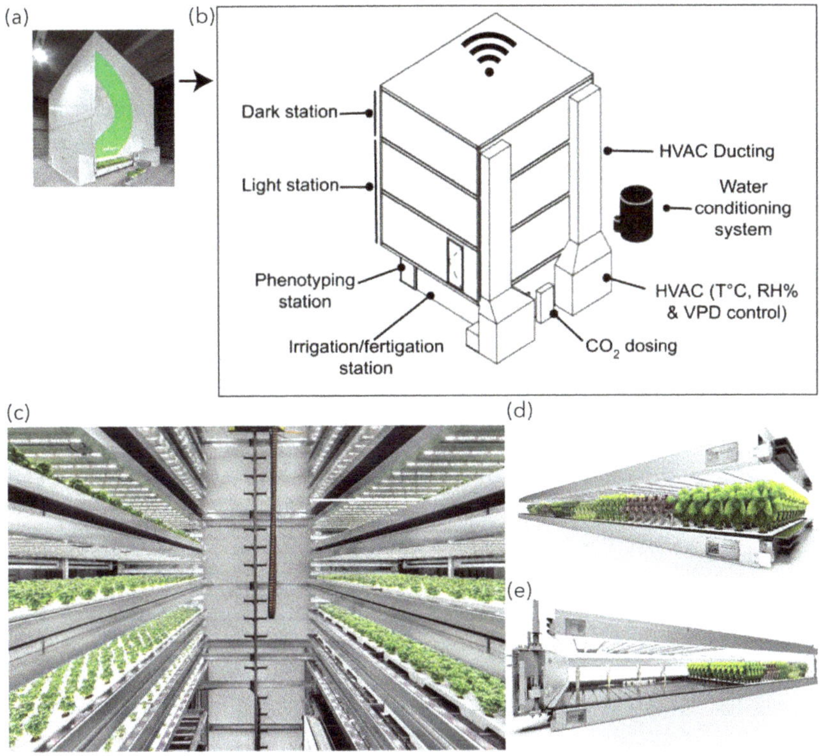


**Figure 3** ACRE growing towers. (a) ACRE tower view from the harvesting/transplanting station side. (b) Distribution scheme of the main components in an ACRE tower as well as the internal distribution of the dark/light, phenotyping, and irrigation stations. Note that the water conditioning system (WCS) that adjusts irrigation water is located outside the tower. One WCS can supply irrigation water to up to four ACRE towers. (c) Internal view of the light stations. Benches are moved between stations through an elevator system located in between two main columns that hold light and dark stations. (d) Closer view of the light station. The height of the light station can vary to accommodate various crops, always maintaining a minimum distance of at least 10 cm between the top of the canopy and the lights. (e) Closer view of the irrigation station. A separate irrigation station design for this type of PFAL supports ebb and flow irrigation. After irrigation, water returns to the WCS to be cleaned through a filter and UVC light system followed by EC and pH adjustment for re-use in follow-up irrigations.

is mobile among stations (light/dark stations, irrigation station, phenotyping station and transplant/harvesting station), and the time each bench stays in a specific station and the settings applied to it are in concordance with the growing recipe of that specific bench. Consequently, many different crops with different requirements can be grown in one ACRE tower within the range of allowed flexibility. In addition, ACRE's design is based on warehouse logistics systems and utilizes the benches and the elevator system to incorporate the specific farming technologies needed for production in a PFAL. Monitoring systems are installed in every ACRE tower to control for environmental settings (i.e. light dimming, T°C, $CO_2$, humidity, irrigation frequency, EC), movement of benches, opening gates and optionally multispectral cameras for crop monitoring.

There are both internal and external modularity considerations in the design and construction of the ACRE tower. Internal components of the ACRE can be multiplied and modified according to the product demand during the planning phase. For example, the light station has been designed as one module to which specific components such as distance to the canopy and LED composition (light spectrum and number of lumineers) can be easily adjusted/changed allowing the system to adapt to many different crops, from leafy greens to fruit crops. External components such as the water conditioning system and the climate control can supply one or more towers depending on the needs per location which facilitates fast expansion and less CapEx in the initial phases when establishing a new production location. The ACRE is a module in itself. Depending on the demand per location, one or more ACREs can be installed in a growing center, so operations are adaptable and scalable.

## 4.2 Software and data

A modular PFAL that can be remotely monitored combined with the digital infrastructure to manage, store, and use information are the backbone of a globally networked PFAL system. Infarm's software platform has been built to be modular, scalable, and secure. The central components of this platform are an IoT (internet of things) layer and the data warehouse. The IoT layer creates an abstraction layer that extracts specific features of the hardware implementation creating basic modules that can be applied to different PFAL types, thus allowing for scalability while keeping the integrity and availability of the monitoring, control, and factory operational systems. The data warehouse is a data infrastructure system that stores, transforms, and enriches every single data point collected from plant health indicators, to PFAL environmental sensor data, to sales performance. On top of these two main pillars, there are three main domains that are modular as well, so they can be used independently or as a whole. As Infarm has full ownership of the design and development of the software domains, each domain can be tailored for any use case, from internal operations, to outsourced operations of hardware, to the support of traditional farming and other PFAL solutions.

### 4.2.1 Farm Software Domain: plant factories with artificial lighting control and monitoring

The farm software domain (FSD) has different digital products that control and monitor all factories to provide global and centralized visibility of the factories' performance. In this domain there are different digital products that provide an alerting system to report factory abnormal behaviors, real-time views on all metrics of the factory environment, apply factory environment settings (light, irrigation, and climate), predict maintenance events, and allow for crop performance monitoring and recipe optimization (Fig. 4). The main aim of this domain is to maximize system uptime to increase harvest performance and reduce maintenance technician time on site. Currently, all digital products of the FSD allow >1360 globally active PFALs to be monitored every 10 s everyday, providing >20 different data points per batch of planted crops related to crop morphology captured with imaging technologies such as plant height, projected leaf area, 3D leaf area, and digital biomass (for further details refer to Manavalan et al., 2021), as well as environmental data (T°C, RH%, EC). Collected information translates into approximately 5 billion/month data points ingestion. The FSD is capable of providing real-time alerts, dispatching >9K alerts/month to field maintenance teams maximizing systems' uptime, and containing an average of >100 critical alerts per day that indicate events likely to negatively affect performance. The data collection and alerting systems can consequently contribute to a networked quality-controlled system that allows the maintenance of homogenous factory productivity across all globally distributed PFALs.

The variety of products offered by Infarm rely on different operational configurations so precise capacity planning is essential to define where and when planting should occur. To do so, the scheduling tools use a unit of area to which any plausible operational configuration is applied. This configuration includes information such as type of crop, density, and environmental settings.

Currently, there is accurate planting capacity across six different types of PFAL configurations, in a total growing area of >9800 $m^2$ with >400 active end product varieties configured and >50 different tray configurations and densities. This leads to >1900 harvest activities which are scheduled on a given day worldwide.

### 4.2.2 Business intelligence

Business intelligence (BI) applications can provide an overview of harvest performance, production center performance, and operational events (topics and issues) in different locations. It provides information at local, regional,

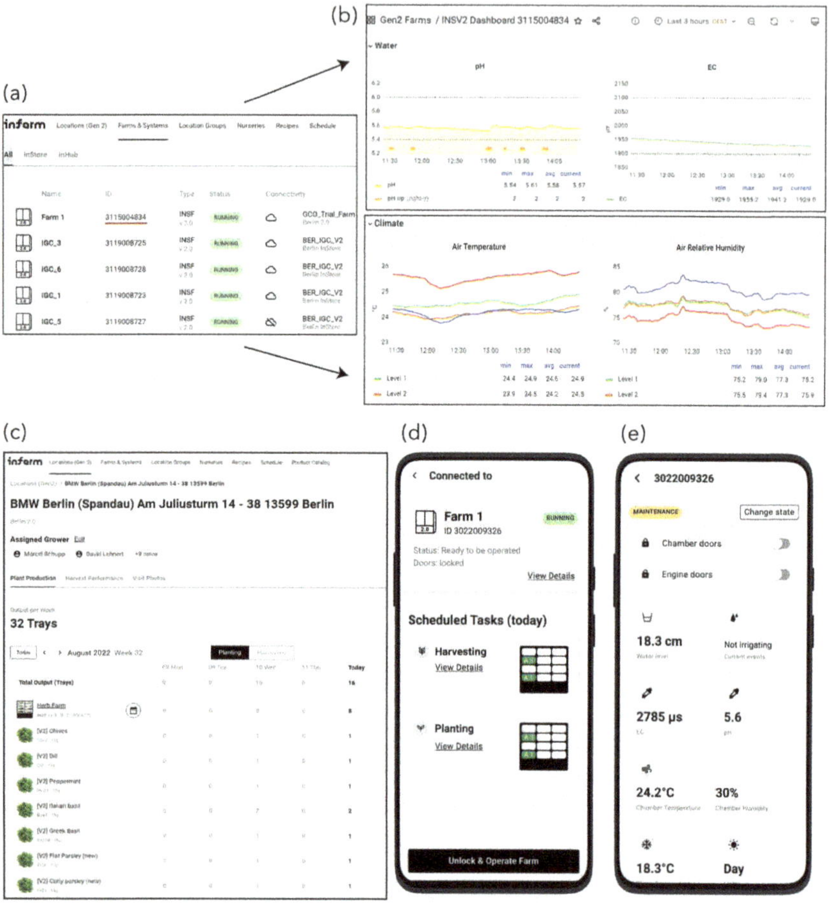


**Figure 4** FSD and OSD digital products. (a) Farmboard: a PFAL management tool. (b) Grafana: a dashboard that provides real-time metrics of each PFAL. As an example, live metrics for a selected PFAL (red underlined) in (a) are displayed in (b) in addition to virtual boards (c), OSD products include various applications for mobile phones used by employees to collect information and apply protocols in production. Apps support the execution of crop-related tasks such as planting, harvesting, and quality monitoring (d) as well as PFALs' hardware and environment maintenance (e).

and global levels on issues such as seeding to harvest reporting, plant factory performance, sales, crop utilization metrics, and sustainability. The power of BI applications relies on the definition of parameters that can provide relevant information comparable at different levels (local to global) and usable by different types of stakeholders. For instance, the number of harvested plants per PFAL can be used to monitor crop performance as well as to understand the efficacy of operations, space used, and energy consumption (Fig. 5 and Table 1).

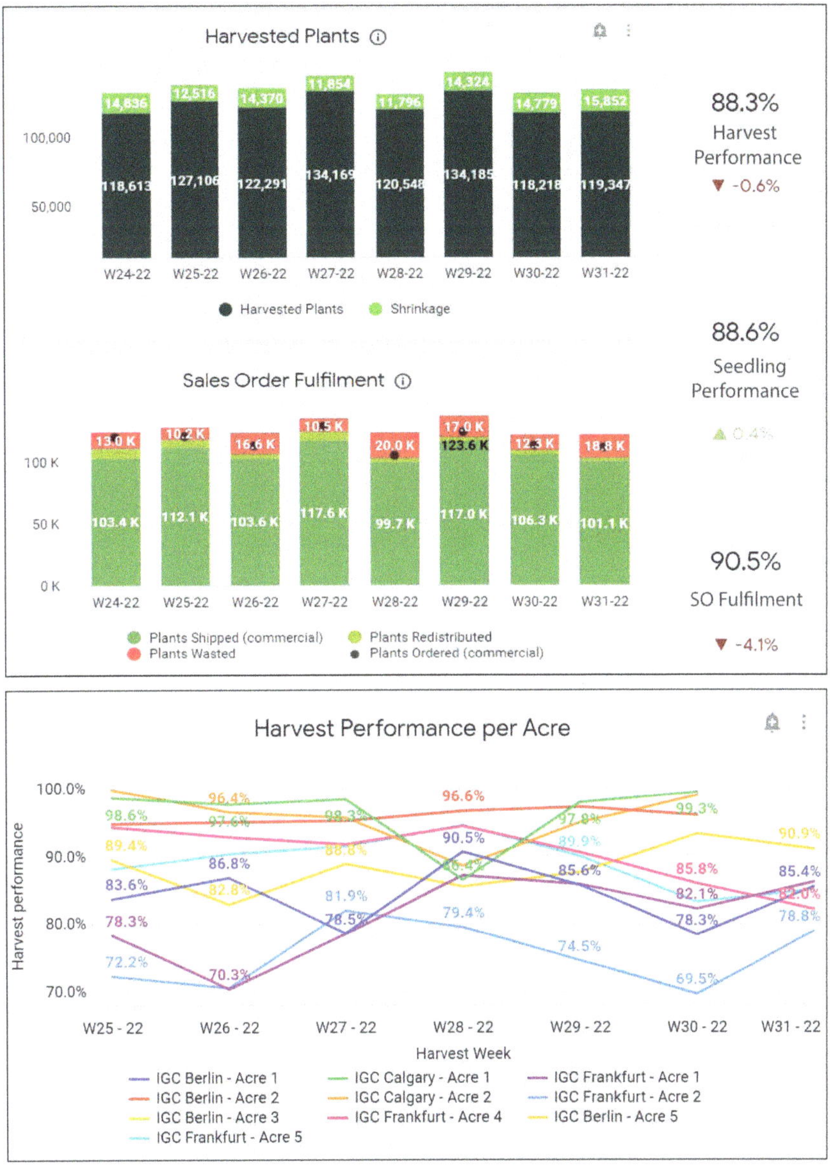


**Figure 5** Representative view of various dashboards of the BI domain. Presented figures compiled from data obtained for 2022 of the production centers of Berlin, Frankfurt, Copenhagen, Paris, and Vancouver. Up to date, there are >300 dashboards that provide business insights ranging from finance to field operations to plant optimization. As an example various parameters such as Harvest Performance (% share of good sellable harvested plants based on the total number of kitted plants), Seedling Performance (% share of good and usable harvested seedlings based on the total number of seeded plants), and SO fulfillment (plants shipped/plants ordered) are shown.

**Table 1** BI table display: Utilization rate per location

| | Utilisation Rate | | | | | | | | |
|---|---|---|---|---|---|---|---|---|---|
| | Week Number | W25-2022 | W26-2022 | W27-2022 | W28-2022 | W29-2022 | W30-2022 | W31-2022 | W32-2022 |
| | Hub Name | Utilisation Rate | Utilisation Rate | Utilisation Rate | Utilisation Rate | Utilisation Rate | Utilisation Rate | Utilisation Rate | Utilisation Rate |
| 1 | Bedford | 44.0% | 44.8% | 48.7% | 51.1% | 52.6% | 57.9% | 62.5% | 64.0% |
| 2 | Berlin | 91.9% | 92.1% | 92.1% | 92.1% | 93.2% | 93.5% | 69.3% | 73.0% |
| 3 | Calgary | 75.6% | 78.7% | 81.1% | 82.1% | 81.3% | 81.3% | 81.3% | 81.3% |
| 4 | Copenhagen | 87.3% | 85.1% | 81.4% | 82.6% | 76.8% | 75.4% | 74.2% | 77.8% |
| 5 | Frankfurt | 89.4% | 95.9% | 97.0% | 98.2% | 97.9% | 84.0% | 85.9% | 87.8% |
| 6 | Hannover | 99.7% | 99.7% | 99.7% | 99.7% | 100.0% | 95.5% | 95.5% | 95.5% |
| 7 | Paris | 74.0% | 70.0% | 74.8% | 80.2% | 80.3% | 76.8% | 72.4% | 72.4% |
| 8 | Toronto | 67.9% | 68.9% | 71.8% | 69.7% | 69.7% | 67.9% | 66.6% | 63.7% |
| 9 | Vancouver | 83.0% | 83.9% | 87.1% | 87.3% | 86.6% | 84.6% | 84.6% | 84.6% |

Values indicate weekly space use efficacy as the percentage of growing space use from growing space available. Real-time data of European and North America growing centers.

The BI insights are essential to driving operational excellence. Used as a tracking system, it has allowed the identification of issues such as overproduction, order fulfillment, and low crop harvest performance. These are then corrected from all plausible perspectives (operational, financial, and R&D) driving the improvement of each process based on real-time data.

# 5 Uniqueness, strengths and weaknesses

The combination of a modular growing system with a FaaS business model has allowed Infarm to expand in an exponential manner. This however relies on a strong R&D effort to develop customized systems. From structural components such as growing stations (LED cassettes, irrigation systems, elevator for benches, etc.) to consumables such as plant trays, many elements have been designed in or for Infarm, in collaboration with expert industrial partners. Once developed, all components are globally available, mostly pre-assembled and certified by international quality standards. Effective internal documentation on processes and guidelines provided by Infarm's digital infrastructure has been crucial for the installation and training of new personnel. Along with data monitoring and tracking, effective documentation has allowed successful installations and establishment of new plant factories in diverse conditions and locations, even with unpredictable logistic issues or traveling restrictions of expert personnel (such as the ones in place worldwide during the COVID-19 pandemic).

The design and development of a unique globally modular PFAL system have been only possible by the acknowledgment of innovation and research as a core principle for business development. In an emerging industry, such ratification relies on visionary investment rather than immediate profit. For Infarm, the development of every new product category requires time and effort and investment that will be returned by its global application.

# 6 The future of networked plant factories

## *6.1 Plant factories with artificial lighting automation*

The future of plant factories lies in automation (van Delden et al., 2021; Saad et al., 2021). Automation facilitates the standardization of processes and reduces the need for personnel to enter the production environment thus minimizing the risk of pests and pathogens compromising production. For full automation of plant production and monitoring significant investments are still required. Whilst the horticultural industry has developed automated solutions for complex steps such as transplanting or harvesting (Jahnke et al., 2016; TTA, 2019), there is still some way to go before humans can be removed from the complete production cycle. One key step to facilitate full-scale automation is the integration of non-invasive high throughput phenotyping throughout the growth cycle. Such phenomics capabilities allow for real-time monitoring of the plants which in turn supports sensor-informed artificial intelligence (AI), leading to optimization of the growth environment. In this process named here biofeedback, the plant itself becomes the environmental sensor which allows for fine-tuning to its particular needs (Fig. 6). A biofeedback system comprises a combination of sensing technology (plant traits and environmental monitoring) to generate self-learning data-driven models that can be used to automate the optimization of growth recipes, thus increasing the resource use efficiency of PFALs (van Delden et al., 2021).

A biofeedback system requires the creation of a digital platform where plant and environmental data are constantly stored, reviewed on quality, and used to feed the ML algorithms connected to the AI environment optimization tools. On a research level, automation of growth recipes through the biofeedback system (using data-driven models) has been shown to successfully maintain crop yield while saving up to 20% of electricity consumption (Infarm-internal trials, 2022). However, there are still various steps needed to apply it and obtain a fully automated PFAL. First, it is necessary to define which phenotyping and environmental monitoring technologies provide the most relevant parameters for canopy monitoring on a weekly/daily/hour scale, and thus the price/benefit of such technology on the production cost per sale unit. For instance, parameters derived from multispectral imaging provide indices that are good indicators of plant stress, however, in a highly controlled environment strong stress conditions are avoided. Consequently, plant growth monitoring is more relevant in a PFAL, and it can be achieved without multispectral imaging. Similarly, appropriate sensors need to be acquired for environmental monitoring. The type of sensors

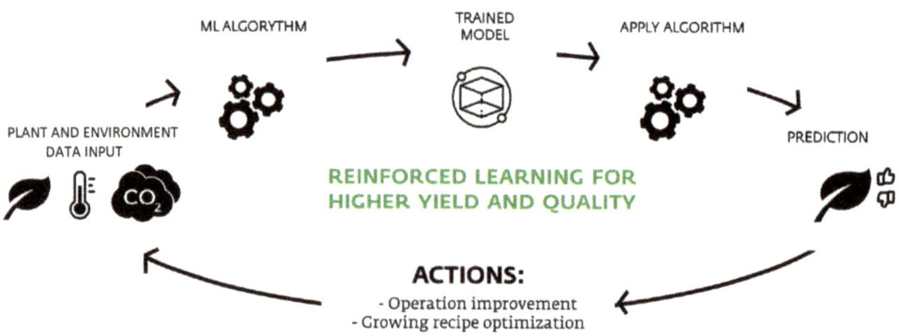


**Figure 6** Automation of plant growth monitoring and recipe adjustment through a data-driven biofeedback system at Infarm, status 2022. Plant and environmental data are collected from seeding throughout the plant's growth cycle and used to predict the final yield of a plant batch. The system can, through a machine learning algorithm (ML), evaluate the growth status of the plant batch daily or every 2 days after transplant. If the prediction does not match the expected yield, the system automatically assigns a specific growth recipe that will increase the likelihood of that batch achieving the desired yield. If, on the other side, the plant batch exceeds the final yield, the recipe is adjusted to reduce resources (light and irrigation input) without compromising the final yield.

required, amount, and sensitivity are determined by the structure/design of the farm as this influences the environmental gradients plants are subjected to in the growing benches. Thus it is important to find a compromise between the number of sensors included in the system, the value of the information provided by those, and the additional cost/benefit added to the farming units. Once the monitoring systems are in place, a second step is to create data analysis pipelines that are transferable and scalable to various types of products: from leafy greens to fruit crops or from single plants to cut products.

Finally, it is important to define the modeling approach to use. Until now, it has been plausible to generate accurate predictions with a data-driven based approach, since having a global network of PFALs can provide large data sets for model training. However, a combination of mechanistic and data-driven models could have the potential to increase predictive power, which remains to be explored.

## 6.2 New crops

To date, production in plant factories has been largely restricted to leafy greens and herbs. This is no accident, these crops combine compact morphology with a harvest index (proportion of the above-ground biomass that can be sold) approaching 1. This combination of traits greatly simplifies their commercialization in a PFAL context. However, if plant factories aim to fulfill their promise to significantly reduce the environmental impact of our food production systems whilst also increasing their resilience they will need to expand their crop portfolio to include a much wider range of crop types (Harris and Kountouris, 2020; Asseng et al, 2020). There are two main challenges associated with this expansion: adapting the hardware and improving the germplasm.

### 6.2.1 Hardware adaptation

Defining the optimal PFAL hardware setup is not simple and comes with many crop-specific adjustments; however, there are some general tenets that should be optimized from a business and sustainability perspective, namely CAPEX and OPEX optimization. When these are placed front and center of the hardware design process the range of environments that can be easily generated is reduced. This set of environments then becomes breeding targets and crops can be divided up into a small set of environmental recipes which simplifies farm design and maintenance. Breeding for the spectra of the most energy and cost-efficient lighting solutions as well as continuous lighting conditions are examples of breeding targets that are specific to plant factories and for which little to no breeding has been done to date (Folta, 2019). Recently several PFAL companies have begun to produce strawberries, this is the first step into fruiting crops (Butturini and Marcelis, 2019); however, this germplasm is far from optimized for PFALs.

### *6.2.2 Improving germplasm*

Recent advances in genomics, genome editing, speed breeding, and statistical genetics have greatly improved both the rate of crop improvement and the precision with which breeding can be conducted. Plant factories in particular are set to benefit from these advances. Unlike conventional farming, breeding for plant factories does not require broadly adapted germplasm which can grow well under a range of conditions but rather highly specialized germplasm optimized for the PFAL environment can be developed without the negative consequences such specialization would incur in the outdoors. In addition, since the production environment is known and can be maintained indefinitely, the breeding target remains static, thus improvements made today will last. This is not always true when farming outdoors where changing climate and soil conditions render older varieties obsolete. Finally, breeding for disease resistance under field or greenhouse conditions takes up a considerable proportion of most breeding program resources. In a fully automated PFAL environment disease will no longer be a problem and breeding efforts can focus on yield and quality, improving sustainability and customer satisfaction.

## 7 Catalyzing a sustainable food production system

The future of our global food production has never faced such challenges and uncertainty. On one hand, our growing global population and improving the standard of living means we will need to produce more food than ever in the coming decades. However, our environment is on the brink of collapse, today we are utterly dependent on this environment for our food supply. PFALs can produce top-quality food regardless of the climate outside. As the climate emergency progresses this resilience of food production in the face of an unpredictable climate is one way in which we can ensure global food supply in the years to come. PFALs don't only provide greater resilience for our food supply but will aid in restoring planetary resilience, freeing up land for nature where ecosystem services and biodiversity can be restored, putting us on track to becoming a truly sustainable civilization living in harmony with nature and mother earth.

## 8 Acknowledgements

The authors would like to thank Pedro Silva (Director of Software and IoT-Infarm), Orie Sofer (Senior Crop Hardware Designer-Infarm), and Serge Caston (Crop Hardware Manager-infarm) for their comments and suggestions on the content of this document.

## 9 Appendix

Appendix 1. Infarm's crop catalog by product name (Status June 2022)

| Aromatic herbs | Lettuces | Leafy greens | Microgreens | Pots | Cut salad mixes (CSM) |
|---|---|---|---|---|---|
| Italian Basil | Crystal | Baby Green | Borage | Italian Basil | 1. Green Leaf, |
| Bordeaux Basil | Lettuce | Pak Choi | Dijon Mustard | Chervil | Red Mustard, |
| Chives | Green | Baby Purple | Garden Cress | Chives | Red Oak Leaf |
| Coriander | Romaine | Pack Choi | Garden | Coriander | 2. Crystal |
| Curly Parsley | Red | Garden | Nasturtium | Curly | lettuce, |
| Dill | Romaine | Cress | Green Pea | Parsley | Spinach, Red |
| Flat Parsley | Red Oak | Greek Cress | Green Radish | Dill | Veined Sorrel |
| Greek Basil | Leaf | Golden Frills | Green Shiso | Flat Parsley | 3. Spinach |
| Green Mint | Red Leaf | Mustard | Leek Doraemon | Green Mint | 4. Green Leaf |
| Lemon Balm | Green Leave | Golden | Red Radish | Lemon | **Mushrooms** |
| Melissa | Salanova | Purslane | | Balm | Shiitake |
| Lemon Basil | Green | Karashina | | Melissa | |
| Lemon Star | Saladuo | Mustard | | Mountain | |
| Mountain | Batavia | Mustard Mix | | Coriander | |
| Coriander | Salatrio | Nero di | | Oregano | |
| Peppermint | | Toscana Kale | | Pea Tops | |
| Peruvian Mint | | Red-Stem | | Peppermint | |
| Rosemary | | Mizuna | | Rosemary | |
| Sage | | Red-veined | | Sage | |
| Thai Basil | | sorrel | | Tarragon | |
| Thyme | | Scarlet Kale | | Thyme | |
| | | Spinach | | Watercress | |
| | | Wasabi | | | |
| | | Rocket | | | |
| | | Wasabi | | | |
| | | Rocket with | | | |
| | | flowers | | | |

## 10 References

Asseng, S., Guarin, J. R., Raman, M., Monje, O., Kiss, G., Despommier, D. D., Meggers, F. M. and Gauthier, P. P. G. (2020). Wheat yield potential in controlled-environment vertical farms. *Proceedings of the National Academy of Sciences of the United States of America* 117(32), 19131-19135.

Butturini, M. and Marcelis, L. F. M. (2019). Vertical farming in Europe: present status and outlook. In: *Plant Factory: an Indoor Vertical Farming System for Efficient Quality Food Production* (2nd edn.). Elsevier Inc. The Netherlands.

Folta, K. M. (2019). Breeding new varieties for controlled environments. *Plant Biology* 21 (Suppl. 1), 6–12.

Harris, Z. M. and Kountouris, Y. (2020). Vertical farming as a game changer for BECCS technology deployment. *Sustainability* 12(19), 1–8.

Jahnke, S., Roussel, J., Hombach, T., Kochs, J., Fischbach, A., Huber, G. and Scharr, H. (2016). phenoSeeder - A robot system for automated handling and phenotyping of individual seeds. *Plant Physiology* 172(3), 1358–1370.

Manavalan, L. P., Cui, I., Ambrose, K. V., Panjwani, S., DeLong, S., Mleczko, M. and Spetsieris, K. (2021). Systematic approach to validate and implement digital phenotyping tool for soybean: A case study with PlantEye. *Plant Phenome Journal* 4(1), 1–17.

Saad, M. H. M., Hamdan, N. M. and Sarker, M. R. (2021). State of the art of urban smart vertical farming automation system: advanced topologies, issues and recommendations. *Electronics* 10(12), 1–40.

TTA (2019). *WWW Document*. Available at: https://www.tta.eu/.

van Delden, S. H., SharathKumar, M., Butturini, M., Graamans, L. J. A., Heuvelink, E., Kacira, M., Kaiser, E., Klamer, R. S., Klerkx, L., Kootstra, G., Loeber, A., Schouten, R. E., Stanghellini, C., van Ieperen, W., Verdonk, J. C., Vialet-Chabrand, S., Woltering, E. J., van de Zedde, R., Zhang, Y. and Marcelis, L. F. M. (2021). Current status and future challenges in implementing and upscaling vertical farming systems. *Nature Food* 2(12), 944–956.

# Chapter 18

# Oishii Farm: gaining the leading edge in the plant factory business and looking ahead

*Hiroki Koga and Kenzo Uchigasaki, Oishii Farm, USA*



## 1 Introduction

The story of indoor vertical farming has its beginnings in Japan. From the early 2000s, indoor vertical farms were researched and then built at scale by both major Japanese conglomerates and smaller-sized enterprises. Having some of the world's leading industrial technology and horticultural techniques, it was no surprise that indoor vertical farming technology began in Japan ahead of the rest of the world. The technique of growing plants indoors without sunlight soon fascinated the public and was touted by the media as the 'future of agriculture'.

However, skeptics wondered whether it was viable to spend such large amounts of capital on the construction and operation of indoor vertical farms when traditional field and greenhouse cultivation were just as reliable. Japan had a well-established market for conventionally grown high-quality fruits and vegetables, weakening the appeal of indoor vertical farms. The first indoor vertical farm companies could not charge the premium prices necessary to cover the initial investment and resource costs of creating such facilities. This resulted in early indoor vertical farms consistently operating at a loss. Eventually, indoor vertical farms were labeled as an 'innovative but not an economically viable industry' and were forgotten by the general public. For the next 10 years, the industry adopted a wait-and-see approach.

Then, around 2015, there was a turning point in agriculture, focusing on the problem of 'sustainability'. Factors such as a stable climate, available land, healthy soil, water, labor, pesticides, and fertilizers – all of which are necessary for modern, intensive agriculture – have begun to enter an era of scarcity. For example, California, which produces most of the fresh vegetables and fruits in the United

http://dx.doi.org/10.19103/AS.2023.0126.25

States, has seen shortages in all these areas. Climate change is an ever-growing threat, especially in California, where wildfires and extreme weather conditions continue to increase in intensity and frequency, affecting land availability, soil health, growing conditions, and water resources. One of the main indicators of climate change is the steadily rising trend in annual temperature (which can be seen in Fig. 1), which has various effects on weather patterns, including extreme weather events and changes in precipitation patterns, which profoundly affect the ability to grow food. In the case of labor, changes in domestic and international politics have reduced the availability of labor e.g. the restrictions on migrant workers from Mexico in the United States during the administration of President Trump. Soil depletion, due to poor soil management (resulting from problems such as compaction, use of monocultures, and reliance on agrochemical inputs), has meant fields in California and other places produce lower yields or are becoming unusable. Concern about the long-term sustainability of conventional farming has resulted in a renewed interest in indoor vertical farming.

At the same time, continuing technological progress has progressively decreased the costs of indoor vertical farming. The price of equipment such as LEDs and air conditioning has continued to fall as they become more energy efficient and are mass-produced (Fig. 2). Operating costs (e.g. labor) have declined due to greater use of automation. Costs are also declining due to cheaper methods of energy production and improvements in facility design. This has led to a situation where some indoor vertical farm can have lower production costs than conventional agriculture. In addition, implementation of water recycling, no need for pesticides, and even removal of a soil medium show the potential of indoor vertical farming at addressing the problem of sustainability.

These improvements have accelerated recent global investment in indoor vertical farms. More than 2 billion dollars e.g. have been raised between the companies Plenty, Bowery, Aerofarms, and Infarm in the United States. This level of investment is many times larger than the amount raised in Japan over the past 20 years. The reason behind such a large investment made in these companies, which still have relatively low turnover, reflects the magnitude of the 'sustainability' problem and a consensus among investors around the world that indoor vertical farms are a key long-term solution.

Indoor vertical farms offer the opportunity for a new era of globalization in agriculture. Unlike other manufacturing industries, regional differences in climate, seasonal conditions, land, and other resources make it impossible to produce exactly the same products globally. It is e.g. challenging to produce strawberries of equivalent quality in regions as diverse as America, Europe, or Africa or even to ensure consistent yields and quality within these regions from season to season. These challenges in ensuring a reliable supply of produce of consistent quality cannot be fully resolved by globalized agri-food supply chains

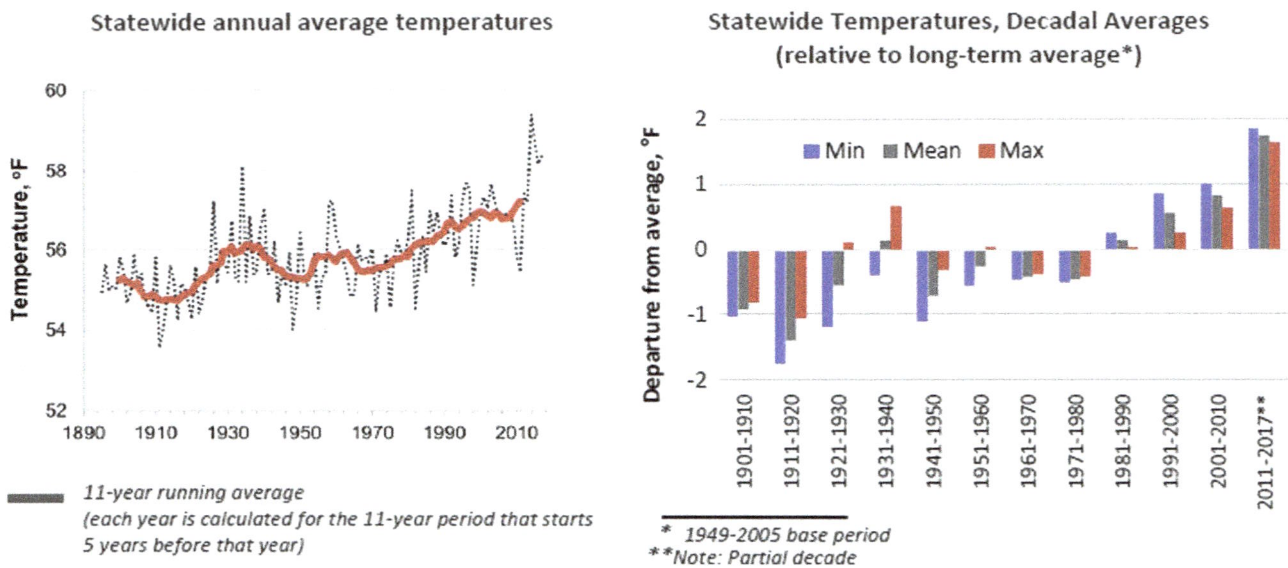


**Figure 1** California annual average temperature and deviation from long-term average. Source: California OEHHA.

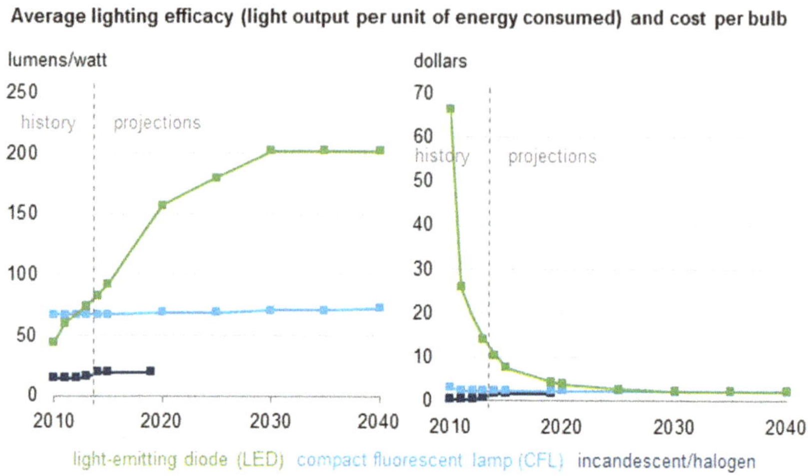


**Figure 2** Changes in LED costs. Source: US Energy Information Agency.

which source produce from different regions around the world. In addition, these supply chains are resource intensive and environmentally damaging.

Indoor vertical farms, on the other hand, can manufacture exactly the same produce in any location consistently without being affected by variables, such as weather, as long as they are well managed and have basic resources such as electricity and a small amount of water. For the first time in the history of agriculture, it will be structurally possible for truly global producers to emerge. In other words, an era can be foreseen where global brands, similar to Tesla, Toyota, and Ford in the car industry, can emerge in the agricultural world. The fresh produce market is worth over a trillion dollars and the company that leads this market could become one of the most valuable companies on the planet. We are now in this revolutionary phase.

What, then, is hindering the realization of this paradigm shift? The industry is making gradual progress, but the answer goes back to achieving viability now. As Tesla did with the electric car, someone needs to prove that indoor vertical farms are no longer a niche technology with uncertain profitability but a truly scalable business model. This is exactly the question Hiroki has been exploring for the last 10 years. Our solution is to focus on strawberries and not just any normal strawberries, but the world's highest-quality Japanese strawberries.

Why strawberries? The crops that indoor vertical farms have produced so far have been mostly leafy greens with low technical hurdles to overcome. However, it has been difficult to place premium prices on these products. The average consumer has no desire to purchase expensive lettuce. Low prices mean the only way to improve business profitability is to cut costs. From the standpoint of looking at this industry as a consultant and investor for 10 years,

Hiroki thought that developing a product with a premium selling price would achieve profitability in the medium to long term more than pursuing cost reductions of 10–20%. Achieving sufficient early profitability makes it possible to invest in increasing productivity and expanding the scale of production. This makes it possible to reduce costs over time and finally reach the point where we can provide the product at a lower price than conventional agriculture. Without this approach, we have seen many examples of companies failing to turn a profit and going bankrupt because they are unable to cut costs sufficiently.

## 2 A brief history of Oishii Farm

Key details:

Founded: December 2016
Business description: production, sales, and research and development of fruits and vegetables in a completely closed indoor vertical farm
Headquarters: Jersey City, State of New Jersey, USA
Farm locations: Jersey City and Kearny (New Jersey), Los Angeles (California)
Number of employees: 140+
Composition: farm + R&D (40%); engineer (40%); business (20%)
Sales channel: Whole Foods and other retailers
Mission:

1. Revolutionize agriculture sustainably through indoor vertical farming.
2. Improve customer experience through a category creating product.

Our goal as a company is to make fresh, quality, safe, and sustainable produce available to everyone around the world, and make indoor vertical farming the new standard for agriculture (Fig. 3). We see a future where everything that can be grown in indoor vertical farms is grown in indoor vertical farms more cheaply than conventionally grown products.

To go back to the beginning, Oishii Farm was established in December 2016 by two co-founders: Hiroki Koga and Brendan Somerville. At the time, Hiroki was enrolled in the MBA course at the University of California – Berkeley while Brendan Somerville was enrolled in the MBA course at the University of California – Los Angeles. They met, hit it off, and established the company while still on their MBA courses. Just before graduating, the two won UC Berkeley's largest accelerator program known as 'Launch' (designed for business start-ups), raised funds from outside investors, moved to New Jersey in 2017, and built the first prototype farm (Fig. 4).

As noted earlier, a fundamental early decision was to focus on growing strawberries. Unlike leafy greens, strawberries can be premium priced. Without

**Figure 3** Oishii Jersey City farm. Source: Oishii.

**Figure 4** Oishii prototype container farm. Source: Oishii.

an indoor vertical farm, it is not possible to supply fresh, delicious, and pesticide-free strawberries of consistent quality all year round, especially in a location like New York City. By meeting this high-value and unmet market demand, Oishii has, 5 years after its founding, become known among investors as the world's most profitable indoor vertical farm.

Another reason for focusing on strawberries was that lettuce production in indoor vertical farms has already existed for more than 20 years. Although there are some differences between producers, the technology for indoor production is well established. It is therefore relatively easy for new entrants to enter the market and accelerate price competition. On the other hand, fruiting crops, such as strawberries and tomatoes, especially flowering crops that require pollination management and integrated pest management, are considered difficult to produce in indoor vertical farms at scale and at profit. By focusing on this area from the beginning and establishing an early lead on research and development, the plan secured an advantage over other companies. Also, as a Japanese-led company, Hiroki's ability to access Japanese facility design technology gave a competitive edge over other companies, especially the latest in Japanese facility design technology which has yet to spread in the United States. As a result, with fewer new entrants, there would be no price collapse and we would have more time to maximize profits and reduce costs. In addition, if more challenging crops such as strawberries could be successfully grown, it would be relatively easy to expand into the production of other fruit vegetables such as tomatoes, melons, cucumbers, and so on. If you are able to learn to grow strawberries indoors, you are in a strong position to grow most other fruiting crops.

Currently, Oishii Farm has been alone in successfully establishing the technology to mass produce strawberries at a profit, but it is only a matter of time before others catch up. A comparison can be made with the electric car industry. Ten years ago, Tesla was the global face of fully electric cars, but there are now more than a dozen companies in the market to challenge Tesla's monopoly. It is self-evident that the same thing will happen in the medium to long term with indoor vertical farms. To continue being the world's number one in the medium to long term, you must also create a 'one-of-a-kind' brand. Before competitors can establish themselves, Oishii needs to secure overwhelming recognition, much like Tesla did in the auto industry. To do that, we need to create an unforgettable product experience. To achieve this, there is no better crop than Japan's high-end strawberries.

The vision for the future is one which, when you enter the supermarket, there is a shelf labeled 'Oishii' where the vegetables and fruits on display are always in-season, pesticide-free, fresh, and delicious. Only if you can't see the product you want, do you move to the next shelf labeled 'organic' or, if it's still not there, go to the 'non-organic' shelf. Oishii will create that vision. In order to build a brand that customers choose first, our products must always 'exceed expectations', so we plan to start by offering crops that are unique to the American market. Strawberries are just the start. Tomatoes and melons are already in development.

Until now, Oishii has been selling directly to consumers. However, by 2022, our mass-production technology has achieved a sufficient volume of output to allow us to sell to retail supermarkets, the main distribution channel for fresh produce. In June 2022, Oishii started distributing its strawberries through the Whole Foods supermarket chain in the Manhattan area in New York City (Fig. 5). By connecting a chain of supermarkets to one of our local farms, the freshest produce can be delivered to the greatest number of consumers.

A unique thing about Oishii is that, at this stage, everything is done in house, from research and development, construction, and propagation to distribution. It is common for indoor vertical farming companies to partner with a separate contractor for construction. However, we built our first mass-production farm from the ground up. Because no company in the world specialized in building indoor vertical strawberry farms, which are significantly more complicated than farms for leafy greens, we realized it would be quicker and cheaper to construct them ourselves. This allowed us to understand the materials and costs involved, make flexible design changes during construction and budget efficiently for future projects. In terms of sales, direct contracts with retail supermarkets, and a top-notch marketing team ensuring effective display and branding, means no need for a separate sales team.

**Figure 5** Whole Foods launch. Source: Oishii.

## 3 R&D and production challenges

Growing strawberries in an indoor setting on a small scale is not that difficult. There is plenty of research on the subject and we have had the opportunity to eat high-quality strawberries at many small-scale facilities in Japan and The Netherlands. The headaches start when these facilities are scaled up. When it comes to stable mass production at a commercial scale, there are so many factors to consider. Hurdles include:

- pollination;
- disease management;
- growing healthy, disease-free seedlings;
- establishing growth conditions to optimize quality;
- precise environmental control; and
- effective monitoring.

These are discussed in more detail in the following paragraphs.

Unlike leafy greens like lettuce, most fruit vegetables, including strawberries, require pollination. In most cases, this is done by bees. It is not difficult to manage bees in outdoor cultivation or greenhouse cultivation, but it is very difficult to control pollination by bees in a completely closed indoor vertical farm. Hand pollination is used in most research and development facilities but is too time-consuming and costly for mass production. We have developed technology that keeps pollination rates by bees above 90% at all times. Even in ordinary greenhouses, pollination rates are only about 70% depending on the crop. Costs will practically double if the pollination rate falls to 50%, so pollination is an incredibly important aspect to manage and improve.

Ensuring a high, consistent pollination rate is achieved by controlling various environmental parameters, collecting data and utilizing AI to ensure that bees act in the same way as they do in a natural environment. Environmental data and image data relating to each plant is collected using a robot that automatically travels round the farm (Fig. 6). Data are transmitted to another internet of things device attached to the bee's hive. AI is then used to create the ideal conditions to ensure that the optimal number of bees can always pollinate within the farm.

Disease management is also a critical aspect of mass production, particularly as many consumers prefer 'pesticide-free' produce and methods of production. In small-scale research or production facilities, it is possible to replace diseased plants or continue production with a slightly lower yield. Crops with a short production cycle, such as lettuce, can also easily be replaced. However, if disease occurs in the mass production of strawberries, which take months to grow from nursery to harvest, entire crops might have to be culled and regrown from scratch with a major impact on cost and profitability.

**Figure 6** Self-driving camera cart. Source: Oishii.

The starting point of any farm is the growth of healthy seedlings which are critical to final quality and yield. If you want to produce completely pesticide-free strawberries, you must prepare disease-free seedlings. However, sourcing externally is too expensive. Growing plants in house from the start is not only more cost effective but also allows our farmers and researchers to observe and understand every new cultivar and the unique traits that come with being grown in an indoor environment. This is invaluable as very few existing cultivars have been grown indoors and certainly not in large-scale commercial conditions.

Propagation is done in house using various techniques to produce virus-free and nearly sterile seedlings, depending on the cultivar (Figs 7 and 8). By establishing in-house propagation units that can supply to all of the company's

**Figure 7** Tissue culture plantlets *in vitro*. Source: Oishii.

production facilities, we ensure a continuous supply of contamination-free, high-quality, tailored planting material that would not be possible using traditional nursery suppliers. By creating a facility design and operation flow that minimizes the risk of bringing in bacteria and viruses from the outside, over the past 5 years, there has never been an outbreak of disease in the production farms. Furthermore, farms are designed to minimize damage in the unlikely event that some kind of disease occurs. After an outbreak of disease, it will often recur in greenhouses no matter how much of the contamination is removed. There is an established proprietary procedure to reset the environment to a completely sterile state if this is needed to remove contamination.

A third key aspect is creating the right growing conditions (recipe) to ensure sufficient quality. The reason why Sembikiya (a Japanese gourmet fruit merchant selling some of the most expensive produce in the world) can set prices ten times higher than supermarket retailers is that it created a system that guarantees the superior quality of the fruits they sell. Strawberries are a

**Figure 8** Tissue culture plantlets rooting. Source: Oishii.

fickle crop and even a slight change in the environment can drastically change the taste. When comparing the same cultivar of strawberries (e.g. Tochiotome in Japan), there is a marked difference in quality between the Tochiotome grown by a dedicated, experienced farmer and the Tochiotome grown by a beginner. While they may look similar, the smell, texture, and taste of one is far superior compared to the other. If Oishii can also supply strawberries of significantly higher quality throughout the year, we can charge much higher prices compared to our competitors. This is especially so in the United States where the culture of very high-quality produce from traditional, small-scale, artisanal growers is not well established.

In Japan, and currently with Oishii, short-day cultivars are the main variety of strawberries grown because of their higher quality in flavor and sweetness. There is a trade-off between yield and quality with short-day cultivars. In conventional Japanese domestic strawberry production, the harvest period is limited by seasonal weather changes and generally runs for about 5 months between December and March–April. Plants are traditionally kept for just one season and then replaced or left to recover through the summer and autumn of the following year.

In contrast, indoor vertical farms can maintain the appropriate environment for growing plants all year round. After flower bud differentiation is set, the harvesting cycle can be continued semi-permanently. However, in order to eliminate stress on plants and stabilize production, seedling management and environmental controls that are unique to indoor vertical farms are required. Since these challenges are new, they have required significant R&D to identify optimum conditions which extends beyond climate to include factors such as fertilizer dosage, pH, and micronutrients.

A fourth factor is environmental control. Factors such as the short growth cycle of leafy greens mean that, even without strict environmental control, little difference can be seen in quality. In some of the indoor vertical farms that we visited, it is common for the temperature to differ by 3°C between the upper and lower shelves. There are farms that are just 5 m tall with a 4–5°C discrepancy between the top and bottom shelves. In contrast, flowering crops such as strawberries take several months to reach harvest and will undergo various phases through growth, including flower bud differentiation. Even small changes in temperature will result in a large difference in overall average temperatures, and the growth rate and traits of seedlings between the upper and lower shelves will differ significantly if similar temperatures are not maintained. If seedling growth or the timing at which flower buds appear is wrong, the necessary environmental requirements will be out of sync for that batch of new plants as they enter different growth periods, disrupting harvesting. Managing a consistent environment is thus key to consistent, high quality.

Our company focused on effective control of environmental conditions at an early stage. In designing an effective heating, ventilation, and air conditioning (HVAC) system, we performed airflow calculations in the farm on a zero basis to optimize air conditioning and exhaust outlets. Compared to leafy plants, differences in day and night temperatures are greater and the amount of heat generated by LEDs is also greater. To design the optimal HVAC system to be able to handle heavy temperature and humidity loads, 3D models of the farm and computer simulations were run and fine-tuned on site after construction. We were able to create an environment with controls within ±0.5°C, allowing for uniform growth, yields and quality.

A key element of a stable environment is an effective monitoring system. This all-encompassing system should be able to monitor and control all the various systems such as air conditioning, irrigation/drainage (including recycling), air mixture ($CO_2$, levels), light intensity, data collection, etc. It must operate continuously and mostly autonomously, automatically adjusting environmental conditions as required or providing an immediate alert if there is a problem. There are a number of companies around the world that provide environmental control equipment for indoor cultivation. However, these systems are mostly designed for greenhouses and are unable to precisely control temperatures within ±0.5°C. Oishii has therefore developed and implemented its own programmable logic controller in house because there was no existing system that centrally manages all the required controls (especially the complicated management of irrigation required for indoor vertical strawberry farms). This system was developed over a number of years and is still being improved day by day but is at the heart of our company's operation.

All of these challenges must be addressed. While it is possible to grow strawberries on a small scale even if some of these components are missing, investing tens of millions of dollars (billions of yen) to build a farm and grow a completely new industry requires that all of these elements are developed and executed perfectly.

## 4 Conclusion and future trends

It is said that there is a global market of about 4 trillion yen or 30 billion dollars for strawberries alone. Assuming that this is constrained by seasonality, it is quite possible that this will more than double if indoor vertical farms are able to produce all year round. Indoor vertical farms could create a market of nearly 10 trillion yen or 73 billion dollars for strawberries alone. If we add in other fruits and vegetables that we know can be produced in indoor vertical farms using our current technology, there is a potential market of more than 1 trillion dollars.

Oishii will continue to release products of 'one-of-a-kind' quality suitable for the Oishii brand. The company already has several farms that produce crops other than strawberries, and is already testing tomatoes and melons that have a sweetness and flavor that consumers have never tasted before. Newer crops will be continuously added to the R&D pipeline to achieve the envisioned future for the Oishii brand. We will use this technology to create a truly global agricultural brand that has not existed before where our products will always be the first choice when consumers go to a supermarket.

In the short term, we will introduce products in the high-end price range by introducing 'one-of-a-kind' products, but behind the scenes, we are working tirelessly to reduce costs. This is not because of a desire to increase profits, but because we want to lower the price so that more people can experience high-quality produce. We are not aiming for a high-end market like Ferrari, but like Tesla, we will improve productivity and scale of production to lower the price and eventually allow everyone to be able access fresher and more delicious produce. If our strawberries (currently 20 dollars) were 5 dollars, there would be no reason to buy poorer-tasting strawberries containing pesticide residues brought in from far away. Only then will we have achieved our mission of achieving a true paradigm shift in the agricultural industry.

Over the past 5 years, we have reduced production costs by more than 50%. Over the next few years, we will continue to reduce costs even more. The three main ways to achieve this are through:

1 increasing yields;
2 cost reduction through operational efficiency; and
3 cost reduction through automation.

Within 10 years, it will be possible to produce strawberries all over the world at a price comparable to that of conventionally grown strawberries. If Oishii-quality strawberries could be bought for a few dollars all year round, people would only buy our strawberries because they were the sweetest, freshest, completely pesticide-free, continuously available and, above all, consistently high-quality strawberries available. Together with the team, we want to create a brand that can gain such overwhelming customer trust for all the crops we produce.

It has often been said that indoor vertical farms will establish a niche in the market, co-existing with a still dominant conventional agriculture. Hiroki's view is that this technology will replace existing agriculture in the long run, becoming the dominant form of agriculture. As noted earlier, current agricultural systems are in crisis. Take the US strawberry market as an example. Most of the country's strawberries are grown in California along the central and southern Pacific coast. Forest fires occur almost every year, extreme weather conditions continue to occur at greater frequencies, land prices soar from having to compete with commercial and residential development, migrant agricultural workers have been deported and wages have increased 3+ times. Input costs such as fertilizers are increasing and governments around the world are putting increasing restrictions on use of agrochemicals such as pesticides and fertilizers, given their environmental impact. Add to the equation a still increasing world population and the current supply chain for fresh food looks increasingly vulnerable in meeting demand.

The four major costs of an indoor vertical farm are labor costs, utility costs, depreciation (construction costs), and land costs. Since automation is progressing more rapidly than in conventional agriculture, labor costs will decrease exponentially in the future. Utility costs are also falling day by day due to improvements in LED and air conditioning performance. As the number of indoor vertical farms increases, the price of equipment will also fall and, as research and development progresses, the space required for cultivation is declining exponentially. All of these trends are making indoor vertical farms ever more competitive compared to conventional agriculture which is currently facing serious upward cost pressures in prices for labor, inputs, and land. In addition, the ability to build indoor vertical farms within or near urban centers reduces logistics costs. In some regions, lettuce can already be grown at a lower cost in an indoor vertical farm than outdoors.

Future areas of research include the following:

- Improving cultivation efficiency;
- Extending the range of crops that can be grown in indoor vertical farms;
- Developing varieties specifically tailored for indoor vertical farming conditions; and
- Developments in automation.

One area of research is improving cultivation efficiency. There is continuing research in areas such as LED irradiation, carbon dioxide concentration, promotion of flower bud differentiation, and control of sugar content which can enhance plant growth and yield. However, these advances will mean little if indoor vertical farming operations are not precisely and consistently controlled so that these refinements can be effectively implemented and reliable data produced to allow further research and improvement. Operational problems such as 'seedlings withered because a part-time worker did not come in' or 'due to system failure, we were not able to promote the growth of seedlings sufficiently during the seedling raising period' will both impact yield and quality and compromise future improvement. If consistency and reproducibility are poor, the market will lose interest.

Another area of development is the range of crops that can be grown in indoor vertical farms. Relatively compact, high turnover and high value-added products such as lettuce, strawberries, and tomatoes have been easier to develop first as businesses, but it will take time for companies to start producing crops like grains (e.g. rice or wheat) or root crops in indoor vertical farms. With improvements in such areas as dwarfing and grafting techniques, the number of crops that can be cultivated will increase further, creating a market potentially comparable in size to the automobile industry.

A third area for improvement is in developing varieties specifically tailored for growth in indoor vertical farming conditions. In general, leafy vegetables and fruit vegetables produced in indoor vertical farms, including our products, are mostly produced by using existing varieties, from which those that grow most efficiently in an indoor vertical farming environment are selected. The next step is the creation of varieties specifically bred for indoor vertical farming production. In the case of strawberries, existing cultivars were originally selected for the specific environmental requirements of particular regions with indoor vertical farms currently recreating these conditions to optimize quality and productivity. However, the greatest advantage of indoor vertical farms lies in the ability to create environments that cannot be realized in the natural world. For example, days and nights no longer have to be split between two 12-h intervals. However, after ascertaining the basic characteristics of crops, we can select them based on the kind of environment that theoretically can maximize yield and quality. In the case of lettuce, some companies have already developed varieties exclusively for indoor vertical farms. To achieve this means partnerships between breeding and production companies which are currently run fairly independently of one another.

A particularly exciting new breeding technique is genome editing which is already being used to accelerate the development of new cereal cultivars, for example. Genome editing promises to play a major role in the process of developing varieties exclusively for indoor vertical farms. Such techniques can be used e.g. in producing dwarf varieties of crops that are currently too large for indoor vertical farming such as blueberries, peaches, coffee, and cocoa.

Lastly, automation is key to reducing the overall cost of indoor vertical farms (Fig. 9). Automation is usually associated with 'labor saving'. Labor costs are, of course, the biggest cost in an indoor vertical farm, but utility costs, construction costs, and rents can all be reduced through automation. If there are no walkways for people to walk on and plants can be grown closer together, the lighting efficiency of LEDs will naturally increase, which will both reduce utility costs and the footprint of facilities. However, indoor vertical farming automation is more challenging than normal factory automation, as the impact

**Figure 9** Tamura fully automated indoor vertical farm. Source: A-Plus.

on plants needs to be considered. Introducing living, breathing, growing, relatively fragile, and heterogeneous components like plants into the equation complicates all aspects of automation. It is essential to use engineers with a farmer's mindset or at least an interest in plants to design automation systems with plants in mind from the outset.

The market for indoor vertically grown produce is there. The agricultural world will change in our lifetime as the climate change forces us to adapt to new environmental conditions. Humans will always need food and to fuel humanity for the future, we will need to evolve how we grow and produce it as humanity has done for millenniums. It will depend on who will take on this challenge and be part of the next step in agriculture. With high-quality strawberries as just the beginning, Oishii will do just that; revolutionize agriculture.

# Chapter 19

## Data-driven operations for a productive and sustainable plant factory

*Katashi Kai and Morio Okabe, Shinnippou Ltd, Japan*



## 1 Introduction

In Japan, an increasing number of companies from various industries have entered the plant factory business owing to the need for a stable supply of agricultural products. However, only few factories are profitable because of the high construction costs and the large amount of energy required for production (Kozai and Niu, 2020).

The brand '808 FACTORY' was established by Shinnippou Ltd., a company that runs pachinko (Japanese pinball), a hotel business, and generates solar power. In 2014, Shinnippou Ltd. constructed a new plant factory, 808 FACTORY, and began lettuce production with up to 10 000 heads per 100 g of conversion per day. In 2017, Shinnippou Ltd. expanded its production capacity to 20 000 lettuce heads per 100 g conversion per day (Shinnippou, 2018).

## 2 Business model

808 FACTORY sells products to supermarkets, retailers, food manufacturers, and restaurants. The business-to-business (B-to-B) and business-to-customer (B-to-C) sales ratios are approximately 50% each. Therefore, we need to differentiate our products from open-field-grown lettuces and competitors to sell at a high price. We repeatedly examined the taste, size, and other

http://dx.doi.org/10.19103/AS.2023.0126.26

characteristics of many lettuce varieties and selected one of them. Moreover, the concept of 'ready to eat without washing' plays an essential role in the differentiation.

Currently, we sell four varieties of B-to-C product types; however, selling a wide variety of B-to-B product types is necessary because the demand for size and shape differs for each company. For example, the boxed lunch industry prefers lettuce varieties with relatively uniform leaf size.

The demand from retailers and supermarkets fluctuates based on the price of open field-grown lettuce. Although B-to-B demands are consistent, they increase during the holiday season. Therefore, we make production plans based on our experience to meet them. We avoid relying on specific partners because unforeseen events such as the COVID-19 pandemic can lead to large fluctuations in demand. Owing to its TV commercials, the 808 FACTORY is familiar to neighboring citizens. Currently, the demand exceeds the supply; thus, lettuce is regularly produced.

## 3 Factory layout

### 3.1 Cultivation rooms

Both factories have a cultivation room separate from the shipping room and an employee entrance separated by more than three compartments to prevent the invasion of insects and dust. All cultivation rooms are airtight and insulated to minimize $CO_2$ loss to the outside environment as well as to reduce the costs associated with air conditioning. Additionally, the rooms are equipped with condensate-recycling facilities to conserve water.

Our experience with the first factory led to several improvements to the second factory. Each cultivation room has an area of ~500 $m^2$. The cultivation rooms in the first and second factories are equipped with 10 and 8 cultivation shelves, providing a total cultivation area of 3600 $m^2$ and 3456 $m^2$ (Hayashi, 2020), respectively. Lettuce is grown using the deep flow technique in the first factory and the nutrient film technique in the second (Fig. 1).

The distance between the tiers of the cultivation shelves in the second factory was one and a half times higher than that in the first factory, allowing lettuce to grow taller. In addition, conveyors for transporting lettuces help workers perform their jobs efficiently in a second factory.

Approximately 5% of the cultivation shelves are idle for test cultivation, cleaning, and other purposes; the remaining 95% are practically 100% operational.

### 3.2 Other rooms in the factories

In addition to the cultivation room, there are washing, seedling cultivation, packing, and shipping rooms. The washing room houses three automatic

**Figure 1** Lettuce cultivation at the second factory.

**Figure 2** An automatic seedling transplanter in operation.

machines that wash the panels and other items used in each cultivation room. Sowing and seedling cultivation are performed in a single sitting in a nursery room, and approximately 25 000 seeds are sown daily using a semi-automated sowing machine. Seedlings are grown for approximately 11 days, and as the second true leaf emerges, they are transplanted into panels using automatic transplanters (Fig. 2) and placed in cultivation rooms. The packing room is maintained at a low temperature of 15°C to preserve the quality

**Figure 3** Packaging machines in operation.

of harvested lettuce. Five packaging machines (Fig. 3) have been installed for each product. Packaging machines are equipped with metal detectors and weight checkers for foreign objects and weights. The shipping room is maintained at a temperature of 10°C. Packaged lettuce is visually inspected for any abnormalities, foreign matter, or package defects, followed by boxing.

## 4 Cultivation room and operations management

A single issue in the early stages of cultivation can have repercussions that last weeks. To reduce such problems, we installed a system that displays cultivation data and correlates the data with each other (Fig. 4). For example, water is supplied and drained simultaneously, accompanied by fluctuations in the electrical conductivity (EC), pH, and water temperature. Subsequently, the nutrient tank automatically provided a sufficient nutrient solution. The purpose of the cultivation data display system is to enable easy checking of the cultivation data and equipment operation. Specifically, because nutrient tubes are prone to clogging, monitoring growth conditions, pH, and other factors are necessary to detect anomalies. The system is displayed on a web platform, usually weekly, to allow everyone to access data from anywhere, not just from their desks. However, comparing a given day with the same day in the previous month or year can also help detect unexpected fluctuations in the cultivation characteristics.

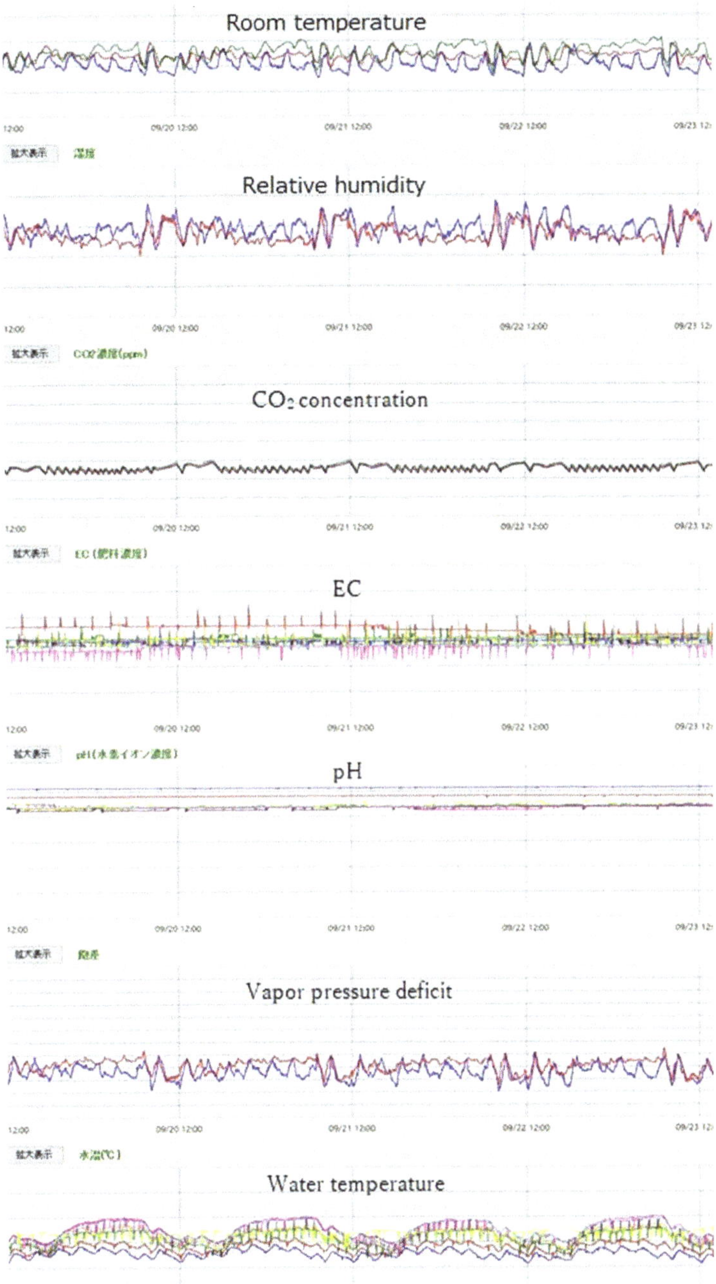


**Figure 4** Cultivation data graph. All cultivation data are displayed on one screen.

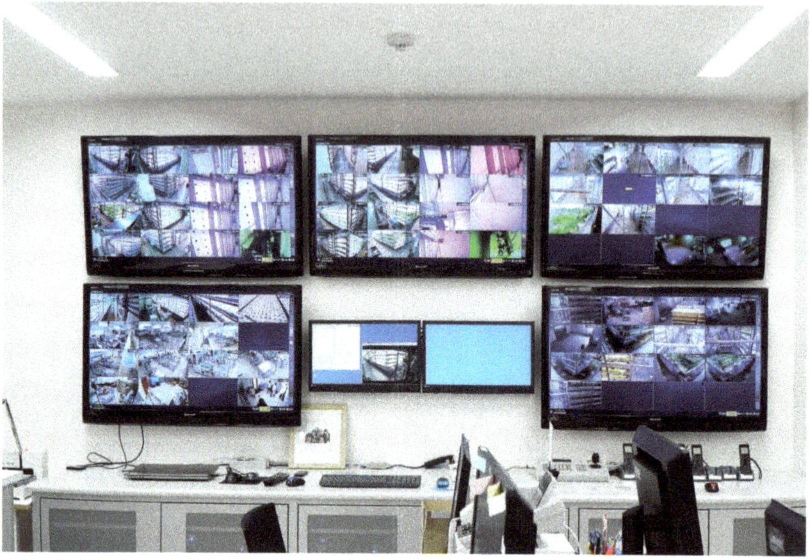

**Figure 5** Images of the webcam list.

All the rooms are equipped with webcams to record each operation. It is possible to ensure the efficiency and safety of the work by reviewing the records (Fig. 5).

## 4.1 Operation management system

Increasing the labor productivity of workers (Fig. 6) is as important as cultivating techniques. Operational management aims to ensure that the work runs smoothly, and that each worker simultaneously completes their tasks. The manager assigns another task if the worker completes the task ahead of the schedule. The workflow in the growing room is in the opposite direction to that for cultivation. After harvest (Fig. 7), the panels are removed, the other cultivation panels are pushed forward from the other side of the cultivation shelf, and the lettuce seedlings are transplanted into new panels. Variations in work speed must be accounted for to prevent mismatches at each process step. Web cameras record the work in every room, allowing us to track the duration of each task. Based on this information, we developed a unique system that assigns work appropriately, considering the work speed and skill level, to ensure that all workers are generally able to finish on time.

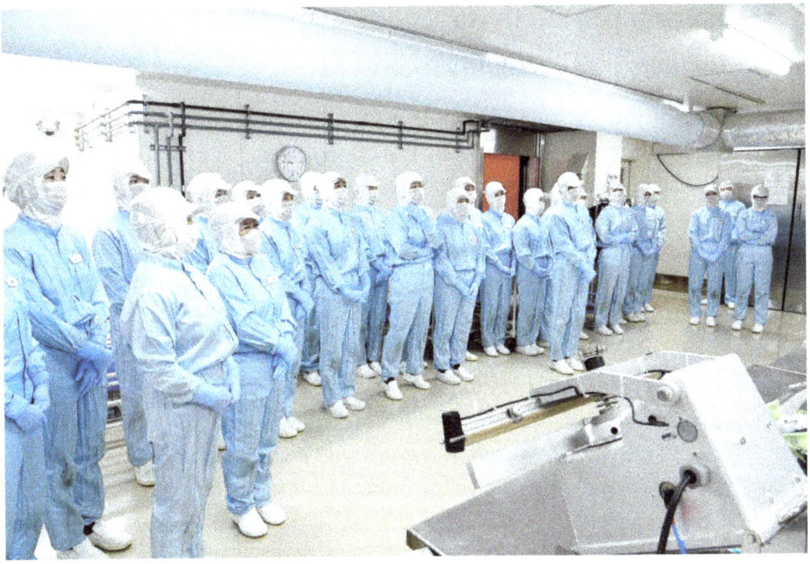

**Figure 6** Morning meeting in the factory.

**Figure 7** Harvesting.

# 5 Measuring productivity

## 5.1 Productivity per unit cultivation area

The varieties of lettuce grown at 808 FACTORY provides options for yield, quality, and size that are preferred by the consumers. For example, the lettuce variety on the left in Fig. 8a has a relatively uniform leaf size and each leaf is not overly large, making the lettuce variety suitable for B-to-B, as in the boxed lunch industry. Its leaves also grow densely, with a high yield per cultivation area (Fig. 8). The variety on the right has a soft texture, spreads widely, and has few leaves, resulting in a low yield per unit cultivation area. However, it is popular among consumers (Fig. 8). Both lettuce heads appear similar in size; however, the frill lettuce head is approximately twice as heavy (94 g vs. 49 g) and has more leaves per plant (Fig. 8). Because overcrowding of the panels with lettuce results in inconsistent overall quality, we grow lettuce with an appropriate spacing within the cultivation panels. Additionally, trimming is performed to maintain quality.

**Figure 8** (a) Left: frill lettuce, right: leaf lettuce. (b) Unfolded leaves of each variety. Both lettuce heads appeared similar in size, but the frill lettuce head was approximately twice as heavy (94 g vs. 49 g) and had more leaves per plant.

### 5.2 Productivity per unit of electric energy consumption

Long photoperiods result in a faster growth rate but lower weight gain per unit of electric energy consumption than short photoperiods do. An optimum photoperiod is specified for each variety, considering the prevention of tip burns and other issues. When calculating the productivity per unit of electric energy consumption, the denominator is the electricity used for lighting and air conditioning in the growing room. The numerator is the weight of the packaged lettuce, excluding the trimmed or discarded lettuce. The amount of electricity consumed in air conditioning changes from summer to winter and thus varies with the season (Table 1).

### 5.3 Productivity per labor hour

Labor productivity was calculated as described previously (Kozai et al., 2022). Labor includes the culturing process and the time involved in packaging, inspecting, and shipping. In B-to-C products, plants are small, and time is spent on packaging and inspection. Including all these times, from cultivation to shipping, the calculation was approximately 5 kg/h. As it is tedious to calculate labor productivity for each product using the 808 FACTORY, we compute it as a whole.

### 5.4 Merchantability ratio

The merchantability ratio was calculated as the weight of shipped lettuce divided by the weight of harvested lettuce. Because 808 FACTORY commercializes high-quality lettuce with a long shelf life, most varieties are trimmed after harvest by approximately 15% of their outer leaves, which deteriorates quickly in quality. Tip burns and diseases rarely occur in standard lettuce varieties, although some occur in certain varieties. Additionally, the merchantability ratio excludes waste resulting from packaging errors and other factors. In terms of

**Table 1** Lettuce productivity per unit of electricity consumption and per unit cultivation area for each B-to-C product type

| Variety | Productivity per unit of electric energy consumption (kg/kWh) | Productivity per unit of cultivation area (kg/m$^2$/day) | Average packaging weight in 808 FACTORY (g/package) |
|---|---|---|---|
| 1 | 0.07-0.08 | 0.18 | 85 |
| 2 | 0.12-0.14 | 0.15 | 60 |
| 3 | 0.05-0.06 | 0.09 | 50 |
| 4 | | 0.09 | 70 |
| Kozai et al. (2022) | 0.11-0.14 | 0.25-0.33 | - |

the weight of lettuce harvested, only approximately 65% of certain varieties are commercialized, whereas others are commercialized between 80% and 85%.

# 6 Controlling cultivation variables

## 6.1 Control of temperature, photoperiod, airflow and $CO_2$ concentration

Many factors are involved in determining the growth conditions. Higher temperatures result in faster growth rates but also render cultivation rooms more prone to mold and insect infestation, which can cause disease and tip burns on lettuce. We first determined the temperature that would keep the entire room clean, without inhibiting lettuce growth. Subsequently, the photoperiod should be maintained under conditions that do not cause tip burns. The length of the photoperiod must be considered because longer photoperiods decrease the rate of weight gain per unit electricity. Controlling airflow is particularly important. Even with airflow control throughout the room, sufficient airflow often did not reach the lettuce inside the cultivation shelves, resulting in tip burns that did not sufficiently increase the growth rate. By applying an airflow of approximately 0.2 m/s to the center of each lettuce plant, we could optimize the growth environment while reducing tip burns and increasing yields by approximately 25% for some varieties (Table 2).

$CO_2$ is maintained at a high concentration because of its low cost. However, photosynthetic activity decreases if the $CO_2$ concentration is excessively high (Yamori, 2020); thus, the $CO_2$ concentration-dependent photosynthetic activity was determined for each variety (Fig. 9). Currently, the concentration is set to approximately 1000–1200 ppm.

## 6.2 Nutrient management

Nutrient management is controlled via EC, and as EC falls below a certain level, fertilizer is automatically applied. As a nitrogen source, ammonia nitrogen, which is efficiently utilized by lettuce, is added in addition to the nitrate-form nitrogen. Additionally, the cultivation conditions are checked regularly, and trace elements are added as needed in case of growth deficiencies.

**Table 2** The percentage increase in yield obtained after the introduction of airflow

| Variety | Yield increase rate (%) |
|---|---|
| 1 | 21 |
| 2 | 26 |
| 3 | 23 |
| 4 | 3 |

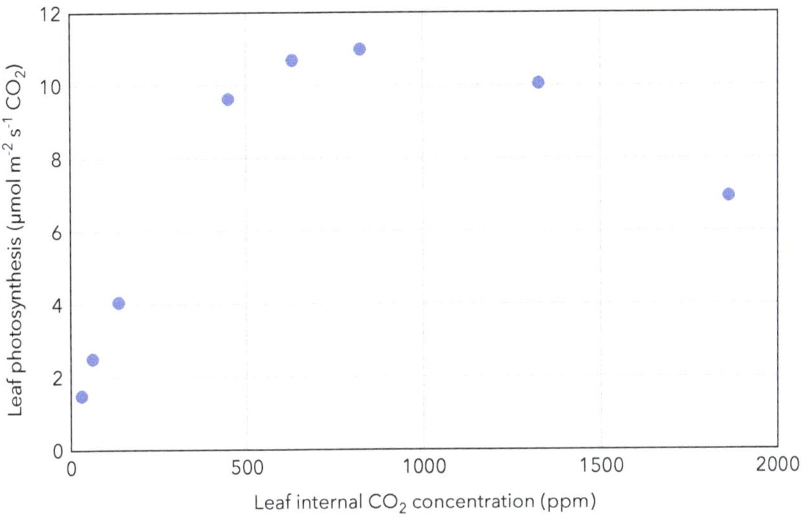


**Figure 9** Correlation between photosynthetic activity and $CO_2$ concentration of a single leaf measured using LI-6400.

## 7 Other technologies and future trends

A 3D printer is used to improve factory design, in which parts may require repair owing to deterioration over time, or there may be a need for connectors to install new equipment. A 3D printer can speed up the creation of the necessary parts to accommodate these requirements. In addition, when combined with 3D modeling, 3D printing can be used to study workability and other factors when designing a plant factory (Fig. 10).

We also collaborated with a venture company to develop seeds for plant factories to improve their variety. The resulting lettuce varieties were tested repeatedly on cultivation shelves and compared with the in-use and available lettuce varieties for growth speed, taste, shape, and variation in the number and area of leaves. Utilizing these tests, new varieties have been introduced every few years.

In addition, we have begun to develop cultivation methods that will allow us to achieve higher productivity per unit of electrical energy consumption in pursuit of our Sustainable Development Goals. For example, we are working to develop cultivation methods that increase cultivation density without compromising quality and cultivation methods that use specific wavelengths to control the deterioration of outer leaves and improve the merchantability ratio. Through these efforts, we plan to increase productivity per unit cultivation area and productivity per unit of electric energy consumption by

**Figure 10** (a) 1/100 scale model created by a 3D printer. (b) Actual cultivation work.

15%. Reflecting on these achievements, we plan to increase productivity by more than 20% in the future. Moreover, we are currently generating solar power on the rooftop and on site, which will be used for cultivation after the feed-in tariff period ends, allowing us to further reduce carbon-derived energy consumption.

## 8 References

Hayashi, E. (2020). Selected PFALs in Japan. In: *Plant Factory (2nd edition)*, eds. Kozai, T., Niu, G. and Takagaki, M. Elsevier, 437–454.

Kozai, T. and Niu, G. (2020). Role of the plant factory with artificial lighting (PFAL) in urban areas. In: *Plant Factory (2nd edition)*, eds. Kozai, T., Niu, G. and Takagaki, M. Elsevier, 7–34.

Kozai, T., Uraisami, K., Kai, K. and Hayashi, E. (2022). Productivity: Definition and application. In: *Plant Factory Basics, Applications and Advances*, Kozai, T., Niu, G. and Masabni, J. (Eds). Elsevier: Amsterdam, The Netherlands, 197–216.

Shinnippou (2018). *Case Study of Commercial Large-Scale PFAL, 808FACTORY*. Lecture at *Five-day Introductory Training Course on Plant Factory with Artificial Lighting (PFAL) at Japan Plant Factory Association* (October 2, 2018). Chiba, Japan.

Yamori, W. (2020). Photosynthesis and respiration. In: *Plant Factory (2nd edition)*, eds. Kozai, T., Niu, G. and Takagaki, M. Elsevier, 197–206.

# Chapter 20

## Design and management of industrial-scale vertical farms: Intelligent Growth Solutions (IGS)

*Gonçalo Neves, Andrew Lloyd, Niall Skinner, Douglas Elder, Dave Scott, Niels Kortstee, Emily Seward, Csaba Hornyik, Alexander Keel-Dwyer and Lawrence Ross, Intelligent Growth Solutions (IGS), UK*



## 1 Introduction

As noted in this book, vertical farming has various potential advantages compared to conventional agriculture in addressing such challenges as climate change and sustainability (Table 1). Advantages include:

- Optimising food quality through environmental control rather than genetic manipulation;
- No pesticide or fungicide use leads to better product safety and reduced environmental impact;
- Highly efficient reuse of water through condensate recovery as well as recirculation of the nutrient solution;
- More efficient use of nutrients due to a closed-loop system;
- Farmers have a year-round growing capability independent of more extreme and unpredictable weather due to climate change, allowing crop diversification and predictable, regular income;
- Repurposing brownfield sites and buildings to increase plant production without losing greenfield and fertile land (with deforestation for agriculture contributing to climate change);

http://dx.doi.org/10.19103/AS.2023.0126.28

**Table 1** Resource-use efficiency comparison between agricultural systems

| Resource efficiency | Vertical farming | Greenhouse farming | Outdoor farming |
|---|---|---|---|
| Water-use efficiency | 1 l/kg lettuce/year | 20 l/kg lettuce/year | 250 l/kg lettuce/year |
| Energy use | 250 kWh/kg/year | 60–180 kWh/kg/year | 0.3 kWh/kg/year |
| $CO_2$ emissions | 158 kg/ton of lettuce | 352 kg/ton of lettuce | 540 kg/ton of lettuce |
| Land-use efficiency | 0.3 $m^2$, 1 kg lettuce/day | 9 $m^2$, 1 kg lettuce/day | 93 $m^2$, 1 kg lettuce/day |
| Food miles | 43 km | 800–1600 km | 3200 km |
| Yield | 80–120 kg/$m^2$/year | 41 kg/$m^2$/year | 3.9 kg/$m^2$/year |

*Source*: Adapted from: Avgoustaki and Xydis (2020).

- Relocalisation of food production with economic benefits to local communities from investment in vertical farms locally;
- Reduced $CO_2$ emissions and waste (compared to conventional remotely located cultivation);
- Reduced food miles with production centres close to retailers and consumers; and
- Higher yields and land-use efficiency.

Scaling up operations is a key challenge for most vertical farming operations in realising this potential. Intelligent Growth Solutions (IGS) can only take advantage of economies of scale to address the two major operational costs, energy and labour, and further improve resource-use efficiency in water, energy, $CO_2$, fertiliser, seeds, labour, space, and time. IGS offers a modular and flexible solution to scaling up operations in vertical farming. One approach is the development of modular self-sufficient growing systems that use technologies such as light-emitting diode (LED) lighting and circulating water systems, a high degree of automation, remote monitoring, and control (Martin and Bustamante, 2021).

## 2 Intelligent Growth Solutions: company profile

Founded in 2013 and headquartered in Scotland, IGS is an innovative, multi-award-winning agri-tech company delivering vertical farming solutions to customers worldwide. IGS is a privately owned business which has benefited from more than £50m of investment from leading international agri-tech and impact investors. The company blends engineering, crop science, and agronomy skills to build the best technology to allow IGS customers to succeed.

A key differentiator from other companies providing technology services is that IGS does not commercially grow crops. IGS enables its customers to produce high-quality crops using economically viable and environmentally sustainable methods. IGS business-to-business model is to supply growers with a year-round capability to produce crops within a highly automated and controlled environment. IGS calls this capability total controlled environment agriculture (TCEA). IGS product is known as a growth tower.

The IGS system is modular with independent subsystems such as fertigation and heating, ventilation, and air conditioning (HVAC) for each growth tower. This makes it possible to fine-tune customer and crop operating conditions. Building in stages and having a highly flexible growing system to grow different crops allows customers to de-risk their business, test their markets, and pivot crops when required.

IGS farmer customers use our growth towers to grow over 100 different crop types, as seen in Figs. 1 and 2, including microgreens, herbs, salad leaves, brassicas, roots, fruits, botanicals (including edible flowers), and forestry seedlings, as seen in Fig. 3. Some crops are grown to maturity for harvest, while others are cultivated as starter plants for onward growth in fields, forests, polytunnels, and greenhouses. Growth towers can support industrial scale operations consisting of hundreds of microclimates, enabling parallel high-volume production, economic efficiency, and operational flexibility.

The ability to cultivate diverse ranges of crops in high volumes and operational flexibility will give IGS farmers a distinct advantage over their

**Figure 1** Multiple crops grown in IGS towers.

**Figure 2** Peppermint, violas, parsley, and micro kale in IGS towers.

**Figure 3** Norway Spruce seedlings.

competitors. The following example demonstrates the versatility of IGS technology and how it can be used to grow more than food crops.

Consider the UK's ambitious target of planting 30 000 ha of new woodland each year (UK Government, 2021). With a density of up to 3000 trees per hectare for conifers (IGS Internal Research, 2022 and Forestry Commission, 2018), this represents up to 90 million new trees planted annually. What's more, The Department for Food and Rural Affairs (DEFRA) estimates the rate of seedling loss in the first five years after planting to be about 20% (IGS Internal Research, 2022). Therefore, to achieve the government's target of 90 million trees annually, traditional glasshouse nurseries will need to grow at least 108 million seedlings annually to account for this loss. Given that the UK's current annual seedling output is about 100 million seedlings (IGS Internal Research,

2022), and there are plans to increase this target to 50 000 ha (ibid.), it seems likely that traditional nurseries will struggle to cope with demand.

Tree species diversity is crucially important to afforestation projects, with The Committee for Climate Change stating, 'it is vital that afforestation continues ... to ensure species and site selection takes account of the need to adapt to current and future impacts from climate change' (Committee on Climate Change, 2020). Furthermore, Fargione et al. recognise that because of climate change, forestation projects in the United States of America may need to 'shift in the composition of genotypes, ecotypes, and species' (Fargione et al., 2021). Significantly, afforestation carbon off-setting projects increasingly require 'additionality' (Internal IGS Research, 2022), which stipulates that 'emission reductions resulting from [a] project [are] higher than the reductions that would have occurred without the project' (McKinsey, 2021). This means that carbon crediting schemes such as the Woodland Carbon Code are now requiring mixed afforestation projects (broadleaf and conifer) (IGS Internal Research, 2022). Simply, crediting bodies are moving away from single-species forests such as Sitka spruce because such forests have an intrinsic timber value, which would justify planting notwithstanding the carbon sequestration potential. In this sense, carbon sequestration projects must be 'additional' to normal timber practices.

In such scenarios, IGS has a considerable advantage over traditional nurseries. Firstly, our towers can significantly scale up the production of tree seedlings. Based on trials run with Forestry and Land Scotland (IGS Internal Research, 2022), IGS knows that one 12-m growth tower can produce 2 million seedlings annually, based on three growth cycles per year (Internal IGS Research, 2022). Importantly, because of the growth tray with light (GTL) independence from one another, it is possible to cultivate different tree species at the same time. In other words, each GTL could run different light and fertigation cycles, allowing for the growth of multiple tree species in a single 12-m tower. This makes IGS towers nimble and ideally suited to requirements of additionality and species diversity. Finally, based on three growth cycles per year, an IGS tower could grow an entirely different crop for a quarter of the year, contributing positively to forestation projects and the global demand for food. With this in mind, IGS technology offers the versatility required to cope with an ever-changing climatic landscape adequately.

IGS growth towers are available to customers as either a capital purchase or a financial leasing model. IGS software, data, and support services are available via an annual subscription. IGS growth towers were commercially launched in mid-2020. IGS has sold more than 200 growth towers to customers across four continents, totalling £100m in sales. At the time of writing, IGS has a diverse workforce of 220 people from 28 nationalities employed across three continents.

## 3 System configuration and description

IGS's 9+ years of R&D have yielded over 20 patents, including our core design and lighting, HVAC, and fertigation technologies. The company improves products continuously, even after deployment, through iterative hardware product development and IGS cloud-based software.

Our growth tower product comes in 6 m, 9 m, and 12 m high variants. All variants sit on the same 41 $m^2$ footprint and feature an identical design protected by more than 22 patents. Each modular growth tower has its bio-secure environment, which includes separate air and fertigation systems. This allows for a highly diverse set of growing conditions within a multi-tower farm with minimal risk of cross-tower contamination. Each growth tower is thus a separate module, including the HVAC and fertigation systems which are small, easy-to-maintain units dedicated to each tower, as seen in Fig. 4. This design approach is like that used in other industries where high reliability and low time to repair are required (e.g. oil rigs). Should problems such as mechanical failure or biocontamination occur, the problem is isolated to a single growth tower rather than affecting the entire site production? The small scale of components means a small maintenance team easily repairs them, whilst modularity means that small stock of spares allows rapid repair of any on-site problem. This approach increases the reliability of a site and reduces the chances of a technical problem impacting the ability of the site to meet orders. Should a

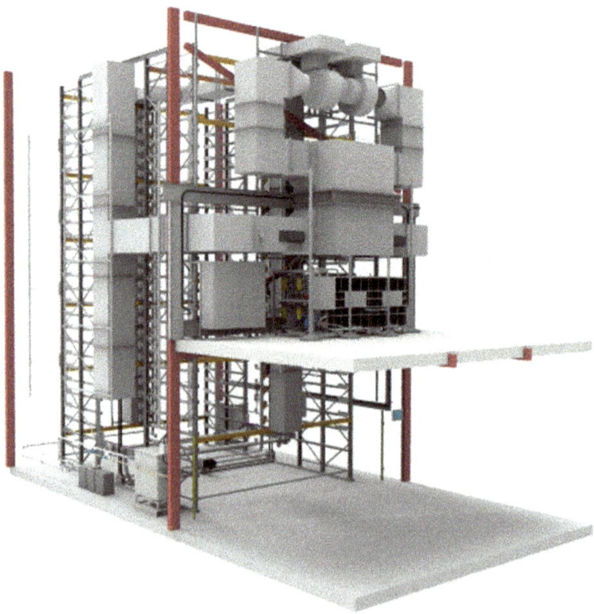

**Figure 4** Three-dimensional view of an IGS tower and its subsystems.

fault not be readily repairable, there are possibilities to cross-connect systems to allow ongoing operation while faults are repaired.

The plant growing area, known as a GTL, is a 6 $m^2$ horizontal hydroponic tray. These GTLs are stacked in two vertical racks, spaced at either 250 mm or 300 mm intervals, and are serviced by a central lift. The central lift performs three core functions:

- GTL movement within the tower;
- Fertigation delivery; and
- Crop imaging.

Each GTL can support a specific microclimate delivering precise levels of light (wavelength and intensity) and irrigation frequency and volume to execute the growth recipe required for each crop type. The grower can adjust recipes to deliver the specific outcome that the grower requires (e.g. taste, nutrient/flavour density, and size). In many designs, it can be common due to the nature of the vertical arrangement of growth areas for climatic variation to occur beyond the mean, which can lead to product waste due to either crop death or lower than target yield at harvest. The GTLs are designed to provide as consistent a growth environment as possible at all points of a GTL. This has been achieved using a combination of modelling tools such as computational fluid dynamics (CFD) and physical measurements from prototypes and working production systems. CFD models are especially useful for design development but require significant calibration of results to real-world measurements. Having real-world data means that future design iterations can be performed more rapidly and accurately using calibrated models.

The growth towers extensively use automated controls to operate safely and reliably round the clock. The grower interacts with the growth towers via secure, cloud-based software (available via a web browser, tablet, or smartphone) to develop growth recipes, plan production, monitor operations, and review crop health.

IGS has developed a novel, patent-protected extra low voltage (ELV) three-phase power transmission system to power the GTLs within the system. The use of three-phase power direct to the GTLs reduces transmission losses and ensures balancing of phase load at scale (a significant safety risk in many large-scale facilities) and significantly reduces the amount of filtering of harmonics required, all of which contribute to energy efficiency.

The system is designed to match the three-phase voltage to the voltage required by the LED drivers once rectified on each GTL. This close matching allows the dimming of the LEDs without significant efficiency loss so that the LEDs can operate at various intensity levels with similar efficiency. This is very different from conventional single-phase LED power, which shows a rapid drop in efficiency away from the optimal setting. Traditionally, when dimming, a tuned pulse width modulation (PWM) is employed to maintain the electrical efficiency of the LED driving circuit; the IGS system can use linear dimming or adjustable PWM (or both simultaneously) to target plant growth and composition as the electrical efficiencies are ensured from the ELV approach. This makes it possible to have a dynamic lighting regime for a particular crop and different lighting regimes for other crops, all using the same LEDs. This offers economies of scale in production without loss of efficiency. An additional benefit of this voltage matching is that the voltage qualifies as ELV, which means the system is much safer for use in potentially wet, humid environments.

The IGS system design allows a completely variable lighting spectrum to be provided to the crop. The light not absorbed by a crop is reflected, and the energy is wasted as heat. IGS continues to develop lighting recipes to minimise applied light while maintaining the desired quality and yield. These offer double-efficiency gains: non-productive light does not have to be generated first, and the waste heat otherwise generated does not need to be cooled.

It is common for HVAC systems to account for over 50% of overall energy use in CEA systems such as advanced greenhouses systems (i.e. Venlo). These systems cover both space cooling/heating and dehumidification. In the case of growing under lights, it is also common for an additional system to be required for luminaire cooling. The IGS system uses a novel, patent-protected HVAC system which uses waste heat from the LEDs and electronics to reheat the air provided to the crop after dehumidification. The system recycles the air in the growth tower air with extremely limited external air required to make up for air lost in leakage through the building. This air can be enriched with $CO_2$ as needed. Reducing external makeup air reduces the energy required (particularly seasonally) to treat that air for use in a growth tower. The air is also sterilised by UV rather than mechanical filtration, reducing energy consumption. These measures mean that HVAC energy use is typically around 30% of the overall system energy use for the IGS system. Key features of IGS HVAC technology include:

- Using recirculated air managed via an airlock system significantly reduces external conditions' effects. The HVAC configuration provides both conditioned air to the crop and cooling of the LEDs to avoid the need for two separate cooling systems;
- IGS use one HVAC system per tower to eliminate the potential for cross-contamination; and
- The crop absorbs around 5% of the water delivered during fertigation. The remainder is recaptured, cleaned, and UV filtered again. Capturing all condensate in the system means that the only water leaving IGS facilities is from the plants.

Control systems can adjust electricity consumption according to local power grid requirements, e.g. by optimising the use of lower-cost and potentially as a virtual power plant providing frequency response services while still addressing plant requirements.

IGS can capture crop images during its lifecycle using crop imaging with lift-mounted cameras. These images feed into computer vision models that enable us to do crop and machinery monitoring automatically. For example, the crop height warning alerts the user if the crop is growing too close to the lights. For instance, Rocket death detection can alert users to wilting and dead rockets and basil, i.e. machinery monitoring (e.g. checking the lights, checking the camera colour calibration, and checking the power supply of the GTL).

Using computer vision to monitor tasks improves resource-use efficiency by automating manual crop and machinery inspections. It is also still in an early stage. Still, IGS is working on the first stages of an energy optimisation tool that can improve the timing of energy consumption to help growers manage/improve their electricity bills.

The IGS system is also designed to capture all wastewater from fertigation and return it to the system for reuse. This includes the condensate water captured from the HVAC dehumidification process. The nutrient solution for the crop is intensively recycled in a system with pumps that run 24 h per day, performing filtration and nutrient top-up operations when not delivering a nutrient solution to crops. Each GTL is irrigated individually rather than using a cascade-type system, which means that both custom fertigation schedules can be set for each GTL and that a large, high-power fertigation system is not required. This further reduces system energy use and provides greater consistency in the solution delivered to each growth area.

## 4 Installation and operation

As already referred, the IGS system is modular and scalable. IGS minimum superstructure installation comprises of two vertical growth towers. Normally, the installation process develops as follows with a few exceptional cases:

- A basic four-tower arrangement plus airlock takes up 307 $m^2$ of land. Each additional pair of towers beyond this minimum adds 104 $m^2$ to the footprint.
- IGS recommends an additional 500 $m^2$ for the service area, which may need to expand with additional towers to create a larger operation.
- IGS works with customers and local contractors to design the facility and build the necessary infrastructure, superstructure, and service areas under IGS supervision.

- IGS installs the growth tower system, including racking and automation, HVAC, and fertigation systems.
- IGS installs its control system based on a programmable logic controller.
- IGS supplies the GTLs with integrated controllable lighting and ventilation for each growth tower.
- Once installed, a comprehensive commissioning checklist is run through and followed by a pre-agreed user acceptance test to ensure the system performs as it should.

In addition, IGS provides full training to customers' staff at IGS Crop Research Centre as seen in Figs. 5–7, in parallel to the project progress so that their operation will run independently.

IGS has full control over growth tower design and deployment. The growth tower design is optimised for easy maintenance and minimal operating cost. For standard parts in a modular system, IGS use off-the-shelf equipment and components instead of proprietary systems for processes such as sowing and harvesting. This means IGS can keep start-up and maintenance costs down by providing items with which farmers are already familiar. If a section of the vertical farming system needs to be replaced, such as a water filter, a lighting panel, or a tray, it can be done with almost no interruption.

The vertical farm offers complete remote control over the environment. Users see a live view of their crops and can adjust settings to better deal with fluctuations in market demand to speed up, slow down, or pause the growth of crops as required. Growth tower operations are managed using a cloud-based management system which is simple to operate and requires no on-site

**Figure 5** IGS CRC (Crop Research Centre) at the James Hutton Institute, UK.

**Figure 6** IGS growth tower.

**Figure 7** IGS growth tower.

IT (Information Technology) management. The controls and data are accessible via IGS Growth Tower Management System (GTMS). GTMS monitors and controls the environment and machinery to maximise the yield and consistency of the crops inside the growth towers. It is essentially the operating system for the growth towers. It autonomously manages all aspects of the crop growth cycle (e.g. lighting, fertigation, imaging, and tray movement). GTMS enables operators to create, modify, and deploy growth recipes to tailor all these aspects for optimal growth for specific (varieties of) crops. A continuously expanding set of new and improved growth recipes are shipped as part of GTMS. The IGS approach was designed with data security and integrity from the outset. All data are secured and encrypted and specific to each customer.

IGS support and guidance vary with IGS clients' crop mix, operating knowledge, market, financing stability, building challenges, etc. IGS advises their customers on how to make decisions about scale while growing excellent quality products, being economically competitive and more sustainable compared to existing food systems. IGS pre-sales team guides their clients throughout the feasibility analyses of their business by using data from IGS crop trials, grower models, and knowledge on executing vertical farming projects worldwide. IGS grower modelling allows customers to test all the operational costs of growing to build a business case that truly reflects the market in which they operate.

More generally, IGS takes a data-driven approach to improve IGS vertical farm system continuously. IGS intends to continue to publish performance-based statistics using actual data from IGS operations. IGS is also committed to furthering the understanding of vertical farming and plant science.

## 5 Conclusion and future trends

IGS focuses on its passion for delighting customers by providing them with market-leading products and support services. IGS investments remain in global expansion and product innovation. Customers should expect to see the introduction of an even more diverse crop range, further efficiency gains, increased levels of automation, and the systems necessary to operate the world's largest vertical farms successfully.

IGS's first customers were new to the market and took the brave step to use our technology. They had to start small to test what would work best. IGS is now seeing those first customers return to us to expand their business with additional towers. At the same time, those entering the market now have confidence in the technology and are looking to buy multiple towers. Many IGS customers plan large deployments of tens or even hundreds of towers. Within these large farms, IGS is increasing levels of automation, which allow automatic GTL movement outside the growth towers and increased use of data science techniques to minimise the manual labour usually required to operate at large scales.

The feasibility of automating the service area and connecting it to the IGS system depends heavily on factors such as the crop mix, region, labour, financing, and building constraints. When designing large-scale facilities, the vertical farming system is only one part. Collaboration and partnerships with automation and machine suppliers are critical. As a rule of thumb, full end-to-end automation from seed to harvest is only feasible economically in projects with a minimum of 10000 $m^2$. At a smaller scale, some supply chain elements can be automated, such as seeding and packaging. These are areas requiring further research and development. Other areas of development include design optimisation for manufacture, transportation and installation as systems grow in scale, as well as developing modules for customised fertigation/HVAC and other parameters to suit crop types. Other broader areas for future research include breeding varieties tailored to vertical farming and advanced low-cost sensing technologies to allow even more precision to address the CapEx/OpEx balance through cost reduction, operational efficiency improvements driven by science, and better yield performance. The core IGS technology has been built to allow plant science to develop in the background and be deployed via software through the TCEA capabilities. Such developments will reduce further wastage, achieving net-zero and full traceability from seed to harvest.

From a research point of view, IGS started with a system focused on optimising leafy greens production. IGS has proven capable of much more, but as the market grows, variants of the machine to suit specific crops will be developed. Variations will include irrigation methods, tower configuration for different crop formats, and customizations of the HVAC and fertigation systems.

IGS is also looking at other types of imaging, e.g. three-dimensional cameras, thermal cameras and collaboration with other sensors that can measure chlorophyll fluorescence.

## 6 Where to look for further information

For further information you can reach out to the IGS team through: https://www.intelligentgrowthsolutions.com/ or +44 131 202 1050. The site includes an introductory guide to technologies and best practices in vertical farming. https://www.intelligentgrowthsolutions.com/growth-intro-guide.

## 7 References

Avgoustaki, D. D. and Xydis, G. 2020. How energy innovation in indoor vertical farming can improve food security, sustainability, and food safety? In Cohen, M. J. (Ed.), *Advances in Food Security and Sustainability* 5, 1-51. doi: 10.1016/bs.af2s.2020.08.002. PMCID: PMC7516583.

Department for Environment, Food & Rural Affairs, Forestry Commission, Natural England, and The Rt Hon George Eustice MP. 2021. Tree planting rate to treble by end of this Parliament. https://www.gov.uk/government/news/tree-planting-rates-to-treble-by-end-of-this-parliament.

Fargione, J., Haase, D. L., Burney, O.T., Kildisheva, O. A., Edge, G., Cook-Patton, S. C., Chapman, T., Rempel, A., Hurteau, M. D., Davis, K. T. and Dobrowski, S. 2021. Challenges to the reforestation pipeline in the United States. Frontiers in Forests and Global Change, 4, 629198.

Forestry Commission. England. 2018. Biomass in a low-carbon economy. Committee on climate change 2018. Annex 1. Sustainable forest management. https://www.theccc.org.uk/wp-content/uploads/2018/11/Annex-1-Sustainable-Forest-Management.pdf.

GIC, McKinsey & Company, Singapore Economic Development Board and Vivid Economics. 2021. Putting carbon markets to work on the path to net zero. How investors can help decarbonise the economy and manage risk-adjusted returns. https://www.mckinsey.com/business-functions/sustainability/our-insights/putting-carbon-markets-to-work-on-the-path-to-net-zero.

Martin, M. and Bustamante, M. J. 2021. Growing-service systems: new business models for modular urban-vertical farming. *Frontiers in Sustainable Food Systems* 5, 787281. https://doi.org/10.3389/fsufs.2021.787281.

# Part 7

## Concluding remarks

# Chapter 21

# Plant factories with artificial lighting (PFALs): Concluding remarks

*Eri Hayashi and Toyoki Kozai, Japan Plant Factory Association, Japan*



## 1 Introduction

This book has introduced the background, concept, and methodology of sustainable plant factories with artificial lighting (PFALs), as well as past, present, and future research and technology in PFALs. As described throughout this book, PFALs can contribute to solving global and local issues in

- food safety and security;
- environmental recovery, improvement, and creation;
- increasing demands for higher quality fresh, functional, pesticide-free foods;
- reducing food mileage and food loss; and
- improving quality of life.

However, in practice, most existing PFALs have not yet achieved the potential that PFALs have. Since the PFAL industry is emerging and still in its infancy, there is significant room for technology development to realize genuinely sustainable PFALs.

## 2 Challenges of sustainable plant factories with artificial lighting

Challenges remain for research and development in achieving sustainable PFALs, including PFALs consisting of one or more cultivation system modules,

http://dx.doi.org/10.19103/AS.2023.0126.31

phenotyping individual plants, controlling three-dimensional environmental factors within a plant canopy, and production of more diverse plant types, such as fruit, root and head vegetables, and medicinal plants, among others.

As highlighted by environmental assessments of PFALs, we need to address reducing carbon emissions from PFALs. Current PFALs consume a significant amount of electricity, chemical fertilizer, plastic products, cement, etc., including emitting $CO_2$ even during the manufacturing of components and PFAL construction (for details, see Chapter 5). To achieve energy-autonomous PFALs, a realistic solution is to use natural energy instead of fossil fuels for electricity generation, which can be one of the significant sources of $CO_2$ emissions.

In promoting energy and other resource use efficiencies in PFALs, plant phenotyping plays an essential role in optimizing the environmental control of PFALs. With the timely advancement of sensing, artificial intelligence, robotics, and other relevant technologies, plant phenotyping technologies in PFALs are expected to advance rapidly and be widely used in the PFAL industry. Indeed, as described in several business case study chapters in this book, phenotyping-related technologies are beginning to be applied to commercial PFALs for growing leafy greens or fruit crops.

Parts 4 and 5 of this book review multiple studies on spectral manipulations for controlling the growth and quality of leafy greens, ornamentals, fruit vegetables, and herbs, which are crucial for expanding the plant types in commercial PFAL production. There are increasing numbers of PFAL-related research studies, particularly on lighting, conducted internationally. Further research with a holistic approach, also considering management practices for improving productivity per electric energy usage and other resources, along with utilizing simulation software, is needed to advance the PFAL industry.

## 3 Opportunities in achieving sustainable well-being societies

As described throughout the book, PFALs are airtight, thermally insulated, and with a high degree of sanitation, making them unique compared to greenhouses or open-field cultivation. Using sensors and actuators, PFALs enable high levels of measurement and controllability of plant productivity, environment, and phenotype, which are affected by the (micro)environment, management, and genotype. Furthermore, PFALs can achieve high predictability and reproducibility with various simulation models. PFALs can thus facilitate data sharing and evolutional research with a multidisciplinary approach for further advancement toward sustainable PFALs.

A PFAL simulator, or a virtual PFAL, connected with a physically existing real PFAL, could be a valuable tool for self-learning, training, and education, along with improving the design and operation of a PFAL. Moreover, virtual PFALs can contribute to science-based entertainment, including games or amusement, or even a novel plant-based communication platform. Scalable PFAL cultivation modules, together with a phenotyping system and PFAL simulators, open up the possibility of sustainable PFALs beyond producing fresh vegetables and high quality transplants (seedlings and cuttings), and selection of transplant for breeding. Visualizing the interaction between phenotypes of both plant individuals and communities, the microenvironment and management in real-time in PFALs based on plant cohort research could also offer valuable lessons for our lives and society besides plant science.

Considering the significant potential of the sustainable PFALs, another form of business model or social business entity might be needed, focusing more on directly coping with global and local issues, not just selling PFAL-grown plants or PFAL-related systems. As PFALs hold the potential to concurrently contribute to solving both global and local issues, including food, environment, energy and resources, and quality of people's lives, the PFAL system can be utilized for multiple purposes depending on the regional situation. Besides producing high-quality plants, plants in a well-insulated and highly airtight PFAL cultivation room could be a powerful tool for purifying water. In achieving inclusive and sustainable societies with improved well-being, it is critical to enhancing the shared value of sustainable PFALs, including fields such as science, engineering, design, and art. Sustainable PFALs have the potential to achieve multiple Sustainable Development Goals (SDGs), including SDG 1: No poverty.

## 4 Conclusion

It is expected that demand for plants or plant-related products will further increase in the coming years, and PFALs are only a part of diverse plant production methods. PFALs, as one of the closed production systems, are similar to our mother planet, i.e. earth, which is also a closed ecological system. Sustainable PFALs can also be integrated with other closed production systems, including insect or fish production, as part of the sustainable social infrastructure directly connected with social citizens.

Adequately integrating various technologies based on PFAL-specific concepts and engineering, in parallel with making full use of past research, multidisciplinary research, and cooperative activities relevant to PFALs, will contribute to realizing the innovative, creative, and evolutional sustainable PFALs.

## 5 Where to look for further information

The following books also provide concepts, technical features, challenges, and opportunities of PFALs:

- Kozai, T. (Eds.) (2018). *Smart Plant Factory: the Next Generation Indoor Vertical Farms*. Springer. 456 Pages.
- Kozai, T., Niu, G. and Masabni, J. (Eds.) (2022). *Plant Factory Basics, Applications and Advances*. Elsevier. 449 pages.

# Index

Milton Keynes UK
Ingram Content Group UK Ltd.
UKHW021124030624
443545UK00001B/4

9 781801 463164